Professor A. W. Bishop's Finest Papers

Professor A. W. Bishop's Finest Papers
A Commemorative Volume

compiled by Laurie Wesley

Whittles Publishing

Published by
Whittles Publishing,
Dunbeath,
Caithness KW6 6EG,
Scotland, UK

www.whittlespublishing.com

ISBN 978-184995-442-6

Printed and bound in Great Britain by Severn, Gloucester

Contents

Introduction

There is a little history behind the creation of this volume which I will explain. Bishop's most influential paper, co-authored with Lauritz Bjerrum, was presented at an ASCE (American Society of Civil Engineers) conference in Boulder, Colorado, in 1960, and although it made a big impression at the time, it is now in danger of disappearing from sight. In addition, two of Bishop's very significant papers were published in the late 1970s in the *Philosophical Transactions of the Royal Society*, not normal reading for the soil mechanics fraternity and thus became known to only a few people. That has remained the case to this day, and suggestions made to ICE (Institution of Civil Engineers) and *Géotechnique* to have these two papers republished have not brought a positive response. So those factors were the initial motivation for creating this volume. In addition, it is nearly 40 years since Bishop retired from his position as professor at Imperial College and a fitting time to remember him with both a biography and a volume of his most important papers. The basis for selecting the papers, apart from the three above, has been twofold: firstly, their importance in the development of soil mechanics and secondly to highlight the nature and range of subjects that Bishop investigated during the thirty seven years of his career.

In addition to 'Bishop's finest papers', there is one paper (Wesley and Pugh) that does not bear his name. The best justification for including it here is perhaps to claim that it may well have been among Bishop's 'finest' papers had he been its lead author. However, there are other justifications for including it. As the reader will note, it is a very long paper and describes a wealth of information coming from very extensive laboratory and field testing. The field testing, especially the construction and instrumentation of full scale trial embankments was very costly and to make the work possible Bishop had obtained a large grant from the Science Research Council. This also financed Richard Pugh and myself as fellow PhD students under Professor Bishop's supervision (in the mid-1970s). Separate papers were to have been written after the completion of our PhDs, with Bishop as the lead author, but because of his illness and early death this didn't happen. Hopefully, the reader will agree that our research was very comprehensive and deserves to be published, both as a tribute to our supervisor and because of its relevance to today's geotechnical profession.

Laurie Wesley

Acknowledgements

Permission from the following organisations to reproduce papers, and provision of the papers as pdf files, is gratefully acknowledged:

The Royal Society (papers 10 and 11)

American Society of Civil Engineers (ASCE) (paper 6)

International Society for Soil Mechanics and Geotechnical Engineering (paper 2)

Norwegian Geotechnical Society (paper 4)

Swedish Geotechnical Society (paper 9)

Teknisk Ukeblad Media (paper 5)

(Publisher's note: The original files for these papers and therefore their resulting print quality vary considerably but we hope that readers will nonetheless find them useful.)

A NEW SAMPLING TOOL FOR USE IN COHESIONLESS SANDS BELOW GROUND WATER LEVEL

A. W. BISHOP, M.A., A.M.I.C.E.

Lecturer in Civil Engineering, Imperial College, University of London

INTRODUCTION

The problem of obtaining undisturbed samples of a sand below ground water level has proved difficult for engineers to solve economically. Even disturbed samples cannot always be satisfactorily obtained owing to segregation and the loss of fines in the usual auger or bailer.

The description " undisturbed " has an even more qualified meaning with sands than with clays. Changes in stress cannot be avoided in withdrawing a sample, and the external forces are not largely replaced by capillary forces as in the case of clays. However, reversible volume changes are usually small in sands and the release of pressure will only have a minor effect on the porosity. It should therefore be possible to obtain samples approximating to the requirements which Hvorslev (1948)* has suggested for undisturbed samples, which are :

(i) The soil structure has not been disturbed.
(ii) There is no change in water content or void ratio.
(iii) There is no change in the constituents and chemical composition of the soil.

Samples may be obtained by sinking either trial pits or lined boreholes. When a trial pit is used and samples are required below ground water level, it is necessary to lower the water level by well points or filter wells, as pumping from a sump is likely to disturb the stability and structure of the sand at the bottom of the pit. Alternatively a shaft may be sunk under compressed air, but both methods are costly and may be impracticable for deep or variable strata.

Sampling from a lined borehole is complicated by the fact that in a clean sand the strength is entirely dependent on the pressure between the grains, and as soon as this is removed, in extracting the sampler, the sand flows out. Samples are sometimes retained in a normal sampling tool, but this is usually because they are slightly cemented or are above the water table and have a small apparent cohesion due to the presence of moisture.

A number of methods of preventing the loss of samples have been developed and are summarized by Hvorslev (1948) and Terzaghi and Peck (1948). They fall under three main headings :

(1) *Mechanical core retainers.* These include various forms of valves and fingers at the lower end of the sampling tube and generally so increase the cross-section of the sampler that an undisturbed sample is difficult to take. Of these, the Denison sampler (Johnson 1940) has been widely used in the United States, but it is stated to be unreliable in clean sands.

(2) *Solidification of the lower end of the sample.* This may be done by (a) injecting chemicals through openings in the sampler shoe to form a plug of insoluble gel, or (b) freezing the lower end of the sample by lowering an annular freezing unit after sampling (Falquist 1941). Of these, the second method has been used with some success, but it is elaborate and expensive.

(3) *Solidification of the sand before sampling.* This has been done by (a) injecting asphalt emulsion, which is removed by a solvent before testing (van Bruggen 1936) or (b) freezing the ground by the use of a cooling mixture in auxiliary pipes, and then sampling (Langer 1939)† or core boring (Brinton 1938).† The first method is slow and the asphalt difficult to remove, and in the second, the danger of expansion during freezing occurs.

A new apparatus developed by Kjellman and Kallstenius (1948) for taking continuous cores up to 60 ft. in length has also been successful in retaining sand strata without the use

*See Bibliography. †Quoted by Hvorslev (1940).

of a core-catcher, but it is not clear from the details published whether it has been used in strata free from lenses of cohesive material.

Another method, that of capping the borehole casing and expelling the water from the sand by compressed air before sampling in order to make use of the slight cohesion of damp sand, was suggested to the author in 1946 by Mr. R. Glossop and formed the starting point of the experimental work carried out at Imperial College which led to the present design. This method is also included by Hvorslev under the third heading as "proposed but not tried out."

Laboratory Tests

The method as originally envisaged was to pass the boring rods through a gland in a cap which could be fitted to the borehole casing, so that compressed air could be used to expel the ground water from the zone to be sampled. The sample of damp sand could then be retained in the sampler by an applied vacuum and withdrawn from the ground before the pressure was released.

The tests were designed to examine whether moist, clean sand would arch satisfactorily across a sampling tube of adequate size, and whether the core could be retained by an applied vacuum.

A fairly coarse uniform sand was used in the tests (Leighton Buzzard sand between the 18 and 25 sieve sizes) as it seemed that this type of sand would present most difficulty. Sampling tubes ranging between 1 in. and 5 in. diameter were tried. The sand was placed in a loose packing under water and was drained just before sampling. When each sample had been withdrawn it was subjected to impact comparable with that received by a sampler being raised from a deep borehole.

The results showed that the method was not very satisfactory, partly because a vacuum sufficient to retain the sample drew air rapidly through the core of sand. It was noticeable, however, that samples taken from the still flooded stratum were retained much more satisfactorily, provided that the lower end of the sample was raised rapidly above water level. This method required a filter in the sampler head to prevent sand passing into the vacuum line and, unless very carefully controlled, the vacuum drew the pore-water through the sample and difficulties due to the passage of air arose again.

The simpler alternative of using a valve to seal off the top of the tube before withdrawing the sampler was also tried. This gave very satisfactory results provided that the valve remained completely free from leakage. The tendency of the sample to slide out of the tube was just balanced by the negative pressure this caused inside the sampler.

When the sample is above water level static equilibrium is possible, but only so long as the surface tension in the water at the base of the sample is able to maintain this negative pressure, and it is therefore surprising that a sample of sand 10 or 15 in. in length can be supported when capillary tension in the pore-water does not exceed a few inches head of water. The explanation is illustrated in Fig. 1.

If a manometer tube is connected to the top of the sampler and a pressure pilot placed in the sand, then provided that there is no air in the system they will both give the same value of pressure head in the sample (showing that there is no flow of water through it) and measure the negative pore-water pressure at the base of the sample, $u = -h_1 = -h_2$, Fig. 1a. With a sample of sand B (Fig. 4) 2·4 in. diameter and 14·6 in. in length, the measured value of u was $-1{\cdot}3$ in. of water. Since the effective stress under a head of 14·6 in. of submerged sand would be approximately equal to 14·6 ins. of water it is obvious that the difference must be due to arching, as in a silo. The maximum negative pressure which the capillary forces can maintain was measured as 7 in. of water, which would not be sufficient to support the sample without the friction developed on the sides of the tube.

With rounded Leighton Buzzard sand (18 to 25 sieves, sand A in Fig. 4), and the same sample dimensions, the value of u varied between $-2{\cdot}0$ and $-2{\cdot}5$ in. of water. The measured maximum capillary pressure was -4 in. of water.

A uniform sand at the upper limit of the coarse sand fraction (7 to 10 sieves) was found to be unstable, but its maximum capillary pressure was only −1·5 in. of water.

The theory of arching (Terzaghi 1943) and preliminary measurements indicate that when the length of the sample exceeds two or three times its diameter no further increase in the base pressure required to support it takes place, but probably a slight decrease. The stability of a sample is therefore controlled by the relation of the diameter of the sampler to the capillary forces in the sand, which are approximately inversely proportional to the effective grain size (of the 10 per cent fraction). The frictional properties of the tube and the sand are also minor variables. The 2·4 in. diameter sampler has been adopted for general purposes, but there is no reason why in fine sand larger cores should not be obtained and, indeed, several 5-in. diameter cores have been taken in the laboratory.

It is important to note that without this small effective pressure at the bottom of the sample the support due to arching is not mobilized and samples left under water flow rapidly out of the sampler.

Design of Field Apparatus

From the foregoing laboratory tests it is clear that the three primary requirements to be incorporated in a field sampler are :

(1) The rapid withdrawal of the sampler to a position where its base is above the water surface.

(2) A completely leak-proof valve at the top of the sampler.

(3) A sample tube diameter for which the capillary forces in the type of sand to be sampled are sufficient to maintain arching.

The first requirement will generally mean an artificial lowering of the water level in the borehole, which leads to piping of sand into the casing unless compressed air or well points are used. A simpler expedient has been found to get over this difficulty, namely, that of having an outer shell which can slide on the rods above the sampler and be used as a " diving bell " into which it is withdrawn after sampling. This expedient appears to be a considerable advance over that of expelling all the water from the borehole, which interrupts the boring operations and may lead to serious piping into the hole on releasing the pressure.

The second requirement is met by using the compressed air needed to expel the water from the bell to close a flexible rubber diaphragm valve over the outlet from the tube after driving.

The dimensions of the apparatus have been chosen so that it can be used with an ordinary portable rig and 6-in. diameter lining tubes, and a motor-car tyre pump is adequate for the compressed air supply (Plate II). The third requirement is satisfied for most sands by the use of a 2$\frac{3}{8}$-in. diameter sampling tube.

The sampler (Fig. 2) is attached to the boring rods by a plain loose-fitting socket, and its weight is taken by the lifting shackle which hooks over a peg on the lowest rod. When the sampler reaches the bottom of the borehole and its weight is removed from the rods the shackle springs off the peg, and enables the rods to be withdrawn separately after the sampler has been forced into the ground. The sampler can then be withdrawn smoothly on a light cable and the delay and jarring due to uncoupling sections of rod are avoided.

The air supply is carried by a small diameter hose to this socket and passes through the hollow rod to the head of the sampler. The rod passes through the weighted head of the bell in a loose-fitting bronze bush, the seal being made by two round Angus sealing rings. The air is led from the rod to the upper side of a flexible rubber disc which it deflects downwards to seal the port in the head of the sampling tube. To ensure a pressure sufficient to close the port the air has to raise a spring-loaded valve adjusted to about 20 lb. per sq. in. before passing into the bell.

The port is $\frac{5}{8}$-in. diameter and stands $\frac{3}{4}$-in. proud of the surface, and has a thin, rounded lip where the rubber diaphragm closes on it. This minimizes the risk of sand preventing the

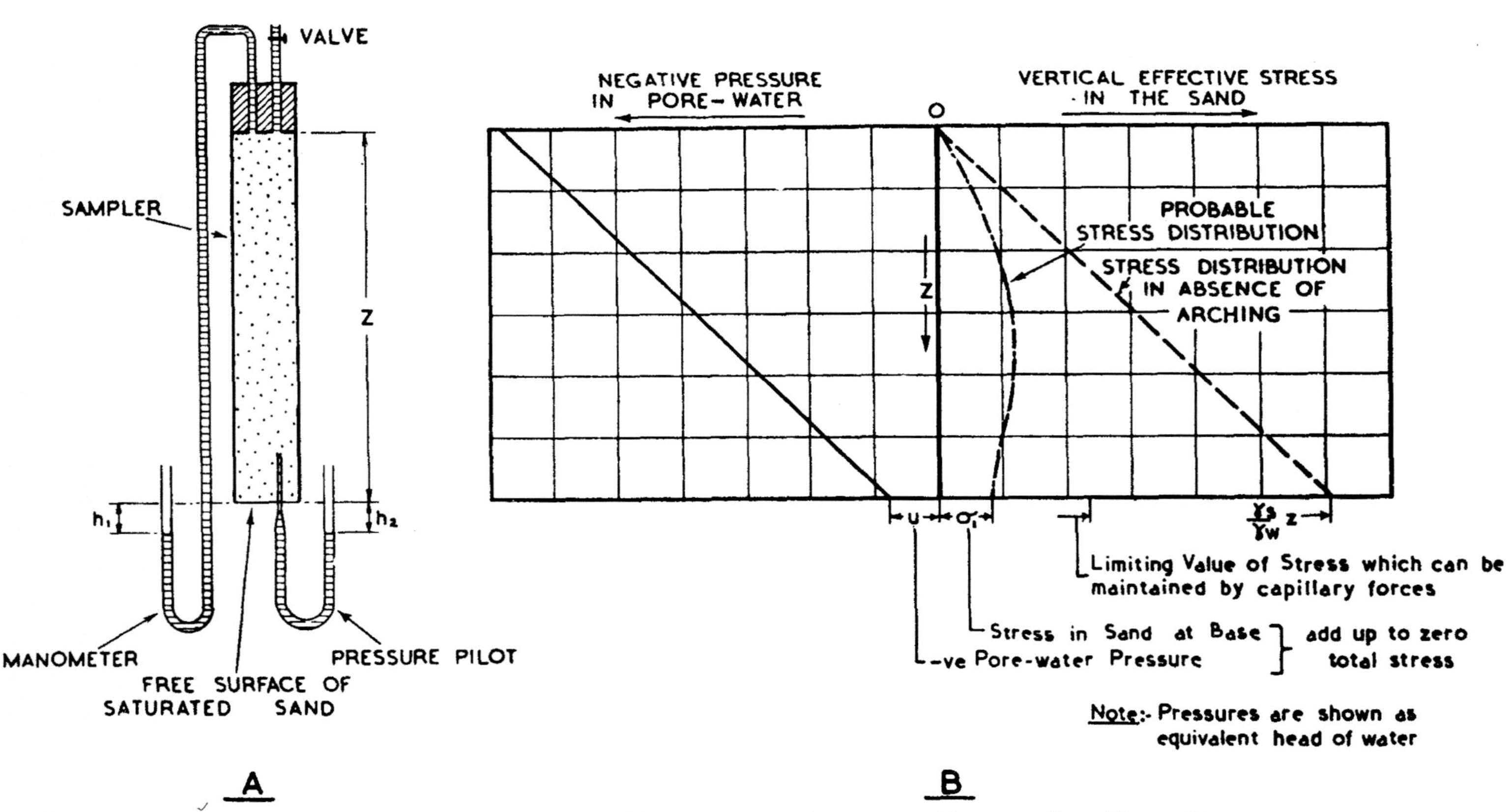

Fig. 1. Experimental Illustration of Maintenance of Arching by Negative Pore-Water Pressure

proper sealing of the opening. The valve has the general advantage of a relatively large port area and no back pressure as in the case of a spring-loaded valve.

The sampling tube is 16-gauge hard-drawn brass $2\frac{3}{8}$-in. internal diameter, with an area ratio* of 11 per cent and no internal release. A round rubber ring below the fixing screws maintains an air-tight seal with the sampling head.

The bell is of 4-in. diameter steam barrel, weighted to prevent it from being lifted by friction as the sampler is withdrawn. Ports are provided just above its foot for the water to be expelled. The weight of the assembled sampler is 86 lb.

The sequence of operations during sampling is illustrated in Fig. 3, and is as follows:

(1) The sampler is lowered on the boring rods, the air valve being open.

*The area ratio is expressed as $\frac{D^2_2 - D^2_1}{D^2_1}$ where D_1 is the internal diameter of the mouth of the sampler and D_2 is the external diameter. It represents the ratio of the cross-sectional area of the sampler to that of the sample it takes. Standard $1\frac{1}{2}$-in. diameter and 4-in. diameter samplers used in this country have area ratios of about 17 per cent and 27 per cent respectively.

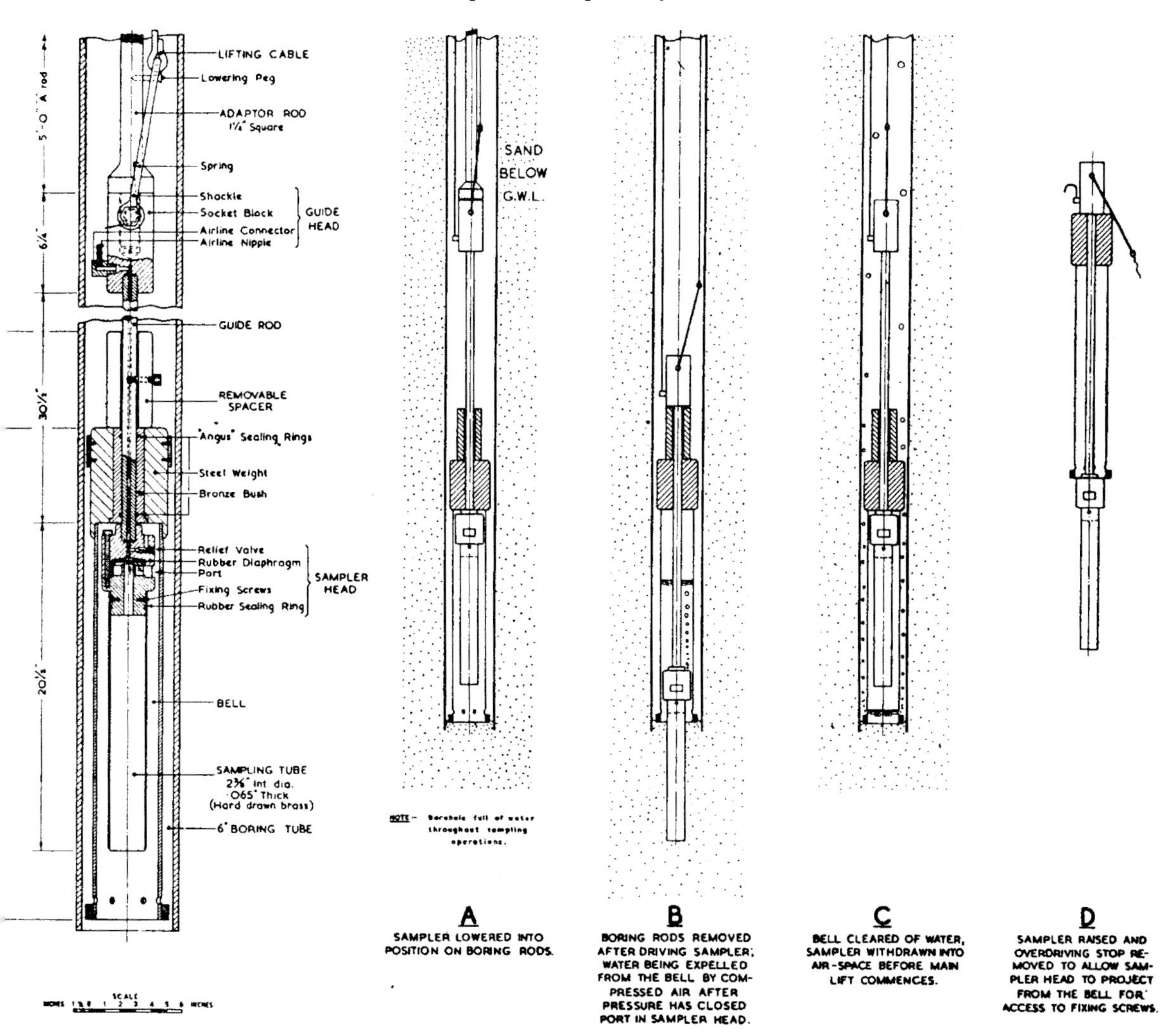

FIG. 2
GENERAL LAYOUT OF SAND SAMPLER

FIG. 3
METHOD OF SAMPLING IN A LINED BOREHOLE

(2) When the bottom is reached the sampler is forced into the ground by a pulley tackle attached to the lining tubes, until the stop on the top of the bell is reached.

(3) The boring rods are withdrawn.

(4) Compressed air is supplied by the pump. This closes the valve and seals off the sample. The excess air expels the water from the bell and bubbles rising in the borehole indicate when this operation is complete.

(5) A steady and rapid pull is applied to the cable, which raises the sampler into the air-space in the bell and then carries the bell up with it to the surface. The pressure gauge should be watched to see that sufficient pressure is maintained to keep the valve closed.

(6) The sampler is placed on a stand (Plate II) and the overdriving stop removed to allow the sampling head to project beyond the bell for removing the tube. A filter plug is held in the sampling tube to prevent loss of the core when the suction is released. Density and permeability measurements can then be made without removing the sample from the tube, which can also be used for transporting it to the laboratory.

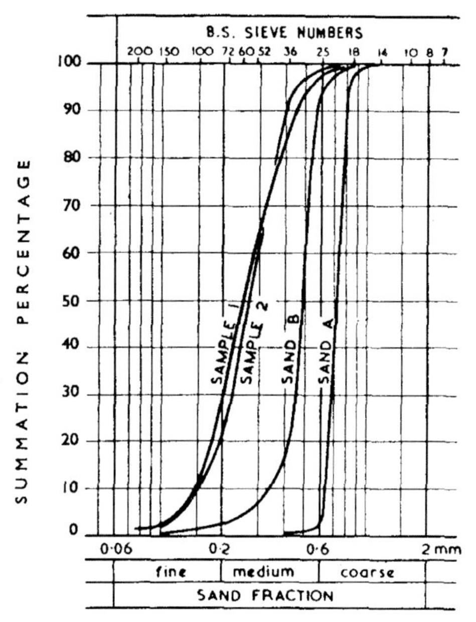

FIG. 4
PARTICLE SIZE DISTRIBUTION OF SANDS IN LABORATORY AND FIELD TESTS

TEST RESULTS

Tests have been carried out with the sampler, both in the laboratory and in the field, and have so far given satisfactory results.

The conditions at the bottom of a borehole are difficult to reproduce exactly in the laboratory; but to obtain an indication of the amount of disturbance and compaction due to the entry of the sampling tube, tests were made in a flooded bed of sand in a bin. Plate II shows a cross-section of a sample from a layered bed of clean medium sand (sand B in Fig. 4) and indicates that the structure is relatively undisturbed.

Another sample of the same sand taken from a bed whose average porosity was 41·5 per cent gave a value of 40·5 per cent in the sampler. The sand was of an open structure (the relative density* was 0·43, the limiting porosities being 46·9 and 32·3 per cent), and it had not been subjected to overburden pressure, so the agreement was quite satisfactory.

Two samples have been taken from a borehole during a site investigation on land reclaimed by pumping river sand (Plate II). The grading curves are given in Fig. 4, and the results of the density measurements will be found in Table I.

TABLE I. MEDIUM GREY SAND

Sample No.	Dry density, lb./cu. ft.	Specific gravity.	Natural porosity, per cent.	Maximum porosity, per cent.	Minimum porosity, per cent.†	Relative* Density.
1	102·4	2·667	38·2	43·2	31·0	0·46
2	103·2	2·686	38·3	42·7	31·0	0·43

*Relative density is $\frac{\epsilon_L - \epsilon}{\epsilon_L - \epsilon_D}$ where ϵ is the natural void ratio of the sand and ϵ_L and ϵ_D are the void ratios in the loosest and densest states respectively. It is therefore a measure of the relative compactness of sands of various gradings. The limiting densities obtained in the dry by techniques Nos. 2 and 6 (KOLBUSZEWSKI 1948) have been adopted as provisional standards.

†Approximate values.

The samples were taken at depths of 12 and 15 ft. below the ground surface, water level being at the depth of 5 ft. 6 in. The sampling tool required considerable force in the latter part of the drive, due probably to arching in the tube in the reverse direction to that which prevents loss of the core during withdrawal. It will be seen that if the drive is rapid there will be little opportunity for compaction in saturated loose sand and little frictional resistance will be mobilized on the sides of the sample, as increases in effective pressure cannot take place. A rapid continuous drive should therefore be expected to produce satisfactory cores. In the case of dense sands the effects of dilatancy may limit the length of core which can be taken, whatever force is applied, owing to the formation of a plug arching across the tube.

A further point which was noted was the necessity of cleaning the lining tube of loose sand and preventing sand piping into the hole before sampling. This can only be done if the head of water in the lining tubes is kept above ground water level. If a " quick " condition is allowed before sampling, either due to unbalanced pressures or shelling, the structure in the zone to be sampled is disturbed.

Summary

Laboratory tests and preliminary field tests have shown that the method of maintaining the stability of samples of non-cohesive sands by a slight negative pore-water pressure gives satisfactory results. The sampling tool is free from most of the elaborations usually required in obtaining sand cores and does not interfere seriously with the ordinary routine of boring,

Provided that care is exercised in cleaning out the borehole to undisturbed material, the sampler itself appears to recover cores in which the structure is preserved and the void ratio relatively unchanged. As the sample is saturated when the tube is driven it is probable that compaction will be less than in drained samples taken in a test pit, especially if the driving is rapid relative to the permeability of the stratum.

Density and permeability measurements can be made on the core without removing it from the sampling tube, and in the case of the permeability test the difficulty of resaturating the sample is avoided. In general, this method offers the possibility of a closer approximation to the requirements for an undisturbed sample than many of the more elaborate procedures.

Acknowledgments

The author wishes to acknowledge the encouragement he has received from the Directors of Messrs. John Mowlem & Co., Ltd., who have made possible the development of the field apparatus, and the assistance of Mr. I. K. Nixon who has supervised many of the practical details.

The laboratory tests were carried out in the Civil Engineering Department of the Imperial College of Science and Technology, and the sampling tool was constructed by the Sanders Precision Engineering Company.

Bibliography

van Bruggen, J. R. (1936). " Sampling and Testing Undisturbed Sands from Boreholes." Proc. 1st Int. Conf. Soil Mechanics. Vol. I, p. 3. *Harvard.*

Falquist, F. E. (1941). " New Methods and Technique in Subsurface Explorations." Journ. Boston Soc. Civil Engrs. Vol. 28, p. 144.

Hvorslev, M. J. (1940). " The Present Status of the Art of Obtaining Undisturbed Samples of Soils." Harvard University.

Hvorslev, M. J. (1948). " Foundation Exploration : A Review of Methods and Requirements." Proc. 2nd Int. Conf. Soil Mechanics. *Rotterdam.*

Johnson, H. L. (1940). " Improved Sampler and Sampling Technique for Cohesionless Materials." Civil Engr. Vol. 10, p. 346.

Kjellman, W., and Kallstenius, T. (1948). " A Method of Extracting Long Continuous Cores of Undisturbed Soil." Proc. 2nd Int. Conf. Soil Mechanics. Vol. I, p. 255. *Rotterdam.*

Kolbuszewski, J. J. (1948). "An Experimental Study of the Maximum and Minimum Porosities of Sands." Proc. 2nd Int. Conf. Soil Mechanics. Vol. I. p. 158. Rotterdam.

Terzaghi, K. (1943). " Theoretical Soil Mechanics." John Wiley. *New York.*

Terzaghi, K. and Peck, R. B. (1948). " Soil Mechanics in Engineering Practice." John Wiley. *New York.*

PLATE II

SAND SAMPLER AND BORING RIG

NOTE. Since this paper was written the sampler has also been successfully used in a bed of fine sand ($D_{10}=0{\cdot}1$ mm). The samples were taken from depths extending to 40 feet below ground level: ground water being 5 feet below the surface.

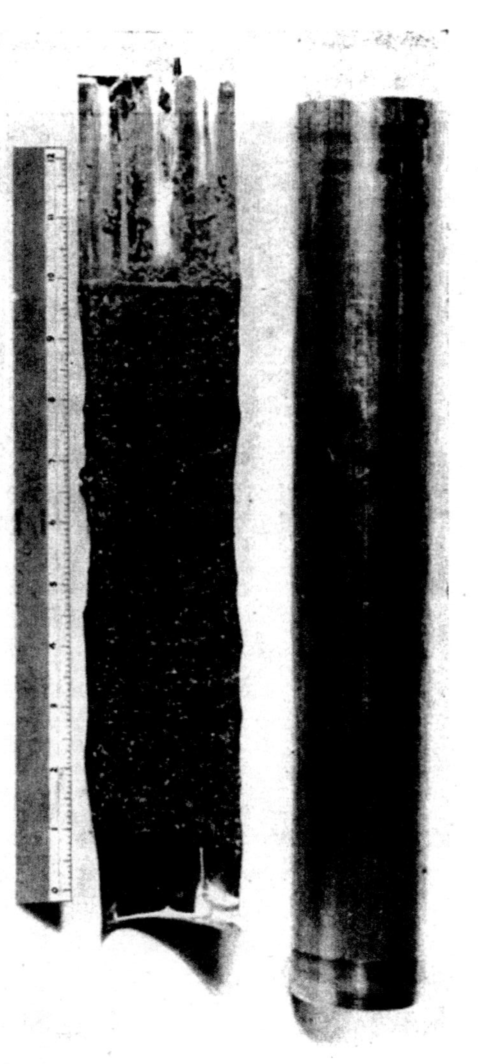

SECTION OF SAND SAMPLE TAKEN IN THE LABORATORY.

SUB-SECTION IV b

DAMS AND LEVEES

V b 1 SOME FACTORS INVOLVED IN THE DESIGN OF A LARGE EARTH DAM IN THE THAMES VALLEY

ALAN W. BISHOP, M.A., A.M.I.C.E.
Lecturer in Civil Engineering, Imperial College, University of London

. INTRODUCTION.

Walton Reservoir No. 1. is situated to the south of the river Thames between Molesey and Walton-upon-Thames. The proposed embankment is approximately 14,300 feet in length and has a maximum height of 50 feet.

The underlying strata consist of flood plain gravel of an average thickness of 15 to 20 feet overlain by a foot or more of top soil and, over one quarter of the length of the embankment, by a bed of brick earth which has a maximum thickness of 6 feet. Within the gravel itself there are lenses of 'bungum', a soft plastic clay, up to 4 feet in thickness and 70 feet in width. Beneath the gravel lies a thick bed of London Clay which has been proved to a depth of 100 feet at two points on the line of the embankment. Normal ground water level is only a few feet below the surface.

The embankment (fig. 1.) is to be constructed of a rolled gravel fill, with a puddled clay core wall which is carried down as a cut-off to key into the London Clay. This stratum forms the impervious bed of the reservoir.

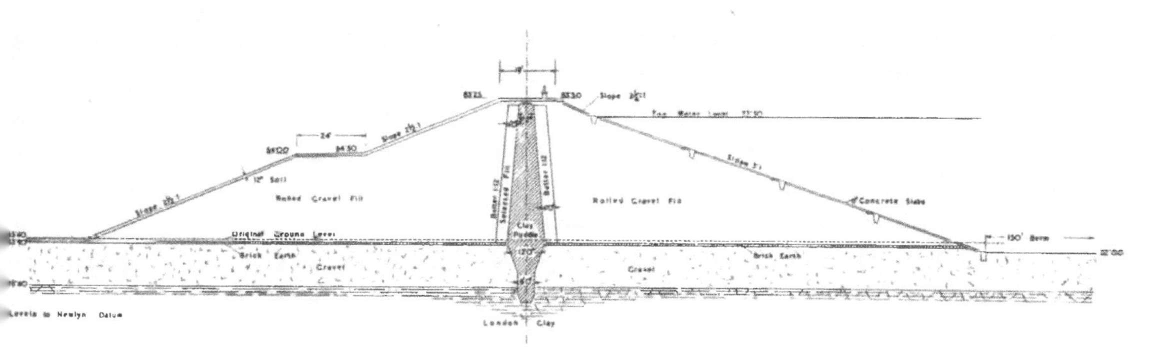

Walton reservoir No I Proposed embankment.cross section 42

FIG.1

2. SITE INVESTIGATION AND LABORATORY TESTS.

The great frictional strength of well graded gravels meant that possible failure zones would largely be confined to the puddle core itself, the thin strata of brick earth and bungum, and the London Clay which borings elsewhere in the Thames Valley had shown to be much softer near the surface than was expected. Data is available from other reservoirs on the shear strength of London Clay puddle after placing, and the natural strata were examined in detail by 19 boreholes, and 5 trial pits, Details of the strength variations in the London Clay have been published elsewhere. (Bishop 1947).

The investigation led to the following conclusions:

a) Puddled Clay.

Values of shear strength and index properties obtained in other investigations (Cooling and Golder, 1942, Bishop 1946) and in unpublished records indicate that during construction, and with the orthodox methods of placing, a shear strength of about 1.4 lb/per sq. inch will be obtained . Consolidation, slight drying and thixotropic hardening may more than double this strength, but as the effect on the factor of safety is small and involves an uncertain time factor, it is omitted from this analysis. The density of London Clay puddle at its normal moisture content (42-45%) is 110 lb per cubic foot.

b) Brick Earth.

This is a stratum composed of layers of dirty sand and brown or mottled sandy clay, and will consolidate rapidly on being loaded. Representative values taken from a cross section in pit 220 are:

Sandy Clay. (Brick Earth Stratum).

L.D. av. = 55% varies from 49% - 68%
P.L. av. = 18% varies from 15% - 21%

From equilibrium tests in shear box,

c = 3.6 lb/per square inch.
ϕ = 20°.

The density is taken as 120 lb per cubic foot.
and submerged density = 57 " " " "
Water Content = 20 - 30%

Samples were obtained from this pit from beneath a section of filling tipped 5 years earlier, and a comparison was made (Figs. 2 a & b) between values of shear strength predicted for that over-burden load from the measured values of c and ϕ, and values measured by immediate triaxial tests on the same samples. It is interesting to note that the actual strengths are in quite good agreement and tend to be slightly higher on the average.

c) Bungum.

This proved to be a difficult stratum to sample, as it occurs only in infrequent lenses in

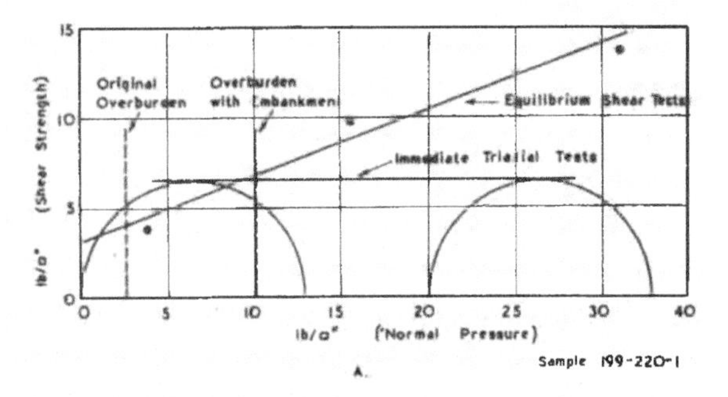

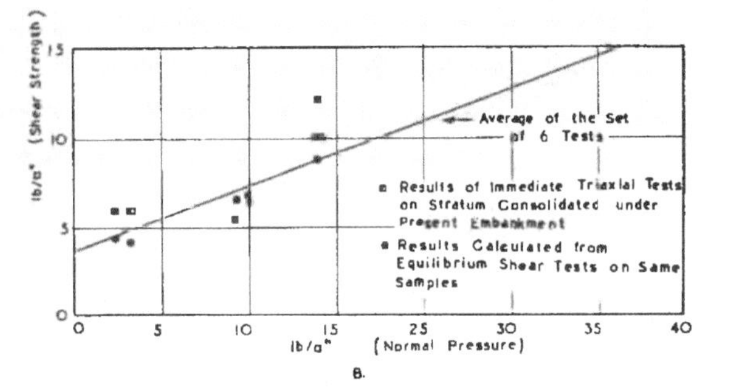

Comparison of actual strengths in brick earth stratum under partially constructed embankment with values calculated from equilibrium shear tests. (pit 199-220)

FIG.2

the gravel and may be thinner than the length of a sampling tube. It is a plastic tenaceous grey bue clay merging into sand on each side

L.L. = 31%
P.L. = 16%
m/c = 16 - 20%
c = 1.8 lb/sq. in
Ø = 25°
(average of 3 samples).

d) London Clay.

This is an overconsolidated silty clay with L.L. of 53-90 and P.L. of 17-34 and water content 21-32, usually termed 'stiff-fissured', although in the upper zones, where a considerable degree of softening has accompanied the pressure release due to the erosion of overlying beds, the compression strength may be as low as 10-20 lb. per square inch. There are considerable variations in strength both horizontally as well as vertically, but these do not obscure the general strength-depth relationship which can be related to the geological history of the stratum. (Bishop 1947) (Fig. 3). It is important to note that the coefficient of variation was almost the same for one pit where the samples lay within 30 feet of each other horizontally as for the whole site. This justifies the use of average values of strength in the stability analysis. The slightly lower strengths in Pit 218 may be accounted for the fact that the clay here was covered by the least thickness of gravel overburden.

As it was clear that a satisfactory embankment could not be constructed on clay of such low strengths as obtained in the upper zone, the effects of consolidation were examined. Samples 4 inches in diameter and 1¼ inches in thickness were consolidated in a large oedometer under loads of 1½ and 3 tons per square foot, and standard 3 inch x 1½ inch diameter test specimens built up of three layers cut from the fully consolidated samples. Immediate triaxial tests were then carried out in the same way as with the natural samples.

This technique was adopted as corresponding most closely to the conditions to which the sample would be subjected. Shear box tests could not be recommended as in fissured clays the results become most erratic. The results are given in Fig. 4 and show that a substantial increase in strength is obtained with consolidation. This fact is of the utmost importance at this site, and it will be seen later that without it the proposed embankment could not be safely constructed.

These samples were cut with their axes vertical, and the failure planes, which were on

TABLE I

	(a) SAMPLES FROM WHOLE SITE			(b) SAMPLES FROM PIT 218 ONLY		
Depth	No.of Samples	Average Strength	Coeff.of Variation	No.of Samples	Average Strength	Coeff.of Variation
Ft.		lb./sq.in			lb./sq.in	
0- 1	38	18	25%	19	16	20%
1- 2	49	23	18%	17	19	26%
2- 4	77	29	38%	19	23	42%
4- 8	65	49	30%	13	45	35%
8-12	86	58	36%			
12-16	46	74	35%			
16-20	32	69	29%			
23-27	12	73	30%			
30-60	43	71	42%			
75-105	23	99	33%			

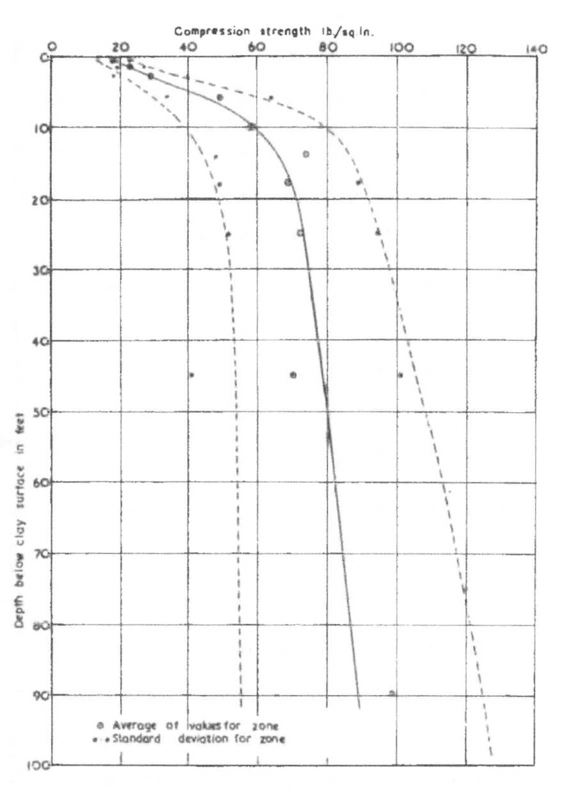

Variation of compression strength with depth
All vertical samples (471 tests)
Immediate triaxial tests-lateral pressure
=30 lb/sq in.

FIG.3

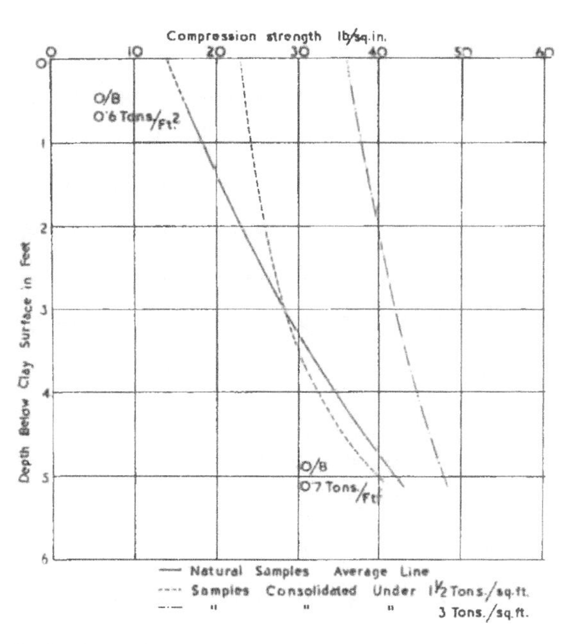

Variation of compression strength with depth-comparison of natural and consolidated vertical samples from pit 218(tests on 13 cores) Immediate triaxial tests-lateral pressure 30 lb/sq.in.

FIG.4

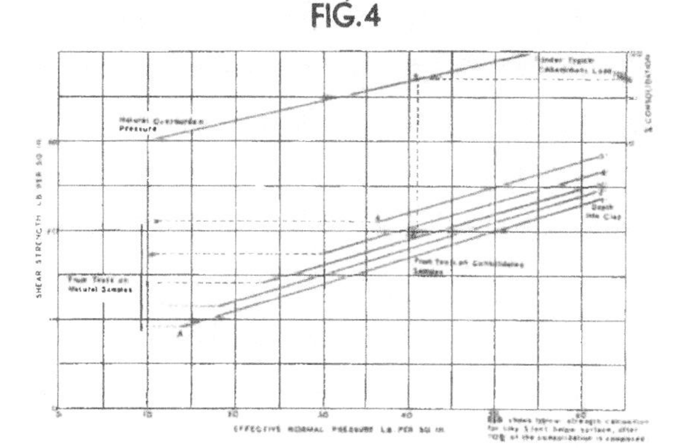

Shear strength plotted against effective pressure for a range of depths.

FIG.5

the average inclined at 56° to the plane of the major principal stress, were therefore also inclined at this angle to the horizontal plane in which the major part of any failure surface would tend to lie. x) As there was the possibility of lower shear strengths in this direction due to horizontal laminations, cores were also taken with their axes inclined to the vertical so as to allow the failure planes to coincide with any horizontal planes of weakness. Those showed a drop in strength of 28% when consolidated under the average weight of the embankment, but as the samples were consolidated in an oedometer this pressure was no longer normal to the laminations, and the strength will give a rather conservative facter of safety. As, however, only a few cores were obtained in this manner, the stability analysis is worked out in detail on the normal test results shown in Fig. 4. and the maximum effect of laminations estimated on a percentage drop in the clay strength.

In Fig. 5. these results are plotted as a relation between shear strength (taken as ½ compression strength) and effective pressure. It will be noted that until the effective pressure. It will be noted that until the effective pressure under which the clay is reconsolidated reaches a value A in Fig. 5. (which varies with its initial strength) no corresponding increase in strength occurs. It is difficult to account for this on disturbance and remoulding effects alone, and it seems to be more consistent with considerable residual lateral pressures produced by pressure release in a vertical direction on a heavily consolidated stratum. On unloading it would be the horizontal stress, now the major principal stress, which would control the strength, and a gain in strength on reconsolidation in an oedometer would only begin when the applied vertical pressure exceeded this. The residual lateral pressure cannot exceed the vertical pressure by a stress greater than the compression strength, and values estimated from Fig. 5 are in agreement with this.

x) The worst plane given by the $\phi = 0$ analysis, although giving the correct factor of safety, does not give the actual failure surface

The coefficient of consolidation is 10 ft^2/year.

e) Gravel.

The gravel stratum, although rather sandy, is fairly well graded. In the embankment good grading will be further ensured by the mixing of any segregated layers in excavation and placing. Density measurements of the natural stratum were not very satisfactory, and gave values which could not be repeated by compaction in the laboratory. Angles of friction were measured in a large shear box (Bishop 1948) for the more sandy material, and values up to 45° for dense packing were obtained. Based on the densities obtained in the laboratory and in rolled gravel elsewhere, the following values are taken:

In situ gravel	ϕ = 43°
Density submerged	= 79 lb/cu.ft.
Above water level	= 130 lb/cu.ft.
Rolled gravel fill	ϕ = 43°
Density as rolled (partly saturated)	= 135 lb/cu.ft.
Density, submerged	= 80 lb/cu.ft.

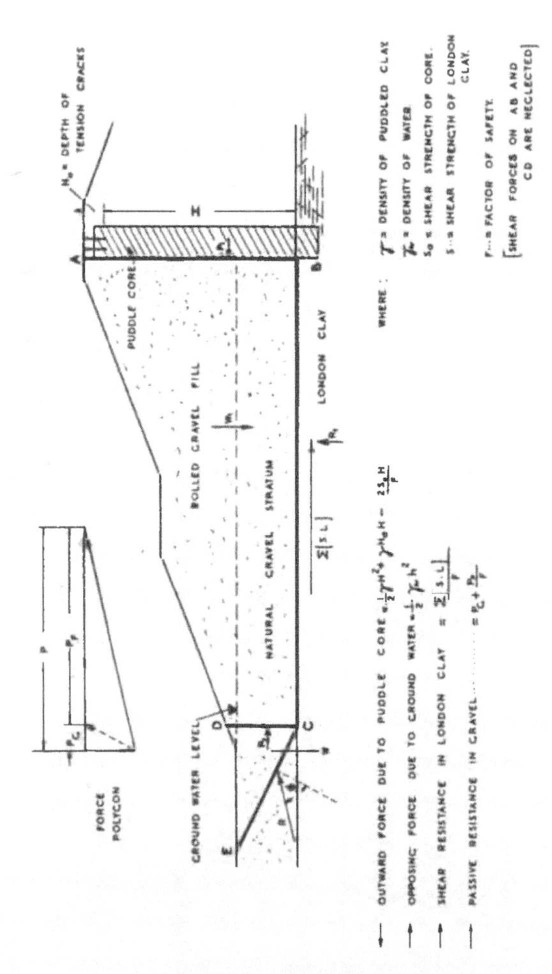

Analysis of failure in London clay surface

FIG.6

3. STABILITY ANALYSIS.

A. Failure in the London Clay.

a) Analysis of Lateral Slide.

The principal type of failure to be considered in design is that due to the pressure of the soft clay puddle core causing the embankment to slide horizontally against the shear resistance of the upper zone of the London Clay and the passive resistance of the gravel at the toe. For the clay core and the London Clay the immediate shear strength is used, i.e. it is assumed that no water content change would take place during failure, and the analysis is therefore a "$\phi = 0$" analysis as described elsewhere. (Skempton 1948).

The active pressure is calculated on the assumption of a tension crack of maximum depth $h_o = \frac{2s_o}{\gamma}$. Integrating the lateral pressure below this depth, and including the factor of safety in the shear strength term, the lateral force P_1 is obtained $= \frac{1}{2}\gamma H^2 + \gamma h_o H - 2 s_o H/F$, where F is the overall factor of safety.

At the toe the passive resistance of the gravel is obtained by finding the plane CE which gives the minimum value to the resistance P. It will be seen from the force polygon that part of this resistance P_G is due to gravity alone, and the remainder P_F is due to friction, and it is only on the latter that a factor of safety should be allowed.

$$\text{i.e. } P_2 = P_G + \frac{P_F}{F}$$

The worst condition will occur when the ground water level rises to the surface under flood conditions, as it reduces the weight of the gravel wedge, which is only partly compensated for by the water pressure $\frac{1}{2}\gamma_w h^2$ on the plane AB.

At any point in the London Clay the increase in pressure due to the embankment can be calculated and from the curves in Figs. 4 and 5 strength - depth profiles can be plotted (Fig.7) i.e. for no consolidation (M_o N_o) and full consolidation (M_1 N_1) in Fig. 7a. To find the degree of consolidation at any depth and time, curves have been plotted (Fig. 8) on the assumption that consolidation goes on in the upper layer as if it were the upper surface of a thick layer with a uniform load increment throughout its depth. (Terzaghi and Fröhlich 1936). Taking the minimum construction time for the embankment as 2 years, with a uniform increment of load, it can be shown that from the completion of construction onwards the percentage consolidation is the average over a period of 2 years from t to t + 2 years where t is the time after the completion of construction. Putting these values in Fig. 7 curves M_1 Q_o N_o etc. give the strength-depth relationship from the end of construction onwards. The minimum value X_o etc., must be used in each case.

The worst plane would actually pass through the points X_o ....., but for simplicity it is assumed to lie at the surface BC in all cases.

The total resisting force is then obtained by summing the forces along BC.

By equating forces we obtain the factor of safety, the position of CD being chosen to give the lowest value. The results may be plotted as a factor of safety - time curve (fig. 9).

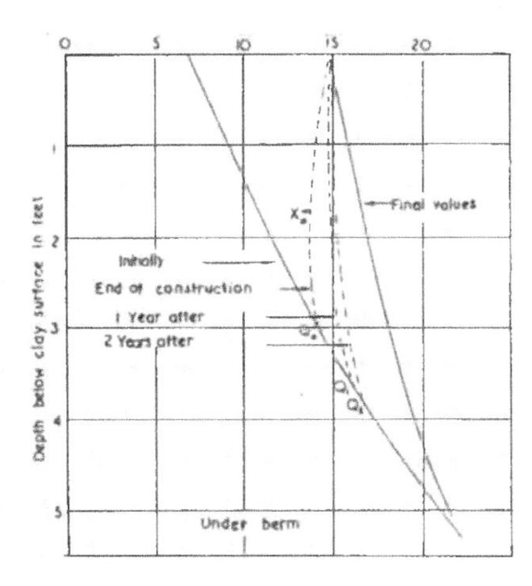

1/2 Compression strength lb/sq.inch.

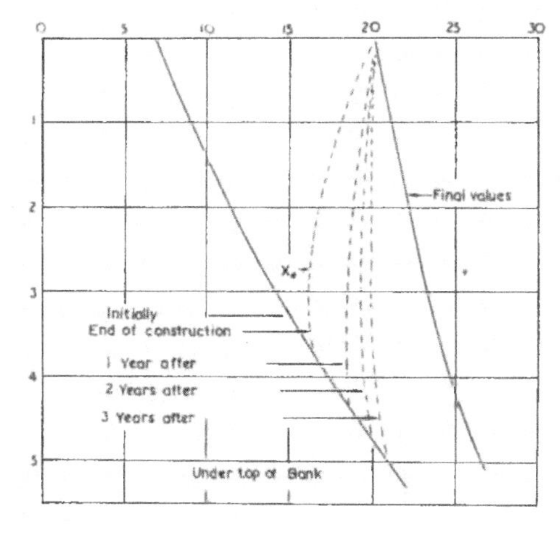

Effect of consolidation on strength of upper clay. Cross section 42 (Outer banks)

FIG.7

b) Discussion of Method.

An analysis of a failure of this type has not been published but a check on its accuracy can be obtained by analysing the stability of a similar type of dam which failed during construction in a soft surface stratum before consolidation had taken place. (See Cooling and Golder 1942 for details of the dam). This gives factor of safety of 0.8, and indicates that the method is conservative. If in this case 50% of the shear strength is assumed to be mobilised up the sides of the core as a form of 'arching', the method gives a factor of safety of 1.0. Without further evidence, however, such an allowance could not be made in design.

The total horizontal force acting outwards on the plane AB can be calculated by other methods for comparison:

I with the full active pressure of the core = 200 x 10^3 lb per ft. run.

II with the active pressure if reduced by 50% side shear = 155 x 10^3 lb per ft. run

III with the active pressure of the gravel + water pressure when the reservoir is full, acting outwards = 134 x 10^3 lb per ft. run

IV the horizontal thrust on the centre line in a homogeneous embankment calculated on elastic theory xa) = 118 x 10^3 lb per ft. run

This indicates that the core-wall itself creates the largest disturbing force, and that in this type of embankment other methods of analysis may be misleading.

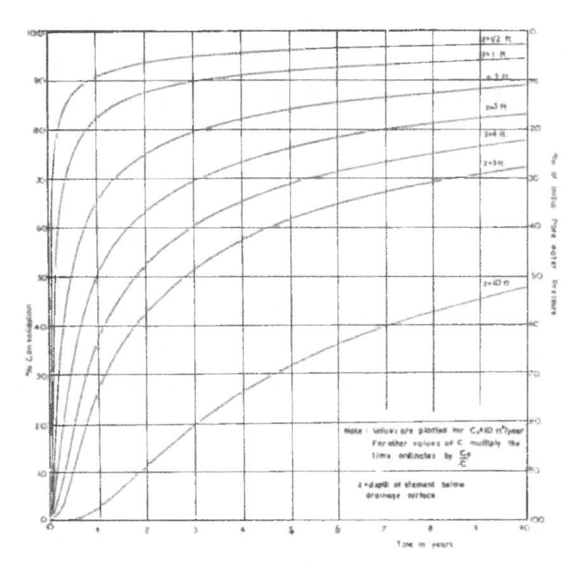

Degree of consolidation(Thick layer with drainage at upper face;load increment uniform throughout depth)

FIG.8

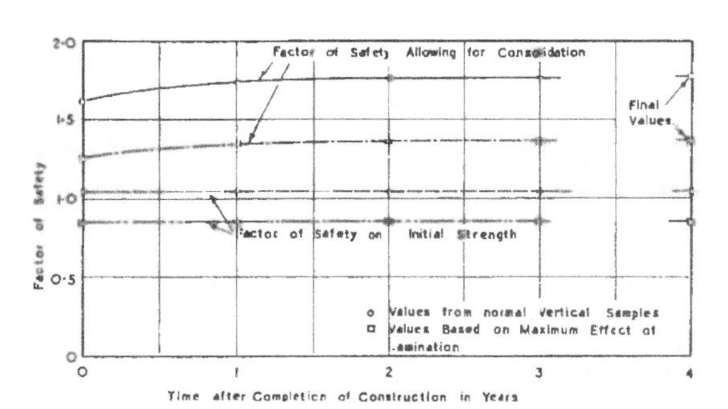

Increase of factor of safety with time (cross section 42 outer bank)

FIG.9

c) Results.

Taking the normal values of strength as given in Figs. 4 and 5, and giving the lower limit values based on the estimate of complete failure on laminations in parentheses, the following results are obtained xb):

<u>OUTER BANK</u>, Cross Section 42 flooded at the toe.

At completion of construction	F = 1.62 (1.27)
After 3 years	F = 1.77 (1.37)
Final	F = 1.78 (1.38)
If no consolidation had been allowed for (See Fig. 9).	F = 1.04 (0.85)

xa) Unpublished work by the Author using Southwell's Relaxation method.

xb) 3 significant figures are given for purpose of comparison. They do not indicate an absolute accuracy.

INNER BANK, Cross section 42.
At completion of construction F = 1.65 (1.28)
If no consolidation allowed for
F = 1.03 (0.83)
Reservoir filled, full softening allowed for at clay surface F = 2.50 (1.94)
Rapid draw-down to level of inner beam, full softening F = 1.33 (1.03)
Rapid draw-down to level of inner beam, 50% softening F = 1.52 (1.16)

d) Conclusions.

The lowest factor of safety is obtained on the inner slope of the embankment when the reservoir is emptied. This is due to the effects of consolidation under the inner slope being reduced by the uplift of the water; some softening would take place, but the amount can only be estimated. In view of the conservative nature of the method, and the fact that such failure would not be catastrophic, the factor of safety is considered to be satisfactory.

B. Failure in the Brick Earth Stratum.

A similar analysis can be made for the brick earth stratum, but allowing full consolidation owing to the greater coefficient of consolidation, and short drainage paths.

Results.

OUTER BANK, Cross Section 42.
Full consolidation F = 2.12

INNER BANK, Cross Section 42.
Full consolidation F = 2.04
Reservoir filled, full softening F = 2.68
Rapid draw-down F = 1.50

C. Failure in a Lense of Bungum.

As their extent and location could only be found with uneconomically close borings, the worst case of a complete lense underlying the embankment is considered, both at the top and bottom of the gravel stratum.

Results.

INNER BANK, Cross Section 42
Rapid draw-down, lense at upper edge of gravel F = 1.44
Rapid draw-down, lense at lower edge of gravel F = 1.46

D. Failure in Gravel Slope due to Rapid Draw Down.

This is unlikely in view of the permeable nature of the fill and the usual rate of draw-off. However taking $\phi = 43^{\circ}$ and $\epsilon = .33$ (Void ratio), and using the relationship that where α is the limiting slope and p is the specific gravity of the particles (= 2.70)

$$\tan\alpha = \frac{p-1}{p+\epsilon}$$

i.e. $\tan\alpha = .524$
i.e. a 1.9.:1 slope is just stable
For the loosely tipped material $\phi = 36^{\circ}$ $\epsilon = .61$
i.e. $\tan\alpha = .373$
and the limiting slope = 2.68 : 1

This indicates that the inner slopes are in any case safe against rapid draw down (at 3 : 1)

4. SUMMARY.

The satisfactory construction of the proposed earth dam for Walton Reservoir was found to depend on the behaviour of a thin soft zone at the top of the London Clay. A site investigation, laboratory tests and a stability analysis have been carried out with this end particularly in view, and indicate that its construction and stability when in use are dependent on the gain in strength due to consolidation under the weight of the embankment itself, but with a controlled rate of construction the factor of safety will be satisfactory. Its stability against failure in the other strata has been examined and is also satisfactory.

ACKNOWLEDGEMENTS.

The work was carried out for the Metropolitan Water Board, and the results are published by kind permission of H.F. Cronin, M.I.C.E. Cief Engineer. Most of the laboratory work was carried out by the Author at the Building Research Station, Watford.

The Author is grateful to G. Andrew Marshall, New Works Engineer, and G.N. Hooper, Resident Engineer at Walton during the investigation, and to Assistant Professor A.W. Skempton and L.F. Cooling, for their encouragement and cooperation.

BIBLIOGRAPHY.

BISHOP, A.W. 1946 "The Leakage of a Clay Core-Wall". Trans.Inst.Water Engineers. Vol.51
BISHOP, A.W. 1947 "Strength Variations in London Clay" Brussels Conf. of Comité Belge pour l'Etude des Argiles.
BISHOP, A.W. 1948 "A Large Shear Box for Testing Sands and Gravels" 2nd Int.Conf.Soil Mechanics.
COOLING, L.F. and GOLDER, H.Q. 1948 "The Analysis of the Failure of an Earth Dam during Construction". Journal Inst.C.E. Vol. 19.
A.W. SKEMPTON, 1948 "The $\phi = 0$ Analysis of Stability and Its Theoretical Basis". 2nd Int. Conf. Soil Mechanics.
TERZAGHI, K., and FROHLICK, O.K. 1936. "Theorie der Setzung von Tonschichten" Deuticke, Vienna.

-o-o-o-o-o-o-

UNDRAINED TRIAXIAL TESTS ON SATURATED SANDS AND THEIR SIGNIFICANCE IN THE GENERAL THEORY OF SHEAR STRENGTH

by

ALAN W. BISHOP, M.A., and GAMAL ELDIN, M.Sc.

SYNOPSIS

In this Article the results of a series of undrained triaxial-compression tests on saturated sand are presented. These indicate :

(1) that, under appropriate conditions, a frictional soil having no true cohesion and a dilatant * structure will exhibit zero angle of shearing resistance and will have the shear characteristics of a purely cohesive material with reference to total stresses ;

(2) that, if a certain value of negative pore-water pressure is reached during shear, these conditions cease to be fulfilled and an apparent angle of shearing resistance is measured.

The factors controlling the changes in pore-water pressure during the application of an all-round pressure are analysed in terms of the relative compressibilities of the soil structure, the pore water, and the soil grains ; and of the areas of contact between the soil grains.

The analysis leads to the conclusion that the angle of shearing resistance in undrained tests (ϕ_u) in saturated soils should have values which are too small to be observed experimentally. This includes soils having a dilatant structure, although, where this is combined with very low compressibility, the angle ϕ_u becomes appreciable.

Initial deviations from full saturation too small to be determined by direct measurement are shown to alter the order of magnitude of the apparent compressibility of the liquid phase to an extent which can cause large angles of undrained shearing resistance to be measured.

Deviations from full saturation of a different character occur even in initially fully saturated samples if the pore-water pressure during shear reaches large negative values, owing to the formation of small bubbles of water vapour and air freed from solution. The resulting angle of shearing resistance is estimated theoretically, and is compared with the values measured in the tests on sand, and with certain other tests on silt.

Dans cet article, on expose les résultats d'une série d'essais de compression triaxiale sans drainage sur du sable saturé. Ces essais montrent que :

(1) sous des conditions appropriées, un sol à frottement n'ayant aucune cohésion vraie et étant d'une structure dilatable * présentera un angle de résistance au cisaillement égal à zéro et aura les caractéristiques de cisaillement d'une matière véritablement cohésive relativement aux contraintes totales ;

(2) si en cours de cisaillement on obtient une certaine valeur de pression négative de l'eau intersticielle, les conditions ci-dessus cessent d'être remplies et on peut mesurer un angle apparent de résistance au cisaillement.

Les facteurs influant sur les changements dans la pression de l'eau intersticielle, lorsqu'on applique la pression dans toutes les directions, sont exprimés en se référant aux compressibilités relatives de la structure du sol, de l'eau intersticielle, et des grains du sol ; et aux surfaces de contact entre les grains du sol.

Cette analyse aboutit à la conclusion suivante : l'angle (ϕ_u) de résistance au cisaillement dans les essais sans drainage sur des sols saturés doit avoir des valeurs trop petites pour pouvoir être observé expérimentallement. Ceci comprend les sols ayant une structure dilatable, quoique, lorsque ceci est combiné avec une très faible compressibilité, l'angle ϕ_u devient appréciable.

On peut se rendre compte que des déviations initiales en ce qui concerne la saturation complète, déviations trop faibles pour pouvoir être déterminés par mesure directe, altèrent l'ordre de grandeur de la compressibilité apparente de la phase liquide, dans une proportion telle que cela peut provoquer la mesure de grands angles de résistance au cisaillement sans drainage.

Des déviations de la saturation complète, d'un caractère différent, peuvent se produire même dans des échantillons pleinement saturés au début de l'essai si la pression de l'eau intersticielle pendant le cisaillement atteint de fortes valeurs négatives, en raison de la formation de petites bulles de vapeur d'eau et d'air libérées de solution. On estime théoriquement l'angle de résistance au cisaillement qui en résulte, et on le compare avec les valeurs mesurées lors des essais sur sable, et avec certains autres essais sur limon.

* That is to say, a soil in which there is a net volume increase, at failure, in a drained test, or a net decrease in pore pressure in an undrained test.

* C'est-à-dire, un sol dans lequel il y a un accroissement net de volume au point de la rupture dans un essai avec drainage, ou une chute nette de la pression des pores dans un essai sans drainage.

Part I: Theoretical Consideration

by

ALAN W. BISHOP

HISTORICAL

The angle of shearing resistance is a measure of the rate of increase of shear strength with applied normal pressure. For the ranges of applied pressure encountered in practice, the shear-strength/pressure relationship is taken to be approximately linear, and can be expressed by Coulomb's empirical law, which was first published in 1776:

$$s = c + \sigma \tan \phi \quad . \quad . \quad . \quad . \quad . \quad . \quad . \quad . \quad . \quad . \quad . \quad (1),$$

where s denotes shear strength;
c ,, apparent cohesion; *
ϕ ,, angle of shearing resistance; *
σ ,, applied pressure normal to the shear surface.

When accurately controlled shear tests began to be carried out about 20 years ago it was soon realized that the angle of shearing resistance obtained from a series of tests on identical samples was, in fact, largely determined by the conditions of drainage from the sample during the test. The first satisfactory experimental evidence was provided by Terzaghi (1932), whose results indicated an angle of shearing resistance of about ½ degree for undrained † triaxial tests on a clay, whilst fully drained tests gave an angle of 23 degrees. Jurgenson obtained similar results on another clay (1934), the angles of shearing resistance for the undrained and the fully drained tests being approximately zero and 32 degrees respectively.

The fact that ϕ is approximately zero in undrained tests on saturated samples has since been confirmed by many investigators, including Terzaghi (1936), who quoted results on both clay and concrete, and Golder and Skempton (1948), who gave the results of tests on a wide range of soft and stiff fissured clays. Golder and Skempton found, however, that the same result was not obtained in tests on silts, or on shales and silt-stones.‡

It is also noteworthy that the fact that the angle of shearing resistance was generally found to be almost zero was not associated with any lack of frictional properties in the soil structure. Hvorslev (1937) and Skempton (1948*a*) determined the angles of true internal friction for a number of typical clays (the values varied between 10 degrees and 32 degrees), and Terzaghi (1936) quotes a value of 34 degrees for concrete.

EXPLANATION OF $\phi = 0$ IN UNDRAINED TESTS

An explanation of the fact that zero angle of shearing resistance is measured in undrained tests on saturated samples was first suggested by Terzaghi (1932, 1936). As generally accepted to-day, the explanation depends on three principles:

* These definitions follow Terzaghi (1943). It should be noted that Taylor (1948) refers to c and ϕ, used in this sense, as "effective cohesion" and "effective frictional angle" respectively, although the quantities refer to total stresses and not to effective intergranular stresses.

† Undrained tests were generally called "immediate tests" in England prior to 1949. In other countries (U.S.A., Netherlands, and Belgium) they are often called "quick" tests, since, owing to the low permeability of clay soils, the same result is obtained if the test is carried out relatively quickly even when drainage is not specifically prevented. Similarly, the fully drained tests are termed "slow" tests owing to the slow testing rate necessary to ensure full dissipation of the pore-water pressure. These terms, however, can be most misleading when used in dealing with soils of widely different permeabilities, since, for example, an undrained test may go on for several months in the study of creep, whilst a drained test on sand may be carried out in ¼ hour.

‡ The quick cell-test as carried out in the Netherlands and Belgium almost always indicates an angle of shearing resistance greater than zero, even in saturated clays. This appears, however, to be due to a method of analysing the results in which a Mohr envelope is drawn to stress circles which do not represent failure conditions, and cannot, therefore, provide relevant data.

(1) The mechanical properties, and hence the strength of a soil, are controlled solely by the intergranular forces.

(2) The effective area of contact between the soil grains is negligible. The pore-water pressure therefore acts equally all round the soil grains, and changes in it do not affect the intergranular forces.

(3) Since water is incompressible compared with the soil structure, a change in applied pressure is carried wholly by the pore water, and the stresses in the soil structure are not changed unless drainage conditions permit a volume change.

EXAMINATION OF BASIC PRINCIPLES

(1)—In studying the intergranular forces in a granular material it is necessary to consider a surface approximating to a plane but passing always through the pore space and the points of contact of the grains. Stresses and areas are then all considered as projected on to this plane.

Taking σ as the total stress on the plane ;
σ' as the average intergranular force per unit area of the plane (= effective stress) ;
u as the hydrostatic pressure in the pore water ; *
and a as the effective contact area between the grains per unit area of the plane ;

then $\sigma = \sigma' + (1 - a)u$

or $$\sigma' = \sigma - (1 - a)u = (\sigma - u) + au \quad \ldots \quad (2)$$

Thus, any strict experimental demonstration that the mechanical properties of a soil are controlled by the intergranular forces requires an independent measurement of the contact-area ratio a to enable the effective stresses to be calculated. This has, in fact, never been done. Since, however, there is a certain amount of evidence that a is very small (this is discussed in section *(2)*, below), σ' has been taken as being approximately equal to $(\sigma - u)$ and, on this basis, the work of Rendulic (1937) and Taylor (1944) provides experimental justification for this first principle in relation to undrained shear tests.

(2)—Direct experimental evidence on the value of the effective contact-area ratio is also limited. Terzaghi (1936) quotes " floating tests " as indicating a value approximating to zero for sands, and two types of buoyancy test as giving values of zero and 5 per cent. for clay samples. Taylor (1944) gives a value of about 3 per cent. from a triaxial test on clay, but only as an indication of its small value.

Terzaghi (1936) also quotes the low angle of shearing resistance in undrained tests on saturated clay and concrete as evidence of negligible values of contact area.

An approximate estimate of the minimum contact area can be made, at least for coarse-grained soils, from the crushing strength of the grains. Bowden and Tabor (1942), in the course of a study of the nature of friction between solids, have shown that real contact between apparently smooth bodies is limited to a large number of small point-contacts.† In the case of metals, the area of these contacts is found to be almost directly proportional to the total force between the surfaces, showing that the actual average pressure at the contact faces remains a constant, which can be related to the strength of the material. (If the contact area had been controlled by the elastic properties of the grains the area would have been proportional to (total force)$^{\frac{2}{3}}$.)

Applying this principle to the contact areas between soil grains :

$$\sigma' = a\sigma_p \quad \ldots \quad (3),$$

where σ_p = the value of the average contact pressure.

---

* This is sometimes called the neutral stress (Terzaghi, 1943), whilst Taylor (1944) uses this term for $(1 - a)u$.

† This is also true in the presence of a liquid.

Now, if the value of σ_p is controlled by the crushing strength S of the grains, then, owing to the presence of the pore-water pressure u,

$$\sigma_p - u = b.S \quad . \quad . \quad . \quad . \quad . \quad . \quad . \quad . \quad . \quad . \quad (4)$$

The constant b will depend on the type of surface failure produced, but its lowest value cannot be much lower than unity. Since the minimum value of b gives an indication of the maximum value of a, take as an example the values $b = 1$ and $S = 20{,}000$ lb. per square inch.*

For an effective stress of 100 lb. per square inch (such as is reached in a normal triaxial test) and $u = 0$, $a = \frac{100}{20{,}000} = 0{\cdot}5$ per cent.

In undisturbed samples the maximum consolidation pressure may control the value of a. In the case of a heavily over-consolidated soil, where $(\sigma - u)$ has an estimated maximum value of 600 lb. per square inch, this gives the value $a = 3$ per cent. For the sand tests quoted in this Article, where the consolidation pressure is only 5·3 lb. per square inch, the value of a is given as 0·03 per cent. approximately.

It is obvious, therefore, that even if this method gives merely the order of magnitude † of a, extremely accurate and systematic tests would be necessary to measure its value experimentally. Test results presented in this Article tend to confirm its very small value.

(*3*)—The third basic principle is that, in the absence of drainage, any change in applied pressure is carried wholly by a corresponding change in the pore-water pressure. This is derived from the following assumptions :

(*a*) that water is incompressible compared with the soil structure ;
(*b*) that since a is zero, the pore-water pressure acts all round the grains and does not affect the intergranular forces ;
(*c*) that, therefore, in the absence of any change in void ratio, the effective stresses remain unchanged ‡ (Terzaghi, 1943).

Direct experimental evidence is provided mainly by Taylor (1944) and Hilf (1948). Taylor measured the effect in clay, and found that the pore-water pressure increased by between 95 per cent. and 100 per cent. of the applied pressure, although only after a considerable time lag (except in cases where the changes in pressure were made during the actual shearing of the sample). Hilf's results implied changes in pore-water pressure equal to those in the total stress once saturation had been reached. The material used was compacted fill for an earth dam.

A theoretical examination of this principle can readily be made in terms of the actual compressibility of water relative to the soil structure and the soil grains. The introduction into the analysis of finite contact areas between the soil grains, however, necessitates physical assumptions about the nature of these contacts, which, in the absence of experimental data, must be considered as tentative.

Fig. 1 represents a magnified section of a single contact between two grains ; there are N of such contacts per unit area. If P is the average force per contact, then the intergranular force per unit area on the plane XX is NP. The total stress on the plane may thus be written :

$$\sigma = NP + (1 - a)u \quad . \quad . \quad . \quad . \quad . \quad . \quad . \quad . \quad . \quad (5)$$

Now if a soil grain is surrounded on all sides by a pressure u, it undergoes cubical

* This is approximately the strength of granites and quartzites under low hydrostatic pressures, and is taken as a constant for the ranges of pressure under consideration.

† Colloidal phenomena in the fine-particle range and chemically formed bonds may produce values of a which cannot be estimated in this way.

‡ This presumes the absence of shearing strain consequent on the change in total stress.

compression only. Hence, it is that part of the total stress which is in excess of u that causes compression of the soil structure * (due to distortion and displacement of the individual grains, Figs 2). If this excess stress be denoted by σ_c, then $\sigma_c = (\sigma - u)$, and hence, from equation (5),

$$\sigma_c = NP - au \quad \quad (6)$$

By definition, the effective stress σ' is the intergranular force per unit area ($= NP$), and so :

$$\sigma_c = \sigma' - au \quad \quad (7)$$

Fig. 1

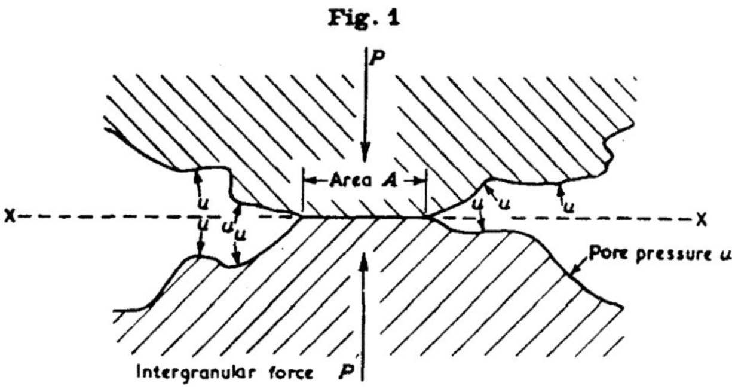

Forces at an intergranular contact

Thus the assumption of finite areas of contact between the grains leads to the conclusion that the volume changes due to the compressibility of the soil structure are controlled not by the effective stress as usually defined, but by $(\sigma' - au)$. The frictional force mobilized during shear is, however, a function of the intergranular force itself, and therefore depends on σ' alone.

Consider now a change $\delta\sigma$ in the total pressure on a fully saturated sample, and let $\delta\sigma'$ and δu be the consequent changes in the effective stress and the pore-water pressure respectively.

Figs 2

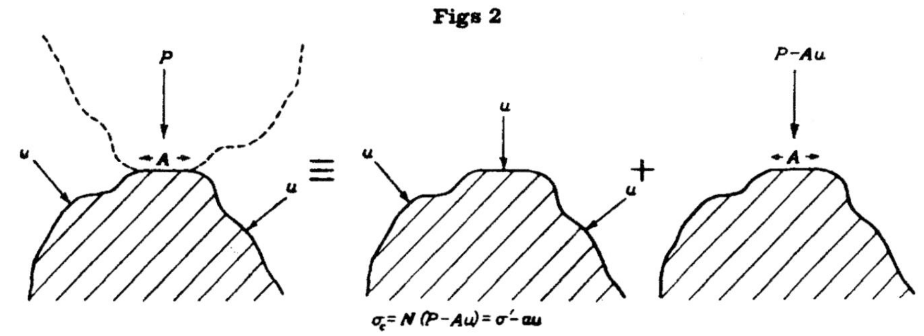

Equivalent forces on a soil grain

The compressibility of the soil structure is taken as C_s, where compressibility is defined as $C_s = -\frac{1}{V} \cdot \frac{\delta V}{\delta \sigma_c}$. If c_u is the compressibility of the pore water, and the soil grains are taken to be incompressible † then, for a sample whose initial volume is V and whose initial porosity is n, the volume changes can be written :

(a) Decrease in volume of the pore water $= C_u \, . \, n \, . \, V \, . \, \delta u$.
(b) Decrease in volume of the soil structure $= C_s \, . \, V \, . \, \delta\sigma_c$.

* This defines compressibility as it is measured in normal test procedure, where the pore-water pressure u is allowed to dissipate and is assumed to be zero in the calculation of the changes in stress.

† The error involved is small, since the cubic compressibility of the grains is of the order 1×10^{-7} to 2×10^{-7} per lb. per square inch, whilst that of water is $3{\cdot}4 \times 10^{-6}$ per lb. per square inch.

B

If drainage is prevented, these decreases must be equal—that is to say:

$$C_u \,.\, nV \,.\, \delta u = C_s \,.\, V \,.\, \delta\sigma_e$$

or

$$nC_u \,.\, \delta u = C_s \,.\, \delta\sigma_e \quad \ldots \ldots \quad (8)$$

Now $\sigma_e = \sigma - u$, and hence equation (8) may be written

$$n\frac{C_u}{C_s} \,.\, \delta u = \delta\sigma - \delta u$$

and thus

$$\delta u = \frac{1}{1 + n \,.\, \dfrac{C_u}{C_s}} \,.\, \delta\sigma \quad \ldots \ldots \quad (9)$$

Equation (9) gives the change in pore-water pressure consequent upon a change in cell pressure, and it will be seen that these changes become equal only when $\frac{C_u}{C_s} = 0$. Contrary to the usual assumption, it is not necessary to take a to be negligible in order to reach this conclusion.

However, in order to examine the effect of this change on the strength, $\delta\sigma'$ must be calculated, and this involves the value of a. Two main cases can be distinguished:

(1) when a is large, due to over-consolidation or chemical bonding, and is independent of stress changes during the test;

(2) when a is determined by the existing effective stress, and is consequently very small, but will undergo variations of the same order of magnitude during the test.

In the first case, $\delta\sigma_e = \delta\sigma' - a\delta u$ from equation (7). With equation (8), this gives $\delta\sigma' = \left(a + n\frac{C_u}{C_s}\right)\delta u.$

Substituting in equation (9),

$$\delta\sigma' = \frac{\left(a + n\dfrac{C_u}{C_s}\right)}{\left(1 + n\dfrac{C_u}{C_s}\right)}\delta\sigma \quad \ldots \ldots \quad (10)$$

The effect on an undrained triaxial test, therefore, of an increase in cell pressure is to cause a slight change in the initial effective stress in the sample, even when it is fully saturated. Subsequent changes in the effective stresses during shear will therefore be those corresponding to a sample consolidated to an effective stress of $(\sigma' + \delta\sigma')$ instead of σ'. If ϕ_{cu}, the angle of shearing resistance for consolidated undrained tests, is known, either from theoretical considerations or from tests, then the angle of shearing resistance for normal undrained tests can be deduced.

It can be seen from Fig. 3 (p. 22) that, if a change in minor principal stress $\delta\sigma_3$ produces a change in the deviator stress at failure of $\delta\sigma_d$, then

$$\sin\phi = \frac{\frac{1}{2}\delta\sigma_d}{\delta\sigma_3 + \frac{1}{2}\delta\sigma_d} \quad \ldots \ldots \quad (11),$$

where ϕ is the angle of shearing resistance,

or

$$\delta\sigma_d = \frac{2\sin\phi}{1 - \sin\phi} \,.\, \delta\sigma_3$$

Hence, in the undrained triaxial test, the change in the strength is given as

$$\delta\sigma_d = \frac{2\sin\phi_{cu}}{1 - \sin\phi_{cu}} \,.\, \delta\sigma' \quad \ldots \ldots \quad (12),$$

and, substituting equations (10) and (12) in equation (11), the angle of shearing resistance with respect to applied stress (ϕ_u) is obtained for the undrained test :

$$\sin \phi_u = \frac{1}{1 + \frac{1 - \sin \phi_{cu}}{\sin \phi_{cu}} \left(\frac{1 + n\frac{C_w}{C_s}}{a + n\frac{C_w}{C_s}} \right)} \quad \ldots \ldots \quad (13)$$

The second case can be solved approximately by assuming that the changes in contact area during the application of the cell pressure are given by equations (3) and (4). The expression obtained for $\sin \phi_u$ is :

$$\sin \phi = \frac{1}{1 + \frac{1 - \sin \phi_{cu}}{\sin \phi_{cu}} \left(\frac{1 + n\frac{C_w}{C_s}}{a + n\frac{C_w}{C_s}\left(1 + \frac{u}{b.S}\right)} \right)} \quad \ldots \ldots \quad (14)$$

The relative importance of the variables may be illustrated by evaluating several numerical examples. Taking values applicable to the tests on sand described in this Article :

$C_w = 3{\cdot}4 \times 10^{-6}$ per lb. per square inch

and $C_s = 570 \times 10^{-6}$ per lb. per square inch at $\sigma' = 5$ lb. per square inch.

$\phi_{cu} = \sin^{-1}(0{\cdot}583) = 35{\cdot}8$ degrees for a porosity $n = 45{\cdot}7$ per cent.

In Table 1 the values of ϕ_u and $\delta u/\delta\sigma$ for $a = 0{\cdot}03$ per cent., as given by equations (14) * and (9), are compared with the values calculated on the assumptions :

(1) that both a and δa (the change in a consequent upon a change in cell pressure) are zero ;

(2) that, due to some other cause, a is large (3 per cent.), and that δa is negligible.

A value of $u = 55$ lb. per square inch is taken as representing the average for the series of tests.

Table 1

	$a = 0{\cdot}03$ per cent.	(1) $a = 0$ $\delta a = 0$	(2) $a = 3$ per cent. $\delta a = 0$
ϕ_u	0·24 degree	0·22 degree	2·50 degrees
$\delta u/\delta\sigma$	99·73 per cent.	99·73 per cent.	99·73 per cent.

The results in Table 1 clearly show that, in undrained tests on a fully saturated loose sand, the angle of shearing resistance should be equal to zero, within the limits of accuracy of experimental measurement. It is also clear that no deductions about the value of the contact-area ratio can be made from the change in pore-water pressure with cell pressure, since this is a function of compressibilities only ; and, furthermore, that unless a reaches much larger values than seem theoretically to be probable, except in very heavily over-consolidated or cemented soils, the effect of a on the value of ϕ_u is small compared with that of the compressibility of the pore water.

In Table 2 the effects of compressibility on ϕ_u and $\delta u/\delta\sigma$ are compared for a number of typical soils, taking both a and δa as being equal to zero.

* The values $b = 1$ and $s = 20{,}000$ lb. per square inch are taken, as before.

Table 2

Soil	Soft clay	Stiff clay	Compact silt	Loose sand	Medium dense sand
n : per cent. ..	60	37	35	45·7	43
ϕ_{cu} : degrees ..	13	17	40	35·8	67
C_s : per lb. per sq. in.	1×10^{-2}	7×10^{-4}	2×10^{-3}	$5{\cdot}7 \times 10^{-4}$	3×10^{-4}
ϕ_u : degrees ..	0·003	0·04	0·06	0·22	3·0
$\delta u/\delta\sigma$: per cent.	99·98	99·82	99·94	99·73	99·51

On the basis of the theory as presented in the foregoing pages it should therefore be impossible, under normal laboratory conditions, to measure an angle of shearing resistance differing appreciably from zero in an undrained test on any fully saturated soil except in the dense-sand range. Here the value of ϕ_u rises rather rapidly owing to an increase in ϕ_{cu}, in addition to a decrease in compressibility. A value can be estimated only very approximately, since the highly dilatant structure causes internal cavitation in undrained tests when normal cell pressures are used, and comparable values of ϕ_{cu} cannot be obtained ; ϕ_u may, however, reach a value of 10 degrees or more.

DEVIATIONS FROM $\phi = 0$

The analysis presented in the previous section of this Article indicates that the angle of shearing resistance in undrained tests on saturated soils should approximate closely to zero for all soils, and, in fact, should be of measurable magnitude only in the case of dense sands.

However, Golder and Skempton (1948) measured angles of shearing resistance varying between 18 degrees and 38 degrees on five different silts, which were in all cases fully saturated. This unexpected result was attributed to dilatancy, although no fully satisfactory explanation could be given,* and this led Skempton (1948*b*) to omit dilatant soils from his theoretical analysis of the changes in pore-water pressure during shear. There is, however, no *a priori* reason for this omission,† and, as will be seen, an explanation of this result can be found in a departure from the conditions assumed to obtain during the test.

It has been shown in the previous section of this Article that deviations from the ($\phi = 0$) result are due primarily to a departure from the constant-volume condition normally assumed in undrained tests on fully saturated samples, and that this departure is due to the compressibility of the pore water. This would not, however, account for angles of from 18 degrees to 38 degrees in relatively compressible soils. Additional volume changes in an undrained test may be due to two causes :

(1) incomplete saturation or entrapped air ;

(2) negative pressures, set up in the pore water during shear, of sufficient magnitude to cause cavitation—that is to say, the formation of voids filled with water vapour and gas freed from solution.

Incomplete saturation.—This is generally assumed to be the cause of appreciable angles of undrained shearing resistance, and is usually associated with a measurable percentage of air voids. Normal density measurements, however, permit the estimation of the air voids to an accuracy of only about 1 per cent., and many so-called " fully saturated " soils may, in fact, contain 1 per cent. or more of air or gas in the voids.

The effect even of this small volume of air is to alter radically the compressibility of the

* This point is also discussed by Kjellman (1948).

† An extension of this theory to include the effects of dilatancy has now been made by Odenstad (1949) and in a more general form by Skempton (awaiting publication).

liquid phase. In order to estimate the magnitude of this effect, an approximate calculation * can be made, based on Boyle's law and Henry's law of solubility. Henry's law states that, at a given temperature, the weight of gas that will dissolve in a given volume of liquid is directly proportional to the pressure (and hence, by Boyle's law, the volume of gas that will dissolve is the same at all pressures).

Let V_v denote the volume of voids ;

$100x$,, the percentage of free air in the voids when the sample is unconfined ;

p_0 ,, the initial pressure of the air in the voids when the sample is unconfined (absolute †) ;

H ,, Henry's coefficient of solubility (= 0·02 cubic centimetre of air per cubic centimetre of water approximately, at room temperature).

Then the total volume of air in the unconfined sample

$$= x \,.\, V_v + (1 - x) \,.\, V_v \,.\, H \qquad (15),$$

and hence, the volume of free air at a new pore pressure p (absolute) is

$$V = \{x \,.\, V_v + (1 - x) \,.\, V_v \,.\, H\}p_0/p - (1 - x)V_v \,.\, H \qquad (16)$$

Defining the additional compressibility of the liquid phase with reference to its initial volume as $C = \frac{-1}{V_v} \cdot \frac{\delta V}{\delta p}$, then

$$C = \{x + (1 - x)H\}\frac{p_0}{p^2} \qquad (17)$$

It is important to note that this additional compressibility is operative only until all the free air is dissolved. From equation (16) it will be seen that, when $V = 0$, $p/p_0 = 1 + \frac{x}{(1 - x)H}$; that is to say, the increase in pore pressure has the value given by the expression

$$p - p_0 = p_0 \cdot \frac{x}{(1 - x)H} \qquad (18)$$

Using equation (9) and assuming that C is large compared with C_u, it is found that the limiting increase in total pressure within which the additional compressibility is operative (σ_s) is given as

$$\sigma_s = \frac{x}{1 - x} \cdot \frac{p_0}{H} + x \cdot \frac{n}{C_s} \qquad (19)$$

The order of magnitude of this effect is illustrated by Table 3, in which the soils referred to in Table 2 are compared on the basis of $(\phi_u)_0$ (the angle of undrained shearing resistance under an initial small increase of pressure) and σ_s, the saturation pressure (above which the ($\phi = 0$) result should be obtained). The value of x is taken as 1 per cent. (the effect of larger values can easily be calculated for comparison) and a is taken as zero. The initial pore pressure is taken as being equal to the atmospheric pressure, but this would not necessarily be true in undisturbed samples from considerable depths.‡

* This neglects the effect of the vapour pressure of water, which is small at room temperature, and also that of the surface tension round the air voids, which becomes important when the bubbles decrease to the size of clay particles (Terzaghi, 1943). A full discussion of these effects is outside the scope of this Article. Calculations based on similar assumptions were made by Bruggeman, Zangar, and Brahtz (1939), and an experimental comparison was made by Hamilton (1939).

† Except where stated, all other stresses are measured above atmospheric pressure, as is usual in laboratory practice.

‡ The negative pore pressure set up in undisturbed samples from deep bore holes may be deduced from expressions obtained by Hansen and Gibson (1949). The constant-volume assumption requires pressures of from − 1½ to − 2 atmospheres in samples from a depth of 100 feet. Thus cavitation (see " Excess negative pore-water pressures," below) may occur in many " saturated " undisturbed samples, and the pressure in the voids thus formed would not in general be equal to atmospheric pressure.

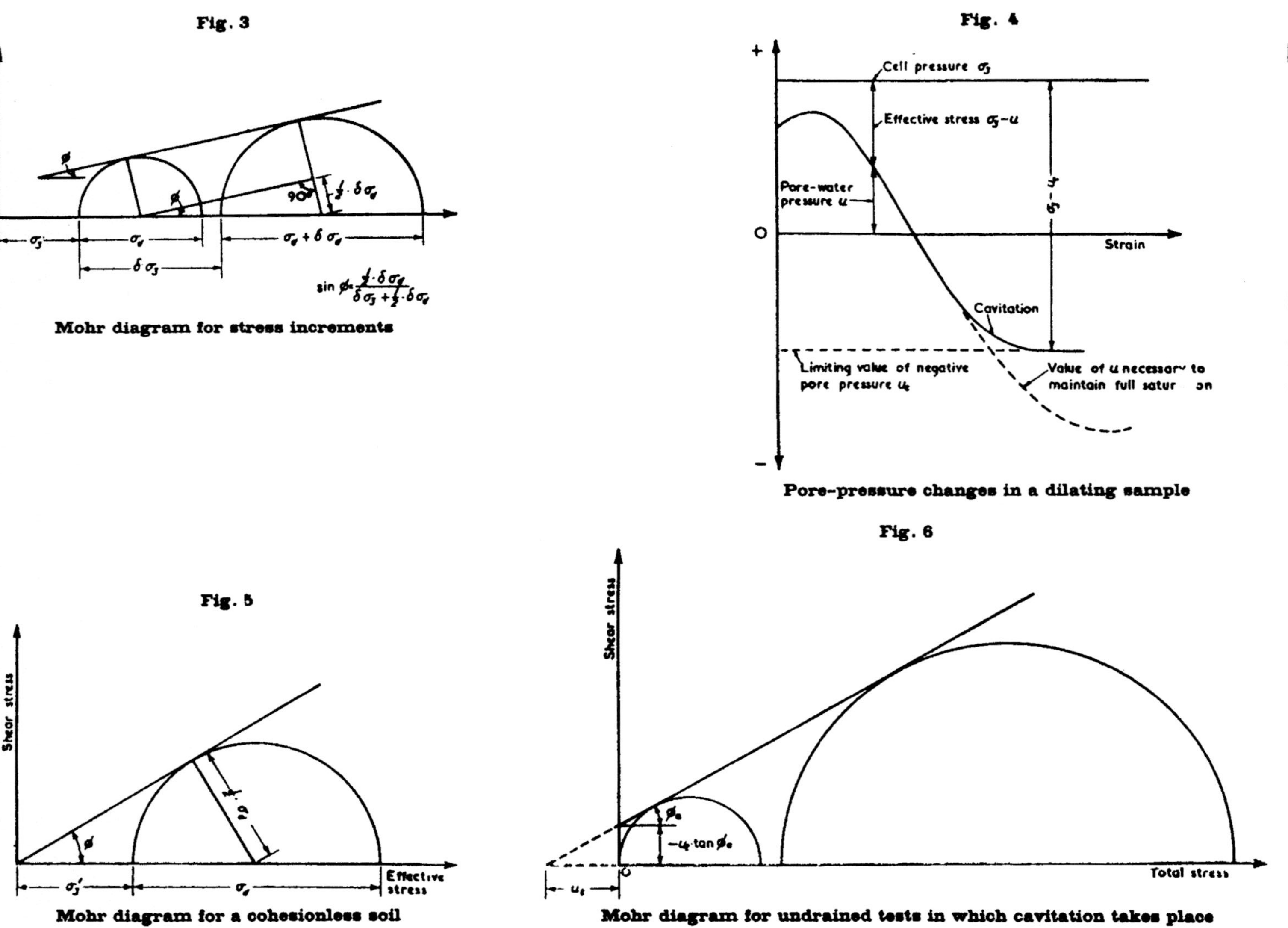

Fig. 3
Mohr diagram for stress increments

Fig. 4
Pore-pressure changes in a dilating sample

Fig. 5
Mohr diagram for a cohesionless soil

Fig. 6
Mohr diagram for undrained tests in which cavitation takes place

Table 3

Soil	Soft clay	Stiff clay	Silt	Loose sand	Medium dense sand
ϕ_{cu} : degrees	15	20	30	32	40
$(\phi_u)_e$: degrees	2	12	12	24	35
σ_s : lb. per square inch ..	8	13	9	15	22

Note : the values of ϕ_{cu} employed in Table 3 are estimated as being those applicable to soils with about 1 per cent. of air voids.

The conclusion may be drawn from Table 3 that the slightest departure from full saturation leads to the measurement of very considerable angles of undrained shearing resistance, even in the case of clays. For all except very incompressible soils, however, this effect disappears under high cell pressures.* It does not, therefore, provide a full explanation of the test results on silts of Golder and Skempton,† although it demonstrates that complete freedom from entrapped air is necessary in undrained tests on sand if satisfactory results are to be obtained.

Excessive negative pore-water pressures.—The effect of excessive negative pore-water pressures is most easily understood by examining the changes in pore-water pressure which take place during shear. These changes are illustrated in Fig. 4. As demonstrated earlier, the effective stress $\sigma_3' = \sigma_3 - u$ to within practical limits of accuracy, where σ_3 = the cell pressure and u = the pore pressure.

In a dilatant soil the pore pressure normally undergoes a slight increase at small strains and then decreases as the structure tends to dilate. This decrease may terminate at a definite strain, controlled by the compressibility of the soil structure, and the strength is then dependent upon the soil structure and almost independent of the cell pressure, as discussed earlier. Alternatively, the drop in pore pressure may be sufficient to produce large negative values, and a point will then be reached at which bubbles of water vapour and gas begin to appear (cavitation). This implies a departure from the fully saturated condition, and a further increase in strain will not reduce the pore pressure below a limiting value at which the growth of bubbles is able to continue.

If u_t is this limiting pore pressure, it will be seen from Fig. 4 that the maximum value of the effective minor principal stress σ_3' is $\sigma_3 - u_t$. In the case of a purely frictional soil, it follows from Fig. 5 that the deviator stress at failure, σ_d, is given by the relation

$$\tfrac{1}{2}\sigma_d = (\sigma_3' + \tfrac{1}{2}\sigma_d) \,.\, \sin \phi_e$$

or

$$\tfrac{1}{2}\sigma_d = \frac{\sin \phi_e \,.\, \sigma_3'}{1 - \sin \phi_e}$$

$$= \frac{\sin \phi_e \,.\, (\sigma_3 - u_t)}{1 - \sin \phi_e} \quad . \; . \; . \; . \; . \; . \; . \; . \; . \; . \; . \quad (20)$$

For a series of tests in which the negative pore pressure reaches the limiting value u_t, the results may be expressed as a Mohr envelope (Fig. 6). Thus it is shown that a departure from the fully saturated condition due to cavitation leads to the measurement of an angle

* It has been known for some time that the unconfined strength is often lower than the average of the strengths at higher cell pressures for saturated undisturbed clay samples, but this has usually been attributed solely to the presence of fissures (Bishop, 1947 ; Golder and Skempton, 1948).

† It is, however, a probable cause of the high angles of shearing resistance measured in clay shales and silt-stones (Golder and Skempton, 1948).

of undrained shearing resistance, with respect to applied stresses, equal to ϕ_e, the angle of true internal friction. In a cohesionless soil an apparent cohesion of $(-u_t \,.\, \tan \phi_e)$ is measured.

It is important to note that the value of u_t does not influence the angle of shearing resistance provided that it is consistent for a given series of samples. The angle therefore remains constant until it reaches a sharp transition to the normal ($\phi = 0$) condition. If the initial consolidation pressure is σ_0' and the angle of consolidated undrained shearing resistance is ϕ_{cu}, then the transition will occur for normally consolidated soils when the deviator stress reaches the value

$$2 \frac{\sin \phi_{cu}}{1 - \sin \phi_{cu}} \,.\, \sigma_0' \qquad \qquad (21)$$

With sand, under laboratory conditions, a value of u_t of about -1 atmosphere would be expected, and hence, taking $\phi = 32$ degrees, a value of $c = 14{\cdot}7 \times 0{\cdot}625 = 9{\cdot}2$ lb. per square inch would be expected, which can be compared with the result of the laboratory tests. It is difficult to estimate the value of c in fine-grained soils, since the limiting negative pressure is affected by the grain size itself, as well as by the amount of dissolved gas and the presence of minute bubbles to act as nuclei; also, the true cohesion of the soils becomes of greater importance than the term $(-u_t \,.\, \tan \phi_e)$.

However, an examination of the tests on silt quoted by Golder and Skempton indicates that the angles of shearing resistance measured correspond quite closely to the values of the angle of true internal friction typical of silts and silty clays (as given by Skempton, 1948*b*). Also, with only one exception, very low values of apparent cohesion were measured (between 3 and 7 lb. per square inch). This form of departure from full saturation, therefore, seems the most reasonable explanation of the results.

SUMMARY OF THEORETICAL TREATMENT

(1) An analysis of the changes in effective stress in a saturated sample of soil during the application of an all-round pressure, on the basis of the relative compressibilities of the soil structure, the pore water, and the soil grains, leads to the following conclusions :—

(*a*) In undrained shear tests an angle of shearing resistance approximating closely to zero should be measured for all soils including those with a dilatant structure, provided that the sample remains fully saturated during the test. The only qualification to this statement is that, in soils such as dense sands where a strongly dilatant structure is combined with a very low compressibility, the combination of the two effects will produce measurable angles of shearing resistance.

(*b*) If the contact-area ratio a is controlled by a limiting pressure related to the strength of the grains, then, at the values of effective stress normally used in the laboratory, the value of the contact-area ratio will be too small to affect the test results. If a has a much larger value due to other causes (such as over-consolidation or colloidal phenomena), the effect on the angle of shearing resistance should be just measurable, but the pore-pressure changes due to changes in the cell pressure will be unaffected since they are a function of the relative compressibilities only.

(*c*) Two types of deviation from full saturation may be responsible for the measurement of large angles of undrained shearing resistance in soils in which conclusion (*a*) indicates that ϕ_u should be zero :—

(i) Slight initial deviations from full saturation outside the limits of direct measurement produce angles of shearing resistance under low pressures even in very compressible soils. The applied pressure below which ϕ_u is not equal to zero increases with a decrease in compressibility, and, in dense soils, is of major importance in interpreting test results.

(ii) Negative pore pressures set up during shear in dilatant soils tested at normal cell pressures may lead to cavitation at a limiting tension in the pore water. This increase in volume results in the measurement of an angle of shearing resistance equal to the angle of true internal friction over the whole range of cell pressures for which this limiting value of pore pressure is reached. Above this point, the conditions required for ϕ_u to be zero will be satisfied.

(2) Published test data provide ample confirmation of the statement that $\phi_u = 0$ for saturated clays under the cell pressures used in normal testing practice. No satisfactory data are available in the case of soils with a dilatant structure, and the following points especially require experimental demonstration :

(*a*) that $\phi_u = 0$ in a soil with a dilatant structure ;

(*b*) that, when $\phi_u \neq 0$ in an initially fully saturated sample, this is due to a limiting negative pore pressure which causes a departure from the conditions assumed to hold in an undrained test.

In the following section of this Article, experimental evidence which confirms these points will be presented.

Part II : Experimental Results

by

Alan W. Bishop and Gamal Eldin

TEST PROCEDURE

The sand used in the tests was the medium-to-fine fraction from a well graded sand of the Folkestone Beds, and was obtained from a deposit being worked adjacent to the River Darent near Brasted, in Kent. The particle-size distribution is shown in Fig. 7 ; the limiting porosities (dry) were found to be 46·2 per cent. and 33·2 per cent. (after Kolbuszewski, 1948).

Preliminary tests and theoretical considerations both showed the necessity of completely eliminating entrapped air if consistent results were to be obtained. This difficulty was largely overcome by boiling the sand in water before depositing it under water through a funnel, as shown in Fig. 8, and by always using freshly boiled water in the apparatus. Considerable care was required to avoid segregation of the particle sizes.

Slight variations of initial porosity were found to be unavoidable, and made the exact repetition of tests difficult. It was therefore decided to increase the number of tests to cover small range of porosities, and plot the test results as deviator stress against porosity. Values for a specific porosity could be read off the curve thus obtained.

Fig. 7

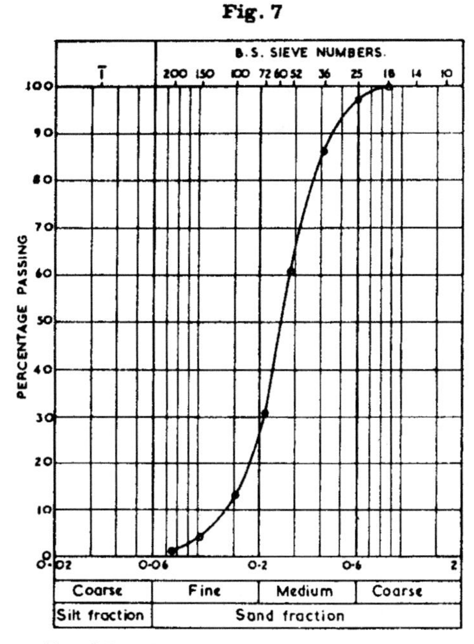

Particle-size distribution : Brasted medium fine sand

Fig. 8

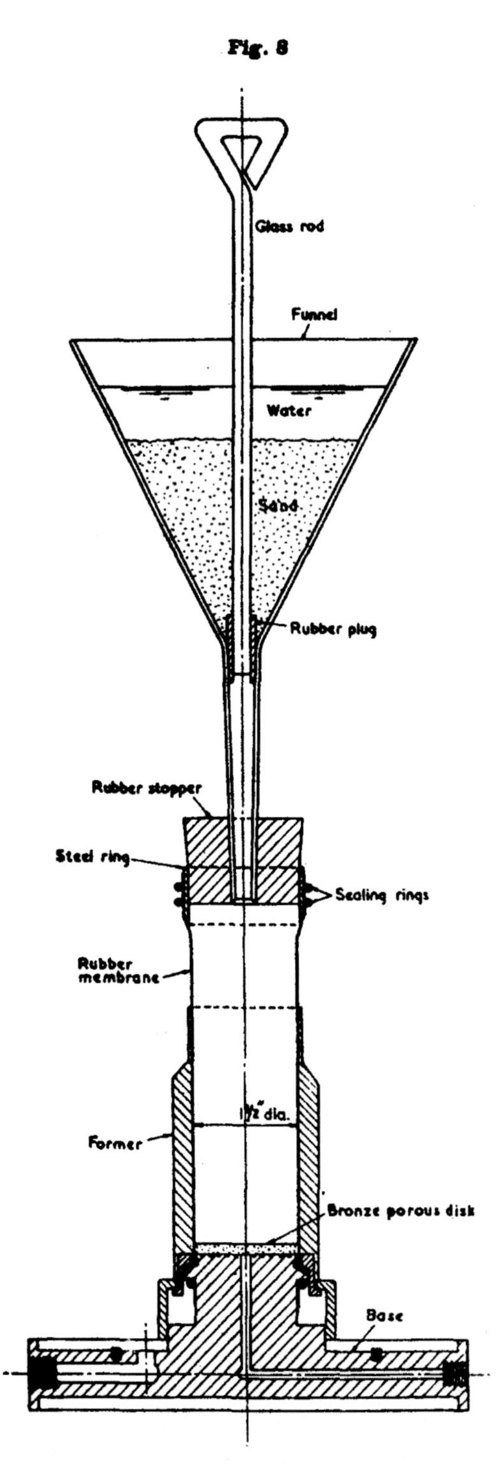

Method of placing sample under water

The lay-out of the apparatus is shown diagrammatically in Fig. 9. The sample is contained in a thin cylindrical rubber membrane (about 0·01 inch in thickness) which is sealed at one end to the base of the cell and at the other end to a duralumin cap. The sample rests on a filter disk of sintered bronze which is connected, through the base of the cell and a flexible copper tube, with the pressure-measuring system. Changes in pore pressure can be measured without any movement of water from the sample by adjusting the screw-controlled piston to maintain a constant mercury-level in the capillary tube D and reading off the changes in pressure from the gauge B. When the pressure becomes negative the readings are continued in the same way on the mercury manometer C. This method has proved to be extremely simple and sensitive in operation, the screw on this apparatus having 20 threads per inch. The piston, which is 1¼ inch in diameter, is fitted with two Gaco rings which give a smooth-running leak-proof seal.

The test procedure followed in a typical test is outlined below. The sample is deposited with a former surrounding the rubber membrane, and the cap is placed under water and sealed. A small negative pressure is applied by lowering the burette F below the level of the sample, the valve E being open. The dimensions of the sample are measured, and the upper section of the triaxial cell is placed in position. The consolidation pressure is now applied by displacing water (whose pressure is measured by gauge A) into the cell by means of compressed air ; F is meanwhile raised to bring the water-level up to the level of the sample, and indicates volume changes during consolidation.

Gauge A, therefore, indicates the consolidation pressure, since the pore-water pressure is zero. Valve E * is now closed and

* This valve must remain completely leak-proof for considerable periods ; a Klinger high-pressure plug-valve has proved extremely satisfactory for this purpose. If the pressure lead is sealed at the connexion to the cell, a pressure change of 100 lb. per square inch on gauge B, produced by the screw control, results only in a small fully recoverable displacement in the mercury-level.

Fig. 9

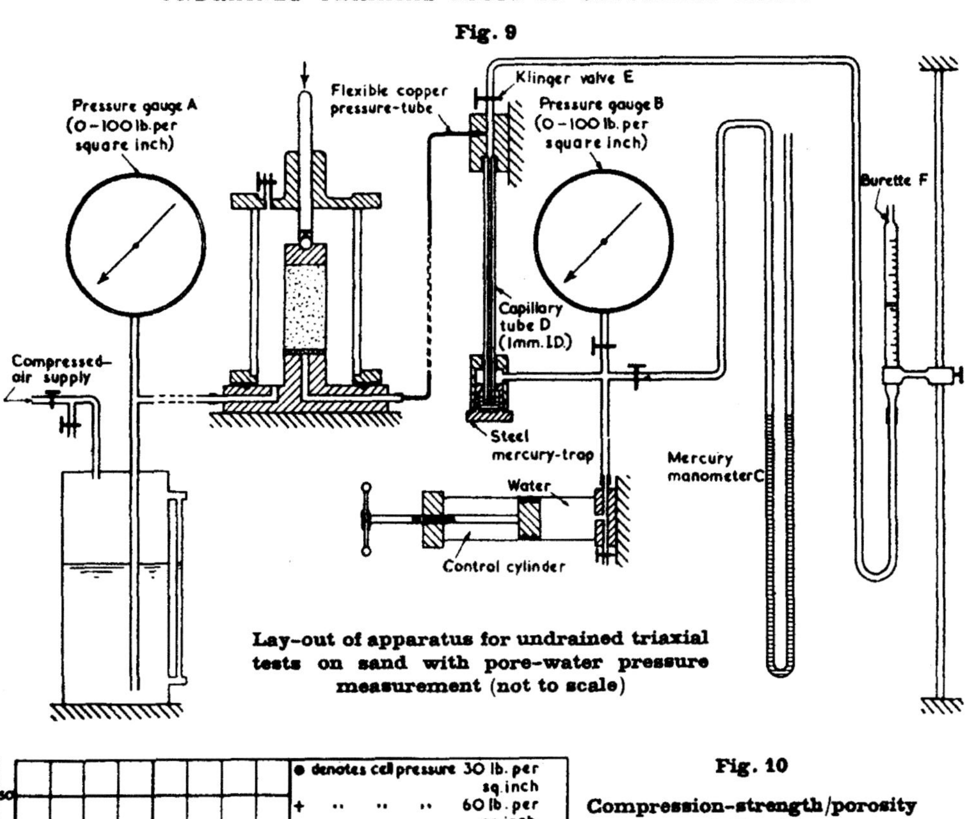

Lay-out of apparatus for undrained triaxial tests on sand with pore-water pressure measurement (not to scale)

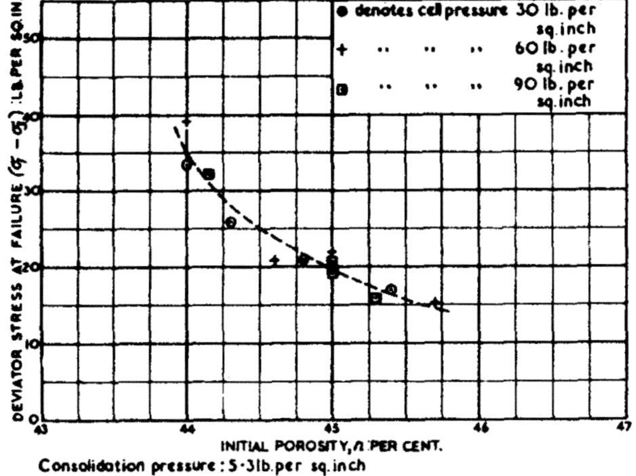

Consolidation pressure : 5·3 lb. per sq. inch

Fig. 10

Compression-strength/porosity relationship

Fig. 11

Mohr diagram for undrained tests on loose sand

SHEAR STRESS : LB. PER SQUARE INCH

TOTAL STRESS : LB. PER SQUARE INCH

Porosity: 44·8 – 45·0 per cent. Consolidation pressure: 5·3 lb. per sq. inch Fully saturated the undrained test can be carried out at any cell pressure required, the pore-pressure changes being given by gauge B.

A rate of strain of approximately ½ per cent. per minute has been used throughout.

TEST RESULTS

(1) In the first series of tests, to examine the effect of dilatancy on the angle of shearing resistance for samples which remained fully saturated during the test, a fairly loose sand was used, since it was intended to avoid excessive negative pore pressures. The initial porosity varied between 44·0 per cent. and 45·7 per cent., and all samples in this series were consolidated to the same initial pressure of 5·3 lb. per square inch.

Figs 12

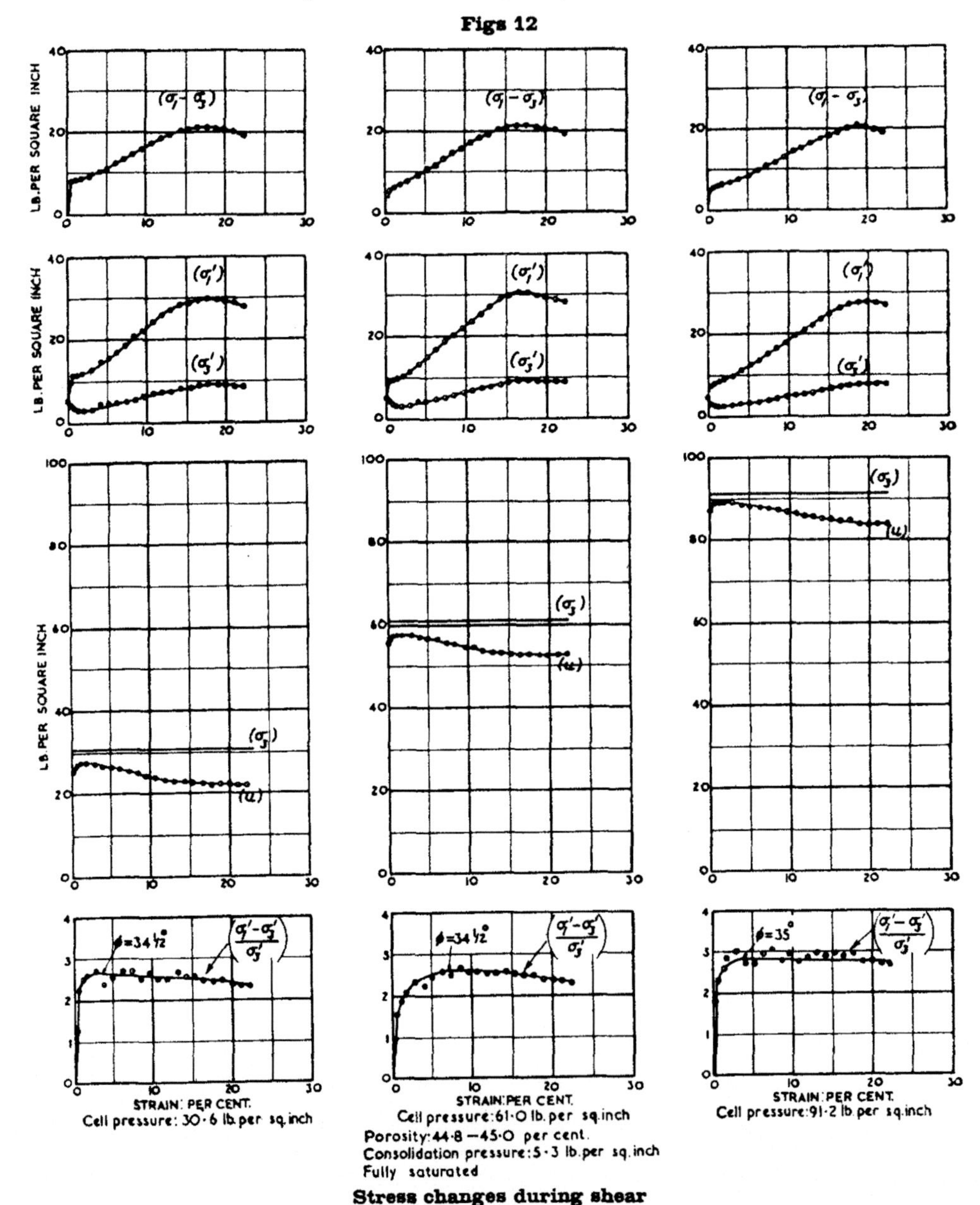

Stress changes during shear

The results are summarized in Fig. 10, where the deviator stress at failure is plotted against the initial porosity for the three values of cell pressure used (30, 60, and 90 lb. per square inch). It will be seen that the results show no significant variation of strength with applied pressure.

Fig. 13

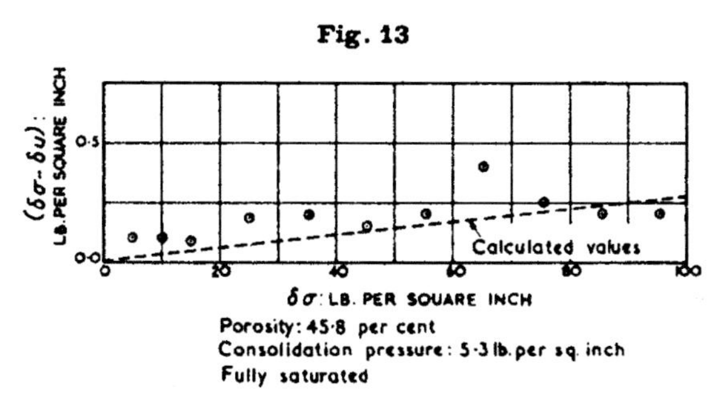

Effect of change in cell pressure on pore-water pressure

In Fig. 11 the Mohr diagram is shown for three of the tests in which the porosities were close to 44·9 per cent. This gives an angle of shearing resistance, with respect to applied stresses, equal to zero to within the limits of experimental accuracy. In Figs 12 the full details of the changes in pore pressure and effective stress for these three tests are given, and it can be seen that they amply confirm the theoretical analysis. The only significant effect of a change in cell pressure is to raise the pore pressure by that amount throughout the test, the relationship between effective stress and strain being the same in each test. The dilatant character of the sand structure is indicated by the fact that the pore pressure at failure has fallen below its value at the beginning of the test.

Fig. 14

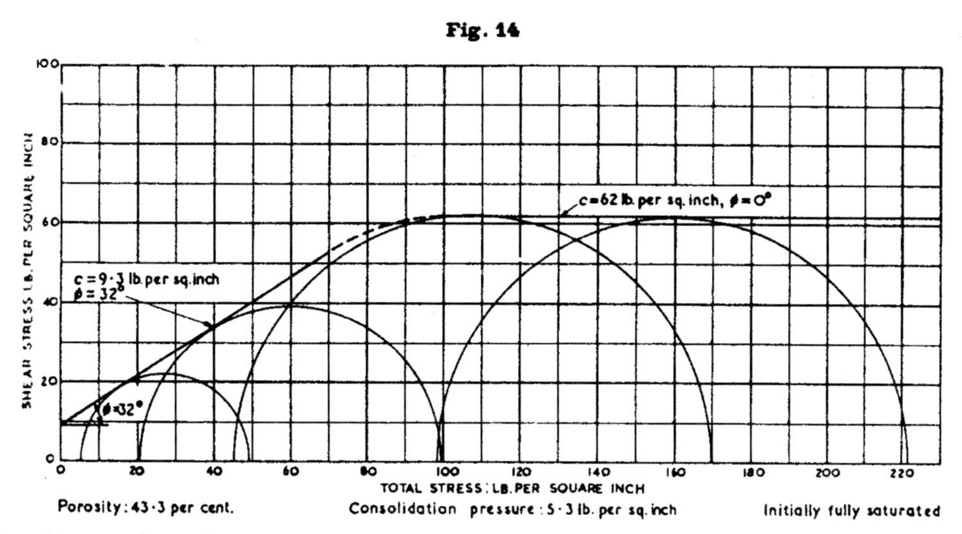

Mohr diagram for undrained tests on sand, showing the effect of cavitation at low cell pressures

The theoretical analysis previously discussed, of the changes in stress during the application of the cell pressure, shows that the change in pore-water pressure should give a direct indication of the effect of the relative compressibilities of the pore water and soil structure. In Fig. 13 the results of a special test to measure this effect are shown. The difference between the change in cell pressure and the consequent change in pore pressure $(\delta\sigma - \delta u)$ is plotted against the change in cell pressure $\delta\sigma$.

In this test the condition of complete absence of drainage from the sample is ensured to a higher degree of accuracy than in the routine tests by adjusting the mercury-level in the capillary tube to allow for the elastic deformation of the connexion to the sample (which is $1{\cdot}0 \times 10^{-4}$ cubic centimetre per lb. per square inch). It will be seen that the results agree quite satisfactorily with the theoretical values calculated from equation (9), using the measured value of the initial compressibility.

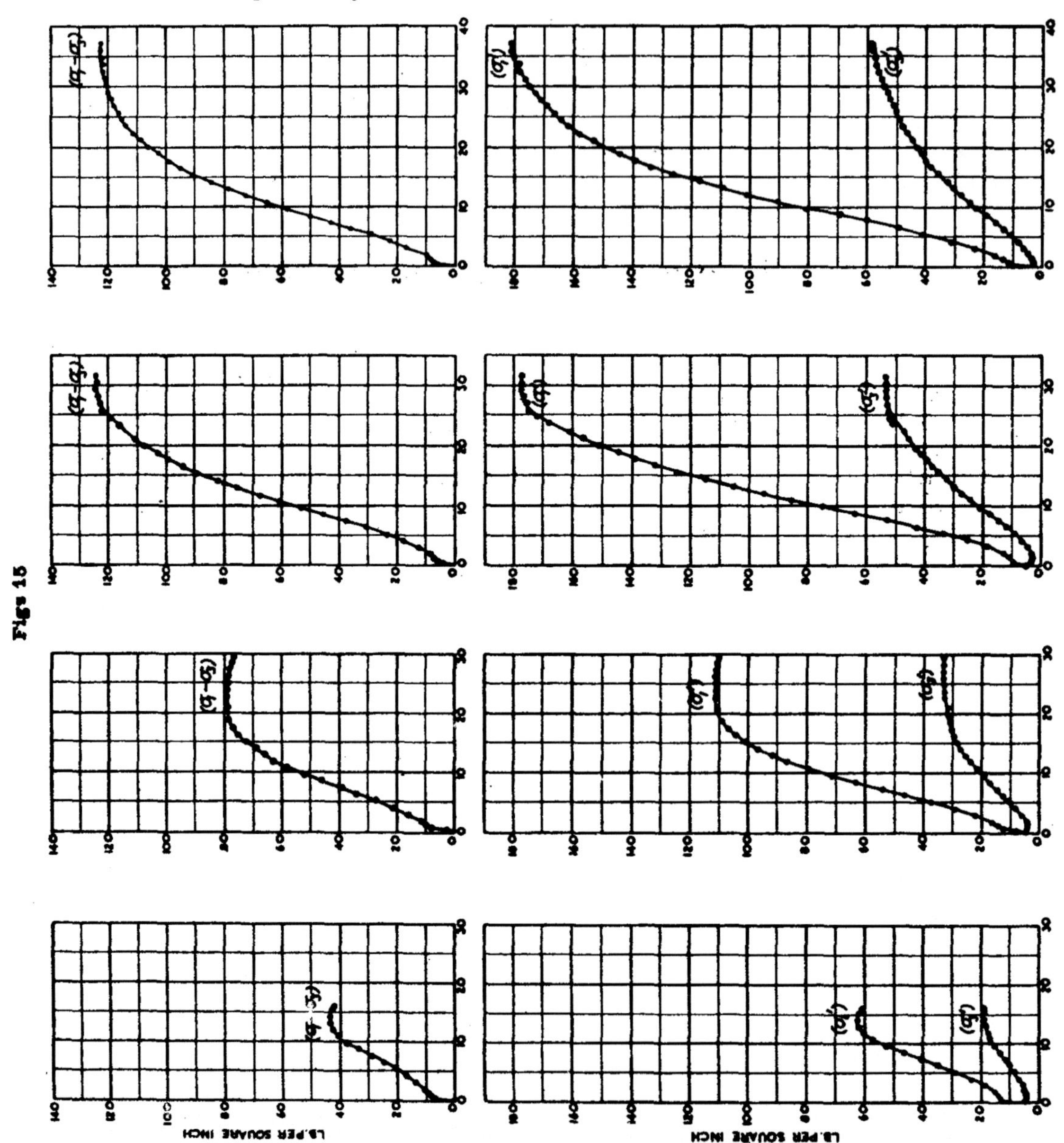

Fig. 15

(2) In the second series of tests, to examine the effect of negative pore pressures on the angle of shearing resistance in initially saturated samples, a lower porosity (about 43·3 per cent.) was used, giving a more strongly dilatant structure. The same consolidation pressure was used (5·3 lb. per square inch).

Fig. 14 shows the Mohr diagram for the results obtained for tests at 5, 20, 45, and 98 lb.

per square inch cell pressure. The first two circles indicate a material with an apparent cohesion of 9·3 lb. per square inch and an angle of shearing resistance of 32 degrees. Above a cell pressure of 45 lb. per square inch the sand behaves as a cohesive material with zero angle of shearing resistance and a " cohesion " of 62 lb. per square inch.

These results are in agreement with the theoretical analysis, and a study of Figs 15, which give the detailed changes in pore pressure and effective stress, confirms the principle invoked.* In the first two tests the effects of dilatancy on the strength were terminated by the pore pressure reaching a limiting negative value of about 14 lb. per square inch (at which cavitation took place in the pore water). In the third test the pore pressure did not fall to this value, and so that test, and the subsequent tests at a higher cell pressure, fulfilled the conditions required for $\phi_u = 0$. The measured value of apparent cohesion is in close agreement with that calculated above.

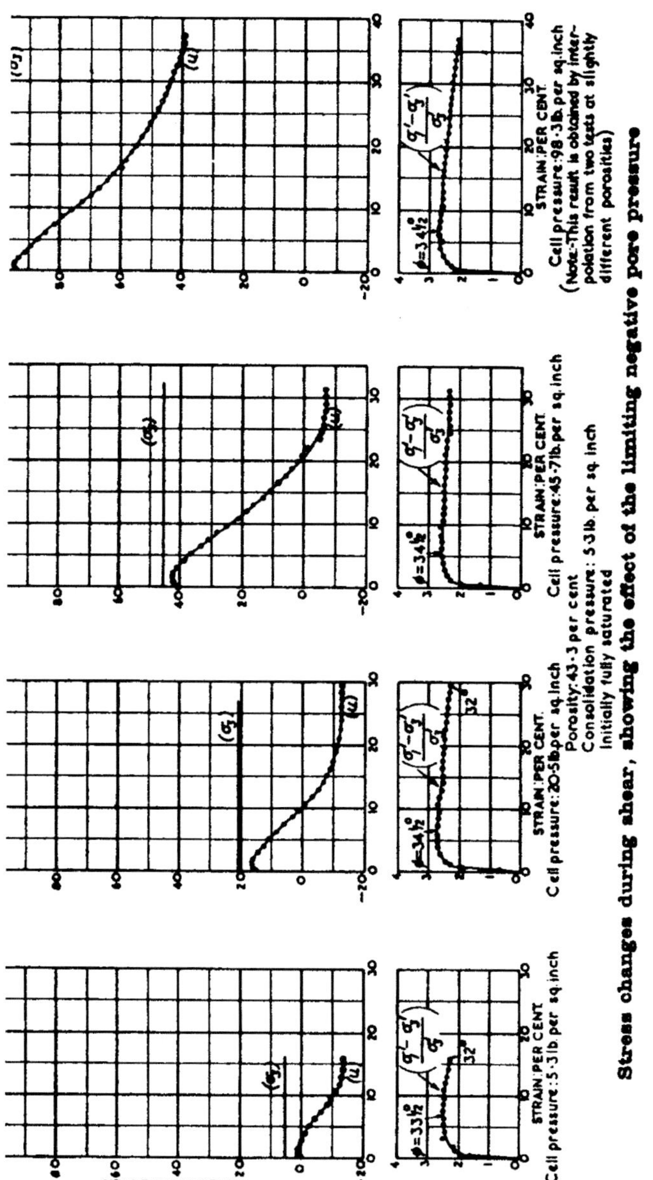

Stress changes during shear, showing the effect of the limiting negative pore pressure

GENERAL CONCLUSIONS

The significance of the test results in the general theory of the shear strength of the soils, and the conclusions of the theoretical treatment, may be summarized as follows:

(1) The angle of shearing resistance measured in a series of undrained tests on saturated samples is, *in general,* close to zero for all types of soil including those possessing a dilatant structure. Soils with a very low compressibility should prove to be the only exceptions to this rule, and appreciable values of ϕ_u may be expected where relatively low compressibility (less than 3×10^{-4} per lb. per square inch, compared with $3{\cdot}4 \times 10^{-6}$ lb. per square inch for water) is combined with a strongly dilatant structure.

(2) It is not inconsistent with this general statement for large values of ϕ_u to be measured on *apparently* fully saturated samples of compressible soils under certain conditions :—

(*a*) Small initial deviations from full saturation due to entrapped air, or to voids formed due to stress release after deep sampling, etc., will cause an angle of shearing resistance to be measured at *low* cell pressures even in very compressible soils.

* This method of pore-pressure measurement cannot be relied on in fine-grained soils in which a negative pressure greater than — 1 atmosphere may be reached.

(*b*) Cavitation caused by excessive negative pore pressures produced during shear by a strongly dilatant structure causes a value of ϕ_u equal to the angle of internal friction to be measured in samples which were initially fully saturated.

Both effects disappear under higher values of cell pressure, where ($\phi_u = 0$) will be measured, but the cell pressures required to produce a zero angle of shearing resistance may sometimes be outside the range employed in normal laboratory practice.

(3) It is not necessary to assume appreciable values of the contact-area ratio a between the soil grains in order to develop a theory consistent with the test results in the cases where ϕ_u is not negligible. Direct determination of the value of a from the changes in pore-water pressure appears to be impossible (*vide* equation (9)). Its indirect determination from the value of ϕ_u can, in the present tests, provide only the negative evidence that here a is zero to within the limits of accuracy of the measurements of shear strength.

ACKNOWLEDGEMENTS

The laboratory tests were carried out in the Civil Engineering Department of the Imperial College, University of London.

The Authors are grateful to Dr A. W. Skempton for his constant interest and discussions during the course of the work.

REFERENCES

BISHOP, A. W., 1947. The London clay. Part III—Strength variations in London clay. *Journées Internationales Consacrées à l'Étude, la Valorisation et la Mise en Œuvre des Roches Argileuses et d'autres Matières Premières Utilisées dans les Industries des Siliçates.* 109–113, Silicates Industriels 13 (9), 1948.

BOWDEN, F. P., and TABOR, D., 1942. Mechanism of metallic friction. *Nature.* 150 : 197–199.

BRUGGEMAN, J. R., ZANGAR, C. N., and BRAHTZ, J. H. A., 1939. Notes on analytic soil mechanics. *Tech. Memo. U.S. Bur. Recl.* No. 592.

COULOMB, C. A., 1776. Essai sur une application des règles de maximis et minimis à quelques problèmes de statique (Essay on the application of maxima-and-minima rules to some statics problems). *Mémoires Académie Royale des Sciences, Paris.* 7 : 353A–353C

GOLDER, H. Q., and SKEMPTON, A. W., 1948. The angle of shearing resistance in cohesive soils for tests at constant water content. *Proc. Second Int. Conf. Soil Mech.* 1 : 185–192.

HAMILTON, L. W., 1939. The effects of internal hydrostatic pressure on the shearing strength of soils. *Proc. Amer. Soc. Test. Mater.* 39 : 1100–1121.

HANSEN, J. B., and GIBSON, R. E., 1949. Undrained shear strengths of anisotropically consolidated clays. *Géotechnique.* 1 : 189–204.

HILF, J. W., 1948. Estimating construction pore pressures in rolled earth dams. *Proc. Second Int. Conf. Soil Mech.* 3 : 234–240.

HVORSLEV, M. J., 1937. Uber die Festigkeitseigenschaften gestoerter bindiger Boeden (On the physical properties of disturbed cohesive soils). *Ingeniorvidenskabelige Skrifter.* A. No. 45. 159 pp.

JURGENSON, L., 1934. The shearing resistance of soils. *J. Boston Soc. Civ. Engrs.* 21 : 242–275.

KJELLMAN, W., 1948. General report on laboratory investigations. *Proc. Second Int. Conf. Soil Mech.* 6 : 76–77.

KOLBUSZEWSKI, J. J., 1948. An experimental study of the maximum and minimum porosities of sands. *Proc. Second Int. Conf. Soil Mech.* 1 : 158–165.

ODENSTAD, S., 1949. Stresses and strains in the undrained compression test. *Géotechnique.* 1 : 242–249.

RENDULIC, L., 1937. Ein Grundgesetz der Tonmechanik und sein experimenteller Beweis (The fundamental law of clay mechanics and the experimental proof). *Bauingenieur.* 18 : 459–467.

SKEMPTON, A. W., 1948*a*. A study of the geotechnical properties of some post-glacial clays. *Géotechnique.* 1 : 7–22.

SKEMPTON, A. W., 1948*b*. A study of the immediate triaxial test on cohesive soils. *Proc. Second Int. Conf. Soil Mech.* 1 : 192–196.

TAYLOR, D. W., 1944. Tenth progress report on shear research. *M.I.T. publication.*

TAYLOR, D. W., 1948. Fundamentals of soil mechanics. *Wiley, New York.* 500 pp.

TERZAGHI, K., 1932. Tragfaehigkeit der Flachgruendungen (Bearing capacity of shallow foundations). *Prelim. Pub. First Cong. Int. Ass. Bridge Struct. Eng.* 659–672.

TERZAGHI, K., 1936. Simple tests determine hydrostatic uplift. *Eng. News-Rec.* 116 : 872–875.

TERZAGHI, K., 1943. Theoretical soil mechanics. *Wiley, New York.* 510 pp.

First Technical Session : General Theory of Stability of Slopes

Tuesday morning, 21 *September*, 1954

Session 1/1

THE USE OF THE SLIP CIRCLE IN THE STABILITY ANALYSIS OF SLOPES

by

ALAN W. BISHOP, M.A., Ph.D., A.M.I.C.E.

INTRODUCTION

Errors may be introduced into the estimate of stability not only by the use of approximate methods of stability analysis, but also by the use of sampling and testing methods which do not reproduce sufficiently accurately the soil conditions and state of stress in the natural ground or compacted fill under consideration. Unless equal attention is paid to each factor, an elaborate mathematical treatment may lead to a fictitious impression of accuracy.

However, in a number of cases the uniformity of the soil conditions or the importance of the problem will justify a more accurate analysis, particularly if this is coupled with field measurements of pore pressure, which is the factor most difficult to assess from laboratory data alone. Two classes of problem of particular note in this respect are :

(i) The design of water-retaining structures, such as earth dams and embankments, where failure could have catastrophic results, but where an over-conservative design may be very costly.

(ii) The examination of the long-term stability of cuts and natural slopes where large scale earth movements may involve engineering works and buildings.

THE USE OF LIMIT DESIGN METHODS

It has been shown elsewhere by a relaxation analysis of a typical earth dam (Bishop, 1952) that, even assuming idealized elastic properties for the soil, local overstress will occur when the factor of safety (by a slip circle method) lies below a value of about 1·8. As the majority of stability problems occur in slopes and dams having lower factors of safety than this, a state of plastic equilibrium must be considered to exist throughout at least part of the slope.

Under these conditions a quantitative estimate of the factor of safety can be obtained by examining the conditions of equilibrium when incipient failure is postulated, and comparing the strength necessary to maintain limiting equilibrium with the available strength of the soil. The factor of safety (F) is thus defined as the ratio of the available shear strength of the soil to that required to maintain equilibrium. The shear strength mobilized is, therefore, equal to s, where :

$$s = \frac{1}{F}\left\{ c' + (\sigma_n - u)\tan\phi' \right\} \qquad (1)$$

where c' denotes cohesion, and ϕ' denotes angle of shearing resistance, } in terms of effective stress.*

σ_n denotes total normal stress,

u denotes pore pressure.

* Measured either in undrained or consolidated-undrained tests with pore-pressure measurement, or in drained tests carried out sufficiently slowly to ensure zero excess pore pressure.

Failure along a continuous rupture surface is usually assumed, but, as the shape and position of this surface is influenced by the distribution of pore pressure and the variation of the shear parameters within the slope, a generalized analytical solution is not possible and a numerical solution is required in each individual case. The rigorous determination of the shape of the most critical surface presents some difficulty (see, for example, Coenen, 1948), and in practice a simplified shape, usually a circular arc, is adopted, and the problem is assumed to be one of plane strain.

MECHANICS OF THE CIRCULAR ARC ANALYSIS

In order to examine the equilibrium of the mass of soil above the slip surface it follows from equation (1) that it is necessary to know the value of the normal stress at each point on this surface, as well as the magnitude of the pore pressure. It is possible to estimate the value of the normal stress by following the method developed by Fellenius (1927, 1936), in which the conditions for the statical equilibrium of the slice of soil lying vertically above each element of the sliding surface are fully satisfied.

A complete graphical analysis of this type is most laborious, and this may result in the examination of an insufficient number of trial surfaces to locate the most critical one. Several simplified procedures have been developed using the friction circle method (Taylor, 1937, 1948 ; Fröhlich, 1951), in which an assumption is made about the distribution of normal stress, but their use is limited to cases in which ϕ' is constant over the whole of the failure surface. More generally the " slices " method is used, with a simplifying assumption about the effect of the forces between the slices.

The significance of this assumption may be examined by considering the equilibrium of the mass of soil (of unit thickness) bounded by the circular arc ABCD, of radius R and centre at O (Fig. 1(a)). In the case where no external forces act on the surface of the slope,

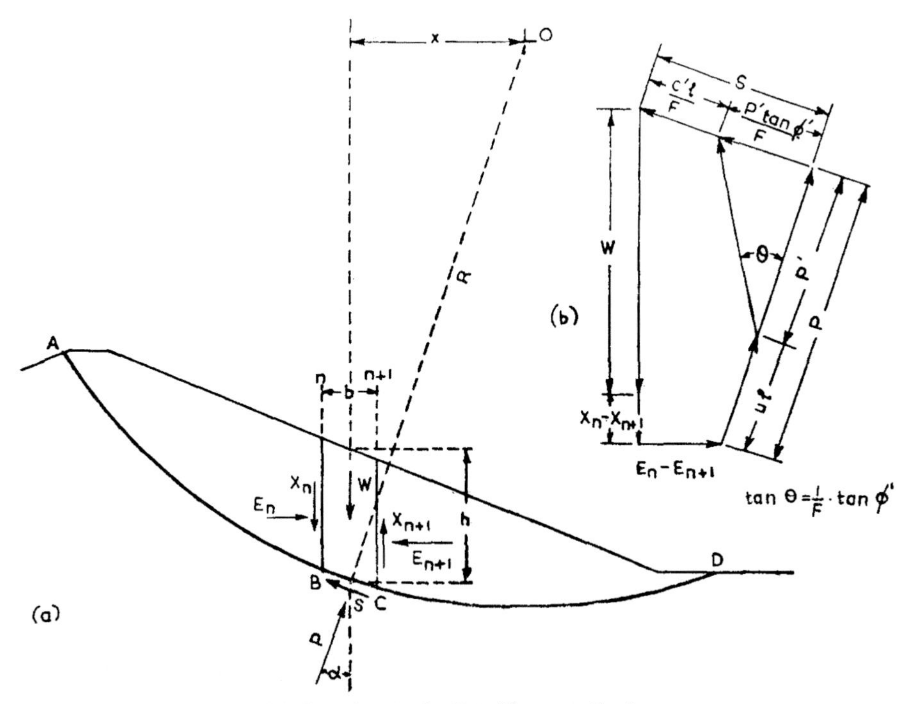

Fig. 1. Forces in the slices method

equilibrium must exist between the weight of the soil above ABCD and the resultant of the total forces acting on ABCD.

Let E_n, E_{n+1} denote the resultants of the total horizontal forces on the sections n and $n + 1$ respectively,

and X_n, X_{n+1} ,, the vertical shear forces,
W ,, the total weight of the slice of soil,
P ,, the total normal force acting on its base,
S ,, the shear force acting on its base,
h ,, the height of the element,
b ,, the breadth of the element,
l ,, the length BC,
α ,, the angle between BC and the horizontal,
x ,, the horizontal distance of the slice from the centre of rotation.

The total normal stress is σ_n, where

$$\sigma_n = \frac{P}{l} \quad \ldots \quad (2)$$

Hence, from equation (1), the magnitude of the shear strength mobilized to satisfy the conditions of limiting equilibrium is s where :

$$s = \frac{1}{F}\left\{ c' + \left(\frac{P}{l} - u\right) \tan \phi' \right\} \quad \ldots \quad (3)$$

The shear force S acting on the base of the slice is equal to sl, and thus, equating the moment about O of the weight of soil within ABCD with the moment of the external forces acting on the sliding surface, we obtain :

$$\Sigma Wx = \Sigma SR = \Sigma slR \quad \ldots \quad (4)$$

It follows, therefore, from equation (3) that :

$$F = \frac{R}{\Sigma Wx} \cdot \Sigma\Big[c'l + (P - ul) \tan \phi'\Big] \quad \ldots \quad (5)$$

From the equilibrium of the soil in the slice above BC, we obtain P, by resolving in a direction normal to the slip surface :

$$P = (W + X_n - X_{n+1}) \cos \alpha - (E_n - E_{n+1}) \sin \alpha \quad \ldots \quad (6)$$

The expression for F thus becomes :

$$F = \frac{R}{\Sigma Wx} \cdot \Sigma\Big[c'l + \tan \phi' . (W \cos \alpha - ul) + \tan \phi' . \Big\{(X_n - X_{n+1}) \cos \alpha - (E_n - E_{n+1}) \sin \alpha\Big\}\Big] \quad \ldots \quad (7)$$

Since there are no external forces on the face of the slope, it follows that :

$$\Sigma(X_n - X_{n+1}) = 0 \quad \ldots \quad (8a)$$

$$\Sigma(E_n - E_{n+1}) = 0 \quad \ldots \quad (8b)$$

However, except in the case where ϕ' is constant along the slip surface and α is also constant (*i.e.*, a plane slip surface), the terms in equation (7) containing X_n and E_n do not disappear. A simplified form of analysis, suggested by Krey (1926) and Terzaghi (1929) and also presented by May (1936) as a graphical method, implies that the sum of these terms

$$\Sigma \tan \phi' . \Big\{(X_n - X_{n+1}) \cos \alpha - (E_n - E_{n+1}) \sin \alpha\Big\}$$

may be neglected without serious loss in accuracy. This is the method at present used, for example, by the U.S. Bureau of Reclamation (Daehn and Hilf, 1951).

1*

Putting $x = R \sin \alpha$, the simplified form may be written :

$$F = \frac{1}{\Sigma W \sin \alpha} \cdot \Sigma \left[c'l + \tan \phi' . (W \cos \alpha - ul) \right] \quad . \quad . \quad . \quad . \quad . \quad (9)$$

In earth dam design the construction pore pressures are often expressed as a function of the total weight of the column of soil above the point considered, i.e.

$$u = \bar{B} \left(\frac{W}{b} \right) \quad . \quad . \quad . \quad . \quad . \quad . \quad . \quad . \quad . \quad . \quad (10)$$

where $\bar{B}$ is a soil parameter based either on field data or laboratory tests.*

In this case, putting $l = b \sec \alpha$, the expression for factor of safety can be further simplified to :

$$F = \frac{1}{\Sigma W \sin \alpha} \cdot \Sigma \left[c'l + \tan \phi' . W(\cos \alpha - \bar{B} \sec \alpha) \right] \quad . \quad . \quad . \quad . \quad (11)$$

This expression permits the rapid and direct computation of the value of F which is necessary if sufficient trial circles are to be used to locate the most critical surface. However, as can be seen from the examples quoted later, the values of F are, in general, found to be conservative, and may lead to uneconomical design. This is especially marked where conditions permit deep slip circles round which the variation in α is large.

To derive a method of analysis which largely avoids this error it is convenient to return to equation (5). If we denote the effective normal force $(P - ul)$ by P' (see Fig. 1(*b*)), and resolve the forces on the slice vertically, then we obtain, on re-arranging :

$$P' = \frac{W + X_n - X_{n+1} - l\left(u \cos \alpha + \frac{c'}{F} \sin \alpha\right)}{\cos \alpha + \frac{\tan \phi' . \sin \alpha}{F}} \quad . \quad . \quad . \quad . \quad . \quad . \quad (12)$$

Substituting in equation (5) and putting $l = b \sec \alpha$ and $x = R \sin \alpha$, an expression for the factor of safety is obtained :

$$F = \frac{1}{\Sigma W \sin \alpha} \cdot \Sigma \left[\left\{ c'b + \tan \phi' . (W(1 - \bar{B}) + (X_n - X_{n+1})) \right\} . \frac{\sec \alpha}{1 + \frac{\tan \phi' . \tan \alpha}{F}} \right] \quad . \quad (13)$$

The values of $(X_n - X_{n+1})$ used in this expression are found by successive approximation, and must satisfy the conditions given in equations (8). In addition, the positions of the lines of thrust between the slices should be reasonable, and no unbalanced moment should be implied in any slice. The factor of safety against sliding on the vertical sections should also be satisfactory, though since the slip surface assumed is only an approximation to the actual one, overstress may be implied in the adjacent soil.

The condition $\Sigma(X_n - X_{n+1}) = 0$ in equation (8*a*) can be satisfied directly by selecting appropriate values of X_n, etc. The corresponding sum $\Sigma(E_n - E_{n+1})$ can be readily computed in terms of the expression used in equation (13). Resolving the forces on a slice tangentially, we obtain the expression :

$$(W + X_n - X_{n+1}) \sin \alpha + (E_n - E_{n+1}) \cos \alpha = S$$

or

$$(E_n - E_{n+1}) = S \sec \alpha - (W + X_n - X_{n+1}) \tan \alpha \quad . \quad . \quad . \quad (14)$$

* In practice the value of B will vary along the slip surface, though for preliminary design purposes it is convenient to use a constant average value throughout the impervious zone.

Now, if equation (13) is written :

$$F = \frac{1}{\Sigma W \sin \alpha} \cdot \Sigma[m] \qquad (15)$$

then

$$S = \frac{m}{F} \qquad (16)$$

and hence

$$\Sigma(E_n - E_{n+1}) = \Sigma\left[\frac{m}{F}\sec\alpha - (W + X_n - X_{n+1})\tan\alpha\right] \qquad (17)$$

The X values must, therefore, also satisfy the condition that :

$$\Sigma\left[\frac{m}{F}\sec\alpha - (W + X_n - X_{n+1})\tan\alpha\right] = 0 \qquad (18)$$

In practice, an initial value of F is obtained by solving equation (13) on the assumption that $(X_n - X_{n+1}) = 0$ throughout. By suitably tabulating the variables a solution can be obtained after using two or three trial values of F. The use of $(X_n - X_{n+1}) = 0$ satisfies equation (8*a*), but not equation (18). Values of $(X_n - X_{n+1})$ are then introduced in order to satisfy equation (18) also. These values can then be finally adjusted, either graphically or analytically, until the equilibrium conditions are fully satisfied for each slice.

From the practical point of view it is of interest to note that, although there are a number of different distributions of $(X_n - X_{n+1})$ which satisfy equation (18), the corresponding variations in the value of F are found to be insignificant (less than 1% in a typical case).

It should also be noted that, since the error in the simplified method given in equations (9) and (11) varies with the central angle of the arc, it will lead to a different location for the most critical circle than that given by the more rigorous method. It will, therefore, generally be necessary to examine a number of trial circles by this latter method.

PARTIALLY SUBMERGED SLOPES

The total disturbing moment is now that of the soil above ABCD, less the moment about O of the water pressure acting on DLM (Fig. 2 (*a*)). If we imagine a section of water bounded by a free surface at MN and outlined by NDLM, and similarly take moments about O, the normal forces on the arc ND all pass through O and the moment of the water pressure on DLM is therefore equal to the moment of the mass of water NDLM about O.

Since the weight of a mass of saturated soil less the weight of water occupying the same volume is equal to its submerged weight, the resultant disturbing moment due to the mass of soil above ABCD and to the water pressure on DLM is given by using the bulk density of the soil above the level of the external free water surface and the submerged density below.

It should be noted that the boundary MN implies nothing about the magnitude of the pore pressures inside the slope and is only used to obtain the statically equivalent disturbing moment.

The disturbing moment is therefore $\Sigma(W_1 + W_2)x \qquad (19)$

where W_1 denotes full weight of the soil in the slice above MN,

W_2 ,, submerged weight of soil in the part of the slice below MN.

It is also convenient to obtain the expression for the effective normal force P' in terms of W_1 and W_2. It can be seen from Fig. 2 (*b*) and (*c*) that, resolving vertically as before, it follows that :

$$P' = \frac{W_1 + W_2 + X_n - X_{n+1} - l\left(u_s \cos\alpha + \frac{c'}{F}\sin\alpha\right)}{\cos\alpha + \frac{\tan\phi' . \sin\alpha}{F}} \qquad (20)$$

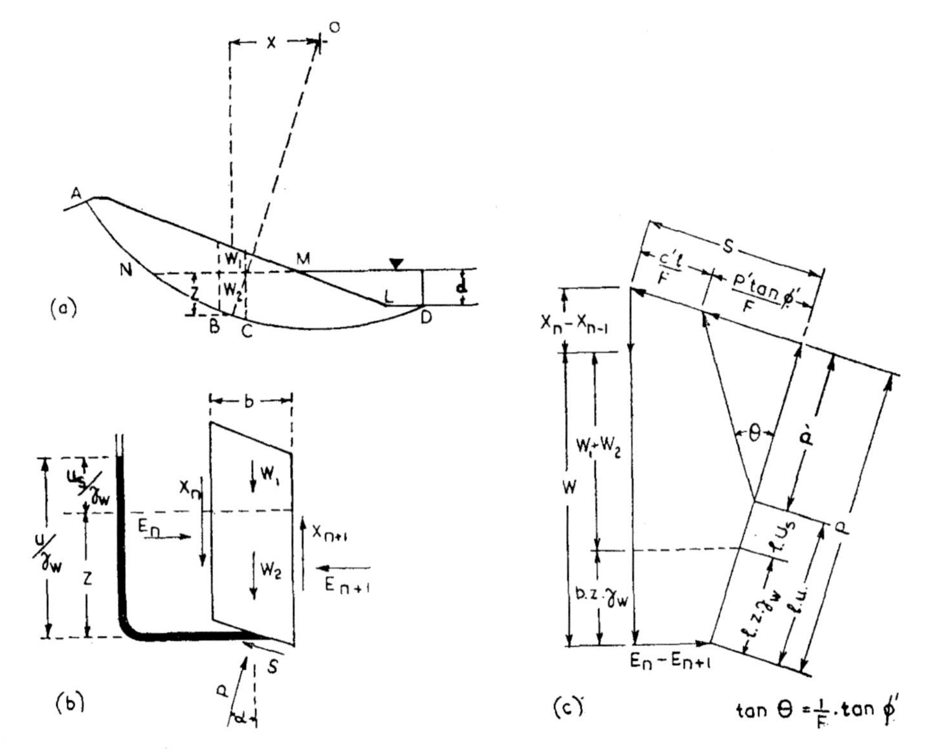

Fig. 2. Partially Submerged Slopes

where u_s is the pore pressure expressed as an excess over the hydrostatic pressure corresponding to the water level outside the slope; that is

$$u_s = u - \gamma_\omega . z \quad \ldots \quad (21)$$

where γ_ω denotes the density of water,
and z ,, the depth of slice below MN. (If no part of the slice is submerged, $u_s = u$.)

It follows, therefore, that for partial submergence the expression for factor of safety * becomes :

$$F = \frac{1}{\Sigma(W_1 + W_2)\sin\alpha} \cdot \Sigma\left[\left\{c'b + \tan\phi'.(W_1 + W_2 - bu_s + \overline{X_n - X_{n+1}})\right\}\frac{\sec\alpha}{1 + \dfrac{\tan\phi'.\tan\alpha}{F}}\right] \quad \ldots \quad (22)$$

The forces between the slices now have to satisfy the conditions :

$$\Sigma(X_n - X_{n+1}) = 0 \quad \ldots \quad (23a)$$

$$\Sigma(E_n - E_{n+1}) = -\frac{1}{2} \cdot \gamma_w d^2 \quad \ldots \quad (23b)$$

* A similar treatment of the simplified method leads to the expression

$$F = \frac{1}{\Sigma(W_1 + W_2)\sin\alpha}\Sigma\left[c'l + \tan\phi'.(\overline{W_1 + W_2}\cos\alpha - u_s b\sec\alpha)\right] \quad . \; . \quad (22a)$$

This treatment of the pore pressure leads to consistent results throughout the full range of submergence (Bishop, 1952).

where d is the depth of water at the toe of the slip, which causes a horizontal water thrust on the vertical section with which the slip terminates.

To obtain an expression for $\Sigma(E_n - E_{n+1})$, we resolve tangentially as before :

$$(W_1 + W_2 + bz\gamma_\omega + X_n - X_{n+1}) \sin \alpha + (E_n - E_{n+1}) \cos \alpha = S$$

or $$(E_n - E_{n+1}) = S \sec \alpha - (W_1 + W_2 + X_n - X_{n+1}) \tan \alpha - \gamma_w zb \tan \alpha \quad . \quad . \quad (24)$$

Writing equation (22) as :

$$F = \frac{1}{\Sigma(W_1 + W_2) \sin \alpha} \cdot \Sigma[m] \quad . \quad . \quad . \quad . \quad . \quad . \quad . \quad . \quad . \quad . \quad (25)$$

then $$S = \frac{m}{F} \quad . \quad . \quad . \quad . \quad . \quad . \quad . \quad . \quad . \quad . \quad . \quad . \quad . \quad . \quad . \quad (26)$$

Hence

$$\Sigma(E_n - E_{n+1}) = \Sigma\left[\frac{m}{F} \sec \alpha - (W_1 + W_2 + X_n - X_{n+1}) \tan \alpha\right] - \Sigma\left[\gamma_\omega zb \tan \alpha\right] \quad . \quad (27)$$

Now $$\Sigma\left[\gamma_\omega zb \tan \alpha\right] = \frac{1}{2}\gamma_\omega \cdot d^2 \quad . \quad . \quad . \quad . \quad . \quad . \quad . \quad . \quad (28)$$

Thus equation (23b) is satisfied when

$$\Sigma\left[\frac{m}{F} \sec \alpha - (W_1 + W_2 + X_n - X_{n+1}) \tan \alpha\right] = 0 \quad . \quad . \quad . \quad . \quad . \quad (29)$$

It can be seen, therefore, that for partially submerged slopes an analogous method is obtained by using submerged densities for those parts of the slices which lie below the level of the external free water surface, and by expressing the pore pressures there as an excess above the hydrostatic pressure corresponding to this water level.

PRACTICAL APPLICATION

Only a limited number of stability analyses have so far been carried out in which both the simplified and the more rigorous methods have been used. From these it appears, however, that the major part of the gain in accuracy can be obtained by proceeding with the solution of equations (13) or (22) only as far as the starting value of F given by taking $(X_n - X_{n+1}) = 0$. This avoids the more time-consuming stages of the solution, and provides a convenient routine method in which the time required for the analysis of each trial circle is from one to two hours or about twice that required using the simplified methods.

This may be illustrated by the two following cases.

(i) *Pore pressures set up in a boulder clay fill during construction.*—The cross-section of the dam and the soil properties used in the analysis are given in Fig 3 (a). From laboratory tests and other data it is considered that an average pore pressure equal to 40% of the weight of the overlying soil might be expected, *i.e.* $\bar{B} = 0{\cdot}4$.

For this condition the simplified method indicates a factor of safety of 1·38. The value given by equation (13), with $(X_n - X_{n+1}) = 0$, is 1·53.

The two methods do not lead to the same critical circle. The circle having a factor of safety of 1·38 by the simplified method gives a value of 1·59 using equation (13) with $(X_n - X_{n+1}) = 0$. The introduction of trial values of $(X_n - X_{n+1})$ makes only a small further change in the factor of safety, a value of 1·60 or 1·61 being obtained depending on the assumed distribution of X_n.

The form of tabulation used for the routine solution with $(X_n - X_{n+1}) = 0$ is given in Table 1.

It is useful from the design point of view to know the influence of possible variations in construction pore pressure on the factor of safety, and for this purpose the factor of safety may be plotted directly against average pore pressure ratio as in Fig. 3 (b).

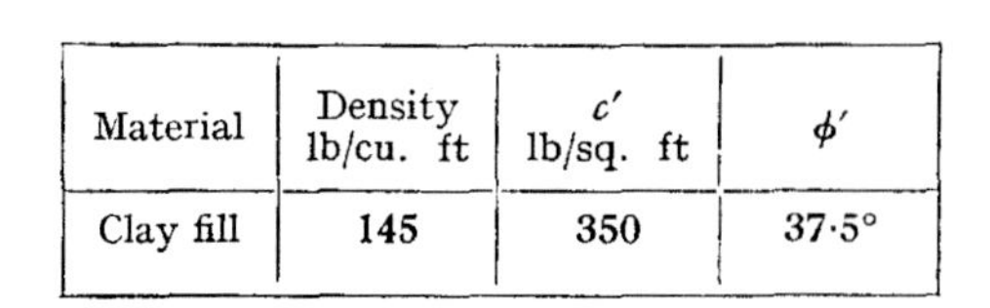

Material	Density lb/cu. ft	c' lb/sq. ft	ϕ'
Clay fill	145	350	37·5°

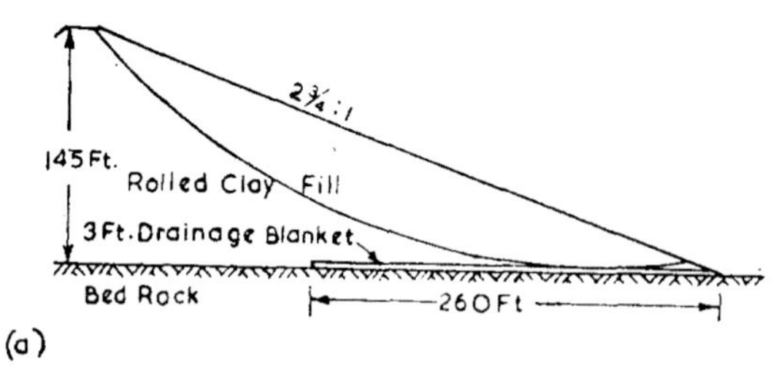

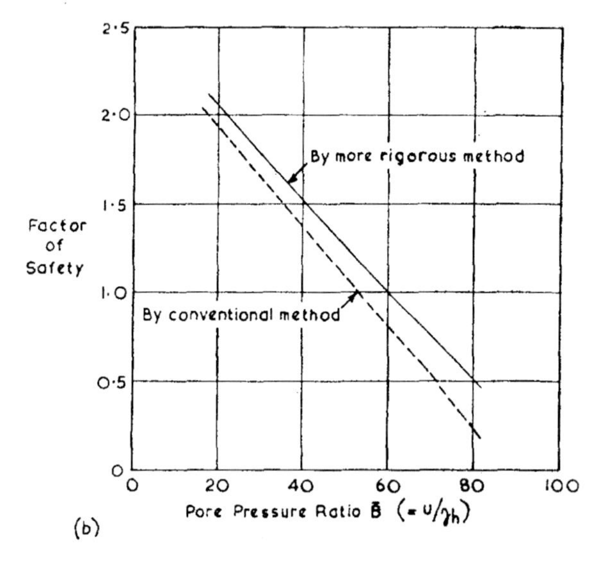

Fig. 3. End of construction case: relationship between factor of safety and pore pressure

(ii) *Pore pressures set up in a moraine fill during a partial rapid drawdown.*—The cross-section of the dam and the soil properties used in the analysis are given in Fig. 4. The basis of the method by which the draw-down pore pressures are calculated has been described by the author elsewhere (Bishop, 1952), and their values are indicated by a line representing stand-pipe levels above different points on the slip surface.

Using the average values of the soil properties (Case I), the simplified method (equation (22*a*)) gives a factor of safety of 1·50, and the more rigorous solution (equation (22), with $(X_n - X_{n+1}) = 0$) gives a value of 1·84.* As before, the two methods lead to different critical circles. The inclusion of trial values of $(X_n - X_{n+1})$ raises the value from 1·84 to 1·92, the value by the simplified method for this particular circle being 1·53. About 80% of the gain in accuracy in this case is thus achieved without introducing the X_n terms.

The form of tabulation for the routine solution in the case of partial submergence is given in Table 2.

CONCLUSION

Errors may be introduced into an estimate of stability by both the sampling and testing procedure, and in many cases an approximate method of stability analysis may be considered adequate. Where, however, considerable care is exercised at each stage, and, in particular, where field measurements of pore pressure are being used, the simplified " slices " method of analysis is not sufficiently accurate.

This is illustrated by the two examples given above. The error is likely to be of particular importance where deep slip circles are involved. This is illustrated by Fig. 5 in which the collected results of the two stability analyses are plotted in terms of the value of the central angle of the arc. It will be seen that the value of the factor of safety given by the simplified method, expressed as a percentage of that given by the more rigorous method [using $(X_n - X_{n+1}) = 0$], drops rapidly as the central angle of the arc increases. This effect is particularly marked for the higher values of excess pore pressure.

* For Case II the corresponding values are 1·14 and 1·48.

Material	Density—lb/cu. ft		Case I		Case II	
	Submerged	Above W.L.	c' lb/sq. ft	ϕ'	c' lb/sq. ft	ϕ'
Moraine	72	135	450	37°	0	40°
Rock	72	118	0	45°	0	40°

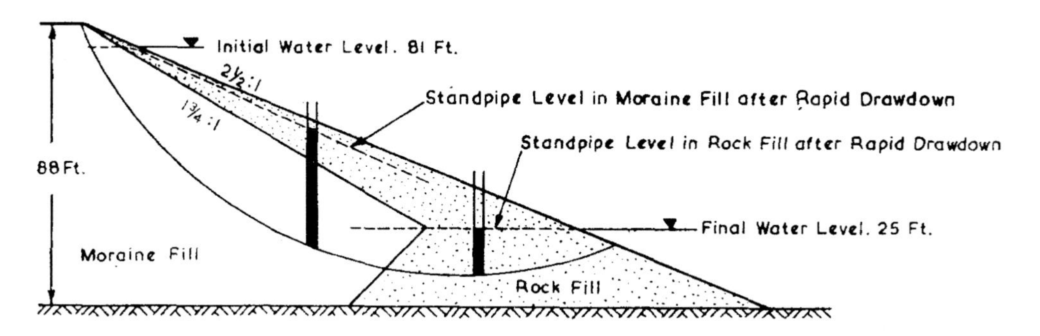

Fig. 4. Drawdown analysis of an upstream slope

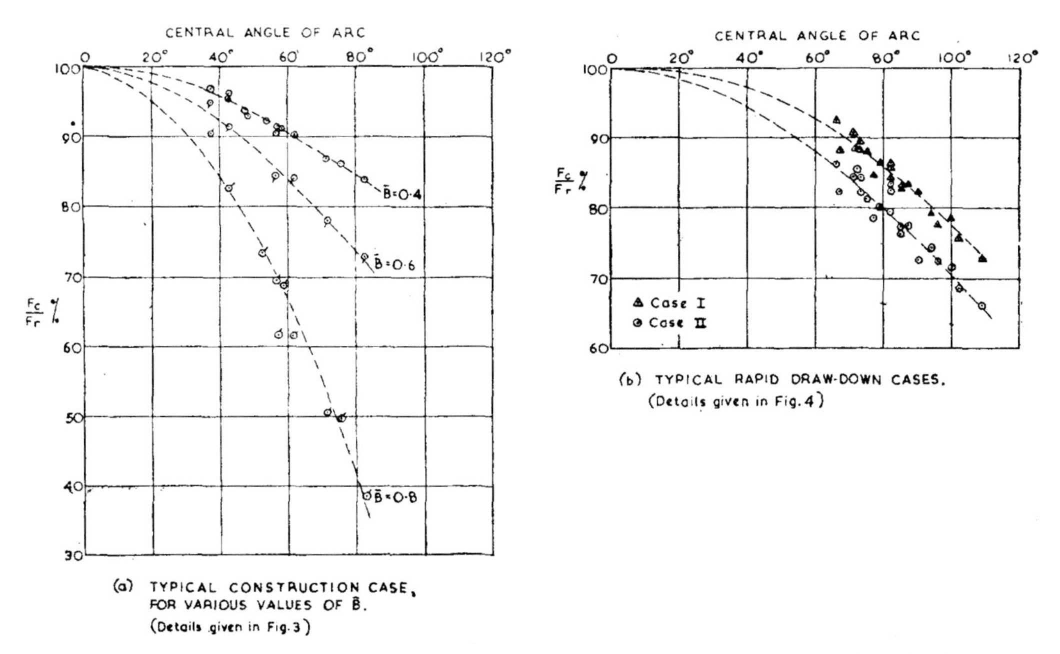

Fig. 5. Influence of central angle of arc on accuracy of conventional method

F_c is factor of safety by conventional method.

F_r is factor of safety by more rigorous method on same sliding surface.

Circle No. 1 **Table 1**

Slice No.	b ft	h ft	W lb. $\times 10^3$	α	$\sin\alpha$	$W \sin\alpha$ (1)	$c'b$ (2)	$W(1-\bar{B})\tan\phi'$ (3)	$(2)+(3)$ (4)	$\sec\alpha$	$\tan\alpha$	$\dfrac{\sec\alpha}{1+\dfrac{\tan\phi' . \tan\alpha}{F}}$ (5)		$(4)\times(5)$ (6)	
												$F=1{\cdot}4$	$F=1{\cdot}6$	$F=1{\cdot}4$	$F=1{\cdot}6$
1	50	23	167	42·2°	·672	112	17	77	94	1·350	·907	·902	·940	85	88
2	50	54	381	34·6°	·568	217	17	175	192	1·215	·690	·883	·912	169	175
⋮	⋮	⋮	⋮	⋮	⋮	⋮	⋮	⋮	⋮	⋮	⋮	⋮	⋮	⋮	⋮
8	40	11	64	−9·2°	−·160	−10	14	29	43	1·013	−·162	1·110	1·097	47	47
						777								1240	1262

$$F = \frac{1240}{777} \quad \frac{1262}{777}$$

$$= 1{\cdot}60 \quad = 1{\cdot}62$$

Circle No. **Table 2**

Slice No.	b ft.	h_r (rock fill)	h_m (moraine)	h_s (submerged)	W_r } W_1	W_m }	W_2	(W_1+W_2) lb. $\times 10^3$	α	$\sin\alpha$	$(W_1+W_2)\sin\alpha$ (1)	h_1 (excess standpipe ht.)	$u_s = \gamma_w . h_1$	$b . u_s$	$(W_1+W_2-b . u_s)$	$(W_1+W_2-b . u_s)\tan\phi'$ (2)	$c' . b$ (3)	$(2)+(3)$ (4)	$\sec\alpha$	$\tan\alpha$	$\dfrac{\sec\alpha}{1+\dfrac{\tan\phi' . \tan\alpha}{F}}$ (5)	$(4)\times(5)$ (6)
1																						
2																						
etc.																						

$$F = \frac{\Sigma(6)}{\Sigma(1)}$$

In these cases the use of the modified analysis outlined above is to be recommended, which, if carried only as far as the $(X_n - X_{n+1}) = 0$ stage, can readily be used for routine work.

Little field evidence is yet available for checking the overall accuracy of the stability calculation except in those cases where the " $\phi = 0$ " analysis has been applicable, and this emphasizes the need for failure records complete with pore-pressure data for both natural slopes and trial embankments.

ACKNOWLEDGEMENTS

The second example is included with the kind approval of the North of Scotland Hydro-Electric Board and their consulting engineers, Sir Alexander Gibb & Partners. The Author is also indebted to Mr D. W. Lamb for carrying out a number of the other calculations.

REFERENCES

BISHOP, A. W., 1952. The Stability of Earth Dams. *Univ. of London. Ph.D. Thesis.**
COENEN, P. A., 1948. Fundamental Equations in the Theory of Limit Equilibrium. *Proc. 2nd Int. Conf. Soil. Mech.* 7 : 15.
DAEHN, W. W., and HILF, J. W., 1951. Implications of Pore Pressure in Design and Construction of Rolled Earth Dams. *Trans. 4th Cong. Large Dams.* 1 : 259.
FELLENIUS, W., 1927. Erdstatische Berechnungen mit Reibung und Kohaesion. Ernst, Berlin.
FELLENIUS, W., 1936. Calculation of the Stability of Earth Dams. *Trans. 2nd Cong. Large Dams.* 4 : 445.
FRÖHLICH, O. K., 1951. On the Danger of Sliding of the Upstream Embankment of an Earth Dam. *Trans. 4th Cong. Large Dams.* 1 : 329.
KREY, H., 1926. Erddruck, Erdwiderstand und Tragfaehigkeit des Baugrundes. *Ernst, Berlin.*
MAY, D. R., and BRAHTZ, J. H. A., 1936. Proposed Methods of Calculating the Stability of Earth Dams. *Trans. 2nd Cong. Large Dams.* 4 : 539.
TAYLOR, D. W., 1937. Stability of Earth Slopes. *J. Boston Soc. Civ. Eng.* 24 : 197.
TAYLOR, D. W., 1948. Fundamentals of Soil Mechanics. *John Wiley, New York.*
TERZAGHI, K., 1929. The Mechanics of Shear Failures on Clay Slopes and Creep of Retaining Walls. *Pub. Rds.*, 10 : 177.

* A discussion of the pore pressures set up in earth dams has since been published in *Géotechnique*, 4 : 4 : 148.

2*

The Principle of Effective Stress

Alan W. Bishop, M. A., D. Sc., Ph. D., A. M. I. C. E.

The principle of effective stress is given, together with an historical review of its development.

The influence of the area of contact between the soil particles is discussed. Theoretical considerations are made, leading to the hypothesis that the area of contact does not influence the effective stress. A series of test results is reviewed, which confirms this hypothesis.

The application of the principle is discussed for the practical case of a two phase fluid in the pore space.

Introduction

Most of the principles and methods now used in Soil Mechanics have roots which extend some way back into the history of the subject. Their development to a position of importance to the civil engineer depends, however, on three requirements. These are:

(1) Laboratory methods to measure the relevant soil properties accurately and easily.
(2) Theories to relate the laboratory measurements to the conditions met with in the actual full scale problems, and suitable design methods based on these theories.
(3) Field measurements to compare the predicted behaviour with actual performance.

With these three requirements in mind, we may re-examine the principle of effective stress, and, in passing, look briefly at its historical development.

The Principle of Effective Stress

It is now being more widely recognised that the principle of effective stress provides not only a satisfactory basis for understanding the deformation and strength characteristics of soil, but also a basis for practical design methods. This is in itself a relatively recent development, as it is only within the last few years that the three requirements referred to above have been satisfied, from the point of view of engineering practice, for all types of soil.

The two simplest implications of the principle of effective stress are:

(1) that volume change and deformation in soils depend not on the total stress applied, but on the difference between the total stress and the pressure set up in the fluid in the pore space. This leads to the expression

$$\sigma' = \sigma - u$$

where σ denotes the total normal stress
u denotes the pore pressure
σ' is termed the *effective stress*.

(2) that shear strength depends, not on the total normal stress on the plane considered, but on the effective stress. This may be expressed by the equation

$$\tau_f = c' + \sigma' \tan \varphi \qquad (2)$$

where τ_f denotes the shear strength
σ' the effective stress on the plane considered
c' the apparent cohesion
φ' the angle of shearing resistance.

The most obvious practical illustration of the volume change aspect of the principle of effective stress is the continued settlement of a foundation under constant load, due to the gradual dissipation of the initial excess pore pressures. Less obvious, however but of considerable importance, are the regional settlements, which have resulted from ground water lowering in compressible soils, either due to pumping for water supply, as in London (Wilson and Grace, 1942) and in Mexico City (Zeevaert, 1953); or due to engineering operations, as in Oslo (Holmsen, 1953). The settlements of shallow foundations in the summer, due to the negative pore pressures set up by drying or by the suction of plant roots, is a further example.

The shear strength aspect of the principle of effective stress is appreciated most easily by comparing the results of undrained tests on a saturated soil with those of drained tests. In the undrained test no increase in strength results from an increase in total normal stress, an increase in pore pressure of equal magnitude being the only result. In the drained test this pore pressure is allowed to dissipate,

An excerpt from a lecture delivered to the Norwegian Geotechnical Society in September 1955.

and an increase in applied stress, which is thus fully effective, is found to result in a corresponding increase in shear strength.

Practical examples of the dependence of shear strength on effective stress are less self-evident than illustrations of deformation. However, the increase in strength of soft clay foundations of earth dams due to pore pressure dissipation has been measured on several occasions (Cooling and Golder, 1942; Bishop, 1948; Skempton and Bishop, 1955); and the gain in strength with depth of normally-consolidated clay strata is another well established fact (Skempton, 1948; Bjerrum, 1954).

Historically, the recognition of the importance of pore pressures may be considered to go back at least to 1886 when Osborne Reynolds demonstrated how a rubber bag full of saturated sand was subject to large negative pore pressures when distorted. Deacon and Hawkesley had pointed out, in connection with the Vyrnwy Dam, that it was the *pressure* set up under a dam by the blocking of an almost imperceptible flow of water, that was a major factor in stability (Deacon, 1896). Uplift pressures within masonry dams were recognised by Levy (1895). The action of water pressure *within* a fine grain soil appears, however, to have first been fully appreciated by Terzaghi (1923) and formed the basis of his theory of consolidation, which dealt with the volume change aspect of effective stress.

The shear strength aspect was recognised, firstly in terms of the difference between undrained and drained test results, in about 1932 by Terzaghi and his co-workers (Terzaghi, 1932). The subsequent work of Rendulic (1937) was of particular importance, as he devised the first practical laboratory method of measuring the pore pressure in a saturated clay. He demonstrated that, however the values of total stress σ and pore pressure u were varied, it was the difference $\sigma - u$ which controlled both deformation and failure conditions.

Investigations into the magnitude of pore pressures set up in partly saturated soils were begun by the U. S. Bureau of Reclamation, and included both theoretical and laboratory work (Brahtz, Bruggeman and Zangar; 1939, Hamilton, 1939), and field measurements (Walker and Daehn, 1948). Experimental investigations into pore pressure changes in saturated soils were also greatly advanced by the work of Taylor (1944, 1948) at the Massachusetts Institute of Technology.

At Imperial College theoretical and laboratory investigations into the magnitude and practical significance of pore pressure changes have taken a central place in our work since the Soil Mechanics section was established in 1946[1]. Emphasis has been given not only to laboratory measurements but also to the provision of a theoretical framework from which rational design methods may be developed to meet the needs of civil engineering practice. This, in turn enables the necessary link to be made between laboratory and field investigations.

More recently the Norwegian Geotechnical Institute under the direction of Dr. Bjerrum has begun to make important contributions in this field.

The Influence of the Area of Contact between the Soil Particles

In order to examine the physical basis of the principle of effective stress it is necessary to consider the forces acting across a surface in the soil which approximates to a plane but passes always through the pore space and points of contact of the soil particles. Normal stress is then equal to the average force perpendicular to this plane, per unit area, and areas are considered as projected onto the plane.

Let σ denote the total normal stress on this plane

σ_i' the average intergranular force per unit area of the plane (often termed the effective stress).

u the hydrostatic pressure in the pore water.

a the effective contact area of the soil particles per unit area of the plane.

It then follows that

$$\sigma = \sigma_i' + (1 - a)\, u \tag{3}$$

Thus

$$\sigma' = \sigma - (1 - a)\, u$$

or

$$\sigma_i' = (\sigma - u) + au \tag{4}$$

It can therefore be seen that effective stress, defined in this way, is not exactly equal to the difference between the total stress and the pore pressure, but is dependent on the contact area between the particles. Although this area may be small, it cannot be zero, as this would imply infinite local contact stresses between the particles. If σ_p is the average component of the local contact stress normal to the plane, then

$$\sigma_i' = a\, \sigma_p$$

or

$$\sigma_p = \frac{\sigma_i'}{a} \tag{5}$$

[1] For example, Skempton, 1948; Bishop 1948; Hansen and Gibson, 1949; Bishop and Eldin, 1950; Bishop, 1952; Bishop and Henkel, 1953; Skempton and Bishop, 1954; Skempton 1954; Bishop 1954.

Thus, although the direct measurement of the magnitude of a is not possible, a lower limit is placed on its value by the crushing strength of the particles, at least for coarse grained soils such as sands[2] (Bishop and Eldin, 1950). For clays and for concrete, in which the influence of uplift is also of considerable practical importance, this limit is less easy to determine and it is necessary to resort to indirect methods of determining a.

However, an important difficulty arises in the interpretation of the results of indirect methods, to which I drew attention in 1950 (loc. cit.). If the effective stress, as defined in equations 3 and 4, controlled volume change, we could observe the effect of several combinations of σ and u on volume change and thus evaluate a.

But if we consider the probable deformations at the contact[3] between two grains under fluid pressure,

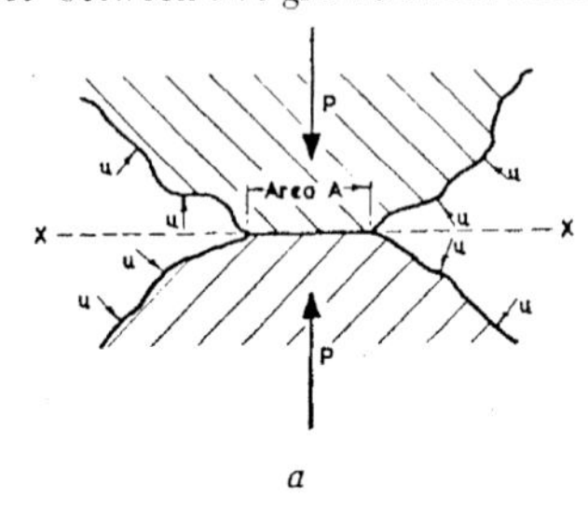

a

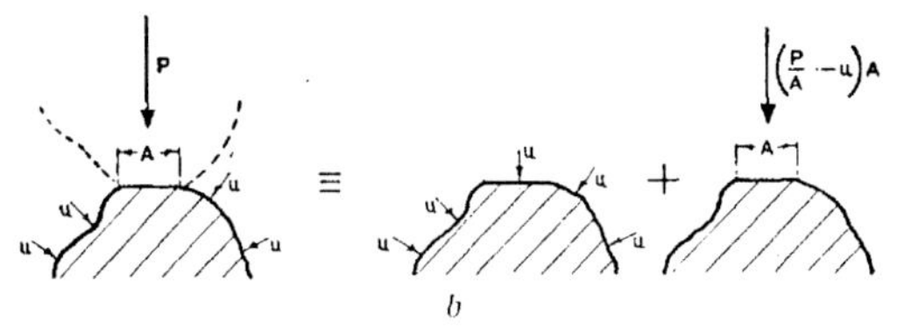

b

Fig. 1. (a) Forces at an intergranular contact. (b) Equivalent forces on a soil grain.

we arrive at an alternative hypothesis about the stress controlling volume change. If P is the average force per contact (Fig. 1) and there are N contacts per unit area, then the intergranular force per unit area of the plane XX is NP and is equal to σ_i'.

Now if a soil particle is subjected to a pressure u over the whole of its surface, it undergoes no distortion[4] but a small decrease in volume. Hence it is only that part of the local contact stress which is in excess of u that causes deformation of the soil structure. This excess stress is equal to $P/A - u$ where A is the area of the particular contact. By summing the corresponding components of intergranular force, an expression is obtained for σ_c', defined as that part of the normal stress which controls volume change due to deformation of the soil structure.

Thus

$$\begin{aligned} \sigma_c' &= N\left(\frac{P}{A} - u\right) A \\ &= NP - uNA \\ &= NP - ua \qquad \text{(since } NA = a\text{)} \\ &= \sigma_i' - au \qquad (6) \end{aligned}$$

and, by substitution from equation (4)

$$\begin{aligned} \sigma_c' &= (\sigma - u) + au - au \\ \text{i.e. } \sigma_c' &= \sigma - u \qquad (7) \end{aligned}$$

This leads to the interesting conclusion that, although the average intergranular force per unit area depends on the magnitude of a, volume changes due to deformation of the soil structure depend simply on the stress difference $\sigma - u$, whatever the value of a. It follows immediately that volume measurement under different combinations of σ and u can provide no data for evaluating a.

Recently Dr. A. S. Laughton carried out some experimental work in the Department of Geodesy and Geophysics at Cambridge to examine this hypothesis, which is of particular importance in predicting the consolidation of ocean sediments under very large hydrostatic pressures. His results provide a confirmation of its validity.

A set of tests were run in a special sealed oedometer in which the magnitude of both the total stress and the pore pressure could be controlled. This permitted the use of combinations of σ and u over a very wide range of stress (up to approximately $\sigma = 1000$ kg per sq. cm). One of the materials tested was lead shot, in which plastic yield at the contact points leaves, on unloading, facets which can be measured under the microscope to give approximate values of a. The test procedure consisted of the application of an increment of total stress together with a known increment of pore pressure, time being allowed for consolidation. The volume change was noted; the applied pore pressure was then reduced to zero without changing the total stress, and, after the same time interval, the volume change was again recorded.

From the results of a series of such increments a curve relating volume change to stress with zero pore pressure can be plotted; and from this curve can be read off the stress $\bar{\sigma}$ which gives the same volume change as each particular combination of σ and u.

According to the hypothesis outlined above the stress $\bar{\sigma}$ is to equal to σ_c', and from equation (7) we have:

$$\sigma - \bar{\sigma} = u$$

[2] Here $\sigma_P - u = b.S$ where b is factor depending on the type of surface contact (probably greater than unity), and S is the crushing strength of the grains, almost independent of the value of u.

[3] The contact between the rigid adsorbed water layers around clay particles may be treated, from this point of view, as a solid contact.

[4] This assumes that the material forming the particie is isotropic.

and hence

$$\frac{\sigma - \bar{\sigma}}{u} = 1 \quad (8)$$

If, on the other hand, the average intergranular force per unit area, σ_i', controls the volume change, as is usually assumed, then $\bar{\sigma}$ is equal to σ_i', and from equation (4) we have:

$$\sigma - \bar{\sigma} = (1 - a)\, u$$

and hence

$$\frac{\sigma - \bar{\sigma}}{u} = 1 - a \quad (9)$$

The results of a test on lead shot are given in Table I.

Table I. *Consolidation Test on Lead Shot (data from Laughton, 1955)*

σ kg/cm²	u kg/cm²	$\frac{\sigma - \bar{\sigma}}{u}$	a[1]
27	8	1.0	0,03
60	16	1.1	—
128	32	1.05	0.11
256	64	0.95	—
512	128	1.0	—
1024	256	0.95	0.95
		av. *1.01*	

[1] The estimate of the magnitude of a under stress involves an approximation as a can only be measured after unloading; values at intermediate stresses are obtained from separate tests.

It is clear from Table I that equation (9), representing the intergranular stress definition of effective stress, is quite incompatible with the experimental results. The results are, however, consistent with equation (8) to within the limits of experimental accuracy. It may therefore be concluded that the hypothesis is correct that $\sigma - u$ controls the volume changes, whatever the contact area between the particles may be.

Tests on a natural sediment also gave consistent values of $(\sigma - \bar{\sigma})/u = 1$. Though in this case the value of a could not be determined directly, it may be inferred from the high stresses used that a differed significantly from zero, and hence that equation (8) is again applicable.

Whether shear strength depends similarly on $\sigma - u$ (and not on a) is still a matter for speculation, as experimental data at the necessarily high stresses is not yet available. In the case of metals the frictional force is found to be proportional to the area of metal to metal contact (Bowden and Tabor, 1942). This area, like the deformation, would be controlled by $\sigma_e (= \sigma - u)$, and hence it is reasonable to infer that σ_e would control shear strength. That the frictional behaviour of soils is analogous to that of metals is less obvious, and alternative views cannot be dismissed at this stage. However, tests such as those of Rendulic (1937), carried out within the relatively small range of stresses encountered in engineering practice, indicate no departure from the hypothesis that $\sigma - u$ is the controlling factor in shear strength. Any remaining uncertainty is of practical importance only at very high pressures or in materials such as concrete in which a may have appreciable values at low stresses.

The Problem of a Two Phase Fluid

In partly saturated soils, the fluid in the pore space consists of both liquid and gaseous phases, existing at different pressures due to surface tension. The simplest case to consider is that of an almost fully saturated soil in which a small quantity of air or other gas exists in the form of isolated bubbles.

The principal effect of the bubbles, from the engineering point of view, is to alter radically the compressibility of the fluid phase; their presence makes little difference to the ability of the fluid to subject each soil particle to an equal all-round pressure. This pressure is equal to that of the liquid phase, and its magnitude would be that measured by a piezometer inserted into the soil. The validity of the effective stress equation is thus unaffected.

This is the case usually assumed in analysingt est results and in performing stability analyses. As the pore pressures are generally only of significant magnitude when the degree of saturation is relatively high[5] (80—100%), this is a reasonable assumption for practical purposes. This view is supported by the fact that no apparent inconsistencies arise in plotting laboratory shear strength data on this basis over a wide range of total stress and pore pressure values.

As the degree of saturation is reduced, a point will be reached when the soil particles will cease to be surrounded by the liquid phase. The pressure in the liquid phase, which is lower than that in the gaseous phase, will then act only over a reduced area. The effective stress equation may then be written:

$$\sigma' = \sigma - u_1 - \chi\,(u_2 - u_1) \quad (10)$$

where u_1 denotes pressure in gas and vapour phase.
u_2 pressure in liquid phase.
χ a parameter which equals unity for saturated soils and decreases with decreasing degree of saturation.

[5] Even if the *initial* degree of saturation were lower, the volume change necessary to create a significant pore pressure would bring the degree of saturation up into this range.

The practical importance of this more general form of the effective stress equation will depend on the range of x values normally encountered and on the magnitude of the surface tension effect $(u_2 - u_1)$ with which they are associated.

Four principal classes of problem may be considered:

(1) *Full saturation.* Here $x = 1$, and the equation reduces to $\sigma' = \sigma - u_2$. This holds true for all saturated soils, whether clays, sands or gravels; and is also true for concrete and porous rocks, at least with the qualification made in the previous section.

(2) *Soil containing a limited quantity of discontinuous air bubbles.* The air may exist in the form of a large number of small bubbles, in which case the term $(u_2 - u_1)$ will be significant. However, x will be close to unity, and hence, to a close approximation, the equation will reduce to $\sigma' = \sigma - u_2$.

An alternative condition may arise a the result of compacting soil which exists in the borrow pits in a saturated state. Here a limited number of larger air bubbles will result, and $(u_2 - u_1)$ will be smaller. The equation will similarly reduce to $\sigma' = \sigma - u_2$.

This class of soil will include compacted earth fills of clay, clay-gravel, moraine, chalk, etc., in which the initial compaction or subsequent volume changes or seepage lead to degrees of saturation of 80% or more. Most problems in which large positive pore pressures are encountered in partly saturated soils may be expected to fall into this category.

(3) *Completely dry soils.* Here $x = 0$ and $u_2 = 0$, and the equation reduces to the form $\sigma' = \sigma - u_1$.

This expression is clearly valid for cohesionless soils, and is supported by test results. For dry cohesive soils, the same result would be anticipated, though from the engineering point of view the result is of limited importance.

(4) *Soils with low degree of saturation.* Here $1 > x > 0$ and $u_1 > u_2$. In general, therefore, the full expression must be used, i. e., $\sigma' = \sigma - u_1 - x\,(u_2 - u_1)$. However, the range of engineering problems in which these conditions are encountered is very limited, at least in temperate climates.

For sands and gravels above the water table the value of the term $x\,(u_2 - u_1)$ is generally not significant compared with the value of σ', as is indicated, for example, by the low unconfined strength of these materials.

It is, therefore only with very fine grain soils at low degrees of saturation that the simple effective stress equation cannot be used. Earth fills of soil in this condition are generally avoided owing to the large changes in volume and shear strength which occur on subsequent saturation. Shallow foundations and pavements may, however, in some cases provide examples of conditions requiring the more complex relationship, and a detailed investigation of the values of x would then be necessary.

References.

Bishop, A. W. (1948) «*Some factors involved in the design of a large earth dam in the Thames Valley*». Proc. 2nd Int. Conf. Soil Mech., *2:* 13—18.

Bishop, A. W. (1952) «*The stability of earth dams*». Ph. D. Thesis, University of London.

Bishop, A. W. (1954) «*The use of pore-pressure coefficients in practice*». Geotechnique *4:* 148—152.

Bishop, A. W. and *Eldin, G.* (1950) «*Undrained triaxial tests on saturated sands and their significance in the general theory of shear strength*». Geotechnique *2:* 13—32.

Bishop, A. W. and *Henkel, D. J.* (1953) «*Pore pressure changes during shear in two undisturbed clays*». Proc. 3rd Int. Conf. Soil Mech. *1:* 94—99.

Bjerrum, L. (1954) «*Geotechnical properties of Norwegian marine clays*». Geotechnique *4:* 49—69.

Bowden, F. P. and Tabor, D. (1942) «*Mechanism of metallic friction*». Nature *150:* 197—199.

Brahtz, J. H. A., Bruggeman, L. R. and *Zangar, C. N.* (1939) «*Notes on analytical soil mechanics*». Tech. Mem. 592, U. S. Bur. Recl.

Cooling, L. F. and *Golder H. Q.* (1942) «*The analysis of failure of an earth dam during construction*». Journ. Inst. Civ. Eng. *19:* 38—55.

Deacon, G. F. (1896) «*The Vyrnwy Works for the water-supply of Liverpool*». Proc. Inst. Civ. Eng. *126:* 24—125.

Hamilton, L. W. (1939) «*The effects of internal hydrostatic pressure on the shearing strength of soils*». Proc. Amer. Soc. Test. Mater. *39:* 1100—1121.

Hansen, J. B. and *Gibson, R. E.* (1949) «*Undrained shear strengths of anisotropically consolidated clays*». Geotechnique *1:* 189—204.

Holmsen, G. (1953) «*Regional settlements caused by a subway tunnel in Oslo*». Proc. 3rd Int. Conf. Soil Mech. *1:* 381—383.

Laughton, A. S. (1955) «*The compaction of ocean sediments*». P. D. Thesis. University of Cambridge.

Levy, M. M. (1895) «*Quelques considerations sur la construction des grands barrages*». Comptes Rendus Acad. Sciences p. 288.

Rendulic, L. (1937) «*Ein Grundgesetz der Tonmechanik und sein experimenteller Beweis*». Bauing. *18:* 459—467.

Skempton, A. W. (1948) «*A study of the immediate triaxial tests on cohesive soils*». Proc. 2nd Int. Conf. Soil Mech. *1:* 192—196.

Skempton, A. W. (1954) «*The pore-pressure coefficients A and B*». Geotechnique *4:* 143—147.

Skempton, A. W. and *Bishop, A. W.* (1954) «*Soils*». Chapter X of Building Materials, their Elasticity and Inelasticity. Amsterdam: North Holland Pub. Co.

Skempton, A. W. and *Bishop, A. W.* (1955) «*The gain in stability due to pore-pressure dissipation in a soft clay foundation*». 5th Congress Large Dams No. 16.

Taylor, D. W. (1944) Tenth progress report on shear research to U. S. Engineers, M.I.T. Publication.

Taylor, D. W. (1948) «*Shearing strength determinations by undrained cylindrical compression tests with pore pressure measurements*». Proc. 2nd Int. Conf. Soil Mech. *5:* 45—49.

Terzaghi, K. (1923) «*Die Berechnung der Durchlässigkeitsziffer des Tones aus dem Verlauf der hydrodynamischen Spannungserscheinungen*». Sitz. Akad. Wissen. Wien Math-naturw. Kl. Abt. IIa, *132:* 125—138.

Terzaghi, K. (1932) «*Tragfähigkeit der Flachgründungen*». Int. Assoc. Bridge Struct. Eng. Prelim. Publ. 659—683.

Walker, F. C. and *Daehn, W. W.* (1948) «*Ten years of pore pressure measurements*». Proc. 2nd Int. Conf. Soil Mech. *3:* 245.

Wilson, G. and *Grace, H.* (1942) «*The settlement of London due to underdrainage of the London Clay*». Journ. Inst. Civ. Eng. *19:* 100—127.

Zeevaert, L. (1953) «*Pore pressure measurements to investigate the main source of surface subsidence in Mexico City*». Proc. 3rd Int. Conf. Soil Mech. *2:* 299—304.

More recent work is summarised in:

A. W. Bishop and D. J. Henkel: «*The measurement of soil properties in the Triaxial Test*». Edward Arnold, London, 1957.

THE RELEVANCE OF THE TRIAXIAL TEST TO THE SOLUTION OF STABILITY PROBLEMS

By Alan W. Bishop[1] and Laurits Bjerrum[2]

1. INTRODUCTION

The purpose of the present paper is to show how the actual properties of cohesive soils measured in the standard undrained, consolidated-undrained and drained triaxial test are applied to the solution of the more important classes of stability problem encountered by the practicing engineer. The failure criteria chosen and the shear parameters by which they are expressed are those found most convenient and most appropriate to the methods of stability analysis used. The relation of the practical shear parameters to the more basic shear parameters proposed, for example, by Hvorslev (1937)[3] is outside the scope of the present paper and is discussed elsewhere (Skempton and Bishop, 1954; Bjerrum, 1954 b).

The practical shear parameters serve to take full account of the principal differences between cohesive soils and other structural materials, such as the dependence of strength on the state of stress and on the conditions of drainage.

Although the purpose of the paper is to present the logical relationship of the various standard tests to the different classes of stability problem, attention must also be drawn to the various limitations of the apparatus in which the triaxial test is usually performed. These include non-uniformity of stress and strain particularly at large deformations, and the inability of the apparatus to simulate the changes in direction of the principal stresses which occur in many practical problems. To enable the quantitative importance of these limitations to be seen in perspective, emphasis is laid on the direct correlation between laboratory tests and field observations of stability (or instability) wherever case records are available.

It may seem to the practicing engineer that many of his problems are too small in scale or are in soils too lacking in homogeneity to apply detailed quantitative methods of stability analysis. However, even for the application of semi-empirical rules it is important to determine into which class the stability problem falls.

2. THE PRINCIPLE OF EFFECTIVE STRESS

One of the main reasons for the late development of Soil Mechanics as a systematic branch of Civil Engineering has been the difficulty in recognising

1. Reader in Soil Mech., Imperial College of Science and Tech., Univ. of London, England.
2. Dir., Norwegian Geotech. Inst., Oslo, Norway.
3. Items indicated thus, Hvorslev (1937), refer to corresponding entries listed alphabetically in the Appendix Bibliography.

that the difference between the shear characteristics of sand and clay lies not so much in the difference between the frictional properties of the component particles as in the very wide difference—about one million times—in permeability. The all-round component of a stress change applied to a saturated clay is thus not effective in producing any change in the frictional component of strength until a sufficient time has elapsed for water to leave (or enter), so that the appropriate volume change can take place.

The clarification of this situation did not begin until the discovery of the principle of effective stress by Terzaghi (1923 and 1932) and its experimental investigation by Rendulic (1937). An examination of current design methods might suggest that the impact of Terzaghi's discovery had yet to be fully felt.

For soil having a single fluid, either water or air, in the pore space, the principle of effective stress may be expressed in relation both to volume change and to shear strength:

(a) The change in volume of an element of soil depends, not on the change in total normal stress applied, but on the difference between the change in total normal stress and the change in pore pressure. For an equal all-round change in stress this is expressed quantitatively by the expression:

$$\Delta V / V = - C_c(\Delta\sigma - \Delta u) \tag{1}$$

where $\Delta V/V$ denotes the change in volume per unit volume of soil,
$\Delta\sigma$ denotes the change in total normal stress,
Δu denotes the change in pore pressure,
and C_c denotes the compressibility of the soil skeleton for the particular stress range considered.

It may be noted in passing that this equation shows that a decrease in pore pressure at constant total stress is as effective in producing a volume change as an increase in total stress at constant pore pressure, a fact which is confirmed by field experience.

(b) The maximum resistance to shear on any plane in the soil is a function, not of the total normal stress acting on the plane, but of the difference between the total normal stress and the pore pressure. This may be expressed quantitatively by the expression:

$$\tau_f = c' + (\sigma - u) \tan \Phi' \tag{2}$$

where τ_f denotes the shear stress on the plane at failure,
c' denotes the apparent cohesion, } in terms of effective
Φ' denotes the angle of shearing resistance, } stress.
σ denotes the total stress on the plane considered,[a]
and u denotes the pore pressure.

In both cases the effective normal stress is thus the stress difference σ - u, usually denoted by the symbol σ'.

The validity of the principle of effective stress has been amply confirmed, for saturated soils, by the experimental work of Rendulic (1937), Taylor (1944), Bishop and Eldin (1950),[b] and Laughton (1955)[b]; and indirectly by the

a. Stresses and pressures are here considered as measured with respect to atmospheric pressure as zero (i.e. gauge pressures). The actual datum does not of course affect the value of the effective stress.

b. A treatment of the influence of contact area is given in these papers.

field records referred to in later sections of this paper. For partly saturated soils, however, a more general form of expression must be used, since the pore space contains both air and water which may be in equilibrium at widely different pressures, due to surface tension. A tentative expression has been suggested for the effective stress under these conditions (Bishop, 1959[b]; 1960), of the form:

$$\sigma' = \sigma - u_1 + x\,(u_1 - u_2) \qquad (3)$$

where u_1 denotes the pressure in the air in the pore space,
u_2 denotes the pressure in the water in the pore space,
and x is a parameter closely related to the degree of saturation S and varying from unity in saturated soils to zero in dry soils.

The parameter x and its values under various soil conditions are discussed in more detail elsewhere (Bishop, 1960; Bishop, Alpan, Blight, and Donald, 1960). It may be noted in passing that for a given soil condition, the value of x measured in relation to shear strength may differ from its value measured in relation to volume change. However, the large positive pore pressures likely to lead to instability in rolled fills will in general only occur if the degree of saturation is high, where x may be equated to unity with little error. The additional complication of observing or predicting pore air pressure may therefore hardly be justified in such cases.

In most stability problems the magnitude of the body forces and of the applied loads is known quite accurately. It is in the magnitude of the shear strength that the main uncertainty lies and it is therefore useful to examine the variables controlling the value of τ_f in equation (2).

The magnitude of the total normal stress σ on a potential slip surface may be estimated with reasonable accuracy from considerations of statics. The shear parameters c' and Φ' are properties which depend primarily on the soil type and to a limited extent on stress history (see Table I in section 6). Provided representative samples are taken and tested in the appropriate stress range, little error need arise in evaluating c' and Φ'. This aspect of any investigation does, however, call for sound judgment and a knowledge of geology.

It is in the prediction of the value of the pore pressure u that in many problems the greatest uncertainty lies. The development of cheap and reliable field devices for measuring pore pressure in soils of low permeability[c] has, however, transformed the situation as far as the practicing engineer is concerned by enabling predictions to be checked and a control to be kept on stability during construction work.

Much of the uncertainty about the pore pressure prediction has arisen from a failure to distinguish between the two main classes of problem[d]:

(a) Problems where pore pressure is an independent variable and is controlled either by ground water level or by the flow pattern of impounded or underground water, for example, and
(b) Problems in which the magnitude of the pore pressure depends on the magnitude of the stresses tending to lead to instability, as in rapid construction or excavation in soils of low permeability.

---

c. See, for example, Casagrande, 1949; U.S.B.R., 1951; Penman, 1956; Sevaldson, 1956; Kallstenius and Wallgren, 1956; Bishop, Kennard, and Penman, 1960.

d. This distinction is discussed in detail by Bishop (1952).

In problems which initially fall into class (b) the pore pressure distribution will change with time and at any point the pore pressure will either decrease or increase to adjust itself to the ultimate condition of equilibrium with the prevailing conditions of ground water level or seepage. The rate at which this adjustment occurs depends on the permeability of the soil (as reflected in its coefficient of consolidation) and on the excess pore pressure gradients, which depend both on the stress gradients and on the distance to drainage surfaces.

The least favourable distribution of pore pressure may occur either in the initial stage or at the ultimate condition or, in special cases, at an intermediate time depending, for example, on whether load is applied or removed, and on other specific details of the problem, as discussed in section 6.

3. PORE PRESSURE PARAMETERS

In slope stability problems the influence of pore pressure on the factor of safety is most conveniently expressed in terms of the ratio of the pore pressure to the weight of material overlying the potential slip surface. This ratio was used by Daehn and Hilf (1951) in the form of an overall ratio of the sum of resolved components of the pore pressure and of the weight of soil, to express the results of the stability analysis of four earth dams, based on the field measurement of pore pressure.

Bishop (1952 and 1954 b) showed that, for a slope in which the ratio of the pore pressure u to the vertical head[e] of soil γh above the element considered was a constant, the value of the factor of safety F decreased almost linearly with increase in pore pressure ratio $u/\gamma h$. Subsequent work by Bishop and Morgenstern (in course of preparation for publication) has shown that, both for pore pressures obtained from flow patterns (class (a) problems) and for those obtained as a function of stress (class (b) problems), the average value of the pore pressure ratio $u/\gamma h$ is the most convenient dimensionless parameter by which to express the influence of pore pressure stability (Fig. 1). The ratio is denoted r_u.

Where the pore pressure is independent of stress, its value is obtained directly from the ground water conditions or flow net and expressed as the average[f] value of r_u. In all other cases (class (b) problems) the ratio must either be obtained from field measurements or predicted from the observed relationships between pore pressure and stress change under undrained conditions and from the theory of consolidation. This in turn necessitates an estimate of the stress distribution within the soil mass.

For the inclusion of the laboratory results in the stability calculation it is convenient to express them in terms of pore pressure parameters. The development of these parameters[g] and their application to practical problems is described in detail elsewhere (Skempton, 1948 b; Bishop, 1952; Skempton, 1954; Bishop, 1954 a; Bishop and Henkel, 1957; Bishop and Morgenstern, 1960).

e. γ is the average bulk density of the soil and h the vertical distance of the surface above the element.

f. Details of the averaging method are given by Bishop and Morgenstern, 1960.

g. Attention was drawn to the possibility of pore pressure changes in clays under the action of a deviator stress before it proved practicable to measure them (Terzaghi, 1925: Casagrande, 1934).

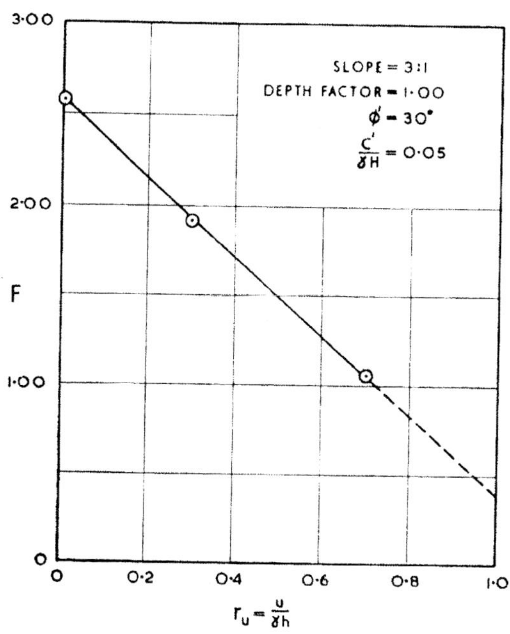

Fig. 1.—The linear relationship between factor of safety and pore pressure ratio for a slope or cut in cohesive soil.

For a change in stress under undrained conditions the change in pore pressure may be expressed as Δu, where

$$\Delta u = B\,[\Delta\sigma_3 + A\,(\Delta\sigma_1 - \Delta\sigma_3)] \tag{4}$$

where $\Delta\sigma_1$ denotes the change in major principal stress,
$\Delta\sigma_3$ denotes the change in minor principal stress,
(in both cases total stresses are considered)
and A and B denote the pore pressure parameters (Skempton, 1954).

Triaxial compression tests show that for fully saturated soils B = 1 to within practical limits of accuracy, and that the value of A depends on stress history and on the proportion of the failure stress applied. This is illustrated in Fig. 2. where the A values for normally and overconsolidated clay are given. The values of A at failure are seen to be very dependent on the overconsolidation ratio (defined as the ratio of the maximum consolidation pressure to which the soil has been subjected to the consolidation pressure immediately before the undrained test is performed).

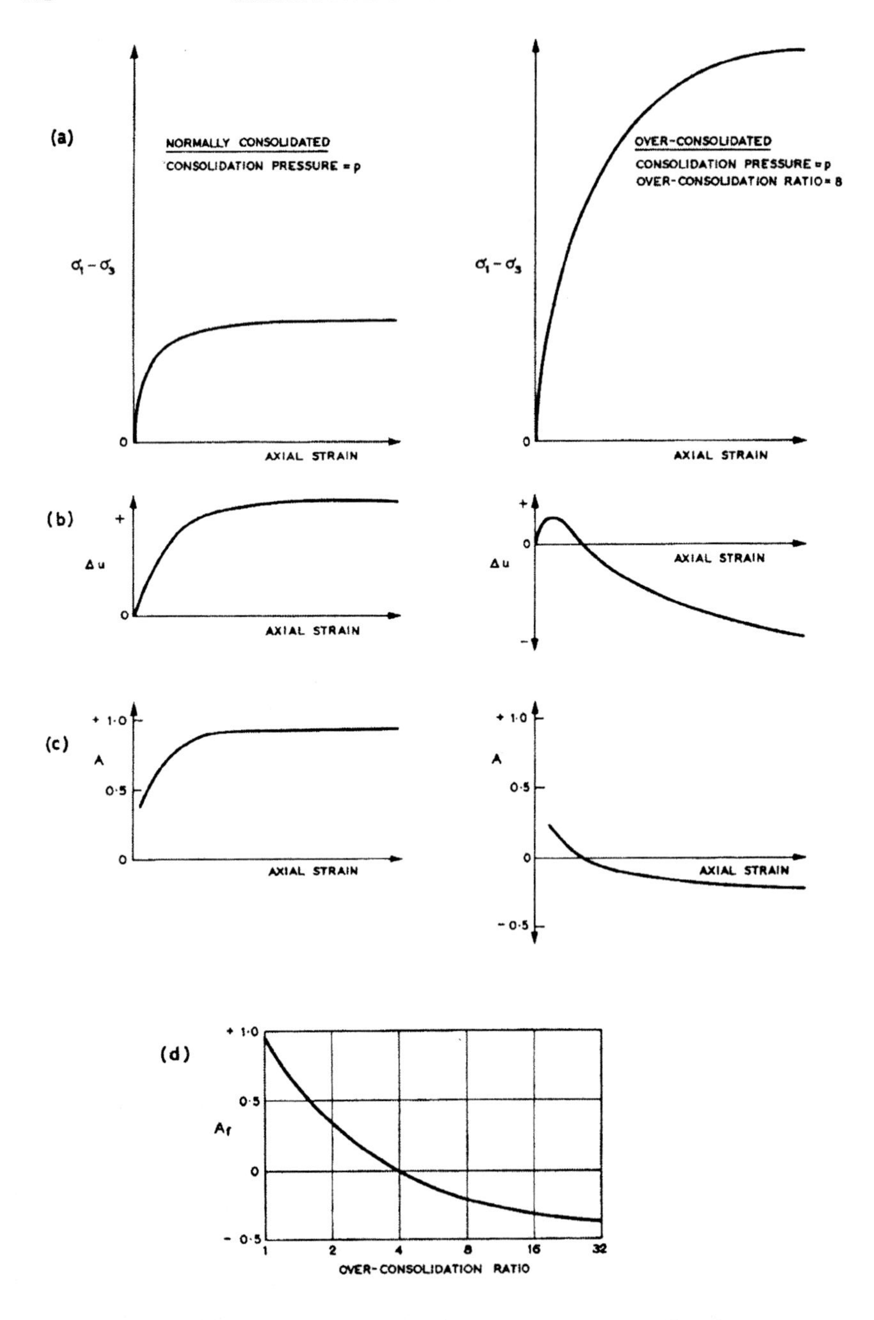

Fig. 2.--The dependence of the pore pressure parameter A on stress history.

For partly saturated soils the value of B lies between 0 and 1 depending on the degree of saturation and the compressibility of the soil skeleton. Typical values of A and B are used in section 6[h].

It should be noted that equation 4 takes no account of the change in the intermediate principal stress $\Delta\sigma_2$, or of possible changes in the directions of the principal stresses. In the majority of stability problems the conditions approximate to plane strain, in which the intermediate principal stress does not equal the minor principal stress as in the standard cylindrical compression test. Theoretical studies (Skempton, 1948 a and b; Hansen and Gibson, 1949; Bishop and Henkel, 1957) indicate that the form of the equation remains the same, but that the triaxial test underestimates the value of A, and the limited amount of test data so far available (Wood, 1958; Cornforth, 1960; Henkel, 1960) supports this view. Little is yet known about the influence of the rotation of the principal stresses on the value of A. The importance of these limitations in practice can at present only be assessed from the overall check with observed pore pressures in the field.

It should also be noted that the principle of superposition can be applied to pore pressure changes in soil only in a very restricted sense. Where the purpose of the test is the accurate prediction of pore pressure at states of stress other than failure, a more accurate result is obtained if the stress increments occurring in practice are closely followed in the test by making simultaneous changes in the values of both σ_1 and σ_3. The test result is then conveniently expressed in terms of the relationship between pore pressure and major principal stress, for the specified stress ratio, using the expression:

$$\Delta u = \bar{B} \,.\, \Delta\sigma_1 \tag{5}$$

The influence of stress ratio on this parameter is illustrated in Fig. 3 for a compacted earth fill.

It should be noted that the value of the parameter $\bar{B}$ only gives the change in pore pressure due to stress change under undrained conditions. The actual pore pressure depends also on the initial value u_0 before the stress change is made (Fig. 4), and is given by the expression:

$$u = u_0 + \Delta u \tag{6}$$

$$\text{i.e.} \quad u = u_0 + \bar{B} \,.\, \Delta\sigma_1 \tag{7}$$

In natural strata u_0 is determined from the initial ground water conditions, being positive below ground water level and negative above. In rolled fill the initial value is usually negative, reaching quite high values in cohesive soils placed at or below the optimum water content (Hilf, 1956; Bishop, 1960; Bishop, Alpan, Blight, and Donald, 1960).

In cases where no dissipation of pore pressure is assumed to occur, the pore pressure ratio r_u used in the stability analysis is directly related to $\bar{B}$:

$$r_u = u/\gamma h = 1/\gamma h \,.\, (u_0 + \bar{B}\Delta\sigma_1) \tag{8}$$

h. The value of B is generally given with respect to changes in pore water pressure. A slightly different value relates the change in air pressure to the change in total stress.

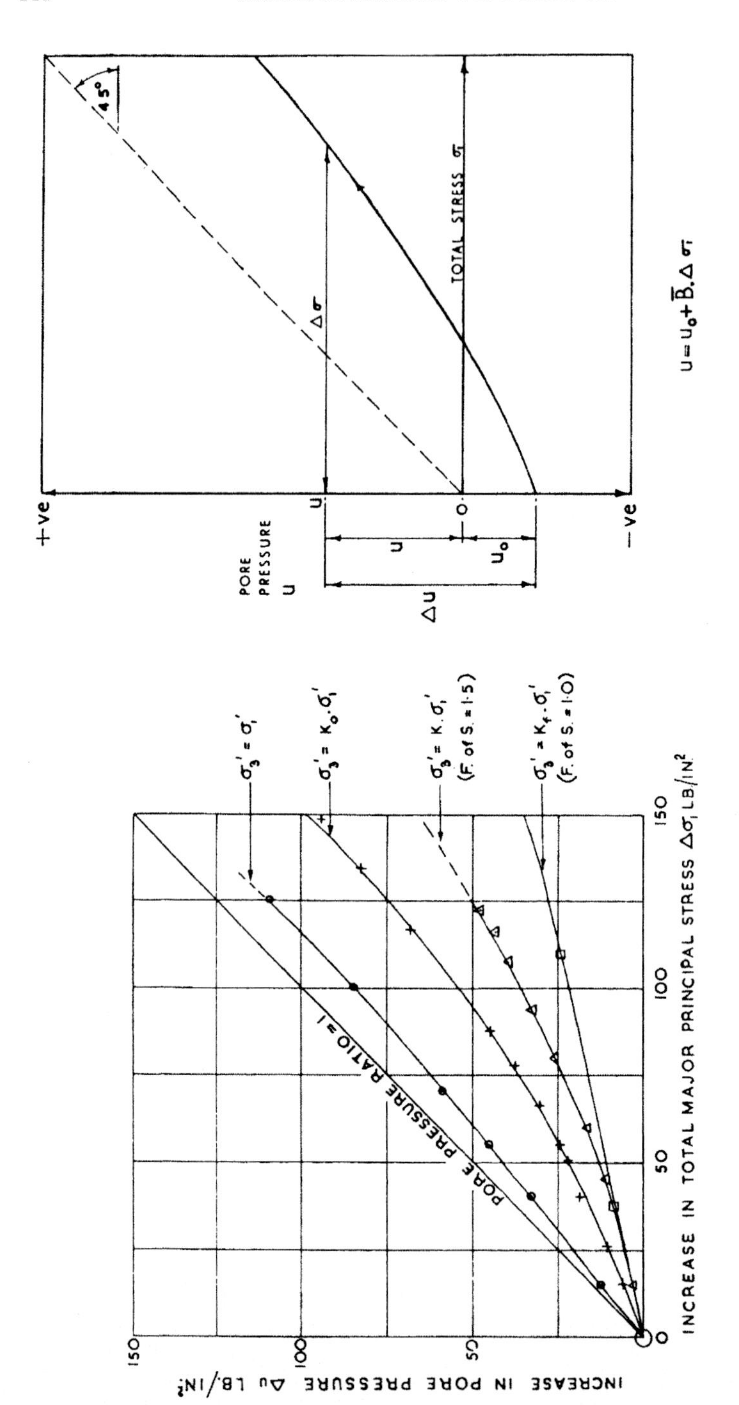

Fig. 3.—The influence of principal stress ratio on the pore pressure parameter $\bar{B}$: boulder clay compacted at optimum + 1%.

Fig. 4.—Pore pressure change under undrained test conditions.

In the special case of the construction of earth fill embankments the average value of $\Delta\sigma_1$ along a potential slip surface approximates to γh (Bishop, 1952), and equation (8) becomes:

$$r_u = \bar{B} + u_0/\gamma h \tag{9}$$

For earth fills of low plasticity placed wet of the optimum the term $u_0/\gamma h$ is small, and a further approximation is sometimes used in preliminary design:

$$r_u = \bar{B} \tag{10}$$

Some typical examples of the use of pore pressure parameters are given in section 6.

4. STANDARD TYPES OF TRIAXIAL TEST

The type of triaxial test most commonly used in research work and in routine testing is the cylindrical compression test. A diagrammatic layout of the apparatus is given in Fig. 5.

The cylindrical specimen is sealed in a thin rubber membrane and subjected to fluid pressure. A load applied axially, through a ram acting on the top cap, is used to control the deviator stress. In a compression test the axial stress is thus the major principal stress σ_1; the intermediate and minor principal stresses (σ_2 and σ_3 respectively) are both equal to the cell pressure.

Connections to the ends of the sample permit either drainage of water or air from the voids of the soil or, alternatively, the measurement of pore pressure under conditions of no drainage.

In most standard tests the application of the allround pressure and of the deviator stress form two separate stages of the test; and tests are therefore classified according to the conditions of drainage obtaining during each stage:

(1) Undrained tests[i].—No drainage, and hence no dissipation of pore pressure, is permitted during the application of the all-round stress. No drainage is permitted during the application of the deviator stress (σ_1 - σ_3).

(2) Consolidated-undrained tests.—Drainage is permitted after the application of the all-round stress, so that the sample is fully consolidated under this stress. No drainage is permitted during the application of the deviator stress.

(3) Drained tests.—Drainage is permitted throughout the test, so that full consolidation occurs under the all-round stress and no excess pore pressure is set up during the application of the deviator stress.

In order to illustrate the inter-relation between results of the different types of test, saturated and partly saturated soils will be considered separately.

i. Alternative nomenclatures have been used both in Europe and the U.S.A. The present terms are considered to be the most descriptive of the test conditions.

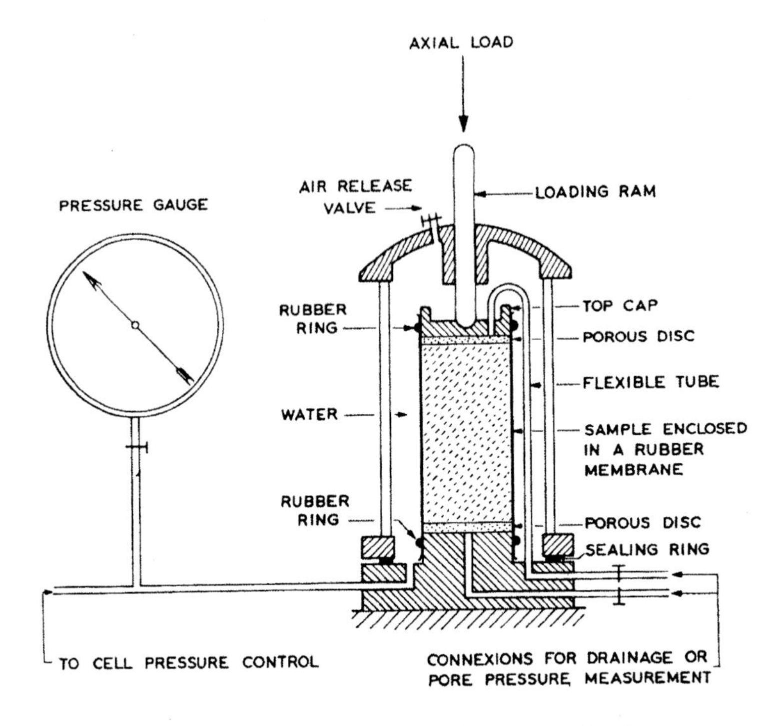

Fig. 5.—Diagrammatic layout of the triaxial cell.

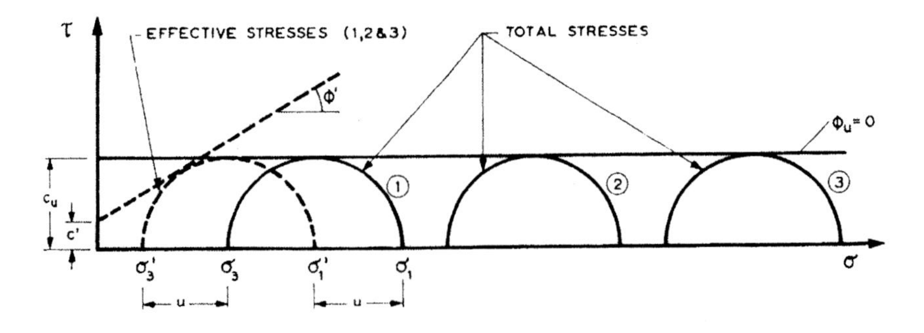

Fig. 6.—Undrained tests on saturated soil: total and effective stress circles.

(a) Undrained Tests on Saturated Cohesive Soils.—

These tests are carried out on undisturbed samples of clay, silt and peat as a measure of the existing strength of natural strata, and on remoulded samples when measuring sensitivity or carrying out model tests in the laboratory.

The compression strength (i.e. the deviator stress at failure) is found to be independent of the cell pressure, with the exception of fissured clays (discussed in section 6) and compact silts at low cell pressures. The corresponding Mohr stress circles are shown in Fig. 6.

If the shear strength is expressed as a function of total normal stress by Coulomb's empirical law:

$$\tau_f = c_u + \sigma \tan \Phi_u \tag{11}$$

where c_u denotes apparent cohesion, Φ_u denotes angle of shearing resistance; in terms of total stress, it follows that, in this particular case,

$$\left.\begin{aligned} \Phi_u &= 0 \\ c_u &= \frac{1}{2}(\sigma_1 - \sigma_3)_f \end{aligned}\right\} \tag{12}$$

The shear strength of the soil, expressed as the apparent cohesion, is used in a stability analysis carried out in terms of total stress, which, for this type of soil, is known as the $\Phi_u = 0$ analysis (Skempton, 1948 a and b). Since the value of c_u may be obtained directly from the unconfined compression test (where $\sigma_3 = 0$), and from the vane test in the field, it is a simple and economical test, but is often used without regard to the class of stability problem under consideration.

For fully saturated soils the increase in cell pressure is reflected in an equal increase in pore pressure and the effective stresses at failure remain unchanged. If pore pressure measurements are made during the test only one effective stress circle is obtained (Fig. 6), and tests at other water contents must be carried out to obtain the failure envelope in terms of effective stress.

In Fig. 7(a) an example is given of the changes in pore pressure during shear in an unconfined compression test and in Fig. 7(b) the Mohr circles are given in terms of total and effective stresses.

The A value measured in the undrained test on a sample of natural ground is very different from the value in situ under a similar change in shear stress. This results from the stress history given to the sample by changes in pore pressure which occur during sampling and preparation due to the removal of the insitu stresses, quite apart from disturbance due to the sampler itself. The release of the deviator stress existing in samples normally consolidated with no lateral yield is a major factor contributing to this effect.

Tests on samples anisotropically consolidated in the laboratory (Bishop and Henkel, 1953) and on undisturbed samples (Bishop, 1960) show that the effective stress in the sample when under an all-round pressure or unconfined can be less than half the effective overburden pressure in situ. Yet when the shear stress is increased to bring the sample to failure, the undrained strength closely corresponds to the in situ strength deduced from stability analysis or from vane tests. This is consistent with the experimental observation that, for a limited range of soil types and stress paths, strength

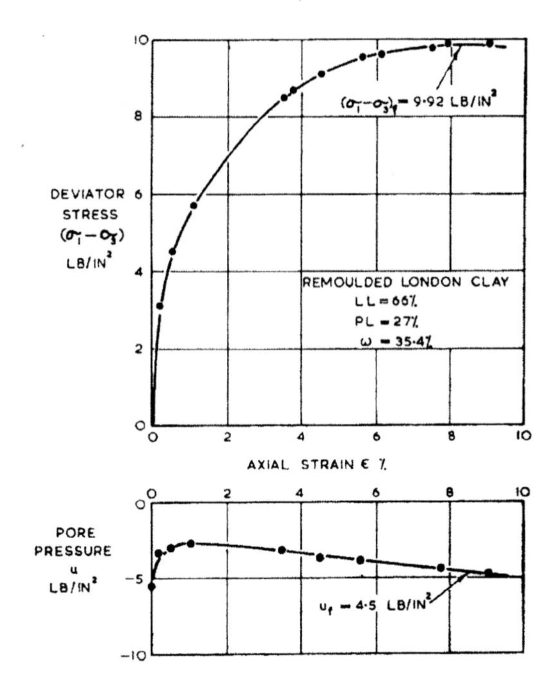

Fig. 7a.—Pore pressure change during shear in an unconfined compression test.

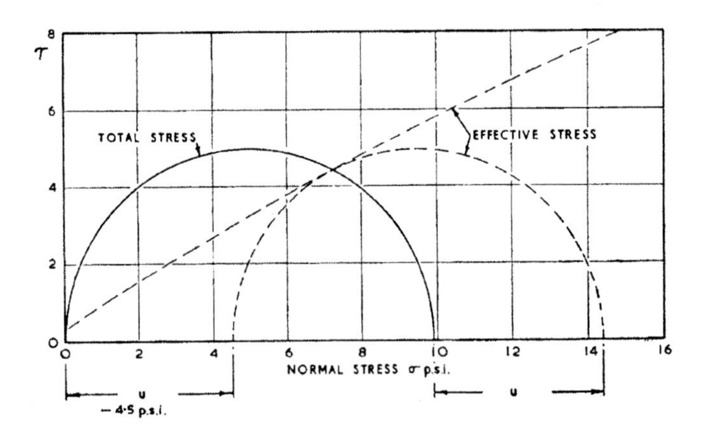

Fig. 7b.—Total and effective stress circles for the unconfined compression test.

and water content are uniquely related (Waterways Experiment Station, 1947; Henkel, 1959).

If it is indeed this fact which provides the empirical justification for the use of undrained compression tests in the $\Phi_u = 0$ analysis, then to reconsolidate the samples in the laboratory under the existing overburden pressure will inevitably lead to an overestimate of the in situ strength of the soil, since reconsolidation is almost always accompanied by a decrease in water content.

(b) Consolidated-Undrained Tests on Saturated Soils.—

These tests are carried out on both undisturbed and remoulded samples of cohesive soils, primarily to determine the values of c' and Φ', but also to determine the values of A and to study the effect of stress history.

In the standard test the sample is allowed to consolidate under a cell pressure of known magnitude (p), the three principal stresses thus being equal. The sample is then sheared under undrained conditions by applying an axial load. As in the case of the undrained test in the previous section, the cell pressure at which the sample is sheared does not influence the strength (except in dilatant silts at low pressures) as illustrated in Fig. 8e. The test result, in terms of total stresses, may thus be expressed by plotting the value of c_u against consolidation pressure p, Fig. 8b.

For normally consolidated soils the ratio c_u/p is found to be constant, its value depending on soil type. However, strengths measured in undrained triaxial tests and vane tests on strata existing in nature in a normally consolidated state, when plotted against the effective overburden pressure, lead to a lower estimate of c_u/p than is found with samples consolidated under equal all-round stress in the laboratory. The difference increases as the plasticity index decreases and appears to be due to two causes:

(i) A naturally deposited sediment is consolidated under conditions of no lateral displacement, and hence with a lateral effective stress considerably less than the vertical stress. The ratio of the effective stresses, termed the coefficient of earth pressure at rest, is generally found from laboratory tests to lie in the range 0.7-0.35, the lower values occurring in soils with a low plasticity index (Terzaghi, 1925; Bishop, 1958 a; Simons, 1958). This cause alone can account in soils of low plasticity for a difference of 50% in the value of c_u/p (for example, Bishop and Henkel, 1953; Bishop and Eldin, 1953).

(ii) Reconsolidation in the laboratory after the stress release associated with even the most careful sampling technique leads to a lower void ratio than would occur in the natural stratum under the same stress. The value of the pore pressure parameter A in particular is sensitive to the resulting modification in soil structure and this in turn leads to a higher undrained strength.

For these reasons the use of the results of consolidated-undrained tests, expressed in terms of total stress either by the parameter c_u/p or by the value of Φ_{cu} (see appendix 1), can be justified in few practical applications. However, if the pore pressure is measured during the undrained stage of the test, the results can be expressed in terms of the effective parameters c' and Φ'. Experience has shown that these parameters can be applied to a wider range of practical problems.

The relationships between the total stress, pore pressure and effective stress characteristics obtained in a typical series of consolidated-undrained

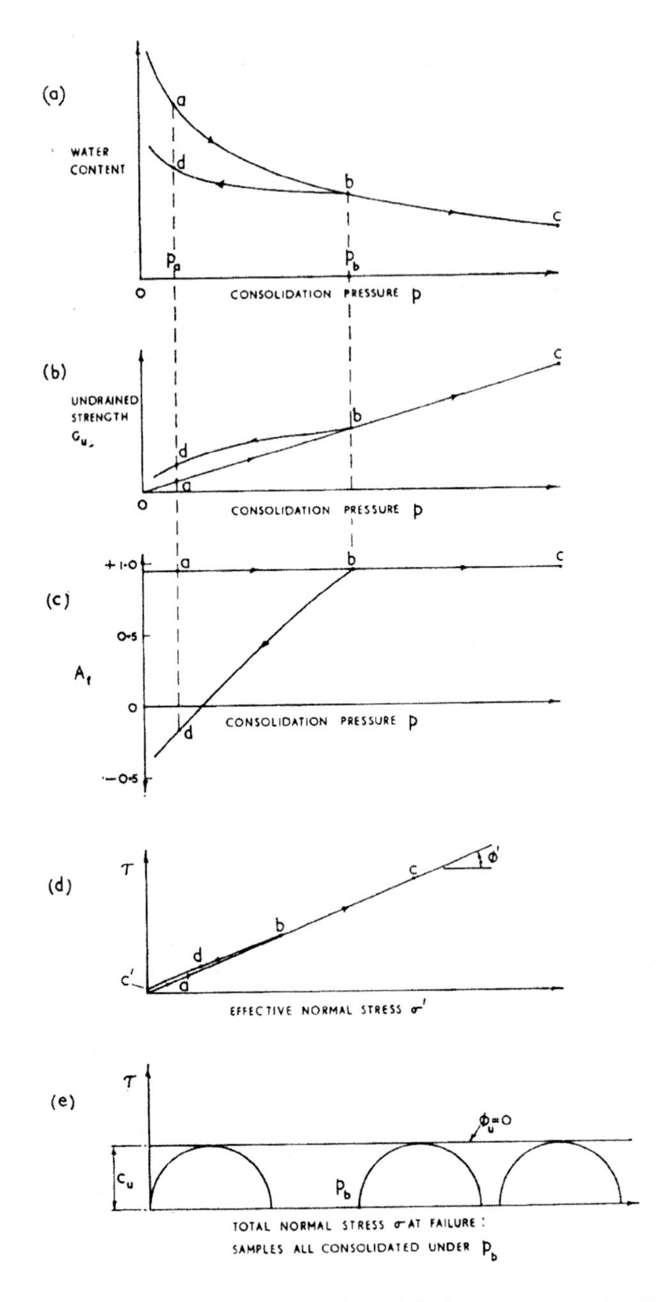

Fig. 8.—The relationships between the total stress, pore pressure and effective stress characteristics for a series of consolidated-undrained triaxial tests on saturated cohesive soil.

triaxial tests are illustrated in Fig. 8. The points, a, b, and c represent normally consolidated samples; the point d represents an over-consolidated sample, the overconsolidation ratio being p_b/p_d, Fig. 8a. For normally consolidated samples the effective stress envelope is a straight line with c' equal to zero (Fig. 8d), the value of Φ' depending on soil type. Over-consolidation results in an envelope lying a little above this straight line; the section of this envelope relevant to any particular practical problem can generally be represented with sufficient accuracy by a slightly modified value of Φ' and a cohesion intercept c'.

The most marked effect of over-consolidation is, however, on the value of A, which, with increasing over-consolidation ratio, drops from a value typically about 1 at failure to values in the negative range (Figs. 2 and 8c). These low A-values are, in turn, largely responsible for the high undrained strength values resulting from over-consolidation (compare point d with point a in Fig. 8b).

Values of c' and Φ' are usually based on the effective stress circles corresponding to maximum deviator stress. However, in some over-consolidated clays in which large decreases in pore pressure during shear are associated with very large failure strains, a slightly larger value of Φ' is obtained by plotting the state of stress at a smaller strain approximating to the point at which the ratio of the principal effective stresses σ_1'/σ_3' reached its maximum value. The difference in the value of Φ' is generally not important from a practical point of view, but in making comparisons between the values of Φ' obtained from consolidated-undrained and drained tests it is necessary to specify which definition of Φ' is being used[j].

(c) Drained Tests on Saturated Soils.—

Drained tests are carried out on both undisturbed and remoulded samples of cohesive soils to obtain directly the shear strength parameters relevant to the condition of long term stability, when the pore pressures have decreased (or increased) to their equilibrium values.

In the standard test the sample is allowed to consolidate under a cell pressure of magnitude p and is then sheared by increasing the axial load at a sufficiently slow rate to prevent any build-up of excess pore pressure. The effective minor principal stress σ_3' at failure is thus equal to p, the consolidation pressure; the major effective principal stress σ_1' is the axial stress. The test results lead directly to the effective stress shear parameters c' and Φ', which for drained tests are often denoted c_d and Φ_d.

The drained tests also provides data on the volume changes which occur during the application of the equal all-round stress and the deviator stress.

(d) Inter-Relationship between the three Types of Test on Saturated Soil.—

Two aspects of this inter-relationship are of practical interest to the engineer concerned with stability problems: (1) The degree of reliability with

j. Whether this difference reflects an actual characteristic of frictional materials or merely the increasing nonuniformity of stress in the cylindrical compression tests at large strains is still open to question. It is, however, clear from tests on sand published by the Waterways Experiment Station (1950) and other unpublished tests at the Norwegian Geotechnical Institute and at Imperial College that for very loose soil structures the maximum deviator stress may occur at smaller strains than the maximum stress ratio, and here the difference in Φ' (of up to 15^0 or so) undoubtedly represents a physical property of the soil.

which the effective stress envelope defined by the parameters c' and Φ' can be assumed to be the same for undrained, consolidated-undrained and drained tests; and (2) the extent to which volume changes in drained tests are an indication of the magnitude of pore pressure changes in consolidated-undrained tests.

In Fig. 9d are compared the results of undrained, consolidated-undrained, and drained tests on clay from the foundation of the Chew Stoke Dam (described by Skempton and Bishop, 1955). The close agreement between the effective stress failure envelopes may be noted. It is also of interest to note from the low values of A_f that an undisturbed sample reconsolidated in the laboratory behaves as though it were 'over-consolidated' even at cell pressures greatly in excess of the in situ pre-consolidation pressure. The intercept c' will in general not be zero for this part of the failure envelope.

That there should be close agreement between the effective stress envelopes for consolidated-undrained and drained tests on normally consolidated samples has been shown theoretically by Skempton and Bishop (1954) using the concept of true cohesion and friction due to Hvorslev (1937). Since Φ' is to some extent time-dependent, it is necessary to use similar rates of testing in making an experimental comparison, and to ensure adequate time for pore pressure measurement in the consolidated-undrained test and for drainage in the drained test. The predicted values of Φ' from the consolidated-undrained test are the higher, but only by 0-1° in typical cases[k].

However, for heavily over-consolidated clays the position is generally reversed[l] and the drained test is usually found to give the higher value, due to the work done by the increase in volume during shear in the drained test, and to the smaller strain at failure.

The volume changes in drained tests have for some time been known to correlate qualitatively with the pore pressure changes in undrained tests. Experimental data on two remoulded clays have recently been presented by Henkel (1959 and 1960) who has described a simple graphical procedure from which the quantitative relationship may be obtained.

(e) Undrained Tests on Partly Saturated Cohesive Soils.—

These tests are most commonly carried out on samples of earth-fill material compacted in the laboratory under specified conditions of water content and density. They are also applied to undisturbed samples of strata which are not fully saturated, and to samples cut from existing rolled fills and trial sections.

The compression strength is found to increase with cell pressure (Fig. 10a), as the compression of the air in the voids permits the effective stresses to increase. However, the increase in strength becomes progressively smaller as the air is compressed and passes into solution, and ceases when the stresses are large enough to cause full saturation, Φ_u the approximating to zero. The failure envelope expressed in terms of total stress is thus nonlinear, and values of c_u and Φ_u can be quoted only for specific ranges of normal stress.

---

k. The sign and magnitude of this difference may change if the failure strains are very dissimilar, as in long term tests reported by Bjerrum, Simons, and Torblaa (1958).

l. If the failure envelopes corresponding to maximum deviator stress are compared.

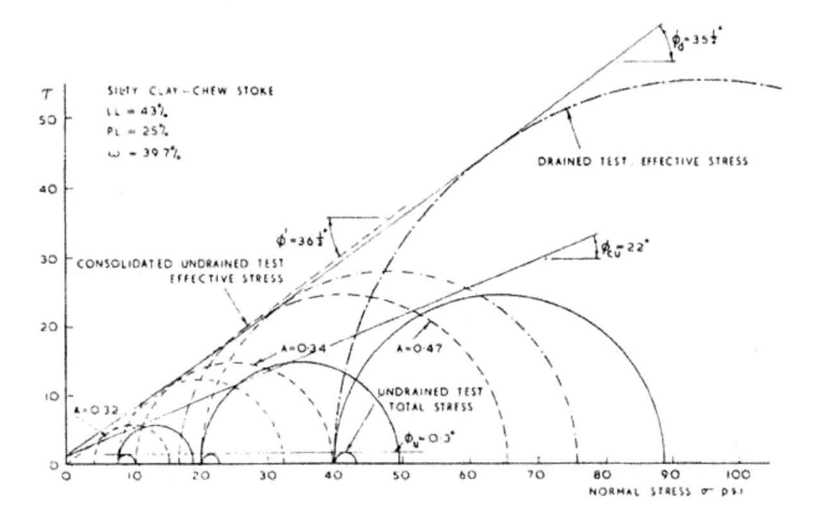

Fig. 9.—Undrained, consolidated-undrained and drained tests on undisturbed samples of Chew Stoke silty clay: maximum deviator stress.

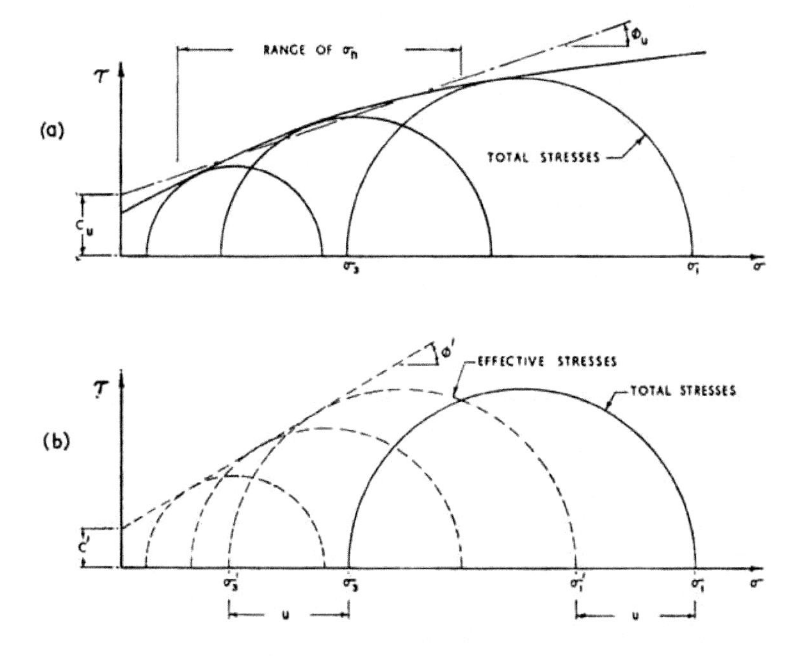

Fig. 10.—Undrained tests on partly saturated cohesive soil (a) in terms of total stress, (b) in terms of effective stress.

If the pore pressure is measured during the test, as is usual where field pore pressure measurements are to be used to check the stability during construction, then the failure envelope can be expressed in terms of effective stress, Fig. 10b. The effective stress envelope is found to approximate very closely to a straight line over a wide range of stress.

However, rather more difficulty arises in defining accurately the effective stress envelope for a partly saturated soil than at first apparent. The first difficulty lies in testing technique, to which attention was drawn by Hilf (1956). This problem is discussed in detail by Bishop (1960) and Bishop, Alpan, Blight, and Donald (1960), where it is concluded that accurate pore water pressure measurements can be made in the triaxial apparatus in partly saturated cohesive soils provided a porous element of very high air entry value is used and provided a considerably reduced rate of testing is accepted.

The second difficulty lies in the form of the expression for effective stress (equation 3), which includes a term for pore-air pressure as well as pore-water pressure for values of the factor x other than unity. The use of the simple expression for effective stress of total stress minus porewater pressure leads to an over-estimate of effective stress of $(1 - x)(u_1 - u_2)$ where $(u_1 - u_2)$ is the difference between pore-air pressure and pore water pressure. Since values of $(u_1 - u_2)$ of up to 40 lb. per sq. inch have already been measured on rolled fill in the triaxial test, and the value of x approximates to the degree of saturation, significant errors in effective stress result from the use of the simpler expression. This is particularly marked near the origin of the Mohr diagram and may lead to the apparent anomaly of a negative 'cohesion' intercept (Bishop, Alpan, Blight, and Donald, 1960). However, pore pressures set up under construction conditions are only critical if the water content of the fill and the magnitude of the stresses lead to almost full saturation, and in this case the error is small enough to be ignored in many practical problems.

(f) Consolidated-Undrained Tests on Partly Saturated Cohesive Soils.—

These tests are carried out on samples of compacted earth-fill material and on undisturbed samples. They may be necessary to determine c' and Φ' when the degree of saturation of the samples is not low enough to result in a sufficient range of strengths in the undrained tests to define a satisfactory failure envelope.

Consolidated-undrained tests in which a backpressure is applied to the pore space to ensure full saturation before shearing are carried out to examine the effect on the values of c' and Φ' of the submergence of fill or foundation strata. Back-pressures of up to 100 lb. per sq. inch are often required to give full saturation on a short term basis.

(g) Drained Tests on Partly Saturated Cohesive Soils.—

Drained tests are carried out on both compacted and undisturbed samples to obtain directly the values of c' and Φ' for the condition of long term stability. Generally a backpressure is applied to ensure full saturation of the sample before the application of the deviator stress, during which the backpressure is held constant.

(h) Inter-Relationship between the Three Types of Test on Partly Saturated Soil.—

Here again two aspects of this inter-relationship are of practical interest to the engineer concerned with stability problems: (1) The comparison of the values of c' and Φ' obtained from the different types of test; and (2) the prediction of pore pressure changes from volume changes.

Tests carried out at Imperial College have generally shown that the difference between the values of Φ' measured in the different types of test are not very significant from a practical point of view. The value of c', however, tends to correlate with water-content at failure. Where all the samples defining a failure envelope show a marked increase in water content in the consolidated-undrained or drained test with a back-pressure, c' is generally reduced. With the lower values of c' obtained by using the improved pore pressure technique described elsewhere (Bishop, 1960; Bishop, Alpan, Blight, and Donald, 1960), the difference in c' obtained in the different tests are less marked, and in some soils are not of practical significance[m] (Fig. 11). The range of soil types so far tested using this technique is, however, rather limited.

It is generally easier to make accurate measurements of pore water pressure under undrained conditions than to make the necessarily very accurate measurements of volume change and degree of saturation on which pore pressure predictions depend. Studies at the Bureau of Reclamation by Bruggeman et al. (1939), Hamilton (1939), Hilf (1948 and 1956) have shown that the change in pore-air pressure can be related to observed volume changes by the use of Boyle's law and Henry's law. However, the magnitude of the difference between pore-air and porewater pressure still has to be found experimentally. For practical purposes, where the pore water pressure is the more significant factor, it is therefore more convenient to measure it directly, particularly if the effect of stress ratio on pore pressure is also to be studied[n].

(i) Advantages and Limitations of the Triaxial Test.—

The advantages and limitations of the triaxial test have been discussed in some detail elsewhere (Bishop and Henkel, 1957), and will be referred to only briefly here.

The principal advantages of the triaxial test as performed on cylindrical specimens are that it combines control of the drainage conditions and the possibility of the measurement of pore pressure with relative simplicity in operation.

The principal limitations are that the intermediate principal stress cannot be varied to simulate plane strain conditions, that the directions of the principal stresses cannot be progressively changed, and that end restraint may modify the various relationships between stress, strain, volume change, and pore pressure.

For most practical purposes the advantages outweigh the limitations, and it will be apparent from section 6 that a very satisfactory correlation does in fact exist between laboratory tests and field observations of stability in many important engineering problems.

---

m. Some difference will in general arise from such factors as the different strains at which 'failure' is taken to occur, the different rates of volume change at failure, and, in soils having true cohesion in the Hvorslev sense, the different water contents of the samples defining the failure envelope.

n. The effect of stress ratio is discussed by Bishop (1952 and 1954 a) and Fraser (1957).

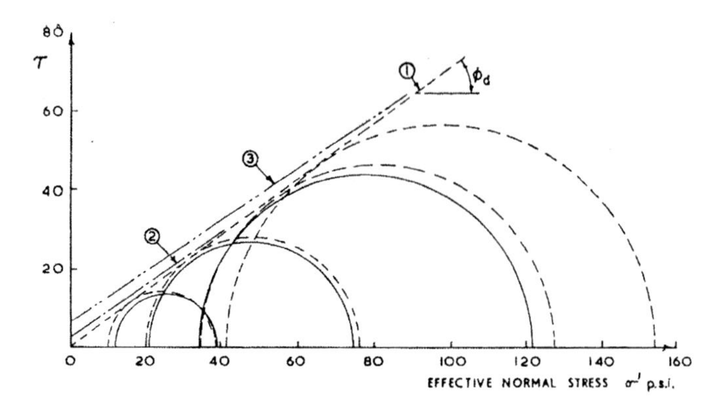

Fig. 11.—Undrained, consolidated-undrained and drained tests on boulder clay compacted at an initial water content 2% dry of optimum: clay fraction 4%.

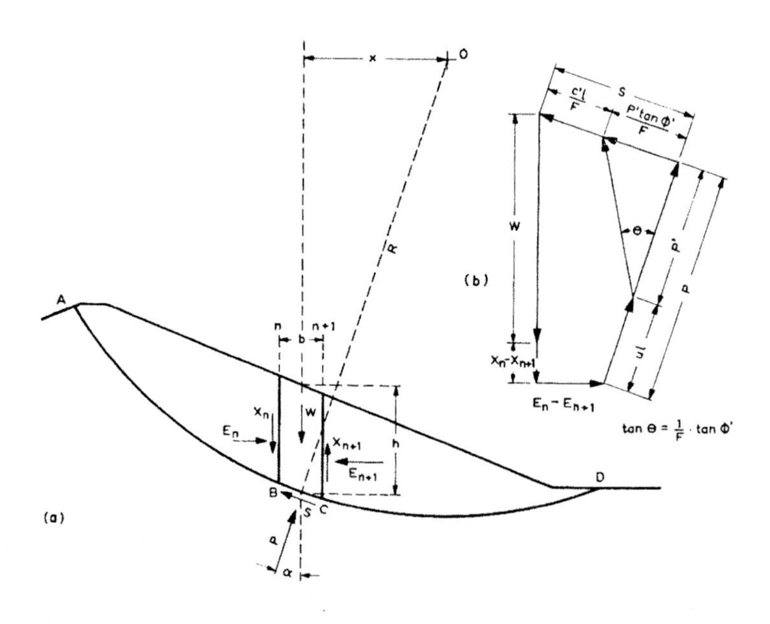

Fig. 12.—Forces in the slices method of stability analysis.

5. METHODS OF STABILITY ANALYSIS

The stability of soil masses against failure under their own weight, or under the action of applied loads, can be examined either by methods based on elastic theory or by methods based on the principle of limit design.

In the first case the stress distribution is calculated and the maximum stresses are then compared with the strength of the soil. As a practical method it is, however, open to several serious objections. Firstly, it is difficult to assess the error resulting from the assumption that the soil mass is a homogeneous elastic material having elastic constants which are independent of the magnitude of the stresses. Secondly, it has been shown that, even if these assumptions were true, local overstress would occur in a typical earth dam section when its factor of safety (by a slip circle method) lay below a value of about 1.8 (Bishop, 1952). The same applies in principle to earth slopes and foundations.

In consequence elastic methods are not applicable to the calculation of the factor of safety when studying observed failures or for design work on embankments and cuts where 1.5 is accepted as a working value for factor of safety. Elastic methods are, however, useful in giving an estimate of the stress distribution studies and for pore pressure prediction.

In most practical stability problems, therefore, the engineer is concerned with the factor of safety against complete failure, rather than against local overstress. The most general definition of factor of safety against complete failure, which can be applied irrespective of the shape of the failure surface, is expressed in terms of the proportion of the measured shear strength that must be mobilized to just maintain limiting equilibrium. The shear strength parameters to which the factor of safety is applied in setting up the equations expressing the condition of limiting equilibrium depend on whether the analysis is carried out in terms of effective stress (c', Φ' analysis) or total stress ($\Phi_u = 0$ analysis). The two cases will be treated separately.

(a) Effective Stress Analysis.—

In the effective stress analysis the proportion of the shear strength mobilized for limiting equilibrium is expressed:

$$\tau = (c'/F) + (\sigma - u)(\tan \Phi'/F) \qquad (13)$$

The value of the factor of safety F is obtained by assuming limiting equilibrium along a trial slip surface (usually the arc of a circle in cross-section), balancing the forces and solving for F. The value of σ is determined from the equilibrium of the soil mass above the failure surface by an appropriate graphical or numerical method. The method of determining the value of u will depend on the class of stability problem.

(I) In class (a) problems, where the pore pressure is an independent variable, the value of u will be obtained from ground water level if there is no flow, or from a flow net if a state of steady seepage exists. The flow net can either be calculated or based on field measurements of pore pressure.

(II) In class (b) problems, where the magnitude of the pore pressure depends on the stress changes tending to lead to instability, the most practical method of approach is that adopted in earth dam design. Here a prediction is made of the actual pore pressure likely to obtain in the stable dam, which should thus check with the field pore pressure measurements usually made during construction. This prediction is based on an approximate stress distribution within the dam, the undrained pore pressure parameter $\bar{B}$ and a calculated allowance for pore pressure dissipation, the value of $\bar{B}$ being re-adjusted if necessary to match the calculated factor of safety.

Where field measurements of pore pressure are available they are of course substituted directly in the analysis.

While any method of stability analysis can be used which correctly represents the statics of the problem, the more complex soil profiles or dam sections involving a number of zones of c' and Φ' and irregular distributions of pore pressure can be handled most readily by a numerical form of the method of slices (Bishop, 1954 b).

As applied to the slip circle analysis (Fig. 12), the method leads to an expression for the factor of safety:

$$F = \frac{1}{\Sigma W \sin\alpha} \cdot \sum \left[\{c'b + \tan\Phi' . [W(1 - r_u) + (X_n - X_{n+1})]\} \frac{\sec\alpha}{1 + \frac{\tan\Phi' . \tan\alpha}{F}} \right] \qquad (14)$$

This expression takes full account of both horizontal and vertical forces between the slices. The vertical shear force term, which cannot be eliminated mathematically, can however be put equal to zero with little loss in accuracy. The method agrees to within about 1% with the modified friction circle method described by Taylor (1948) in two cases which have been checked.

The programming of the digital electronic computer 'DEUCE' for the numerical method by Little and Price (1958) has given it an additional and overwhelming advantage, since any specified pattern of slip circles can be analysed at a rate of about 5 seconds per circle using about 30-50 slices. This leaves the engineer free to investigate the effect of varying his assumptions about soil properties and pore pressure, and to modify his design, without the heavy burden of computation previously involved.

The extension of the slices method to noncircular surfaces has been undertaken by Janbu (1954 and 1957) and Kenney (1956) and it is at present being programmed for the computer.

(b) Total Stress Analysis.—

In the total stress analysis the proportion of the shear strength mobilized is expressed, for the Φ_u = 0 condition, as:

$$\tau = c_u / F \qquad (15)$$

In the notation of Fig. 12, the expression for the factor of safety using the slip circle analysis becomes:

$$F = \Sigma c_u l / \Sigma W \sin\alpha \qquad (16)$$

When Φ_u = 0 the inter-slice forces enter into the calculation only if a non-circular slip surface is used.

For saturated soils the apparent cohesion c_u is equal to one half of the undrained compression strength (equ. 12) and its value is obtained from undrained tests on undisturbed samples or from vane tests. The value of c_u usually varies with depth and appropriate values must be used around the trial failure surface.

It should be noted that the use of this method is correct only where the field conditions correspond to the laboratory tests conditions, i.e. where the shear stress tending to cause failure is applied under undrained conditions[o]. It cannot in general be applied using undisturbed samples from slopes, for example, where the water content has had time to adjust itself to the stress changes set up by the formation of the slope.

The validity of the $\Phi_u = 0$ method is in fact restricted to saturated soils[p] and to problems in which insufficient time has elapsed after the stress change considered for an increase or decrease in water content to occur. It is therefore an 'end of construction method'. Whether the factor of safety subsequent to construction will have a lower value depends on the sign and magnitude of the stress changes. The particular cases are discussed in Section 6.

The use of total stress methods in which Φ_u is not zero, or in which the angle of consolidated undrained shearing resistance Φ_{cu} is used, is, in the opinion of the authors, to be avoided except in special cases, owing to the difficulty of determining the physical significance of the factor of safety thus obtained.

(c) Relationship between Total and Effective Stress Methods of Stability Analysis.—

Since the failure criterion and the associated method of stability are only convenient means of linking the stability problem with the appropriate laboratory test, a soil mass in limiting equilibrium should be found to have a factor of safety of 1 by whichever method the analysis is performed. As total stress methods can only be applied under undrained conditions, it is convenient to demonstrate this point by a simplified analysis of a vertical cut in saturated clay immediately after construction (Fig. 13).

To simplify the mathematics of the problem it is assumed that the undrained strength c_u does not vary with depth, and that the effective stress failure envelope is represented by $c' = 0$ and a constant value of Φ'. The failure surface is assumed to approximate to a plane without tension cracks.

The critical height H under these conditions is known to be equal to $4c_u/\gamma$ where γ is the density of the soil. The factor of safety of the soil adjacent to a vertical cut of depth H can be calculated in terms of either total or effective stress:

(I) Total Stress.—From equation 16:

$$\begin{aligned} F &= (\Sigma\, c_u\,.\, l)/(\Sigma\, W \sin\alpha) \\ &= (c_u\,.\, H \operatorname{cosec}\alpha)/(\tfrac{1}{2}\,.\,\gamma H^2\,.\,\cot\alpha \sin\alpha) \\ &= (2c_u)/(\gamma\,.\, H\cos\alpha\sin\alpha) \end{aligned} \tag{17}$$

Putting $dF/d\alpha = 0$ to obtain the value of α giving the lowest value of F we obtain $\alpha = 45^\circ$.

Substituting in equ. 17 gives:

$$F = 4c_u/\gamma H \tag{18}$$

Substituting $4c_u/\gamma$ for H we obtain $F = 1$.

---

o. The error introduced by the fact that the principal stress directions in most practical problems differ from those in the laboratory is discussed by Hansen and Gibson (1949).

p. Stiff fissured clays under a reduction of normal stress are an exception.

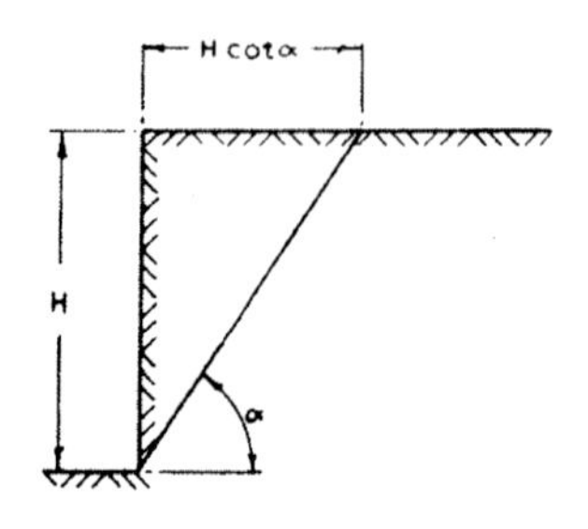

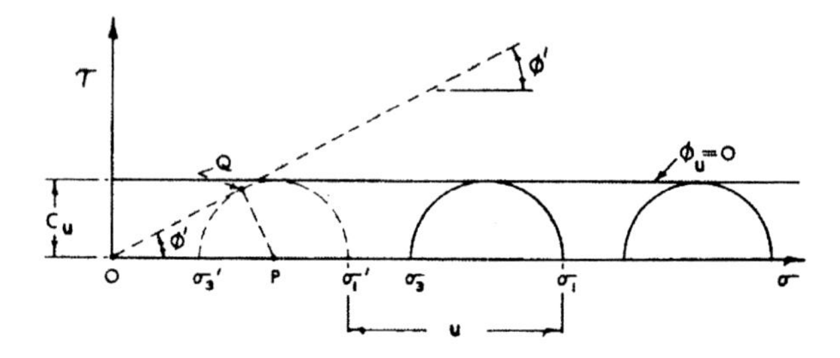

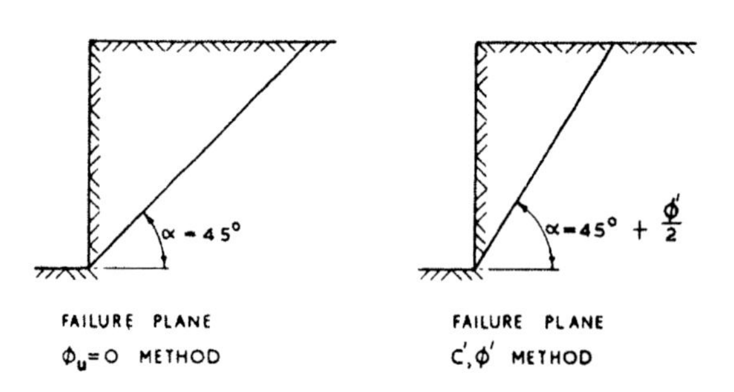

Fig. 13.—Simplified analysis of the stability of a vertical cut in saturated cohesive soil immediately after excavation, using both total and effective stress methods.

(II) Effective Stress.—From the Mohr diagram in Fig. 13 we can obtain the pore pressure in an element of soil at failure in terms of the major principal stress. It follows from the geometry of the triangle OPQ that:

$$(\sigma_1 - \sigma_3)/2 = [(\sigma_1 - u) - \{(\sigma_1 - \sigma_3)/2\}] \sin \Phi' \quad (19)$$

Putting $(\sigma_1 - \sigma_3)/2 = c_u$ and rearranging we obtain:

$$u = \sigma_1 - c_u \cdot [(1 + \sin \Phi')/(\sin \Phi')] \quad (20)$$

For a plane slip surface, and with $c' = 0$ the expression for F given in equ. 15 simplifies to the form:

$$F = [1/(\Sigma\, W \sin \alpha)]\,.\,\Sigma[\tan \Phi'(W \cos \alpha - ul)] \tag{21}$$

For failure on a plane, the state of stress corresponds in this case to the Rankine active state, and thus the major principal stress σ_1 is equal to γh, the vertical head of soil above the element.

Substituting in equ. 21 the value of u given by equ. 20, and putting $c_u = \gamma H/4$, we obtain the expression for F:

$$F = \tan \Phi' (\cot \alpha - [1/\sin 2\alpha]\,.\,[(\sin \Phi - 1)/(\sin \Phi)]) \tag{22}$$

Putting $dF/d\alpha = 0$ we now find that the minimum value of F is given by the inclination $\alpha = 45^0 + \Phi'/2$. Substituting this value in equation 22 and expressing the angles in terms of $\Phi'/2$, we find that the expression again reduces to $F = 1$.

This comparison illustrates two important conclusions. Firstly, both total and effective stress methods of stability analysis will agree in giving a factor of safety of 1 for a soil mass brought into limiting equilibrium by a change in stress under undrained conditions. Secondly, although the values of factor of safety are the same, the position of the rupture surface is found to depend on the value of Φ used in the analysis. The closer this value approximates to the true angle of internal friction, the more realistic is the position of the failure surface[q], and this is confirmed by the analysis of the Lodalen slide in terms of c' and Φ' (Sevaldson, 1956 and Section 6).

The choice of method in short term stability problems in saturated soil is thus a matter of practical convenience and the $\Phi_u = 0$ method is generally used because of its simplicity, unless field measurements of pore pressure are to be used as a control. It should be noted, however, that for factors of safety other than 1 the two methods will not in general give numerically equal values of F. In the effective stress method the pore pressure is predicted for the stresses in the soil, under the actual loading conditions, and the value of F expresses the proportion of c' and $\tan \Phi'$ then necessary for equilibrium. The total stress method on the other hand implicitly uses a value of pore pressure related to the pore pressure at failure in the undrained test. The high factor of safety shown for example in the $\Phi_u = 0$ analysis of a slope of over-consolidated clay in which the pore pressure shows a marked drop during the latter stages of shear will therefore not be reflected in the effective stress analysis in such a marked way.

It cannot be too strongly emphasized that a comparison between effective stress and total stress methods can only be logically made when the shear stress tending to cause instability has been applied under undrained conditions. The use of the $\Phi_u = 0$ method under other conditions cannot be justified theoretically and in practice often leads to very unrealistic results (see Table V).

---

q. This point is discussed more fully by Terzaghi, 1936 b; Skempton, 1948 a; and Bishop, 1952.

6. THE APPLICATION OF STABILITY ANALYSIS TO PRACTICAL PROBLEMS

In this section the stability analysis of a number of typical engineering problems will be examined. The purpose of the examination is in the first place to obtain a clear qualitative picture of what happens to the variables controlling stability during and after the construction operation or load change under consideration. The second purpose of the examination is to indicate the most dangerous stage from the stability point of view and to select the appropriate shear parameters and method of stability analysis.

It is not possible to generalise about the solution of practical problems without considering the principal properties of the soil in each case. It will have been apparent from section 2 that the permeability of the soil has an important bearing on the way in which the stability problem is treated. In the more permeable soils (e.g. sands and gravels) the pore pressure will be influenced by the magnitude of the stresses tending to lead to instability only under conditions of transient loading. Both end of construction and long term problems will fall into class (a) in which pore pressure is an independent variable. Only in the less permeable soils do the relative merits of alternative methods of analysis have to be considered in most practical cases.

In Table I are listed representative values of the shear strength parameters of some typical soils arranged in order of decreasing permeability. The wide range of permeability values will be noted, and it will be apparent that it is here that the largest quantitative difference between the soil types lies.

Table I.—Permeability and Shear Strength Parameters of Typical Soils. (* Signifies Undisturbed Samples).

Material	Plasticity Index P I %	Permeability K cm/sec. (Approx.)	c′ lb./sq.ft.	Φ' Degrees
Rock fill: tunnel spoil	–	5	0	45
Alluvial gravel: Thames Valley	–	5×10^{-2}	0	43
Medium sand: Brasted	–		0	33
Fine sand	–	1×10^{-4}	0	20–35
Silt: Braehead	–	3×10^{-5}	0	32
Normally consolidated clay of low plasticity – Chew Stoke*	20	1.5×10^{-8}	0	32
Normally consolidated clay of high plasticity – Shellhaven*	87	1×10^{-8}	0	23
Over-consolidated clay of low plasticity – Selset boulder clay*	13	1×10^{-8}	170	$32\frac{1}{2}$
Over-consolidated clay of high plasticity – London clay*	50	5×10^{-9}	250	20
Quick clay*	5	1×10^{-8}	0	10–20

The more important problems are:

(a) Bearing Capacity of a Clay Foundation.—

This problem may be illustrated most simply in terms of the construction of a low embankment on a saturated soft clay stratum with a horizontal surface. In Fig. 14a is shown diagrammatically the variation with time of the factors which govern stability, i.e. the average shear stress along a potential sliding surface and the average pore pressure ratio.

The excess pore pressure set up in an element of clay beneath the embankment is given by the expression:

$$\Delta u = B[\Delta\sigma_3 + A(\Delta\sigma_1 - \Delta\sigma_3)] \quad (4)$$

For points beneath the embankment Δu will in general be positive and have its greatest value at the end of construction, since B = 1 and A is positive for normally or lightly overconsolidated clay. Unless construction is slow or the clay contains permeable layers, little dissipation of pore pressure will occur during the construction period. After construction is completed the average value of r_u will decrease as redistribution and dissipation of the excess pore pressures occur, until finally the pore pressures correspond to ground water level.

The factor of safety given by the effective stress analysis will thus show a minimum value at or near the end of construction, after which it will rise to the long term equilibrium value[r]. For the long term stability calculation it is obviously appropriate to take the values of c' and Φ' from drained tests. For the end of construction case the same values may also be used, for, though it is more logical to take the value from undrained and consolidated-undrained tests expressed in terms of effective stress, the error is on the conservative side and is likely to be small.

The use of the effective stress method for the end of construction case means, however, that the pore pressures must be predicted or measured in the field. Typical field measurements of pore pressure under an oil storage tank are illustrated in Fig. 14b (after Gibson and Marsland, 1960). However, field measurements are usually limited to the more important structures and to earth dams, and the application of this method to the end of construction case will in other instances have to depend on estimated pore pressure values. As this estimate involves an assumption about the stress distribution (which is influenced by how nearly limiting equilibrium is approached) and the determination of the value of A, it is usually avoided by going directly to the $\Phi_u = 0$ analysis which is applicable to the end of construction case with zero drainage.

The undrained shear strength to be used in the $\Phi_u = 0$ analysis is obtained from undrained triaxial tests (or unconfined compression tests) on undisturbed samples, or from vane tests in the field. In the majority of problems involving foundations on soft clay, where it is quite clear that the long term factor of safety is higher than the value at the end of construction, there is then no need for the more elaborate testing and analysis required by the effective stress method. However, if appreciable dissipation of pore pressure is likely to occur during construction, it is uneconomical not to take advantage of it in calculating the factor of safety and the effective stress method is then required.

The failure of a bauxite dump at Newport (reported by Skempton and Golder, 1948), may be taken as an example of the use of the $\Phi_u = 0$ analysis for end of construction conditions (Fig. 15). After relatively rapid tipping,

r. The position of the most critical slip surface will of course change as the pressure pattern alters.

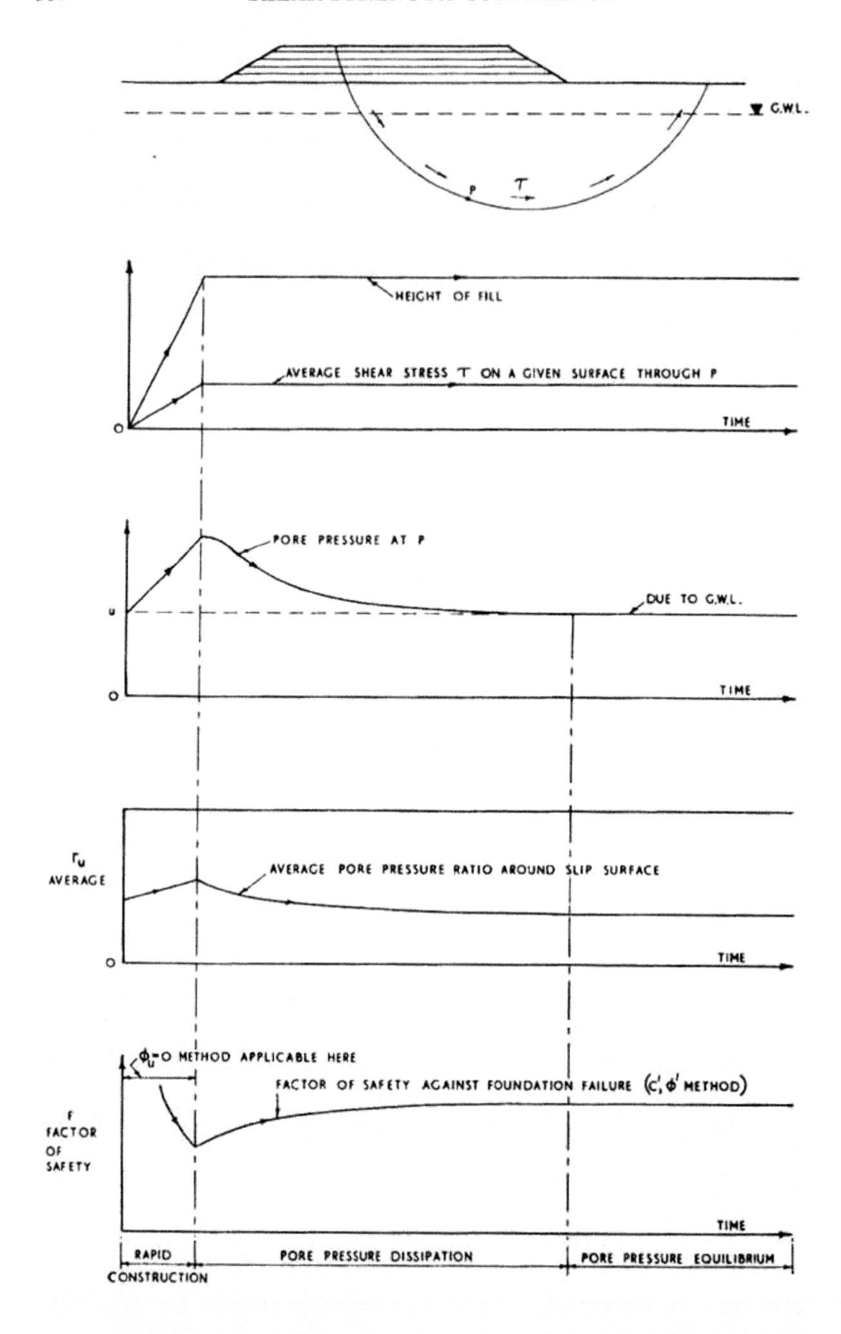

Fig. 14a.—Variation with time of the shear stress, local and average pore pressure, and factor of safety for the saturated clay foundation beneath a fill.

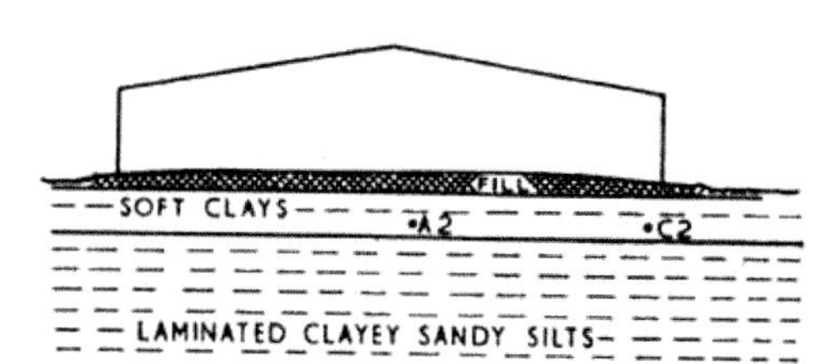

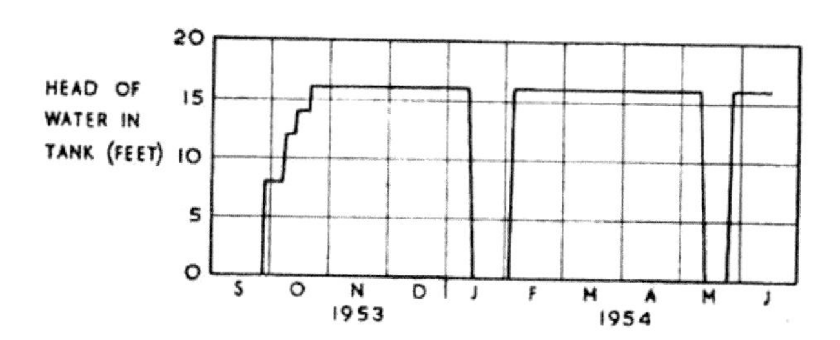

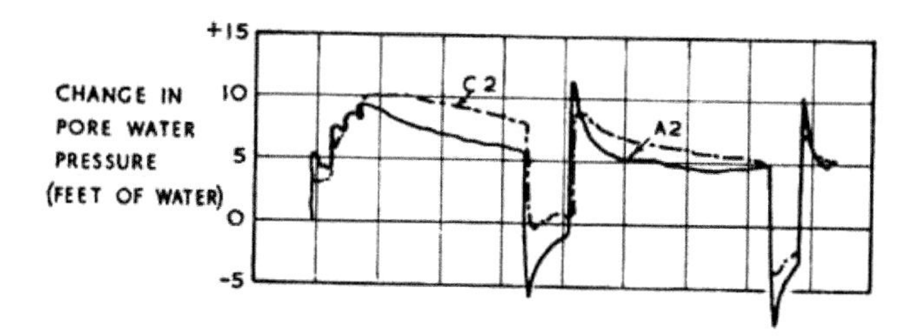

Fig. 14b.—Pore pressure changes in a soft clay foundation on filling and emptying a storage tank (after Gibson and Marsland, 1960).

failure occurred at a height of 25 feet; the factor of safety by the $\Phi_u = 0$ analysis was subsequently found to be 1.08, which can be accepted as agreement to within the limit of experimental accuracy.

In this case the fill was a granular material and its contribution to the shearing resistance was small. In cases where the fill is a cohesive material of high undrained strength, the use of the full value of this strength in the $\Phi_u = 0$ analysis gives misleading results. The explanation appears to be that shear deformations set up in the soft clay foundation under undrained loading

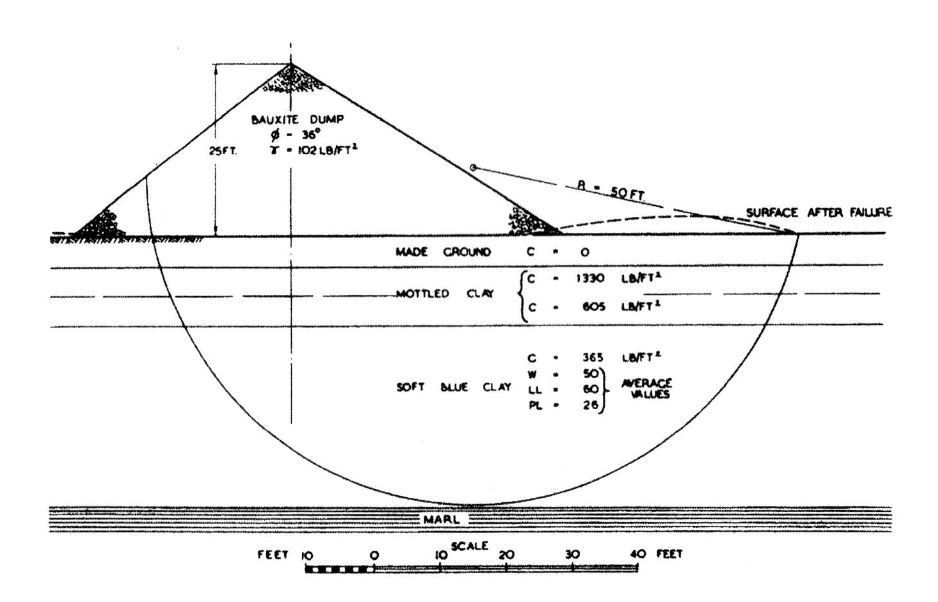

Fig. 15.—Failure of a bauxite dump at Newport (after Skempton and Golder, 1948).

set up tensile stresses in horizontal direction in the more rigid fill above and result in vertical cracks. A description of cracks wider at the bottom than at the top and passing right through the fill is given by Toms (1953a).

As the factor of safety rises with time, no long term failures can be quoted in this category.

The calculation of the ultimate bearing capacity of a structural foundation on a saturated clay is in principle the same as the problem treated above. However, the bearing capacity is not calculated by assuming a circular sliding surface, but is computed from the theory of plasticity for both the total and effective stress analyses, the results being expressed directly as bearing capacity factors.

For large foundations on soft clay the ultimate bearing capacity will increase with time after loading. For small, shallow foundations on stiff clay the ultimate bearing capacity will decrease with time, but in most cases settlement considerations will govern the design.

An example of the $\Phi_u = 0$ analysis of an end of construction foundation failure has been given by Skempton (1942). Here a footing 8 feet by 9 feet founded on a soft clay with $c_u = 350$ lb./sq. ft. failed at a nett foundation pressure of 2500 lb./sq. ft. (Fig. 16). Using a bearing capacity factor of 6.7 for this depth to breadth ratio, a factor of safety of 0.95 is obtained.

Two series of loading tests on the stiff fissured London clay may also be mentioned, where the $\Phi_u = 0$ analysis has led to factors of safety about 1.02 (Skempton 1959). These tests are particularly interesting as showing that the effect of fissures, which can lead to serious difficulties with end of construction problems in open excavations (see section 6b) does not prevent the successful use of the $\Phi_u = 0$ analysis in bearing capacity calculations under end of construction conditions.

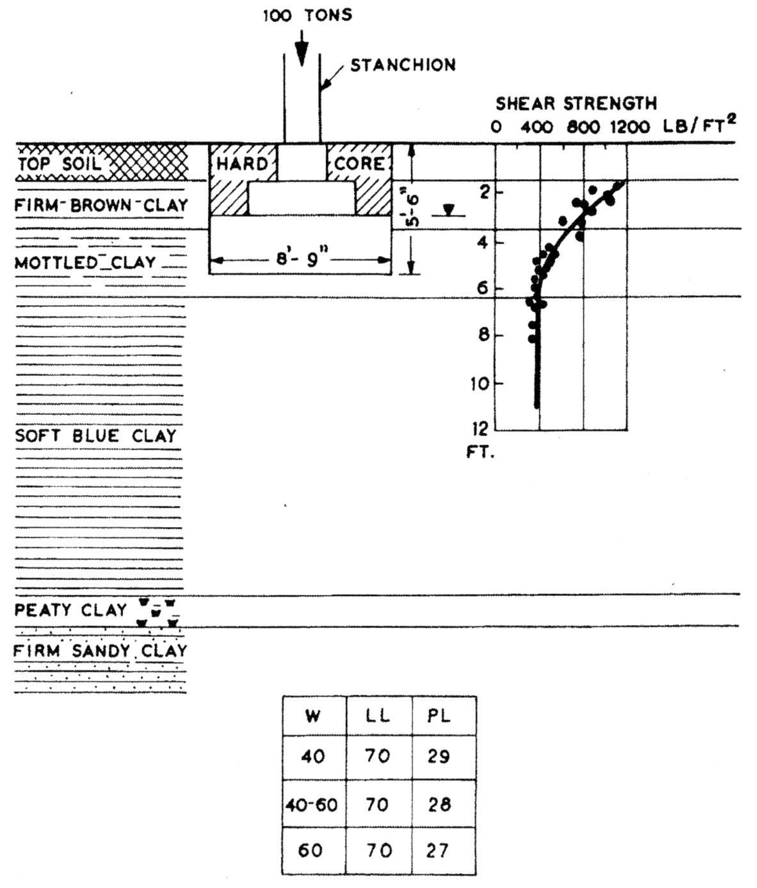

W	LL	PL
40	70	29
40-60	70	28
60	70	27

Fig. 16.—Failure of a foundation on soft clay at Kippen (after Skempton, 1942).

An example of the long term failure of a small heavily loaded foundation of stiff clay is more difficult to find. The long term failure of tunnel arch footings described by Campion (1951) probably falls in this category.

It may however be concluded that with few exceptions the end of construction conditions is the most critical for the stability of foundations and that for saturated clays this may be examined more simply by the $\Phi_u = 0$ analysis. From the field tests and full scale failures tabulated in Table II it is apparent that an accuracy of $\pm$ 15% can be expected in the estimate of factor of safety. One of the exceptions is dealt with in section 6(i). Where partial dissipation of pore pressure occurs during construction an analysis in terms of effective stress is used, and examples of the analysis are discussed in section 6(h).

Table II.—End of Construction Failures of Footings and Fills on a Saturated Clay Foundation: Φ_u = 0 Analysis.

1. Footings, loading tests.

Locality	Data of clay					Safety factor $\Phi_u = 0$ analysis	Reference
	W	LL	PL	PI	W-PL / PI		
Loading test, Marmorerá	10	35	15	20	-0.25	0.92	Haefeli, Bjerrum
Kensal Green	–	–	–	–	–	1.02	Skempton 1959
Silo, Transcona	50	110	30	80	0.25	1.09	Peck, Bryant 1953
Kippen	50	70	28	42	0.52	0.95	Skempton 1942
Screw pile, Lock Ryan	–	–	–	–	–	1.05	Morgan 1944, Skempton 1950
Screw pile, Newport	–	–	–	–	–	1.07	Wilson 1950
Oil tank, Fredrikstad	45	55	25	30	0.67	1.08	Bjerrum, Øverland 1957
Oil tank A, Shellhaven	70	87	25	62	0.73	1.03	Nixon 1949
Oil tank B, Shellhaven	–	–	–	–	–	1.05	Nixon (Skempton 1951)
Silo, U.S.A.	40	–	–	–	–	0.98	Tschebotarioff 1951
Loading test, Moss	9	–	–	–	–	1.10	NGI
Loading test, Hagalund	68	55	20	35	1.37	0.93	Odenstad 1949
Loading test, Torp	27	24	16	8	1.39	0.96	Bjerrum 1954 c
Loading test, Rygge	45	37	19	18	1.44	0.95	Bjerrum 1954 c

2. Fillings

Locality	W	LL	PL	PI	W-PL / PI	Safety factor $\Phi_u = 0$ analysis	Reference
Chingford	90	145	36	109	0.50	1.05	Skempton, Golder 1948
Gosport	56	80	30	50	0.48	0.93	Skempton 1948 d
Panama 2	80	111	45	66	0.53	0.93	Berger 1951
Panama 3	110	125	75	50	0.70	0.98	Berger 1951
Newport	50	60	26	34	0.71	1.08	Skempton, Golder 1948
Bromma II	100	–	–	–	1.00	1.03	Cadling, Odenstad 1950
Bocksjön	100	90	30	60	1.17	1.10	Cadling, Odenstad 1950
Huntington	400	–	–	–	–	0.98	Berger 1951

Table III.—End of Construction Failures in Excavations: Φ_u = 0 Analysis.

Location	Soil type	Data of Clay					Factor of safety: $\Phi_u = 0$ analysis	Reference
		W	LL	PL	PI	W-PL / PI		
Huntspill	Intact clay	56	75	28	47	0.6	0.90	Skempton, Golder, 1948
Congress Street	Intact clay	24	33	18	15	0.4	1.10	Ireland, 1954
Skattsmanso I	Intact clay	101	98	39	59	1.05	1.06	Cadling, Odenstad, 1950
Skattmanso II	Intact clay	73	69	24	45	1.09	1.03	Cadling, Odenstad, 1950
Bradwell	Stiff-fissured clay	33	95	32	63	0.02	1.7	Imperial College, 1959 (Skempton, La Rochelle)

(b) The Stability of Cuts and Free-Standing Excavations in Clay.—

The changes in pore pressure and factor of safety during and after the excavation of a cut in clay are illustrated in Fig. 17.

The change in pore pressure can conveniently be expressed by putting B = 1 and re-arranging equ. (4) in the form:

$$\Delta u = [(\Delta\sigma_1 + \Delta\sigma_3)/2] + (A - 1/2)(\Delta\sigma_1 - \Delta\sigma_3) \tag{23}$$

The reduction in mean principal stress will thus lead to a decrease in pore pressure, and the shear stress term will also lead to a decrease in pore pressure unless A is greater than 1/2, if the unknown effect on pore pressure of changing the directions of the principal stresses is neglected. An estimate of the stress distribution can be made from elastic theory if the initial factor of safety of the slope is high, or from the state of limiting equilibrium round a potential slip surface if the factor of safety is close to 1.

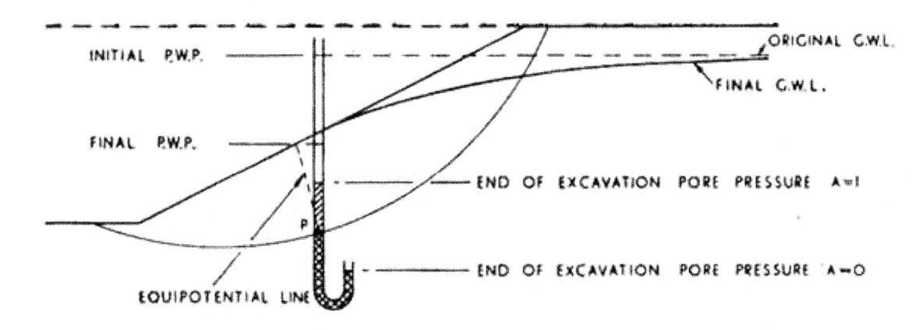

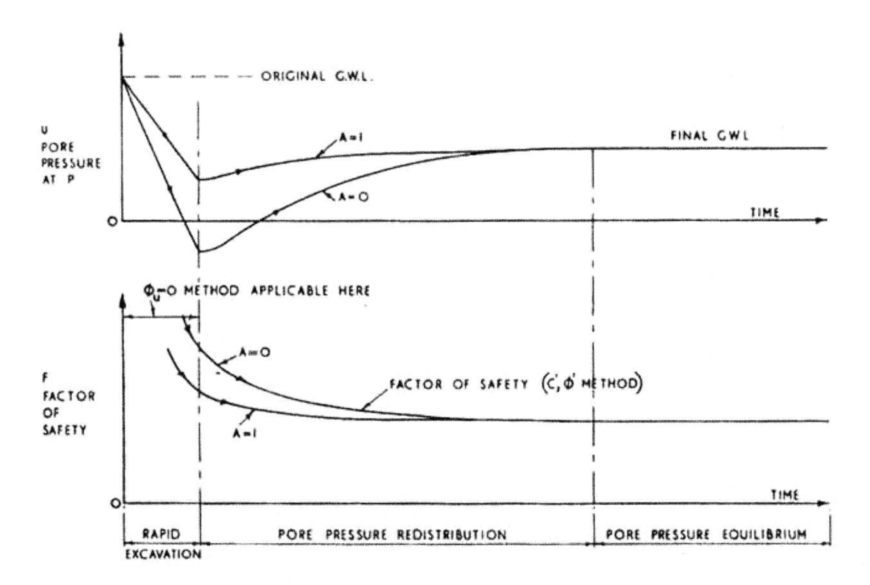

Fig. 17.—The changes in pore pressure and factor of safety during and after the excavation of a cut in clay.

In Fig. 17 the changes in pore pressure at a representative point are shown for the values A = 1 and A = 0. The final equilibrium values of pore pressure are taken from the flow pattern corresponding to steady seepage[s].

Using values of c' and Φ' from drained tests or consolidated-undrained tests expressed in terms of effective stress the factor of safety can be calculated at all stages from equation 14. In the majority of cases, unless special drainage measures are taken to lower the final ground water level, the factor of safety reaches its minimum value under the long term equilibrium pore pressure conditions.

An example of the investigation of a long term failure of a cut in terms of effective stress has been given by Sevaldson (1956). The slide took place in 1954 in a clay slope at Lodalen near Oslo, originally excavated about 30 years earlier (Fig. 18). Since the slide occurred without any apparent change in external loading, it can be considered to be the result of a gradual reduction in the stability of the slope. Extensive field investigations and laboratory studies were carried out to determine the pore pressure in the slope at the time of failure and the shear parameters of the clay.

Triaxial tests gave the values c' = 250 lb./sq. ft. and Φ' = 32°. An effective stress analysis using equation 14 gave a factor of safety of 1.05, and confirms the validity of the approach to within acceptable limits of accuracy.

Where the final pore pressures are obtained from a flow net not based on field measurements, allowance should be made for the fact that the permeability of a water laid sediment is generally greater in a horizontal direction (Sevaldson 1956). The highest wet season values obviously represent the most critical conditions.

The excavation of cuts in stiff fissured and weathered clays presents some special problems which have been discussed in detail by Terzaghi, 1936 a; Skempton, 1938 c; Henkel and Skempton, 1955; Henkel, 1957; etc. The reduction in stress enables the fissures to open up and they will then represent weak zones which a sliding surface will tend to follow. The fissures will also increase the bulk permeability of the clay (an increase of about 100 times is reported by Skempton and Henkel, 1960) so that the pore pressure rise leading to the long term equilibrium state will occur more rapidly.

The presence of fissures is reflected in the factors of safety obtained using the effective stress analysis with field values of pore pressure and values of c' and Φ' measured in the laboratory on 1-1/2" diameter samples. An analysis of three long-term cutting failures in London clay by Henkel (1957), using the laboratory values of c' = 250 lb./sq. ft. and Φ' = 20°, gave factors of safety of 1.32, 1.35, and 1.18. Putting c' = 0 gave values of 0.78, 0.81, and 0.82 respectively and obviously underestimated the factor of safety. If the value of c' required to give a factor of safety of 1 is plotted against time after construction (Henkel, 1957; De Lory, 1957), it is found that the value of c' shows a definite correlation with time (Fig. 19) and, as will be seen in section 6(c), appears to approach zero in natural slopes on a geological time scale.

An effective stress analysis based on a value of c' related empirically with time for each clay type will obviously give a close approximation to the correct factor of safety. For large scale work and for remedial measures on active slips where the large strains tend to reduce c', it is prudent to ensure a factor of safety of at least 1 with c' = 0.

s. This method has also been given since 1956 by Skempton in his lectures at Imperial College.

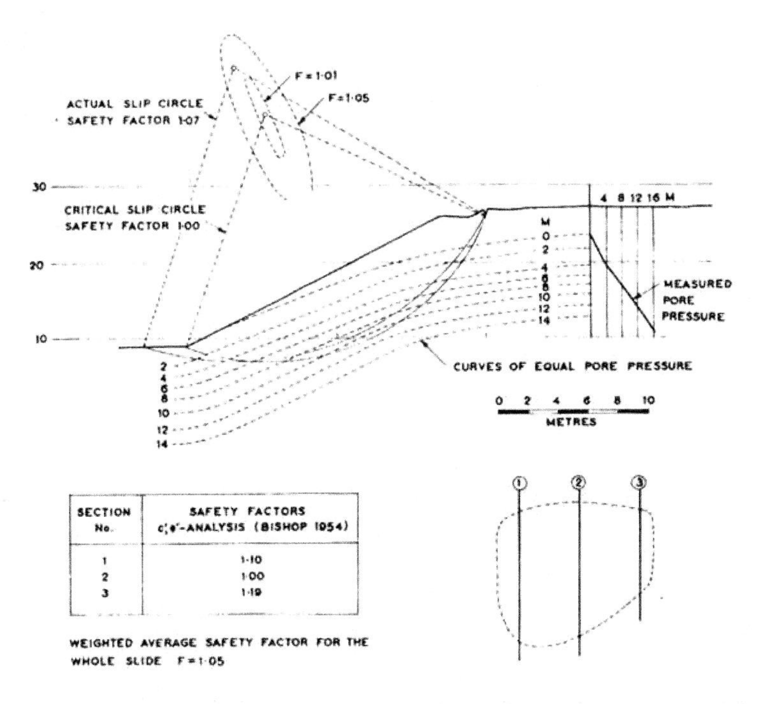

SECTION No.	SAFETY FACTORS c',φ'-ANALYSIS (BISHOP 1954)
1	1·10
2	1·00
3	1·19

Fig. 18.—Long term failure in a cut at Lodalen (after Sevaldson, 1956).

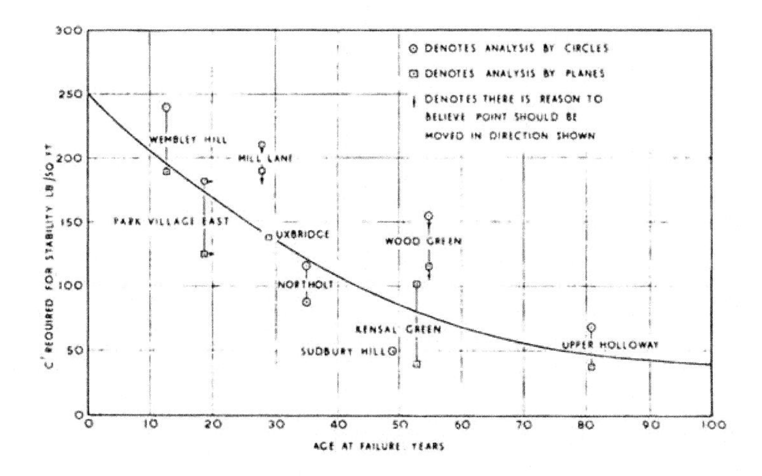

Fig. 19.—Long term failures in stiff-fissured London clay: Correlation of apparent cohesion c' required in effective stress analysis with age of cut at failure (after De Lory, 1957).

The mechanism of the drop in c' before a failure is initiated is not clearly understood, but may be associated with stress concentrations due to the presence of fissures, the progressive spread of an overstressed zone in a soil which tends to dilate and absorb water on shear, and the effect of cyclical fluctuations in effective stress due to seasonal water level changes. On the limited evidence so far available from Lodalen and Selset (see section 6(c)) it does not appear to occur in any marked way in non-fissured clays.

In temporary work, where the end of construction condition is of primary interest, the factor of safety may obviously be calculated by using the $\phi_u = 0$ analysis and the undrained shear strength. This method may also be used with advantage where it is necessary to check that the initial factor of safety is not lower than the long term value, as it avoids the necessity of explicitly determining the stress distribution and pore pressure values at the end of construction. Four examples of its use are given in Table III.

Also included in Table III is an example of the use of the $\phi_u = 0$ method for end of construction conditions in a cut in London clay, which led to an overestimate of the factor of safety by 70%. Whether this is simply a consequence of the opening of fissures due to stress release, or due to changes in pore pressure even in the short period of excavation due to the high bulk permeability, is not yet clear. The reduction in strength at low stresses is apparent in stiff fissured clay even in undrained tests in the triaxial apparatus (Fig. 20), but hardly appears adequate to account for the 70% error. A conservative factor of safety must obviously be used in similar cases, and the rapid adjustment to the equilibrium pore pressure condition must be allowed for in prolonged construction operations.

(c) Natural Slopes.—

Natural slopes represent the ultimate long term equilibrium state of a profile formed by geological processes. The pore pressures are controlled by the prevailing ground water conditions which correspond to steady seepage, subject to minor seasonal variations in ground water level. Natural slopes therefore fall into class (a) in which the pore pressure is an independent variable.

In principle the analysis is the same as that of the long term equilibrium of a cut or excavation. However the pore pressures will have already reached their equilibrium pattern which can be ascertained from piezometer measurements in the field; and the natural processes of softening, leaching etc., will have already reached an advanced stage. An analysis based on laboratory tests of this material would therefore be expected to lead to close agreement with observed slopes in limiting equilibrium.

Relatively few natural slopes in limiting equilibrium have yet been analysed in terms of effective stress, but some representative examples are collected in Table IV. The two cases involving intact clays, Drammen and Selset, are being examined in greater detail, but the preliminary values of factor of safety of 1.15 and 1.03 respectively show that the method can be used with reasonable confidence. Had c' tended to zero the Selset slope would have shown a factor of safety of less than 0.7, for example, which is outside the limit of experimental error.

However, in stiff-fissured clays special account has to be taken of the progressive reduction in the value of c', which appears eventually to approach zero in the failure zone, since the shear strains and water content change associated with failure are very localized and tests on the bulk of the soil do not reveal the decrease in c'.

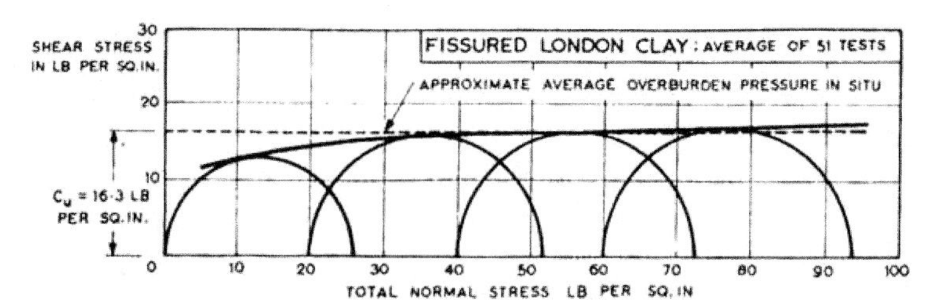

Fig. 20.—The strength of stiff-fissured London clay in undrained tests (after Bishop and Henkel, 1957).

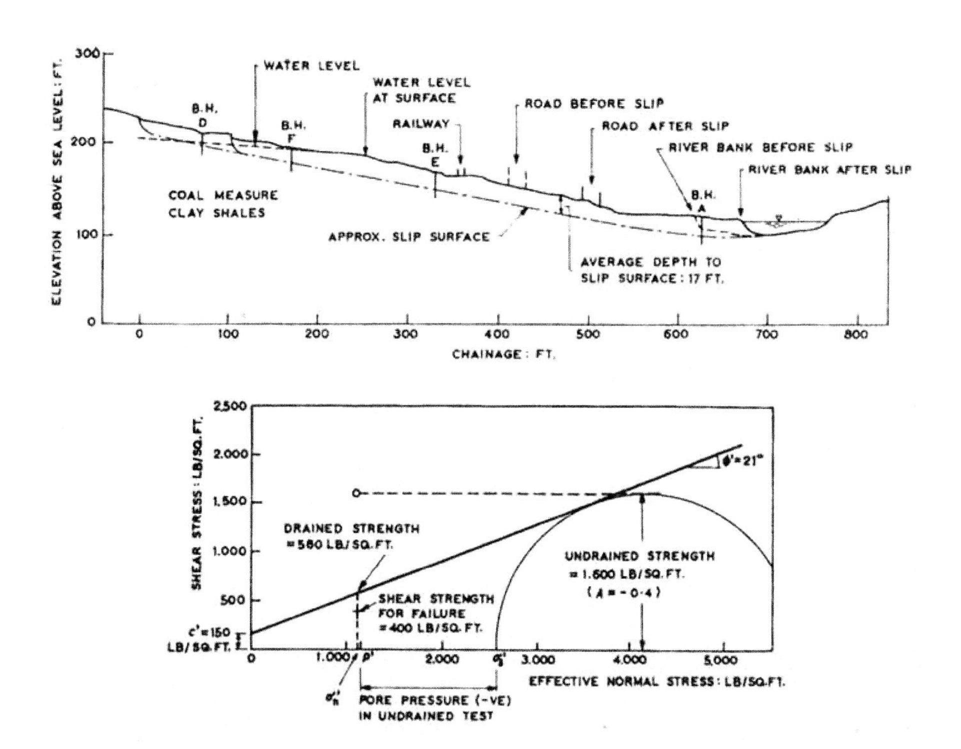

Fig. 21.—Natural slope failure in stiff-fissured clay at Jackfield (after Henkel and Skempton, 1955): Cross-section and shear strength data. Shear strength for limiting equilibrium, 400 lb./sq. ft. (F = 1.0). Undrained shear strength (c' = 150 lb./sq. ft. Φ' = 21° A = -0.4), 1.600 lb./sq. ft. (F = 4.0). Drained shear strength (c' = 150 lb./sq. ft. Φ' = 21°), 580 lb./sq. ft. (F = 1.45). Drained shear strength (c' =0 Φ' = 21°), 430 lb./sq. ft. (F = 1.07).

Table IV.—Failure of Natural Slopes: Effective Stress Analysis.

Location	Soil type	Data of Clay					Factor of safety: c', Φ' analysis	Reference
		W	LL	PL	PI	$\frac{W-PL}{PI}$		
Drammen	Normally Consolidated (Intact)	31	30	19	11	1.09	1.15	Bjerrum and Kjærnsli 1957
Selset	Overconsolidated (Intact)	13	26	13	13	0	1.03	Imperial College
Jackfield	Overconsolidated (stiff-fissured)	20	45	20	25	0	1.45 (1.07 with $c' = 0$)	Henkel and Skempton 1955

Table V.—Long Term Failures in Cuts and Natural Slopes: $\Phi_u = 0$ Analysis (after Bjerrum and Kjaernsli, 1957).

1. Overconsolidated, fissured clays

Locality	Type of slope	Data of clay					Safety factor, $\Phi_u = 0$ analysis	Reference
		W	LL	PL	PI	$\frac{W-PL}{PI}$		
Toddington	Cutting	14	65	27	38	-0.34	20	Cassel, 1948
Hook Norton	Cutting	22	63	33	30	-0.36	8	Cassel, 1948
Folkestone	Nat. slope	20	65	28	37	-0.22	14	Toms, 1953 b
Hullavington	Cutting	19	57	24	33	-0.18	21	Cassel, 1948
Salem, Virginia	Cutting	24	57	27	30	-0.10	3.2	Larew, 1952
Walthamstow	Cutting	–	–	–	–	–	3.8	Skempton, 1942
Sevenoaks	Cutting	–	–	–	–	–	5	Toms, 1948
Jackfield	Nat. slope	20	45	20	25	0.00	4	Henkel/Skempton, 1955
Park Village	Cutting	30	86	30	56	0.00	4	Skempton, 1948 c
Kensal Green	Cutting	28	81	28	53	0.00	3.8	Skempton, 1948 c
Mill Lane	Cutting	–	–	–	–	–	3.1	Skempton, 1948 c
Bearpaw, Canada	Nat. slope	28	110	20	90	0.09	6.3	Peterson, 1952
English Indiana	Cutting	24	50	20	30	0.13	5.0	Larew, 1952
SH 62, Indiana	Cutting	37	91	25	66	0.19	1.9	Larew, 1952

2. Overconsolidated, intact clays

Tynemouth	Nat. slope	–	–	–	–	–	1.6	Imperial College
Frankton, N.Z.	Cutting	43	62	35	27	0.20	1.0	Murphy, 1951
Lodalen	Cutting	31	36	18	18	0.72	1.01	N.G.I.

3.* Normally consolidated clays

Munkedal	Nat. slope	55	60	25	35	0.85	0.85	Cadling/Odenstad, 1950
Säve	Nat. slope	–	–	–	–	–	0.80	Cadling/Odenstad, 1950
Eau Brink cut	Cutting	63	55	29	26	1.02	1.02	Skempton, 1945
Drammen	Nat. slope	31	30	19	11	1.09	0.60	N.G.I.

The landslide at Jacksfield provided a good example of the application of the effective stress analysis to a natural slope in stiff-fissured clay (Henkel and Skempton, 1955) and is illustrated in Fig. 21. The slope of the hillside is 10.5°, and when the slip took place in the winter 1951-52, a soil mass 600 feet by 700 feet and 17 feet in thickness moved gradually downward about 100 feet.

The calculated average shear stress in the clay was about 400 lb./sq. ft. Drained tests on undisturbed samples gave $c' = 150$ lb./sq. ft. and $\Phi' = 21°$, which with the observed pore pressures gave a shear strength of 580 lb./sq. ft. and a factor of safety of 1.45. Putting $c' = 0$ gave a shear strength of 430 lb./sq. ft. and a factor of safety of 1.07.

For natural slopes in stiff fissured clays it therefore appears necessary to use $c' = 0$ in the effective stress analysis. This is confirmed by observations made by Skempton and De Lory (1957) on the maximum stable natural slope found in London clay, and by Suklje (1953 a and b), and Nonveiller and Suklje (1955) in other fissured materials. It is interesting to speculate on whether the drop in c' is due to the fissures, or whether both are due to some more fundamental difference in the stress-strain-time relationships between the fissured and intact clays.

A second class of soil which gives rise to special problems includes very sensitive or quick clays. These clays show almost no strength in the remoulded state, and they will therefore tend to flow as a liquid if a slide occurs. A small initial slip in a slope may therefore have catastrophic consequences as the liquified clay will flow away and will not form a support for the exposed clay face, with the result that the whole of an otherwise stable slope may fail in a series of retrogressive slips taking place under undrained conditions.

A factor which affects the quantitative analysis in the case of quick clays is the influence of sample disturbance on the values of c' and Φ' measured in the laboratory. Soft clays of low plasticity are very sensitive to disturbance and reconsolidation in the triaxial test is always accompanied by a reduction in water content. Particularly where the initial water content is above the liquid limit laboratory tests appear to overestimate the value of Φ'. The investigation of a recent slide in quick clay in Norway has given a value of Φ' calculated from the statics of the sliding mass which is less than 50% of the value measured in the triaxial test.

The occurrence of quick clays is limited to certain well defined geological conditions, and where they are encountered special precautions in sampling, testing and analysis are always necessary (Holmsen, 1953; Rosenqvist, 1953; Bjerrum, 1954a and 1955c).

The application of laboratory tests to the stability of natural slopes raises two general matters of principle. The time scales of the load application are so different in the laboratory and in the field that it is perhaps surprising that satisfactory agreement between the results can be obtained at all. Laboratory results quoted by Bishop and Henkel (1957), and Bjerrum, Simons and Torblaa (1958) indicate that under certain conditions Φ' may have a lower value at low rates of loading. This effect may be partly offset by the fact that the worst ground water conditions which touch off the slip are only of seasonal occurrence; and by factors such as the effect of plane strain on the value of Φ' and the omission of 'end effects' in the stability analysis. It is also well known that considerable creep movements occur in slopes still classed at stable. However, with the exceptions noted, the overall correlation between laboratory and field results is quite acceptable from a practical point of view.

Secondly, it may well be asked why the factor of safety cannot be calculated with equal accuracy using the $\Phi_u = 0$ analysis and the undrained strength of samples from the slope where pore pressure and water content equilibrium have been attained. A large number of case records of slides in both natural slopes and cuttings are summarized in Table V and it is evident that as a practical method it is most unreliable, giving values of factor of safety ranging between 0.6 in sensitive clays to 20 in heavily overconsolidated clays.

The fundamental reason for the difference is that in the undrained test the pore pressure is a function of the stress applied during the test, and it is not necessarily equal to the pore pressure in situ. To obtain a factor of safety of 1 for a slope in limiting equilibrium using undrained tests would require that the same pore pressure should be set up in the sample when the in situ normal and shear stresses were replaced. This is in general prevented by the irreversibility of the stress-strain characteristics of the soil and by the changes in the principal stress directions. The latter occur even with in situ tests.

The position is made worse by the fact that the water content changes both in overconsolidated and in sensitive clays are very localized at failure and samples which do not pick up these layers can have little bearing on the stability analysis. A sample from the 2 inch thick slip zone at Jackfield in which large strains had occurred was found to have a water content 10% above the adjacent clay and an undrained strength within 12% of that required for a factor of safety of 1 (Henkel and Skempton, 1955). This layer was difficult to find and sample, and as clay outside the failure zone had an undrained strength nearly four times as great the method has little predictive value.

(d) Base Failure of Strutted Excavations in Clay.—

During excavation in soft clay, base failure sometimes occurs accompanied by settlement of the adjacent ground. Failures of this type[t] have occurred in excavations for basements, in trenches for water and sewage pipes, and in the shafts for deep foundations.

The construction of temporary excavations is generally carried out sufficiently rapidly for pore pressure changes to be ignored. The change in stress thus occurs under undrained conditions and the stability can be calculated using the $\Phi_u = 0$ analysis and undrained tests.

The factor of safety F against base failure can be derived from the familiar bearing capacity theory, considering the excavation as a negative load. This leads to the expression (Bjerrum and Eide, 1956):

$$F = N_c \cdot c_u/(\gamma D + q) \tag{24}$$

where D denotes depth of excavation

γ » density of the clay

c_u » the undrained strength of the clay beneath the bottom of the excavation

q » the surface surcharge (if any)

N_c » dimensionless bearing capacity factor depending on shape and depth of excavation.

---

t. Bottom heave failures can also occur in clay if a pervious layer containing water under sufficient head lies close beneath the excavation. For example see Garde-Hansen and Thernöe 1960, and Coates and Slade 1958.

The analysis of the failure of seven excavations is given in Table VI. The results indicate that in practice an accuracy of within ± 20% can be expected.

It should be noted that this type of failure is not caused by inadequate strutting, but the loads and distortion after its occurrence may initiate a more general collapse.

Table VI.—Base Failure of Strutted Excavations in Saturated Clay: $\Phi_u = 0$ Analysis (after Bjerrum and Eide, 1956).

Site	Dimensions $B \times L$: m	Depth D : m	Surcharge p : tons/sq. m	Density γ : tons/cu. m	Shear strength S : tons/sq. m	Sensitivity	B/L	D/B	N_c theoretical	Safety factor F	Average safety factor
1. Pumping station, Fornebu, Oslo	5.0×5.0	3.0	0.0	1.75	0.75	50	1.0	0.60	7.2	1.03	0.96
2. Storehouse, Drammen	4.8×∞	2.4	1.5	1.90	1.2	5–10	0.0	0.50	5.9	1.16	
3. Pier shaft, Göteborg	ø 0.9	25.0	0.0	1.54	3.5	20–50	1.0	28.0	9.0	0.82	
4. Sewage tank, Drammen	5.5×8.0	3.5	1.0	1.80	1.0	20	0.69	0.64	6.7	0.93	
5. Test shaft (N) Ensjøveien, Oslo	ø 1.5	7.0	0.0	1.85	1.2	140	1.0	4.7	9	0.84	
6. Excavation, Grev Vedels pl., Oslo	5.8×8.1	4.5	1.0	1.80	1.4	5–10	0.72	0.78	7.0	1.08	
7. "Kronibus shaft", Tyholt, Trondheim	2.7×4.4	19.7	0.0	1.80	3.5	40	0.61	7.3	8.5	0.84	

(e) Earth Pressures on Earth Retaining Structures.—

If the displacement of an earth retaining structure is sufficient for the full development of a plastic zone in the soil adjacent to it, the earth pressure will be a function of the shear strength of the soil. This condition is apparently satisfied in much temporary work and in many permanent structures. The distribution of pressure is a function of the deformation of the structure and the soil, and can only be predicted after detailed consideration of the movements involved.

For temporary excavations in intact saturated clay it is generally sufficient to calculate the total earth pressure using the $\Phi_u = 0$ analysis and the undrained shear strength. Justification for this procedure is to be found in the field measurements published by Peck (1942), Skempton and Ward (1952), and Kjaernsli (1958). More recent measurements in soft clay carried out by the Norwegian Geotechnical Institute however indicate that the total earth pressure may exceed the value determined by the $\Phi_u = 0$ analysis and that the ratio of the actual to the calculated load increases with the number of struts used to carry it.

The long term earth pressure is logically computed using the effective stress analysis with values of c' and Φ' taken from drained tests or consolidated-undrained triaxial tests with pore pressure measurements together with the least favourable position of the water table. This will in most cases represent a rise in earth pressure. Few examples are available to confirm this analysis other than of gravity retaining walls which are themselves founded in the same clay stratum. The problem is then in effect one of overall stability since the slip surface passes beneath the wall which is then of little more consequence than one 'slice' in the slices method of analysis (Fig. 22).

A number of failures of this type in stiff fissured London clay have been analysed by Henkel (1957), and here consistent active and passive pressures on the walls have been obtained using the effective stress analysis and observed water levels, together with the reduced values of c' shown in Fig. 19. It should be noted that where the excavation in front of the wall is deep, the presence of the passive pressure is insufficient to prevent the occurrence of progressive softening.

The behaviour of gravity retaining walls can of course throw little light on the end of construction earth pressures in fissured clays. Measurements of strut load have however been made by the Norwegian Geotechnical Institute in a trench in the weathered stiff fissured crust overlying a soft clay stratum (Di Biagio and Bjerrum, 1957; Bjerrum and Kirkedam, 1958). Here the softening appeared to proceed more rapidly than in cuts in London clay, for after only a few months the strut loads corresponded to the value given by the effective stress analysis with $c' = 0$.

The evidence from this cut and the indirect evidence from the Bradwell slip (section 6(b)) indicates that the $\Phi_u = 0$ analysis does not correctly represent the behaviour of stiff fissured clays under decreasing stresses even shortly after excavation. The rapid dissipation of negative pore pressures due to the presence of open fissures is obviously an important factor in temperate climates. Under long term conditions the $\Phi_u = 0$ method is also inapplicable for the reasons given in sections 6(b) and (c).

(f) The Stability of Earth Dams.—

It is impossible to deal adequately with all the stability problems arising in earth dam construction in one short section. However, the most important

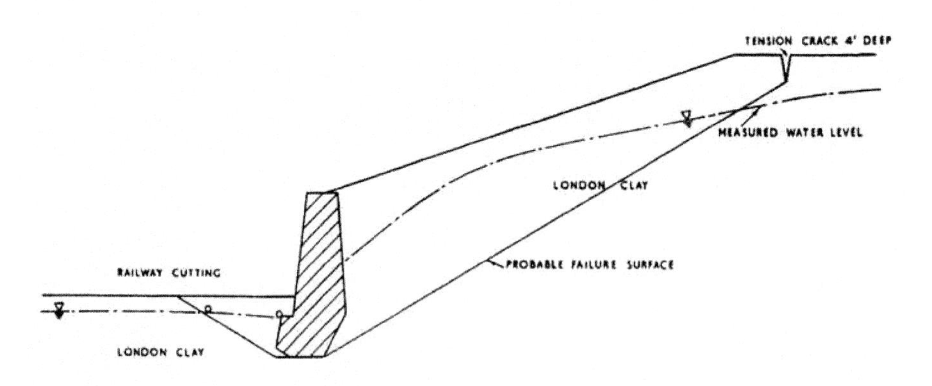

Fig. 22.—Retaining wall failure in stiff-fissured London clay (after Henkel, 1957).

principles may be illustrated by considering the stability of a water retaining dam built mainly of rolled earth fill (Fig. 23).

The stability of the slopes and foundation of an earth dam against shear failure will generally have to be considered under three conditions:

(1) During and shortly after construction,
(2) With the reservoir full (steady seepage), and
(3) On rapid drawdown of the impounded water.

Additional considerations arising from the possibility of failure in a clay foundation stratum are outlined in section 6(h). In this section attention will be limited to the fill.

The stability may be calculated for all three conditions in terms of effective stresses. This involves the measurement of c' and Φ' in the laboratory and an estimate of the pore pressure values at each stage. The use of explicitly determined pore pressures in the analysis enables the field measurements of pore pressure which are made on all important structures to be used as a direct check on stability during and after construction. It also enables the design estimates to be checked against the wealth of pore pressure data now becoming available from representative dams—for example the extensive work of the U.S.B.R. recently summarized by Gould (1959) and special cases such as the Usk dam (Sheppard and Aylen, 1957) and Selset dam (Bishop, Kennard and Penman, 1960).

To work directly with undrained test results expressed in terms of total stress may be unsafe in low dams as it implies dependence on negative pore pressures which will subsequently dissipate; and uneconomical in high dams in wet climates as no account is taken of the dissipation of excess pore pressure during the long construction period.

For the effective stress analysis the values of c' and Φ' are generally obtained from undrained triaxial tests with pore pressure measurement. In earth fill compacted at water contents well above the optimum a series of consolidated-undrained tests with pore pressure measurements may have to be used instead, in order to obtain a sufficient range of effective stresses to define a satisfactory failure envelope.

For the analysis of the condition of long term stability under steady seepage and for the case of rapid drawdown it is necessary to consider the effect of saturation on the values of c' and Φ'. As mentioned in section 4(h), test results show that, in general, the value of Φ' remains almost unchanged. Where c' has an appreciable value in the undrained tests this will decrease. However, tests using the improved techniques described by Bishop (1960) and Bishop, Alpan, Blight, and Donald (1960) have failed to reproduce the high cohesion intercepts previously reported in undrained tests. Provided c' has been accurately measured in either type of test, the differences may only be of significance in important works where the margin of safety is small. Whether c' is likely to become zero in rolled fill on a really long term basis is discussed by Bishop (1958 b) and Terzaghi (1958). Evidence so far does not appear to point to such a reduction.

The principal factors controlling the pore pressure set up during construction are:

(i) The placement moisture content and amount of compaction, and hence the pore pressure parameters;
(ii) The state of stress in the zone of the fill considered, and
(iii) The rate of dissipation of pore pressure during construction.

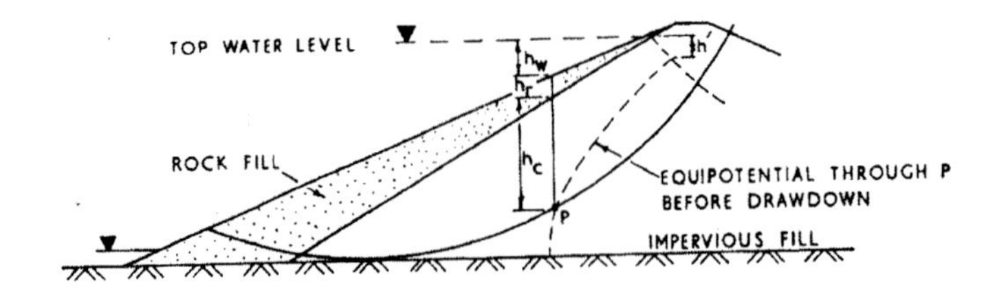

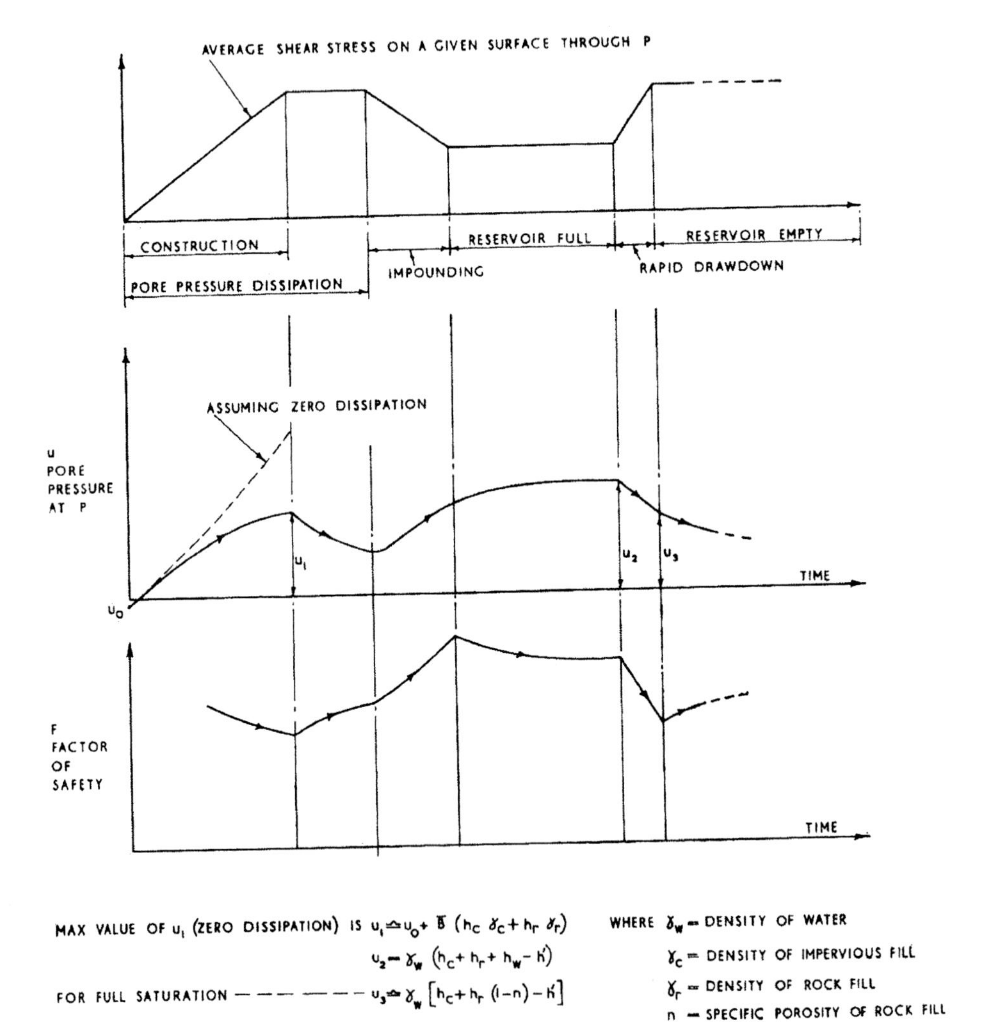

Fig. 23.—The changes in shear stress, pore pressure and factor of safety for the upstream slope of an earth dam.

In section 3 it was shown that the pore water pressure set up under undrained conditions can be expressed in the form:

$$u = u_0 + \bar{B} . \Delta\sigma_1 \quad (7)$$

In Fig. 24 the values of u_0 and $\bar{B}$ are plotted against water content for a series of samples prepared with the compactive effort used in the standard compaction test. This clearly shows the sensitivity of the value of the initial pore pressure to the placement water content; the importance of this effect, both in design and construction, cannot be overemphasised.

It is usually assumed that the value of σ_1 is equal to the vertical head of soil γ. h above the point considered, although the direction in which σ_1 acts is not necessarily vertical. This is a reasonably satisfactory assumption when averaged around a complete slip surface, but tends to overestimate the pore pressure in the centre of the dam and underestimate it near the toes (Fig. 25, after Bishop, 1952). It enables the pore pressure ratio required for the stability analysis to be expressed, under undrained conditions as

$$r_u = \bar{B} + u_0/\gamma h \quad (9)$$

or, at the higher water contents where $\bar{B}$ is large and u_0 small, more simply as

$$r_u = \bar{B} \quad (10)$$

However, in most earth fills a considerable reduction in the average pore pressure results from dissipation even during the construction period. A numerical method of solving the practical consolidation problem with a moving boundary has been given by Gibson (1958). It should be noted that in many almost saturated soils even a small amount of drainage has a marked effect on the final pore pressure, since it not only reduces the pore pressure already set up, but also reduces the value of $\bar{B}$ under the next increment of load. The theoretical basis of this reduction is discussed by Bishop (1957), and it is confirmed by field results from the Usk dam (Fig. 26a).

As an example of the distribution of pore pressure at the end of construction the contours from the Usk dam are illustrated in Fig. 26b. The effectiveness of the drainage layers placed to reduce the average pore pressure in the fill will be apparent.

The average pore pressure ratio along a potential slip surface at the end of construction may be kept within safe limits either by restricting the size of the impervious zone, by strict placement water content control or by special drainage measures (as at the Usk and Selset dams). Which is the more economical procedure will depend on the climatic conditions and the fill materials available.

For reservoir full conditions the pore pressure distribution may be predicted from the flow net corresponding to steady seepage. Accuracy is difficult to obtain owing to the non-uniformity of the rolled fill and differences in the ratio of horizontal to vertical permeability, so conservative assumptions should be made. However, with properly placed drainage zones the average value of r_u for the downstream slope is generally less than during construction, except in low dams or with rather dry placement.

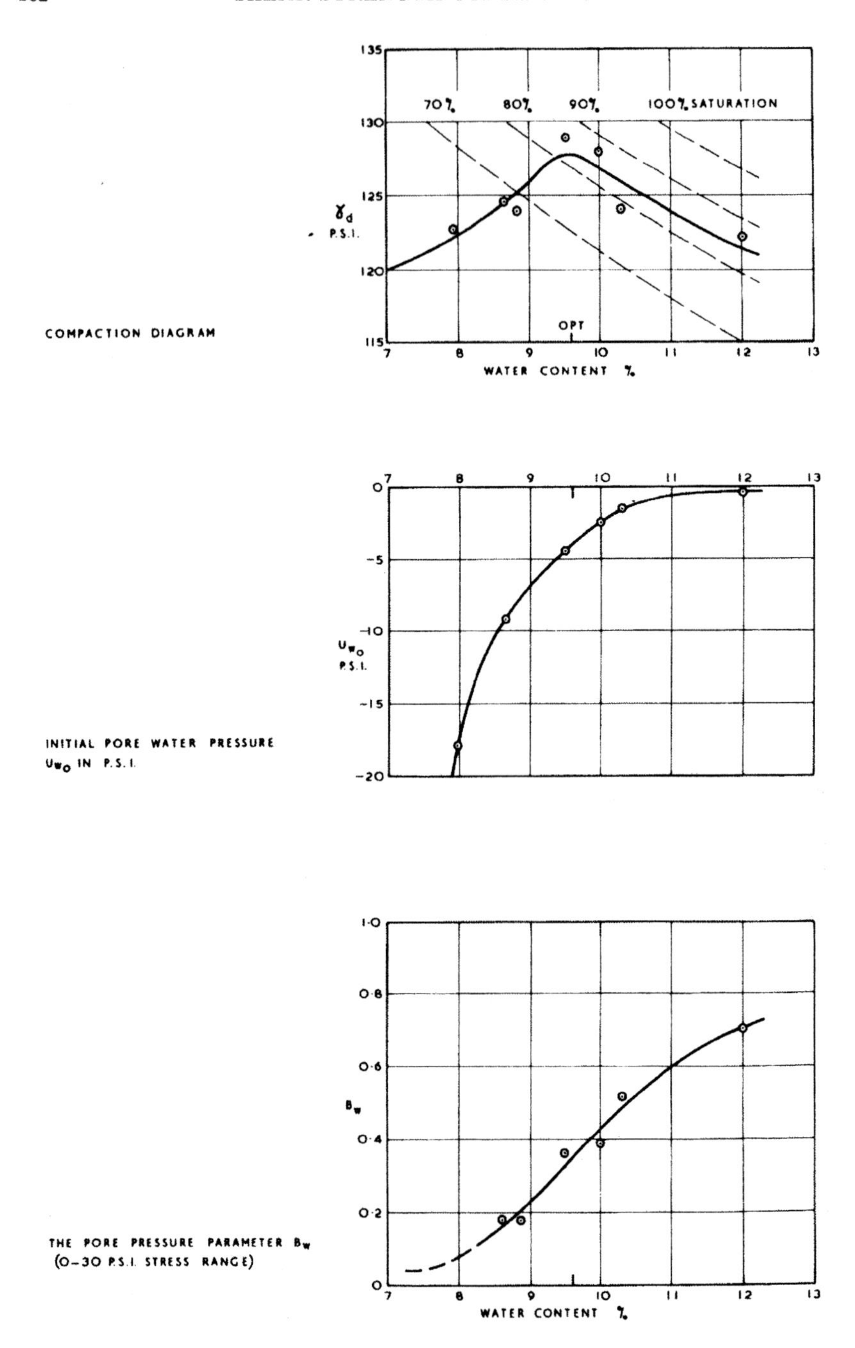

Fig. 24.--Represents Values from Standard Compaction Test under Equal All Round Pressure Increase: Compacted Boulder Clay, Clay Fraction 6%

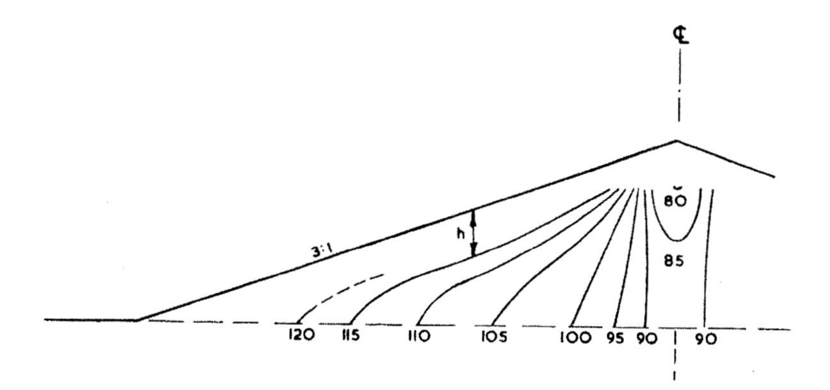

Fig. 25.—Major Principal Stress σ, as a Percentage of γ . h, the Vertical Head of Soil Above the Point Considered (After Bishop, 1952).

A method of predicting the excess pore pressures resulting from rapid drawdown has been proposed by Bishop (1952 and 1954a). In this method the change in pore pressure on drawdown is assumed to take place under undrained conditions and is deduced from the stress change and the pore pressure parameters (see Fig. 23). For saturated fills the value of $\bar{B}$ is taken as 1; the change in the value of σ_1 is due to the removal of the water load from the face of the dam and the drainage of water from the voids of the rockfill.

This method shows reasonable agreement with the results of the field measurements on the Alcova dam (Glover, Gibbs and Daehn, 1948). Two recent cases of very rapid drawdown soon after completion are not in such good agreement, but both involve complicating factors (Bazett, 1958; Paton and Semple, 1960).

Fig. 23 shows diagrammatically the variation in pore pressure and factor of safety for the various phases in the life of the upstream slope of the embankment calculated as described above. The lowest values of factor of safety are usually reached at the end of construction and on rapid drawdown.

For the downstream slope, end of construction and steady seepage are the two critical stages. However, during steady seepage the danger is generally not so much from the pore pressures, which are easily controlled by drainage measures, but from the possibility of piping and internal erosion in the foundation strata, and from crack formation in the fill.

For small earth dams built largely of saturated soft clay, the stability during the construction period can be calculated by the $\Phi_u = 0$ analysis using the undrained strength (Cooling and Golder, 1942). However, where the fill is much stronger than the clay foundation strata satisfactory results are not obtained (for example, Golder and Palmer, 1955) for the reasons given in section 6(a). The long term stability can of course be determined only by the effective stress analysis.

(g) Stability of Slopes in Sand and Gravel on Drawdown.—
In relatively pervious soils of low compressibility the distribution of pore pressure on drawdown is controlled by the rate of drainage of pore water

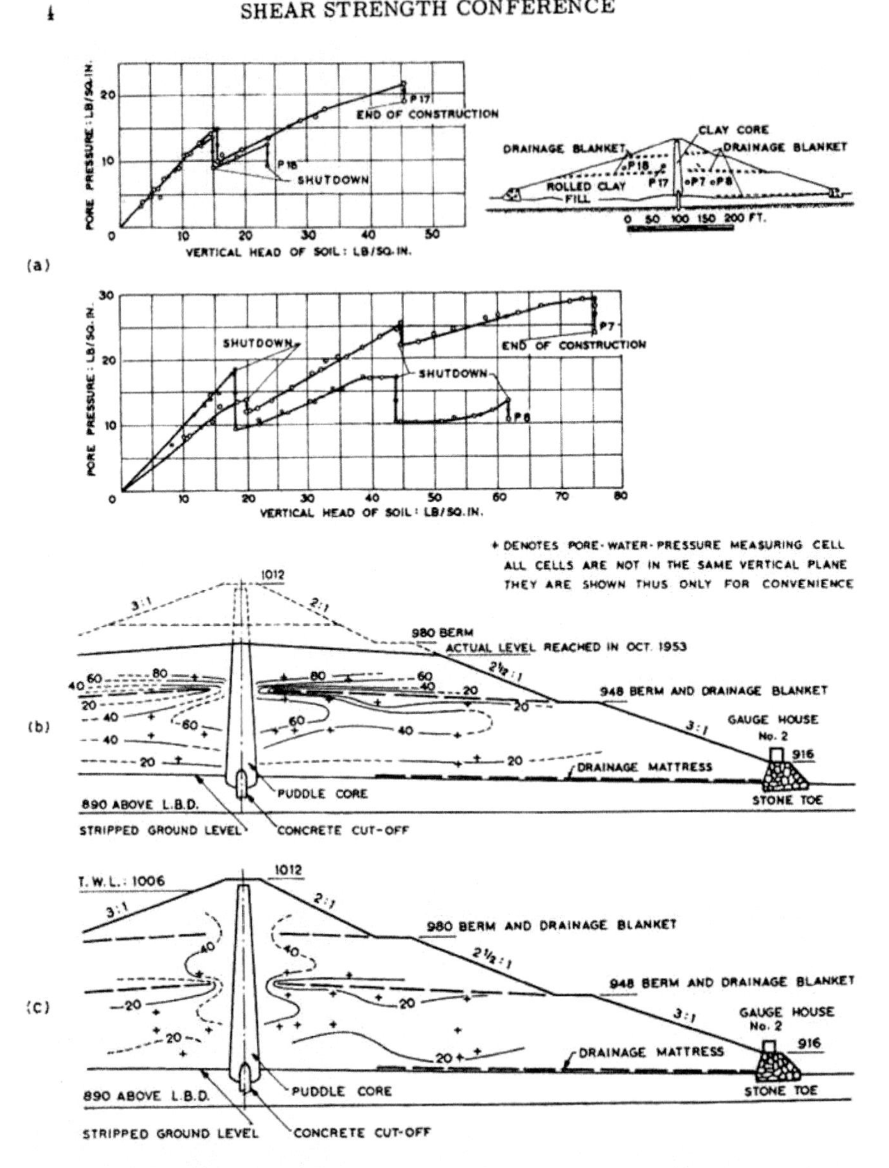

Fig. 26.—Field measurements of pore pressure from the Usk dam: (a) The reduction in pore pressure and in the value of $\bar{B}$ due to pore pressure dissipation (after Bishop, 1957); (b) and (c) contours of pore pressure expressed as a percentage of the vertical head of soil in October, 1953 and October 1954 (after Sheppard and Aylen, 1957).

from the soil. This condition can be represented by a series of flownets with a moving boundary as shown by Terzaghi (1943) and Reinius (1948).

The flow pattern is a function of the ratio of drawdown rate to permeability and the values of the pore pressures to be used in the stability analysis can be taken from the appropriate flow-net. The influence of the greater permeability in the horizontal direction is considerable, but, in one case examined, tended to increase rather than reduce the factor of safety.

The values of c' and Φ' are obtained from drained tests, c' approaching zero for free-draining materials.

An example of a drawdown failure in Thames gravel is shown in Fig. 27. The initial slope of the gravel was 33^0 and the permeability about 0.05 cm/sec. The value of Φ' in the loose state was 36^0. Failure occurred when the pool was lowered at a rate of about 1 foot per day (Bishop, 1952).

(h) The Stability of a Clay Foundation of an Embankment where the Rate of Construction Permits Partial Consolidation.—

It is not uncommon in earth dam construction to encounter geological conditions in which the foundation strata include a soft clay layer at or near the surface, of sufficient extent to be likely to lead to failure in an embankment having conventional side slopes (for example Cooling and Golder, 1942; McLellan, 1945; Bishop, 1948; Skempton and Bishop, 1955; Bishop, Kennard and Penman, 1960). It is then necessary to assess the economics and practicability of a number of alternative solutions. The soft layer may be excavated, if its depth and ground water conditions permit; or an embankment with very flat slopes may be accepted, its factor of safety being calculating using the $\Phi_u = 0$ method which assumes zero drainage. Alternatively, account may be taken of the dissipation of pore pressure which occurs due to natural drainage (for example, Bishop, 1948) or due to special measures, such as vertical sand drains, designed to accelerate consolidation (for example, Skempton and Bishop, 1955; Bishop, Kennard and Penman, 1960). In this case an effective stress analysis is used.

An expression for the initial excess pore pressure in a saturated soft clay layer where B = 1 has been obtained by Bishop (1952):

$$\Delta u = \Delta p + p_o \, . \, [(1-K)/2] + (2A-1)\sqrt{p_o^2[(1-K)/2]^2 + \tau^2} \tag{25}$$

where Δp denotes change in total vertical stress due to the fill,

τ » shear stress along the layer set up by the fill,

p_o » initial vertical effective stress,

and Kp_o » initial horizontal effective stress in clay layer.

This expression illustrates the dependence of pore pressure on the change in shear stress as well as on the change in vertical stress, though the latter predominates in most practical cases. To avoid the error in the estimate of A which may arise from the change in void ratio on reconsolidation of undisturbed samples, the value of A may be deduced from the relation between the undrained strength and the effective stress envelope, using an assumed value of K. The relationship is given by Bishop (1952).

The estimate of the rate of dissipation of pore pressure is based on the theory of consolidation. It is here that the greatest uncertainity arises, especially in stratified deposits, and field observations of pore pressure are advisable on important works.

Fig. 27.—Drawdown failure in Thames Valley gravel (after Bishop, 1952).

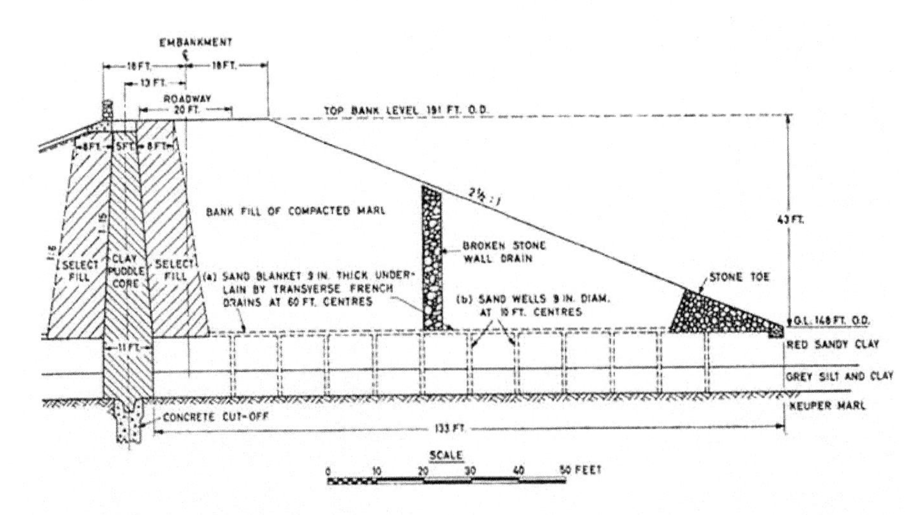

Fig. 28.—Downstream slope of the Chew Stoke dam showing vertical sand drains to accelerate dissipation of pore pressure in soft clay foundation (after Skempton and Bishop, 1955).

The values of c' and ϕ' are taken from drained tests or consolidated undrained tests with pore pressure measurement.

The downstream slope of the Chew Stoke dam (Fig. 28) which had a factor of safety against a foundation failure of 0.8 using the $\phi_u = 0$ analysis was safely constructed using a sand drain spacing designed to give a factor of safety of 1.5 (Skempton and Bishop, 1955). Field observations of pore pressure indicated that the actual factor of safety was rather higher than 1.5 owing to the greater horizontal permeability resulting from stratification of the clay. The Selset dam, founded on a boulder clay with little apparent stratification, showed a smaller difference between predicted and observed pore pressure values (Bishop, Kennard and Penman, 1960).

(i) Some Special Cases.--

In the examples described above the variation in safety factor with time was either a steady increase or a steady decrease during the period from the end of construction until the pore pressures reached an equilibrium condition. The following examples will illustrate that under certain conditions we may temporarily encounter a lower factor of safety at an intermediate stage.

Such cases are obviously very dangerous, as a failure might well occur some weeks or months after the completion of construction, in spite of the fact that it had been ascertained that the factor of safety was adequate in both the initial and final stages. The basic reason in each case is that the redistribution of excess pore pressure which occurs during the consolidation process may lead to a temporary rise in pore pressure outside the zone where the load is applied.

An interesting example is the stability of a river bank in a clay stratum under the action of the excess pore pressure set up by pile driving for a bridge abutment in the vicinity (Bjerrum and Johannessen, 1960). The changes in pore pressure with time are shown diagrammatically in Fig. 29a for two points, one in the centre of the pile group and one outside it, but beneath the slope. The excess pore pressures set up by pile driving will dissipate laterally as well as vertically, particularly in a water-laid sediment where the horizontal permeability tends to be greater than the vertical.

The exact magnitude of the effect is difficult to predict theoretically, and in this case field measurements of pore pressure were used to regulate the progress of pile driving. The most critical distribution of pore pressure occurred shortly after piling was completed (Fig. 29b). The factor of safety of the slope was calculated using the effective stress analysis with c' = 200 lb./sq. ft. and $\phi' = 27°$, and dropped from an initial value of 1.4 to 1.15 after piling, assuming low water in the river in each case.

A case in which the spread of pore pressure led to an embankment failure some time after construction has been analysed in detail by Ward, Penman and Gibson (1955). The clay foundation on which the embankment was built included two horizontal layers of peat. Due to the relatively higher permeability of the peat, the redistribution of pore pressure after the end of construction resulted in temporary high pore pressures in the peat on both sides of the embankment, where initially the pore pressures were low. Because of the difference between the void ratio pressure curves for consolidation and swelling, the reduction in effective stress due to the presence of an additional volume of pore water is much greater than the increase in effective stress in the zone from which this volume has migrated. An overall decrease in effective stress along a critical slip path can thus occur (Fig. 30) at an

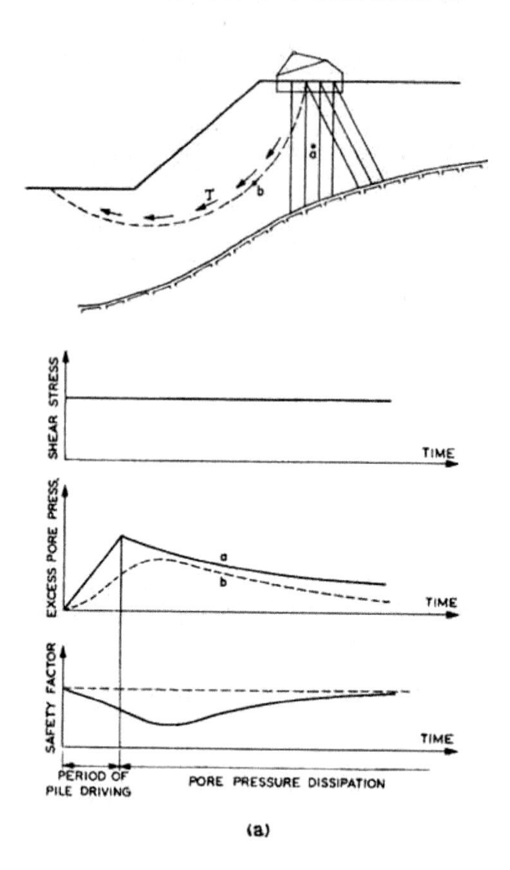

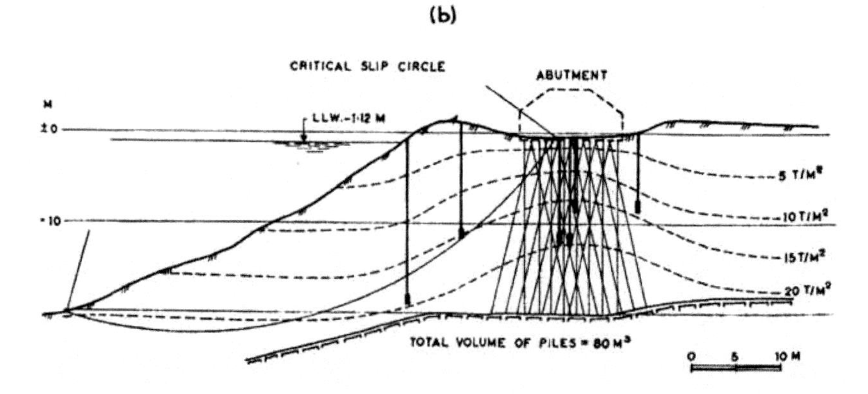

Fig. 29.—The effect of pore pressure set up by pile driving on the stability of an adjacent clay slope: (a) Changes in pore pressure and factor of safety with time: diagrammatic, (b) observed pore pressures just after the completion of pile driving (after Bjerrum and Johannessen, 1960).

intermediate stage, although in the long term equilibrium state the bank foundation would have been stable.[u]

Other engineering operations may result in a similar danger, such as the rapid construction of an embankment or stockpile even some way back from a river bank, cut or quay wall close to limiting equilibrium, and the driving of ordinary piles or screw piles through the clay slopes of rivers or harbours. In such cases an awareness of the danger will either lead to a modification of the operation, or to the use of field measurements of pore pressure as a control while it is carried out.

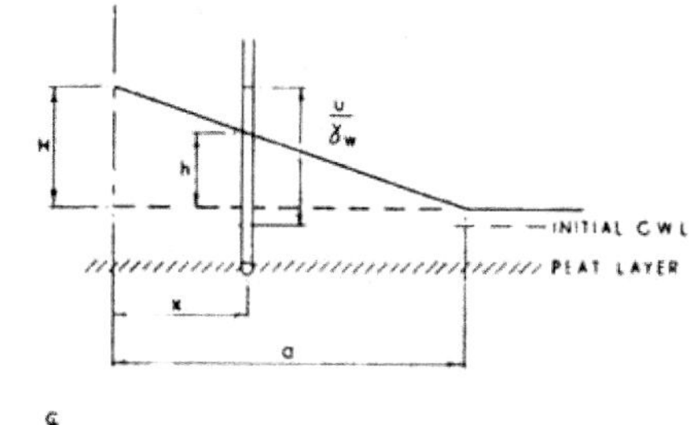

(a) Embankment on a clay foundation with a horizontal peat layer—diagrammatic;

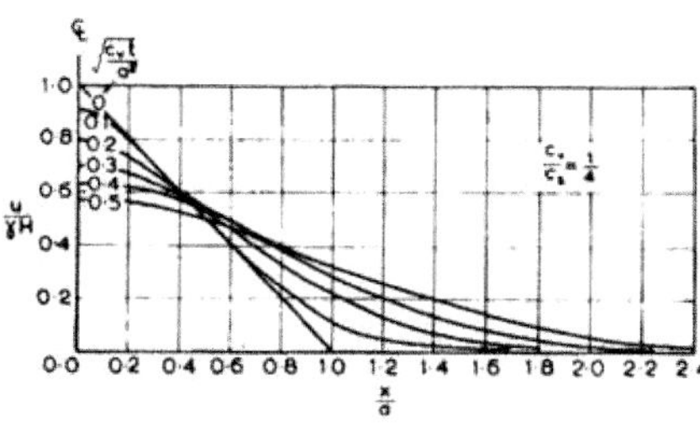

(b) changes in pore pressure with time, as a fraction of $\gamma . H$ where γ is density of fill;

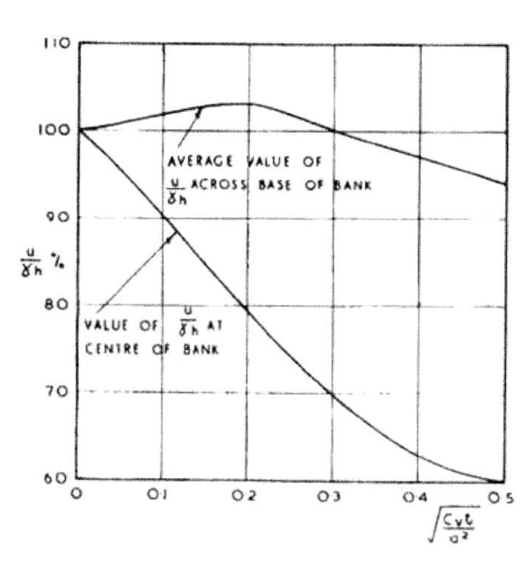

(c) effect of redistribution on maximum and average pore pressure for $c_v / c_s = 1/4$ where c_v is coefficient of consolidation and c_s is coefficient of swelling (after Ward, Penman, and Gibson, 1954).

Fig. 30.

u. Other failures of this type are described by Terzaghi and Peck (1948).

It is probable that a number of failures attributed to 'creep' could more correctly be attributed to the redistribution of pore pressure which occurs after construction.

(j) Design in Earthquake Areas.—

The analysis of the stability of a structure or dam in an area subject to earthquakes raises special problems which are outside the scope of this paper.

It is, however, known that a transient load will leave residual excess pore pressures which may be positive or negative depending on the void ratio and stress history of the soil (Bishop and Henkel, 1953). A possible way of evaluating the stability under earthquake conditions may therefore be to use a series of consolidated-undrained tests in which the stress ratio during consolidation is chosen to represent the conditions prevailing in the field before the earthquake. The sample is then subjected to a series of small variations in deviator stress under undrained conditions corresponding to the additional seismic stresses. The magnitude of the residual pore pressure and the additional strain will indicate the likelihood of failure under field conditions.

A discussion of the additional shear stresses likely to be set up in earthquake areas is given by Ambraseys (1959).

7. CONCLUSIONS

The discussion and case records presented in this paper point to four main conclusions:

(I) The effective stress analysis is a generally valid method for analysing any stability problem and is particularly valuable in revealing trends in stability which would not be apparent from total stress methods.

Its application in practice is limited to cases where the pore pressures are known or can be estimated with reasonable accuracy. These include all the class (a) problems, such as long term stability and drawdown in incompressible soils, where the pore pressure is controlled by ground water conditions or by a flow pattern. It is also applicable to both class (a) and class (b) problems where field measurements of pore pressure are available.

Those class (b) problems where the magnitude of the pore pressure has to be estimated from the stress distribution and the measured values of the pore pressure parameters can often be solved more simply by the $\Phi_u = 0$ analysis. However, this alternative gives no indication of the long term stability and does not enable account to be taken of dissipation of pore pressure during construction, which may contribute greatly to economy in design.

(II) Where a saturated clay is loaded or unloaded at such a rate that there is no significant dissipation of the excess pore pressures set up, the stability can be determined by the $\Phi_u = 0$ analysis, using the undrained strength obtained in the laboratory or from in-situ vane tests.

This method is very simple and reliable if its use is restricted to the conditions specified above. It is essentially an end of construction method, and in the majority of foundation problems, where the factor of safety increases with time, it provides a sufficient check on stability. For cuts, on the other hand, where the factor of safety generally decreases with time, the $\Phi_u = 0$ method can be used only for temporary work and the long term stability must be calculated by the effective stress analysis.

(III) The two methods of analysis require the measurement of the shear strength parameters c' and ϕ' in terms of effective stress on the one hand and the undrained shear strength c_u under the stress conditions obtaining in the field on the other.

For saturated soils the values of c' and ϕ' are obtained from drained tests or consolidated undrained tests with pore pressure measurement, carried out on undisturbed samples. The range of stresses at failure should be chosen to correspond with those in the field. Values measured in the laboratory appear to be in satisfactory agreement with field records with two exceptions. In stiff fissured clays the field value of c' is lower than the value given by standard laboratory tests; in some very sensitive clays the field value of ϕ' is lower than the laboratory value.

For partly saturated soils the values of c' and ϕ' are obtained from undrained or consolidated-undrained tests with pore pressure measurement, or from drained tests. Provided comparable testing procedures are used the differences between the values of ϕ' obtained appear not to be significant from a practical point of view. The values of c' will be slightly influenced by moisture content differences resulting from the different procedures.

The undrained shear strength c_u is obtained from undrained triaxial tests on undisturbed samples (or from unconfined compression tests, except on fissured clays) and from vane tests in situ. It cannot be obtained, without risk of error on the unsafe side, from consolidated-undrained tests where the sample is reconsolidated under the overburden pressure. The error is serious in normally consolidated clays of low plasticity, and though it can be minimised by consolidating under the stress ratio obtaining in the field, the effect of reconsolidation on the void ratio cannot be avoided.

For this same reason it is probably more realistic to calculate the value of the pore pressure parameter A for undisturbed soil from the relationship between the undrained strength of undisturbed samples and the values of c' and ϕ', rather than to measure it in a consolidated-undrained test.

(IV) The reliability of any method can ultimately be checked only by making the relevant field measurements when failures occur or when construction operations are likely to bring a soil mass near to limiting equilibrium. The number of published case records in which the data is sufficiently complete for a critical comparison of methods is still regrettably small.

ACKNOWLEDGMENTS

The work of K. Terzaghi, Arthur Casagrande, A. W. Skempton and the late D. W. Taylor has contributed so much to the background of any study of shear strength and stability that specific references in the text are inadequate acknowledgment. The authors would also like to express their gratitude to their colleagues at Imperial College and the Norwegian Geotechnical Institute for valuable comments on the manuscript.

APPENDIX I.—THE USE OF THE PARAMETER ϕ_{cu}

In section 4b reference has been made to errors likely to arise in applying in the field the relationship between undrained strength and consolidation pressure obtained in the laboratory from the consolidated-undrained test.

Two inherent errors have been referred to: The effect of reconsolidation after sampling on the void ratio and on the value of the pore pressure parameter A; and the error arising from consolidation under a stress ratio different from that obtaining in the ground. A further error may arise from the way in which the results are introduced into the stability analysis.

This point is illustrated in Fig. 31 (after Bishop and Henkel, 1957). The test is usually performed by consolidating the sample under a cell pressure p, and then causing failure under undrained conditions by increasing the axial stress. The total minor principal stress at failure (σ_3) is thus equal to p; the total major principal stress is $(\sigma_1)_{cu}$. The slope of the envelope to a series of total stress circles obtained in this manner (Fig. 31a) is denoted Φ_{cu}, the angle of shearing resistance in consolidated undrained tests, and is about one half of the slope of the effective stress envelope (denoted by Φ') for normally consolidated samples. This relationship between shear strength and total normal stress can only be used in practice if the identity between consolidation pressure and total minor principal stress imposed in the test also

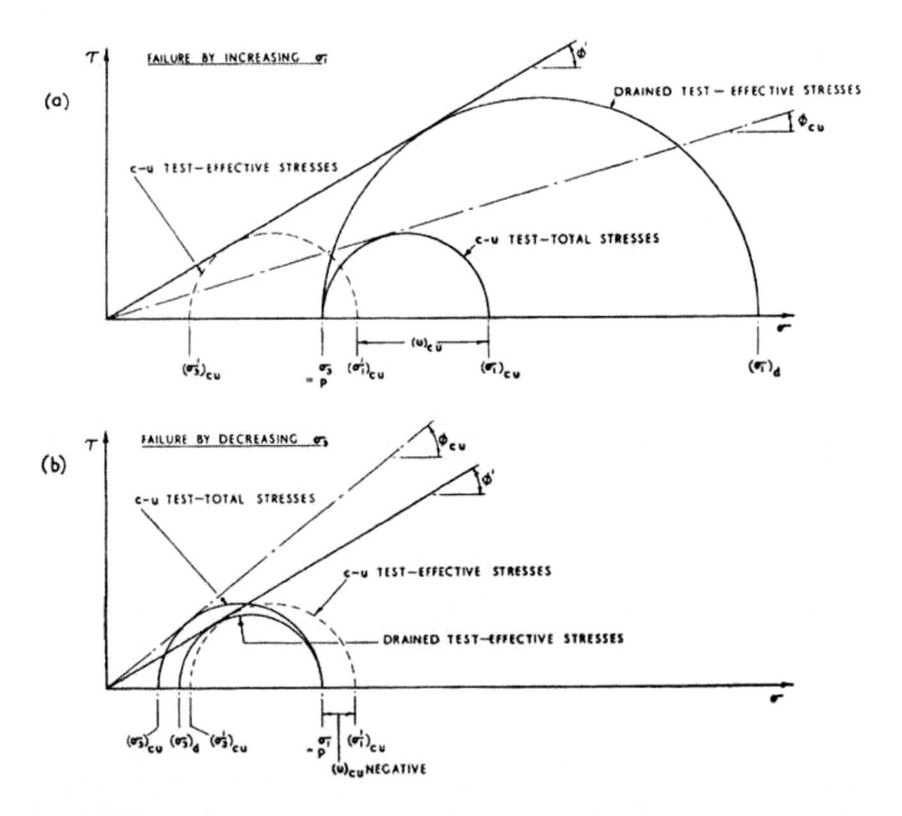

Fig. 31.—The consolidated-undrained test on a saturated cohesive soil in terms of total and effective stresses: (a) Failure by increasing major principal stress σ_1; (b) failure by decreasing minor principal stress σ_3 (after Bishop and Henkel, 1957).

applies around the slip surface considered. Passive earth pressure appears to be the only case in which this is approximately true.

Had the failure been caused by holding σ_1 constant and equal to p and decreasing the total minor principal stress σ_3, the undrained strength would have remained the same, and a radically different value of Φ_{cu} would have been obtained, Fig. 31b. The relationship between shear strength and total normal stress would then approximate to the case of active earth pressure.

The general use of Φ_{cu} (defined in Fig. 31a) as an angle of shearing resistance in conventional stability analyses is therefore likely to lead to very erroneous results, even if the samples are anisotropically consolidated. If σ_3 increases during the undrained loading (as in foundation problems) the factor of safety will be overestimated; if σ_3 decreases, as in the excavation of a cutting, the error may lead to an under-estimate of the factor of safety.

The most logical solution appears to be to plot contours of undrained strength in terms of the consolidation pressure in the ground prior to the undrained loading to be examined, and then to use the $\Phi_u = 0$ analysis. This method is of course limited to the end of construction analysis, in which it is assumed that insufficient time has elapsed for consolidation or swelling to occur.

In rapid drawdown analyses suggested by Terzaghi (1943) and Lowe and Karafiath (1959) the undrained strength is related to the effective normal stress on the potential failure plane before drawdown. However, unless the samples are failed by reducing the stresses, there is a danger of overestimating the undrained strength of compacted samples which are difficult to saturate fully in the laboratory.

APPENDIX II.—BIBLIOGRAPHY ON SHEAR STRENGTH AND STABILITY

1. Ambraseys, N. N. (1959), The seismic stability of earth dams. Thesis. (University of London). London. 2 vol.

2. Bazett, D. J. (1958), Field measurement of pore water pressures. Canadian Soil Mechanics Conference, 12. Saskatoon. Proceedings, p. 2-15.

3. Berger (1951), Unpublished report.

4. Bishop, A. W. (1948), Some factors involved in the design of a large earth dam in the Thames valley. International Conference on Soil Mechanics and Foundation Engineering, 2. Rotterdam. Proceedings, Vol. 2, p. 13-18.

5. Bishop, A. W. (1952), The stability of earth dams. Thesis. (University of London). London. 176 p.

6. Bishop, A. W. (1954 a), The use of pore pressure coefficients in practice. Geotechnique, Vol. 4, No. 4, p. 148-152.

7. Bishop, A. W. (1954 b), The use of the slip circle in the stability analysis of slopes. European Conference on Stability of Earth Slopes, Stockholm. Proceedings, Vol. 1, p. 1-13. Geotechnique, Vol. 5, No. 1, 1955, p. 7-17.

8. Bishop, A. W. (1957), Some factors controlling the pore pressures set up during the construction of earth dams. International Conference on Soil Mechanics and Foundation Engineering, 4. London. Proceedings, Vol. 2, p. 294-300.

9. Bishop, A. W. (1958 a), Test requirements for measuring the coefficient of earth pressure at rest. Brussels Conference on Earth Pressure Problems. Proceedings, Vol. 1, p. 2-14.

10. Bishop, A. W. (1958 b), Discussion on: Terzaghi, K. Design and performance of the Sasumua dam. Institution of Civil Engineers. Proceedings, Vol. 11, November, p. 348-352.

11. Bishop, A. W. (1959), The principle of effective stress. Teknisk ukeblad, Vol. 106, No. 39, p. 859-863. (Norwegian Geotechnical Institute. Publ., 32.)

12. Bishop, A. W. (1960), The measurement of pore pressure in the triaxial test. Pore Pressure and Suction in Soil Conference, London, p. 52-60.

13. Bishop, A. W., Alpan, J., Blight, G. and Donald, V. (1960), Factors controlling the strength of partly saturated soils. Research Conference on Shear Strength of Cohesive Soils. Proceedings.

14. Bishop, A. W. and Eldin, G. (1950), Undrained triaxial tests on saturated sands and their significance in the general theory of shear strength. Geotechnique, Vol. 2, No. 1, p. 13-32.

15. Bishop, A. W. and Eldin, A. K. G. (1953), The effect of stress history on the relation between Φ and porosity in sand. International Conference on Soil Mechanics and Foundation Engineering, 3. Zürich. Proceedings, Vol. 1, p. 100-105.

16. Bishop, A. W. and Henkel, D. J. (1953), Pore pressure changes during shear in two undisturbed clays. International Conference on Soil Mechanics and Foundation Engineering, 3. Zürich. Proceedings, Vol. 1, p. 94-99.

17. Bishop, A. W. and Henkel, D. J. (1957), The measurement of soil properties in the triaxial test. London, Arnold. 190 p.

18. Bishop, A. W., Kennard, M. F. and Penman, A. D. M. (1960), Pore pressure observations at Selset dam. Pore Pressure and Suction in Soil Conference, London, p. 36-47.

19. Bishop, A. W. and Morgenstern, N. (1960), Stability coefficients for earth slopes. In preparation.

20. Bjerrum, L. (1954 a), Geotechnical properties of Norwegian marine clays. Geotechnique, Vol. 4, No. 2, p. 49-69. (Norwegian Geotechnical Institute. Publ., 4).

21. Bjerrum, L. (1954 b), Theoretical and experimental investigations on the shear strength of soils. Thesis. Oslo. 113 p. (Norwegian Geotechnical Institute. Publ., 5).

22. Bjerrum, L. (1954 c), Stability of natural slopes in quick clay. European Conference on Stability of Earth Slopes, Stockholm. Proceedings, Vol. 2, p. 16-40. Geotechnique, Vol. 5, No. 1, 1955, p. 101-119. (Norwegian Geotechnical Institute, Publ., 10).

23. Bjerrum, L. and Eide, O. (1956), Stability of strutted excavations in clay. Geotechnique, Vol. 6, No. 1, p. 32-47. (Norwegian Geotechnical Institute. Publ., 19).

24. Bjerrum, L. and Johannessen, I. (1960), Pore pressures resulting from driving piles in soft clay. Pore Pressure and Suction in Soil Conference, London, p. 14-17.

25. Bjerrum, L. and Kirkedam, R. (1958), Some notes on earth pressure in stiff fissured clay. Brussels Conference on Earth Pressure Problems. Proceedings, Vol. 1, p. 15-27. (Norwegian Geotechnical Institute. Publ., 33.)

26. Bjerrum, L. and Kjaernsli, B. (1957), Analysis of the stability of some Norwegian natural clay slopes. Geotechnique, Vol. 7, No. 1, p. 1-16. (Norwegian Geotechnical Institute. Publ., 24).

27. Bjerrum, L., Simons, N. and Torblaa, I. (1958), The effect of time on the shear strength of a soft marine clay. Brussels Conference on Earth Pressure Problems. Proceedings, Vol. 1, p. 148-158. (Norwegian Geotechnical Institute. Publ., 33.)

28. Bjerrum, L. and Øverland, A. (1957), Foundation failure of an oil tank in Fredrikstad, Norway. International Conference on Soil Mechanics and Foundation Engineering, 4. London. Proceedings, Vol. 1, p. 287-290. (Norwegian Geotechnical Institute. Publ., 26).

29. Bruggeman, J. R., Zangar, C. N. and Brahtz, J. H. A. (1939), Notes on analytic soil mechanics. Denver, Colo. (Department of the Interior, Bureau of Reclamation. Technical memorandum, 592).

30. Cadling, L. and Odenstad, S. (1950), The vane borer. Sthm. 87 p. (Royal Swedish Geotechnical Institute. Proceedings, 2).

31. Campion, F. E. (1951), Part reconstruction of Bo-Peep tunnel at St. Leonards-on-Sea. Institution of Civil Engineers. Journal, Vol. 36, p. 52-75.

32. Casagrande, A. (1934), Discussion of Dr. Jürgenson's papers, entitled "The application of the theory of elasticity and theory of plasticity to foundation problems" and "Research on the shearing resistance of soils." Boston Society of Civil Engineers. Journal, Vol. 21, p. 276-283. Boston Society of Civil Engineers. Contributions to soil mechanics (925-940. Boston 1940, p. 218-225.

33. Casagrande, A. (1949), Soil mechanics in the design and construction of the Logan airport. Boston Society of Civil Engineers. Journal, Vol. 36, p. 192-221. (Harvard University. Graduate School of Engineering. Publ., 467—Soil mechanics series, 33).

34. Casagrande, A. and Albert, S. G. (1930), Research on the shearing resistance of soils. Cambr., Mass. Unpubl. (Massachusetts Institute of Technology. Report).

35. Cassel, F. L. (1948), Slips in fissured clay. International Conference on Soil Mechanics and Foundation Engineering, 2. Rotterdam. Proceedings, Vol. 2, p. 46-50.

36. Coates, R. H. and Slade, L. R. (1958), Construction of circulating-water pump house at Cowes Generating Station, Isle of Wight. Institution of Civil Engineers. Proceedings, Vol. 9, p. 217-232.

37. Cornforth, D. (1960), Thesis. (University of London). In preparation.

38. Cooling, L. F. and Golder, H. Q. (1942), The analysis of the failure of an earth dam during construction. Institution of Civil Engineers. Journal, Vol. 19, p. 38-55.

39. Daehn, W. W. and Hilf, J. W. (1951), Implications of pore pressure in design and construction of rolled earth dams. International Congress on Large Dams, 4. New Delhi. Transactions, Vol. 1, p. 259-270.

40. De Lory, L. A. (1957), Long-term stability of slopes in over-consolidated clays. Thesis. (University of London). London.

41. Di Biagio, E. and Bjerrum, L. (1957), Earth pressure measurements in a trench excavated in stiff marine clay. International Conference on Soil Mechanics and Foundation Engineering, 4. London. Proceedings, Vol. 2, p. 196-202. (Norwegian Geotechnical Institute. Publ., 26).

42. Fraser, A. M. (1957), The influence of stress ratio on compressibility and pore pressure coefficients in compacted soils. Thesis. (University of London). London.

43. Garde-Hansen, P. and Thernöe, S. (1960), Grain silo of 100,000 tons capacity, Mersin, Turkey. CN Post (Cph.), No. 48, p. 14-22.

44. Gibson, R. E. (1958), The progress of consolidation in a clay layer increasing in thickness with time. Geotechnique, Vol. 8, No. 4, p. 171-182.

45. Gibson, R. E. and Marsland, A. (1960), Pore-water observations in a saturated alluvial deposit beneath a loaded oil tank. Pore Pressure and Suction in Soil Conference, London, p. 78-84.

46. Glover, R. E., Gibbs, H. J. and Daehn, W. W. (1948), Deformability of earth materials and its effect on the stability of earth dams following a rapid drawdown. International Conference on Soil Mechanics and Foundation Engineering, 2. Rotterdam. Proceedings, Vol. 5, p. 77-80.

47. Golder, H. Q. and Palmer, D. J. (1955), Investigation of a bank failure at Scrapsgate, Isle of Sheppey, Kent. Geotechnique, Vol. 5, No. 1, p. 55-73.

48. Gould, J. P. (1959), Construction pore pressures observed in rolled earth dams. Denver, Colo. 97 p. (Department of the Interior. Bureau of Reclamation. Technical memorandum, 650).

49. Hamilton, L. W. (1939), The effects of internal hydrostatic pressure on the shearing strength of soils. American Society for Testing Materials. Proceedings, Vol. 39, p. 1100-1121.

50. Hansen, J. B. and Gibson, R. E. (1949), Undrained shear strengths of anisotropically consolidated clays. Geotechnique, Vol. 1, No. 3, p. 189-204.

51. Henkel, D. J. (1957), Investigations of two long-term failures in London clay slopes at Wood Green and Northolt. International Conference on Soil Mechanics and Foundation Engineering, 4. London. Proceedings, Vol. 2, p. 315-320.

52. Henkel, D. J. (1959), The relationships between the strength, pore-water pressure, and volume-change characteristics of saturated clays. Geotechnique, Vol. 9, No. 3, p. 119-135.

53. Henkel, D. J., (1960), The strength of saturated remoulded clay. Research Conference on Shear Strength of Cohesive Soils. Proceedings.

54. Henkel, D. J. and Skempton, A. W. (1955), A landslide at Jackfield, Shropshire, in a heavily over-consolidated clay. Geotechnique, Vol. 5, No. 2, p. 131-137.

55. Hilf, J. W. (1948), Estimating construction pore pressures in rolled earth dams. International Conference on Soil Mechanics and Foundation Engineering, 2. Rotterdam. Proceedings, Vol. 3, p. 234-240.

56. Hilf, J. W. (1956), An investigation of pore-water pressure in compacted cohesive soils. Denver, Colo. 109 p. (Department of the Interior. Bureau of Reclamation. Technical memorandum, 654).

57. Holmsen, P. (1953), Landslips in Norwegian quick-clays. Geotechnique, Vol. 3, No. 5, p. 187-194. (Norwegian Geotechnical Institute. Publ., 2).

58. Hvorslev, M. J. (1937), Über die Festigkeitseigenschaften gestörter bindiger Böden. Kbh., (Gad). 159 p. (Ingeniørvidenskabelige skrifter, A 45).

59. Ireland, H. O. (1954), Stability analysis of the Congress street open cut in Chicago. Geotechnique, Vol. 4, No. 4, p. 163-168.

60. Janbu, N. (1954), Application of composite slip surfaces for stability analysis. European Conference on Stability of Earth Slopes, Stockholm. Proceedings, vol. 3, p. 43-49.

61. Janbu, N. (1957), Earth pressure and bearing capacity by generalized procedure of slices. International Conference on Soil Mechanics and Foundation Engineering, 4. London. Proceedings, Vol. 2, p. 207-212.

62. Kallstenius, J. and Wallgren, A. (1956), Pore water pressure measurement in field investigations. Sthm. 57 p. (Royal Swedish Geotechnical Institute. Proceedings, 13).

63. Kenney, T. C. (1956), An examination of the methods of calculating the stability of slopes. Thesis. (University of London). London.

64. Kjaernsli, B. (1958), Test results, Oslo subway. Brussels Conference on Earth Pressure Problems. Proceedings, Vol. 2, p. 108-117.

65. Larew, H. G. (1952), Analysis of landslides. Wash. D. C. 39 p. (Highway Research Board. Bulletin, 49).

66. Laughton, A. S. (1955), The compaction of ocean sediments. Thesis. (University of Cambridge). Cambr.

67. Little, A. L. and Price, V. E. (1958), The use of an electronic computer for slope stability analysis. Geotechnique, Vol. 8, No. 3, p. 113-120.

68. Lowe, J. and Karafiath, L. (1959), Stability of earth dams upon drawdown. Panamerican Conference on Soil Mechanics and Foundation Engineering, 1. Mexico. Paper 2-A, 15 p.

69. McLellan, A. G. (1945), The Hollowell reservoir scheme for Northampton. Water and water engineering, Vol. 48, p. 7-26.

70. Morgan, H. D. (1944), The design of wharves on soft ground. Institution of Civil Engineers. Journal, Vol. 22, p. 5-25.

71. Murphy, V. A. (1951), A new technique for investigating the stability of slopes and foundations. New Zealand Institution of Engineers. Proceedings, Vol. 37, p. 222-285.

72. Nixon, J. K. (1949), $\Phi = 0$ analysis. Geotechnique, Vol. 1, No. 3, 4, p. 208-209, 274-276.

73. Nonveiller, E. and Suklje, L. (1955), Landslide Zalesina. Geotechnique, Vol. 5, No. 2, p. 143-153.

74. Odenstad, S. (1949), Stresses and strains in the undrained compression test. Geotechnique, Vol. 1, No. 4, p. 242-249.

75. Paton, J. and Semple, N. G. (1960), Investigation of the stability of an earth dam subject to rapid drawdown including details of pore pressure recorded during a controlled drawdown test. Pore pressure and Suction in Soil Conference, London, p. 66-71.

76. Peck, R. B. (1942), Earth pressure measurements in open cuts, Chicago (Ill.) subway. American Society of Civil Engineers. Proceedings, Vol. 68, p. 900-928. American Society of Civil Engineers. Transactions, Vol. 108, 1943, p. 1008-1036.

77. Peck, R. B. and Bryant, F. G. (1953), The bearing capacity failure of the Transcona elevator. Geotechnique, Vol. 3, No. 5, p. 201-208.

78. Penman, A. D. M. (1956), A field piezometer apparatus. Geotechnique, Vol. 6, No. 2, p. 57-65.

79. Peterson, R. (1952), Studies—Bearpaw shale at damsite in Saskatchewan. N. Y. 53 p. (American Society of Civil Engineers. Preprint, 52).

80. Reinius, E. (1948), The stability of the upstream slope of earth dams. Sthm. 107 p. (Swedish State Committee for Building Research. Bulletin, 12).

81. Rendulic, L. (1937), Ein Grundgesetz der Tonmechanik und sein experimenteller Beweis. Bauingenieur, Vol. 18, No. 31/32, p. 459-467.

82. Rosenqvist, I. T. (1953), Considerations on the sensitivity of Norwegian quick-clays. Geotechnique, Vol. 3, No. 5, p. 195-200. (Norwegian Geotechnical Institute. Publ., 2).

83. Sevaldson, R. A. (1956), The slide in Lodalen, October 6th, 1954. Geotechnique, Vol. 6, No. 4, p. 1-16. (Norwegian Geotechnical Institute. Publ., 24).

84. Sheppard, G. A. R. and Aylen, L. B. (1957), The Usk scheme for the water supply of Swansea. Institution of Civil Engineers. Proceedings, Vol. 7, paper 6210, p. 246-274.

85. Simons, N. (1958), Discussion on: General theory of earth pressure. Brussels Conference on Earth Pressure Problems. Proceedings, Vol. 3, p. 50-53. (Norwegian Geotechnical Institute. Publ., 33.)

86. Skempton, A. W. (1942), An investigation of the bearing capacity of a soft clay soil. Institution of Civil Engineers. Journal, Vol. 18, p. 307-321.

87. Skempton, A. W. (1945), A slip in the West Bank of the Eau Brink cut. Institution of Civil Engineers. Journal, Vol. 24, p. 267-287.

88. Skempton, A. W. (1948 a), The $\Phi = 0$ analysis of stability and its theoretical basis. International Conference on Soil Mechanics and Foundation Engineering, 2. Rotterdam. Proceedings, Vol. 1, p. 145-150.

89. Skempton, A. W. (1948 b), A study of the immediate triaxial test on cohesive soils. International Conference on Soil Mechanics and Foundation Engineering, 2. Rotterdam. Proceedings, Vol. 1, p. 192-196.

90. Skempton, A. W. (1948 c), The rate of softening in stiff fissured clays, with special reference to London clay. International Conference on Soil Mechanics and Foundation Engineering, 2. Rotterdam. Proceedings, Vol. 2, p. 50-53.

91. Skempton, A. W. (1948 d), The geotechnical properties of a deep stratum of post-glacial clay at Gosport. International Conference on Soil Mechanics and Foundation Engineering, 2. Rotterdam. Proceedings, Vol. 1, p. 145-150.

92. Skempton, A. W. (1950), Discussion on: Wilson, G. The bearing capacity of screw piles and screwcrete cylinders. Institution of Civil Engineers. Journal, Vol. 34, p. 76.

93. Skempton, A. W. (1951), The bearing capacity of clays. Building Research Congress, London. Papers, division 1, part 3, p. 180-189.

94. Skempton, A. W. (1954), The pore pressure coefficients A and B. Geotechnique, Vol. 4, No. 4, p. 143-147.

95. Skempton, A. W. (1959), Cast in-situ bored piles in London clay. Geotechnique, Vol. 9, No. 4, p. 153-173.

96. Skempton, A. W. and Bishop, A. W. (1954), Soils. Building materials, their elasticity and inelasticity. Ed. by M. Reiner with the assistance of A. G. Ward. Amsterdam, North-Holland Publ. Co. Chapter X, p. 417-482.

97. Skempton, A. W. and Bishop, A. W. (1955), The gain in stability due to pore pressure dissipation in a soft clay foundation. International Congress on Large Dams, 5. Paris. Transactions, Vol. 1, p. 613-638.

98. Skempton, A. W. and DeLory, F. A. (1957), Stability of natural slopes in London clay. International Conference on Soil Mechanics and Foundation Engineering, 4. London. Proceedings, Vol. 2, p. 378-381.

99. Skempton, A. W. and Golder, H. Q. (1948), Practical examples of the $\Phi = 0$ analysis of stability of clays. International Conference on Soil Mechanics and Foundation Engineering, 2. Rotterdam. Proceedings, Vol. 2, p. 63-70.

100. Skempton, A. W. and Henkel, D. J. (1960), Field observations on pore pressures in London clay. Pore Pressure and Suction in Soil Conference, London, p. 48-51.

101. Skempton, A. W. and Ward, W. H. (1952), Investigations concerning a deep cofferdam in the Thames estuary clay at Shellhaven. Geotechnique, Vol. 3, No. 3, p. 119-139.

102. Suklje, L. (1953 a), Discussion on: Stability and deformations of slopes and earth dams, research on pore-pressure measurements, ground-water problems. International Conference on Soil Mechanics and Foundation Engineering, 3. Zürich, Proceedings, Vol. 3, p. 211.

103. Suklje, L. (1953 b), Plaz pri Lupoglavu v ecocenskem flisu. (Landslide in the eocene flysch at Lupoglav.) Gradbeni vestnik, Vol. 5, No. 17/18, p. 133-138.

104. Taylor, D. W. (1944), Cylindrical compression research program on stress-deformation and strength characteristics of soils; 10 progress report. Cambr., Mass. 46 p. Publ. by Massachusetts Institute of Technology. Soil Mechanics Laboratory.

105. Taylor, D. W. (1948), Fundamentals of soil mechanics. N. Y., Wiley. 700 p.

106. Terzaghi, K. (1923), Die Berechnung der Durchlässigkeitsziffer des Tones aus dem Verlauf der hydrodynamischen Spannungserscheinungen. Akademie der Wissenschaften in Wien. Mathematisch-naturwissenschaftliche Klasse. Sitzungsberichte. Abteilung II a, Vol. 132, No. 3/4, p. 125-138.

107. Terzaghi, K. (1925), Erdbaumechanik auf bodenphysikalischer Grundlage. Lpz., Deuticke. 399 p.

108. Terzaghi, K. (1932), Tragfähigkeit der Flachgründungen. International Association for Bridge and Structural Engineering. Congress, 1. Paris. Preliminary publ., p. 659-683, final publ., 1933, p. 596-605.

109. Terzaghi, K. (1936 a), Stability of slopes of natural clay. International Conference on Soil Mechanics and Foundation Engineering. 1. Cambr., Mass. Proceedings, Vol. 1, p. 161-165.

110. Terzaghi, K. (1936 b), The shearing resistance of saturated soils and the angle between the planes of shear. International Conference on Soil Mechanics and Foundation Engineering, 1. Cambr., Mass. Proceedings, Vol. 1, p. 54-56.

111. Terzaghi, K. (1943), Theoretical soil mechanics. N. Y., Wiley. 510 p.

112. Terzaghi, K. (1958), Design and performance of the Sasumua dam. Institution of Civil Engineers. Proceedings, Vol. 11, November, p. 360-363.

113. Terzaghi, K. and Peck, R. B. (1948), Soil mechanics in engineering practice. N. Y., Wiley. 566 p.

114. Toms, A. H. (1948), The present scope and possible future development of soil mechanics in British railway civil engineering construction and maintenance. International Conference on Soil Mechanics and Foundation Engineering, 2. Rotterdam. Proceedings, Vol. 4, p. 226-237.

115. Toms, A. H. (1953 a), Discussion on: Conference on the North Sea floods of January 31st-February 1st, 1953. Institution of Civil Engineers, Publ., p. 103-105.

116. Toms, A. H. (1953 b), Recent research into coastal landslides at Folkstone Warren, Kent, England. International Conference on Soil Mechanics and Foundation Engineering, 3. Zürich. Proceedings, Vol. 2, p. 288-293.

117. Tschebotarioff, G. P. (1951), Soil mechanics, foundations, and earth structures; an introduction to the theory and practice of design and construction. N. Y., McGraw-Hill. 655 p.

118. U. S. Department of the Interior. Bureau of Reclamation (1951), Earth manual; a manual on the use of earth materials for foundation and construction purposes. Tentative ed. Denver, Colo. 332 p.

119. Ward, W. H., Penman, A., and Gibson, R. E. (1954), Stability of a bank on a thin peat layer. European Conference on Stability of Earth Slopes, Stockholm. Proceedings, Vol. 1, p. 122-138, Vol. 3, p. 128-129. Geotechnique, Vol. 5, No. 2, 1955, p. 154-163.

120. Waterways Experiment Station, Vicksb., Miss. (1947), Triaxial shear research and pressure distribution studies on soils. Vicksb., Miss. 332 p.

121. Waterways Experiment Station, Vicksb., Miss. (1950), Potamology investigations. Triaxial tests on sands, Reid Bedford Bend, Mississippi river. Vicksb., Miss. 54 p. (Report, 5-3).

122. Wilson, G. (1950), The bearing capacity of screw piles and screwcrete cylinders. Institution of Civil Engineers. Journal, Vol. 34, p. 4-73.

123. Wood, C. C. (1958), Shear strength and volume change characteristics of compacted soils under conditions of plane strain. Thesis. (University of London). London.

Paper No. 6589

SELSET RESERVOIR: DESIGN AND PERFORMANCE OF THE EMBANKMENT

by

Alan Wilfred Bishop, M.A., D.Sc.(Eng.), A.M.I.C.E.

Reader in Soil Mechanics, Imperial College of Science and Technology

and

Peter Rolfe Vaughan, B.Sc.(Eng.)

Imperial College of Science and Technology, lately of Edward Sandeman, Kennard and Partners

Second of two Papers for discussion at an Ordinary Meeting on Tuesday, 13 March, 1962 at 5.30 p.m., and for subsequent written discussion

---

SYNOPSIS

The 130-ft high embankment at Selset Reservoir has been successfully constructed of clay of low permeability in an area of high rainfall by using closely spaced drainage blankets to reduce the high pore-water pressures set up. Sand drains have been used to accelerate the consolidation of zones of soft clay in the foundation.

Extensive field observations of pore-water pressure, both in the fill and in the foundation, have been made as a control measure during construction. The estimates of pore pressure made at the design stage are reviewed in the light of the results.

The rise in groundwater level downstream of the cut-off on impounding has also been recorded. The effect of the resulting uplift on the stability of the downstream slope and south abutment, and its control by relief wells, are also discussed.

INTRODUCTION

THE geology of the site and the relatively high rainfall at Selset have presented an interesting combination of problems in applied soil mechanics. The field measurements of pore-water pressure, settlement and foundation seepage serve to indicate the value, and limitations, of predictions made at the design stage.

2. The four principal problems have been the design of a large embankment of clay of low permeability in an area where about 20 in. of rain are likely to fall during the construction season; the construction of the embankment on a foundation containing extensive areas of soft clay; the stabilizing of the valley side downstream of the dam where landslides had been occurring before construction; and the control of uplift pressures downstream of a grouted cut-off.

A more detailed description of the engineering works carried out is given in the accompanying Paper by J. Kennard and M. F. Kennard.[1]

3. The initial geological investigation, made under the direction of Mr E. Morton in 1953, revealed the sequence of the foundation strata and the index properties of the clay in the borrow area.

4. The embankment height was to be 130 ft, and because of the stability problems which had arisen at Muirhead (height 70 ft),[2] Knockendon (height 90 ft)[3], and Usk (height 110 ft)[4] in boulder clay fills of similar or more favourable properties, Professor A. W. Skempton and the senior Author were consulted about the embankment design.

5. Samples from the borrow pit area were taken in the summer of 1954 and tested at Imperial College, the results serving as the basis for the design of the embankment section. A detailed investigation of the foundation by means of trial pits could not be undertaken until the commencement of the contract, and took place in the summer of 1955.

6. Improvements in testing technique have occurred during the period of construction, and additional tests on representative batches of fill taken during construction are at present being carried out at Imperial College. The implications of the results are discussed in Parts 2 and 3.

7. Developments in methods of stability analysis also occurred within this period, particularly in the application of the effective stress analysis to the non-circular failure surface, which is potentially the most critical case for an embankment having a soft clay foundation or subdivided by drainage layers. This method has been used to re-check the design and to evaluate the effect of the observed construction pore-water pressures on stability.

PART 1. THE INFLUENCE OF CLIMATIC CONDITIONS ON EMBANKMENT DESIGN

8. In slope stability problems the influence of pore-water pressure on the factor of safety is most conveniently expressed in terms of the ratio of the pore pressure to the weight of soil overlying the potential slip surface.[5, 6] It has been shown by the senior Author[6, 7] that, for a slope in which the ratio of the pore pressure u to the vertical head* of soil γh above the element of soil considered is a constant, the value of the factor of safety F decreases almost linearly with increase in pore-pressure ratio $u/\gamma h$, (Fig. 1). Subsequent work[8] has shown that, for the distributions of pore pressure encountered in relatively homogeneous slopes, the *average* value of the pore pressure ratio $u/\gamma h$ can be used in place of a uniform constant value with little loss in accuracy.

9. The factor of safety may thus be expressed by an equation of the form

$$F = m - n.r_u \quad . \quad . \quad . \quad . \quad . \quad . \quad . \quad . \quad (1)$$

where r_u denotes the ratio $u/\gamma h$.

10. If the cohesion intercept c' is expressed in terms of the dimensionless number $c'/\gamma H$, where H is the height of the slope, then the values of the stability coefficients m and n may be tabulated for typical soil properties and slopes.[8] The influence of pore pressure on the factor of safety in any particular case can then be obtained directly.

---

[1] The references are given on p. 345.

* Here γ denotes unit weight of soil and h the vertical distance of the element from the surface of the slope.

11. For a 4 to 1 slope and values of ϕ' and $c'/\gamma H$ approximating to those of the Selset embankment it can be seen from Fig. 1 that a change in average pore-pressure ratio from 0·33 to 0·56 reduces the factor of safety from 1·5 to 1·0. This change from an acceptable factor of safety to a value corresponding to limiting equilibrium can be brought about, in a boulder clay, by a change in average compaction water content of as little as 1%.

12. The relationship between pore-pressure ratio and the slope required to give a particular value of the factor of safety may also be obtained from the

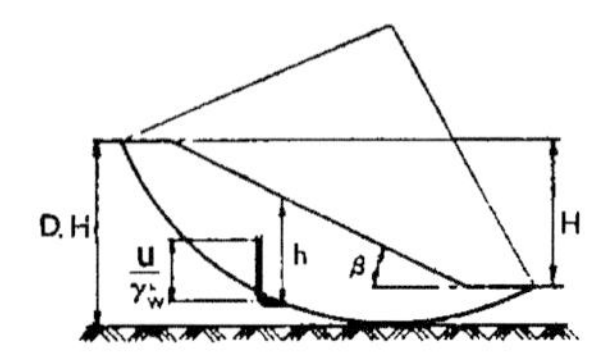

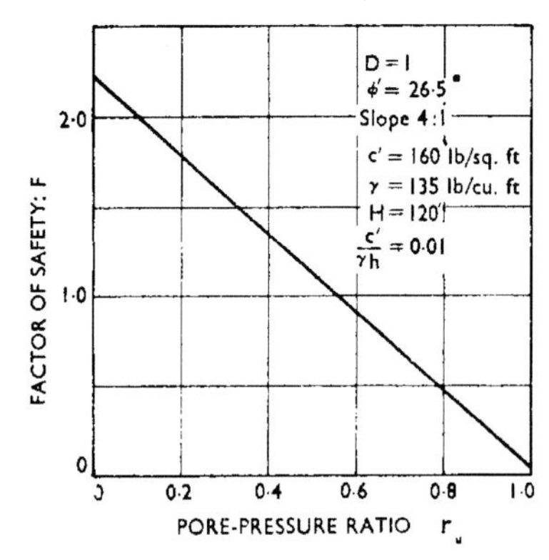

FIG. 1.—RELATION BETWEEN FACTOR OF SAFETY AND PORE-PRESSURE RATIO

stability coefficients (Fig. 2). For the value of $c'/\gamma H = 0{\cdot}01$ and $\phi' = 26{\cdot}5°$ a slope of 4 to 1 is required to give a factor of safety of 1·5 if the pore-pressure ratio is 0·33. But for this same pore-pressure ratio the slope could be steepened to 2·65:1 before the factor of safety dropped to 1·0. Little sign of distress would be apparent until the limiting value of 1·0 was approached. At a slope of 3:1 this soil might stand indefinitely if the pore-pressure ratio and values of c' and ϕ' were applicable to the condition of long-term loading. If its low factor of safety were not known, the slope of 3:1 might therefore be taken, on an empirical basis, as a precedent for economical design.

13. This places the engineer in a dilemma. To require a factor of safety of 1·5 involves the use of about 50% more fill than would be just adequate to ensure stability if all the design assumptions were fulfilled. Is this requirement over-conservative? Or is a factor of safety of even 1·5 too low for a water-retaining

structure whose failure during construction is expensive and during operation might be catastrophic?

14. The answer given in any particular case will be influenced by the engineer's opinion on the thoroughness of the initial investigation of soil properties, on the type of fill being used and on the control likely to be exercised by the site staff during construction. In the Authors' view insufficient consideration has yet been given to the requirements to be called for in new construction and to the extent to which these requirements should be applied in the inspection of existing embankments.

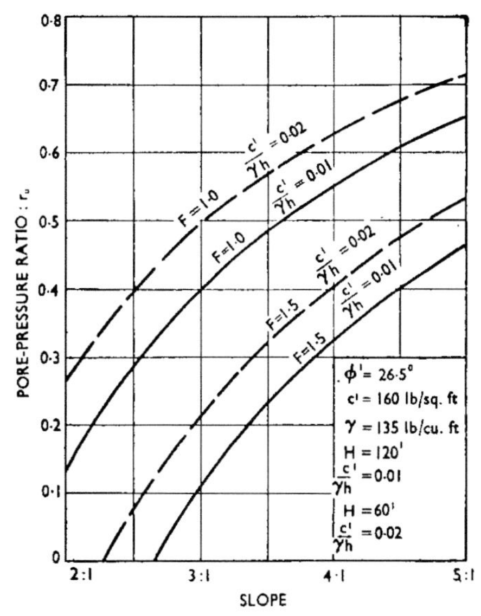

FIG. 2.—RELATION BETWEEN PORE-PRESSURE RATIO AND SLOPE STABILITY

15. The principal uncertainty in the determination of factor of safety lies in the estimate of pore pressure represented by the term r_u. This estimate has to be made for three separate sets of conditions; construction, reservoir full, and drawdown. The initial estimate will have to be based on laboratory tests, theoretical considerations and field observations made on structures previously constructed under similar conditions. The estimate may be checked during construction and operation by field observations of pore-water pressure, and this is now done on most important structures. But if full advantage is to be taken of these observations, some latitude must be allowed both in the design and in the contract for varying the cross-section or the spacing of drainage layers or zones of pervious fill to suit the actual conditions revealed during construction.

16. Where the rainfall during the construction season is high the most critical condition, for clay fill, is represented by the pore-water pressures set up during construction, unless special drainage measures are adopted, as at Selset. The average value of r_u which would have been encountered in the Selset

embankment under conditions of zero drainage is plotted in Fig. 3 against placement water content. Fill placed more than 1% above optimum water content would lead, with no drainage, to r_u values in excess of 0·6, which would require a slope flatter than 5:1 to give a factor of safety of 1·5. Experience on the Usk dam[4] had indicated that under the rainfall conditions prevailing at many upland sites in Great Britain much of the fill might in fact be placed up to 2·5% wetter than the optimum water content. The reduction of excess pore pressure by drainage layers then becomes the only economical solution.

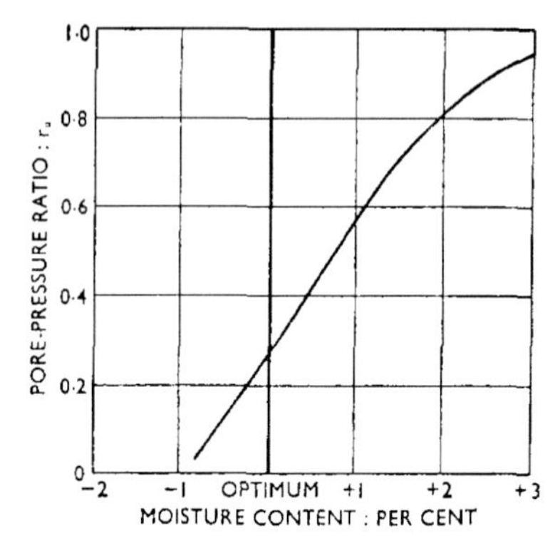

FIG. 3.—TYPICAL VARIATION OF r_u WITH MOISTURE CONTENT FOR SELSET EMBANKMENT

17. The substitution of rockfill or other free-draining fill for the clay might at first sight seem a more economical alternative provided the side slopes could be steepened to balance the greater unit cost of these materials. However, where a clay core is used and the foundation consists of clay, the strength of these two zones largely determines the safe slope and full advantage cannot be taken of the higher strength of the free-draining fill, except possibly for low dams.

PART 2. PROPERTIES OF THE CLAY FILL AT SELSET

18. The initial geological investigation of the south borrow pit area gave the index properties and the particle size distribution of the clay. Bulk samples were taken in 1954 at four levels in the 40 ft cliff face cut by the river Lune at the edge of the borrow pit. The three lower samples were mixed to form a representative batch for a series of triaxial tests with pore-pressure measurement to determine the strength characteristics in terms of effective stress and the rate of dissipation of pore pressure under one-dimensional drainage. This batch of material had the properties given in Table 1A. The triaxial tests were carried out by Hansen.[9]

19. The results of the shear strength tests are given in Table 1B and the results of the dissipation tests are shown plotted in Fig. 10.

20. As a result of the shear tests a cohesion intercept c' of 500 lb/sq. ft and an angle of shearing resistance ϕ' of 24° were used in the design. The density of the fill was taken to be 135 lb/cu. ft.

21. At the time these tests were carried out, relatively coarse saturated porous disks were being used in the triaxial apparatus for the measurement of pore-water pressure. Recent work has shown that with partly saturated soils this can

result in a large over-estimate of the cohesion intercept, the air pressure in the pore space probably being recorded instead of the pore-water pressure. Since the pore air pressure is higher due to surface tension, this leads to an under-estimate of effective stress. The consequent displacement of the failure envelope on the Mohr diagram leads to the overestimate of the cohesion intercept. Fine-grain porous disks having a high air entry value are now being used to avoid this error. A detailed discussion of the testing technique is given elsewhere.[10, 11, 12, 13] The results of more recent tests are given in Table 1B.

22. It will be noted that the samples saturated by a back pressure showed values of $c'=0$ and $\phi'=26{\cdot}5°$. Values obtained from tests on the partly saturated soil using fine porous disks and a low rate of testing gave $c'=160$ lb/sq. ft and $\phi'=26{\cdot}5°$.

23. Most of the triaxial tests were carried out on material from which stones exceeding the $\frac{3}{8}$-in. sieve size were removed, and test specimens 4-in. dia. were used. For the samples in which full saturation had to be ensured, specimens 1·5 in. dia. were used, and material exceeding the $\frac{1}{8}$ in. sieve size was excluded. The effect on the values of c', ϕ', and c_v is probably not very significant, but is on the conservative side. The effect on the optimum density and water content is most marked, and field densities and water contents need to be corrected for stone content in making comparisons.

24. The compressibility of the fill is shown in Fig. 4. The results of tests to examine the permeability and coefficient of consolidation of the boulder clay after passing through the pug mill for use in the puddle core are shown in Fig. 5. It will be noted that the value of the coefficient of permeability k decreases with increasing consolidation, and that its value obtained from direct measurement agrees well with that deduced from the coefficient of consolidation c_v and the compressibility.

PART 3. DESIGN OF EMBANKMENT

25. For the reasons outlined in Part 1 the design was based from the outset on the principle that the danger to stability from the high pore pressures due to wet placement conditions would be eliminated by dividing the fill into layers by horizontal drainage blankets whose spacing could be chosen to meet even the most adverse conditions (Fig. 6). This principle had been applied successfully at the Usk dam[4] on the advice of Professor Skempton and the senior Author.

26. The conditions at Selset were, however, much more severe, as the permeability of the clay was one tenth of that at Usk and the embankment was to be some 20 ft higher.

27. The rate of consolidation of a clay layer sandwiched between two drainage surfaces is given by Terzaghi's theory of consolidation.[14] The minimum dissipation of pore pressure occurs in the middle of the layer and this value, together with the average degree of dissipation of pore pressure, is plotted in Fig. 7 against time factor T, defined by the expression:

$$T = \frac{c_v . t}{d^2} \qquad . \quad . \quad . \quad . \quad . \quad . \quad . \quad . \quad (2)$$

where c_v denotes the coefficient of consolidation
t denotes the time after the application of the load
and $2d$ the distance between the layers.

TABLE 1A.—TYPICAL INDEX PROPERTIES OF SELSET BOULDER CLAY

	Original design assumptions	Range measured during construction	Typical more plastic material	Typical less plastic material
Clay fraction: per cent. (on material less than $\frac{3}{8}$ in.)	18	16–28	23	16
Liquid limit: per cent	30	22–39	35	22
Plastic limit: per cent	16	12–19	17	14
Plastic index: per cent	14	8–20	18	8
Proctor optimum moisture content: per cent (on material less than $\frac{3}{8}$ in.)	10·7	8·5–11·5	11	9
Proctor optimum dry density: lb/cu. ft (on material less than $\frac{3}{8}$ in.)	126	128–121	123	127

Notes: 1. The range of specific gravities measured was 2·65–2·70.
2. The clay from the north borrow pit tended to be less plastic than the clay from the south borrow pit.

TABLE 1B.—UNDRAINED-TRIAXIAL COMPRESSION TESTS ON COMPACTED BOULDER CLAY

Type of test	Moisture content: per cent	Max. particle size included in samples: inch	Size of samples: inches	Measurement or pore pressure	Approx. time to failure: hours	c': lb/sq. ft²	ϕ'
			ORIGINAL TESTS				
Undrained (partly saturated)	8·9	$\frac{3}{8}$	8 × 4 dia.	Coarse stone	5	2,300	26°
Undrained (partly saturated)	10·8	$\frac{3}{8}$	8 × 4 dia.	Coarse stone	5	1,800	24°
Undrained (partly saturated)	12·5	$\frac{3}{8}$	8 × 4 dia.	Coarse stone	5	450	28°
Consolidated undrained (partly saturated)	13·5	$\frac{3}{8}$	8 × 4 dia.	Coarse stone	5	150	28°
			MORE RECENT TESTS				
Consolidated undrained (saturated with back pressure) .	11·6	$\frac{1}{8}$	3 × 1½ dia.	Coarse stone	80	0	26½°
Undrained (partly saturated)	11·6	$\frac{3}{8}$	8 × 4 dia.	Fine stone	120	160	26½°

Notes: 1. All samples were prepared with standard compactive effort.
2. Tests on more plastic material.

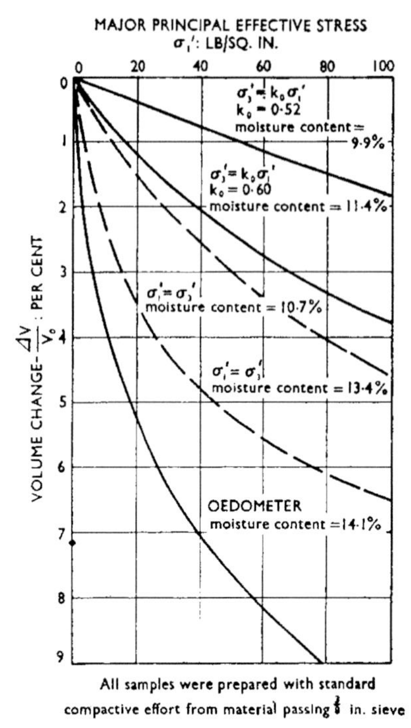

FIG. 4.—COMPRESSIBILITIES OF COMPACTED BOULDER CLAY AS MEASURED ON 4-IN.DIA. SAMPLES IN THE TRIAXIAL APPARATUS AND ON A 9-IN.-DIA SAMPLE IN THE OEDOMETER

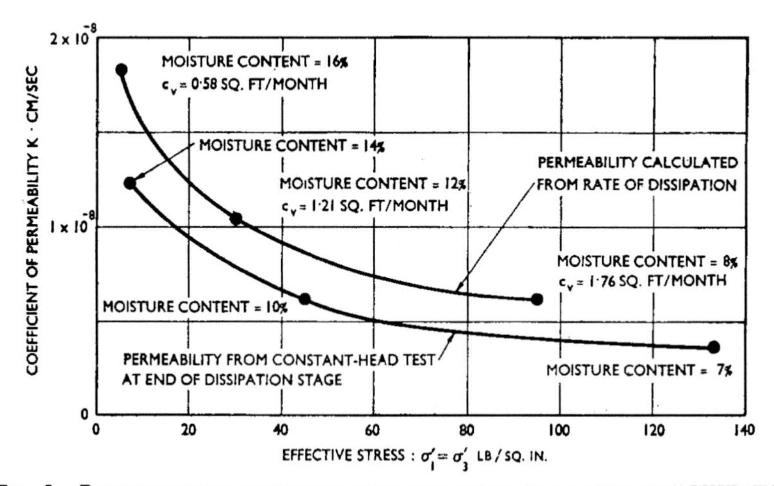

FIG. 5.—PERMEABILITY OF THE PUDDLE CLAY AS MEASURED IN A LABORATORY DISSIPATION TEST ON A 4-IN.-DIA. SAMPLE

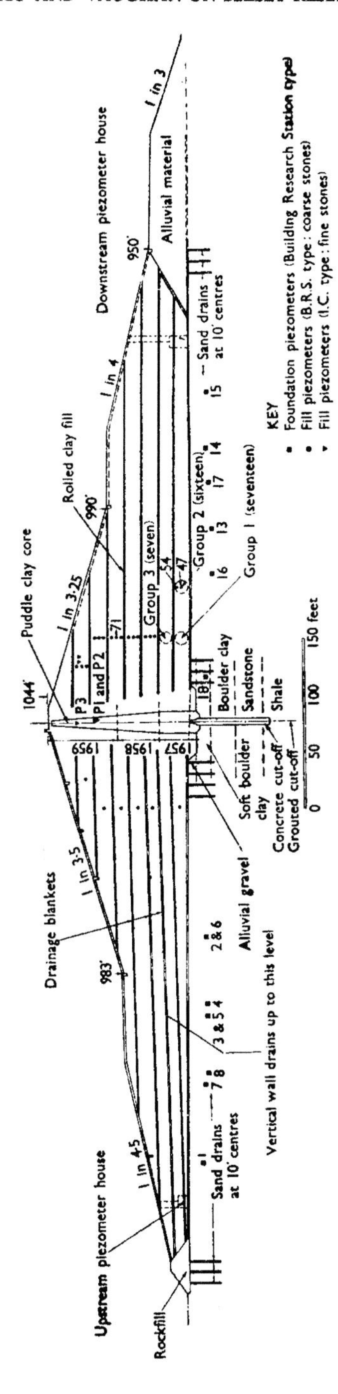

FIG. 6.—TYPICAL EMBANKMENT CROSS-SECTION SHOWING IN PROJECTION THE POSITIONS OF THE PIEZOMETERS

28. Where little consolidation occurs in any one layer of clay before its completion up to the next drainage blanket this solution can be applied directly.

29. Thus from Fig. 7 an average dissipation of 50% of the excess pore pressure could be obtained in a 9-month period by drainage layers spaced 16 ft apart for a value of c_v of 1·5 sq. ft/month.

30. A more detailed analysis based on the proposed construction programme of 3 years for the embankment led to a provisional spacing of 15 ft between the centres of drainage layers 9 in. thick, which raised no major practical or economic difficulties. This may be compared with the spacing of 40 ft used at the Usk dam, where the value of c_v was 11 sq. ft/month.

31. The cross section was therefore worked out on the assumption that the

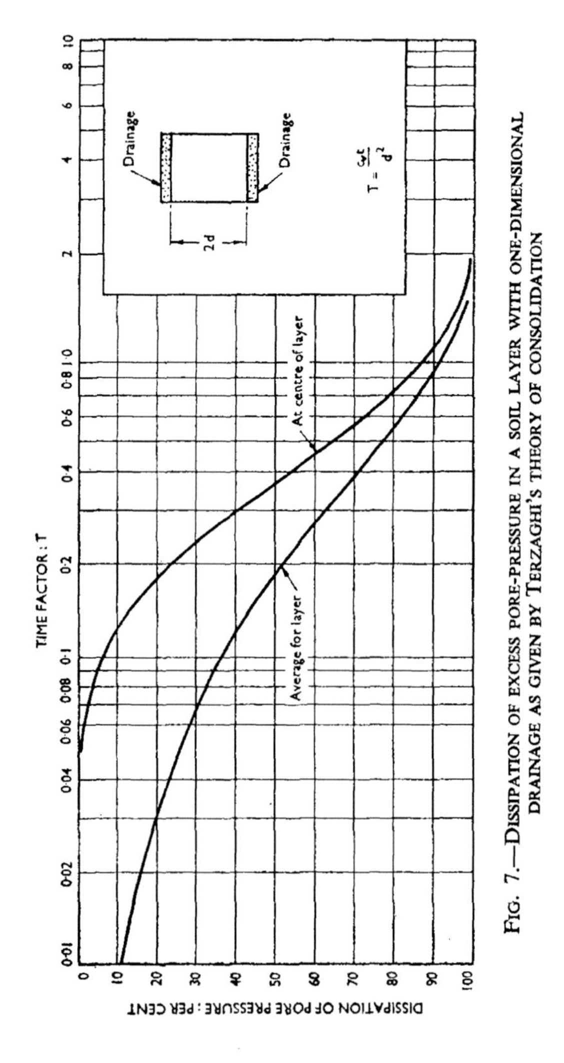

FIG. 7.—DISSIPATION OF EXCESS PORE-PRESSURE IN A SOIL LAYER WITH ONE-DIMENSIONAL DRAINAGE AS GIVEN BY TERZAGHI'S THEORY OF CONSOLIDATION

average value of r_u in the fill would be 0·4 (50% of an initial value of 0·8) in the zones where drainage blankets were used.

32. The decision to use a central puddle core of the same boulder clay had already been made and in this zone and in the adjacent clay fill from which drains would be excluded a value of $r_u = 0{\cdot}8$ was assumed. Up to that time no measurement of excess pore pressure in a puddle core had been made. Earlier theoretical studies made by the senior Author[6] and field measurements made in Sweden[15] had indicated that side shear (or "arching") would reduce the value of vertical stress in the core below that expected in a homogeneous fill. The consequent reduction in pore pressure in the core would be partly counter-balanced by the effect of load transfer on to the undrained and possibly rather wet fill adjacent to it. It was therefore felt necessary to use a high average value for this combined central zone in the preliminary design. Subsequent field measurements (Part 4) have shown this to be over-conservative.

33. At this stage the circular arc stability analysis was used (including the horizontal forces between the slices)[7] with values $c' = 500$ lb/sq. ft and $\phi' = 24°$. Values of $c' = 0$ and $\phi' = 24°$ were used for the puddle clay core. The cross section in Fig. 6 showed a factor of safety between 1·4 and 1·6 for both deep slip circles and circles coming out at the berms. For soils having a significant value of the cohesion intercept c' the term $c'/\gamma H$ increases with decreasing height. It is apparent from Fig. 2 that for a given value of r_u this would permit the slope to be progressively steepened towards the crest. This is taken into account by the use of berms whose levels were chosen to fit into the topography of the valley.

34. The stability of the downstream slope would have its minimum value at the end of construction after which the pore pressures would drop to a relatively low value. Owing to the underdrain and drainage blankets it was considered that impounding would have little long term effect on the pore pressures within the downstream slope and the factor of safety would thus considerably exceed 1·5.

35. The upstream slope had, however, to be stable under drawdown conditions. In a saturated clay of low permeability the pore-pressure on drawdown was estimated to correspond to a stand-pipe level equal to the height of fill above the point considered[16, 17, 18]. This corresponds to an r_u value equal to γ_w/γ_f, where γ_w is the unit weight of water and γ_f is the unit weight of fully saturated fill, and is 0·46 for the Selset material. To reduce the pore-pressure ratio much below this value at the end of construction does not lead to any economy of material in the upstream slope, if a factor of safety of at least 1·3 is required under drawdown conditions.

36. Two details of the section shown in Fig. 6 are of practical importance. The lowest drainage layer beneath the upstream slope has a fall towards the centre of the dam, and provision was therefore made in the design for a sump at the inner end of this layer to be pumped until the first 15 ft of fill had been placed.

37. Material stripped from the valley floor, other than peat or organic material, was placed as a wide berm at the downstream toe, where in spite of its low strength it contributed to the stability of the embankment.

38. An analysis of some of the pore-pressure data from the Usk dam suggested that the rate of drainage in rolled fill might be greater in the horizontal direction than the vertical, and that vertical wall drains might be more effective than the

horizontal blankets. On practical grounds a minimum spacing of 50 ft between vertical drains 18 in. thick was proposed. For the vertical drains at this spacing to be more effective than horizontal blankets in obtaining the required factor of safety an average permeability ratio of at least 6 was necessary.

39. As this could only be proved, for the particular fill and rolling equipment used, by full scale experiment, both horizontal and vertical drains were included in the first season's fill and a detailed investigation made by means of field pore-pressure measurements. The contract was framed to permit a variation in the type and number of drains to suit the field conditions and in fact, as a result of the pore-pressure measurements, the vertical drains were discontinued after the first season.

40. To simplify construction a single layer of filter material was used. The specifications required that its grading should lie within the limits given by the Terzaghi filter rule,[17] and that its permeability should be at least 50,000 ft/year to ensure free flow of water from the layers, which had a maximum width of about 450 ft.

41. The use of drainage blankets accelerates the settlement of the fill and reduces the long term settlement after completion. As the necessary settlement allowance could be estimated more accurately in terms of the observed values of pore pressure remaining to be dissipated at the end of construction, a provisional estimate only was made at the design stage. The final values were given to the contractor in the last season, and all fill placed outside the nominal cross section as a long term settlement allowance was confined to the upper 54 ft of the embankment and paid for as a separate item. The settlement allowance above the puddle core was 2·97 ft at the maximum section.

42. Developments in the application of the effective stress analysis to non-circular slip surfaces (due to Janbu[19] and Kenney[20]) subsequently made it possible to recheck the design using slip paths corresponding more closely to the critical paths in a layered system. Kenney re-calculated the pore-pressure distribution corresponding to the proposed construction programme, assuming a placement water content 3% above the optimum throughout construction and a value of c_v of 1 sq. ft/month. On this basis the most critical non-circular path would pass through fill placed in the second season when the final season's fill was placed (Fig. 8). The low factor of safety of 1·06 obtained was partly due to the wide undrained zone originally planned upstream of the core, and this zone was therefore reduced, as shown by the dotted lines, in the actual construction giving an increase in factor of safety of about 25 per cent. It is interesting to note that the circular arc analysis gave 1·24 in place of the value of 1·06 for this case.

43. As it was not known whether pore pressures as severe as assumed in this analysis would occur under the actual site conditions no change in cross section or drainage spacing was recommended at this stage. The observed pore pressures in fact proved to be lower and no change was necessary.

Part 4:—Performance of embankment

44. Three principal means were used to check the performance of the embankment:

1. The field measurement of pore-water pressure within the fill,
2. The settlement of reference points buried within the fill,

3. The settlement and lateral displacement of reference points at the surface of the embankment.

45. Of these the last, and apparently simplest method, proved the least reliable, owing to the displacement of reference points by heavy plant and to the difficulty of carrying the necessarily very accurate survey over long distances in adverse weather conditions. Details of the instruments used for measuring pore-water pressure have been given elsewhere.[1, 21] Settlement within the fill was measured by four water-level type instruments and by two electrical sounding devices developed jointly with the Building Research Station, which recorded the levels of a series of steel plates buried at intervals in the central core around vertical plastic tubes.

46. The recorded values of excess pore-water pressure, (Fig. 9) provided an immediate check on the values used in the stability analysis and thus on the factor of safety at any stage. A more detailed analysis of the rate of build-up and dissipation of pore pressure was also made to confirm that the individual assumptions on which the estimate of pore pressure was based were realized in practice. This applied in particular to the field value of the coefficient of consolidation c_v, which largely determined the blanket spacing, and on which the only other full scale check had been made at the Usk dam.

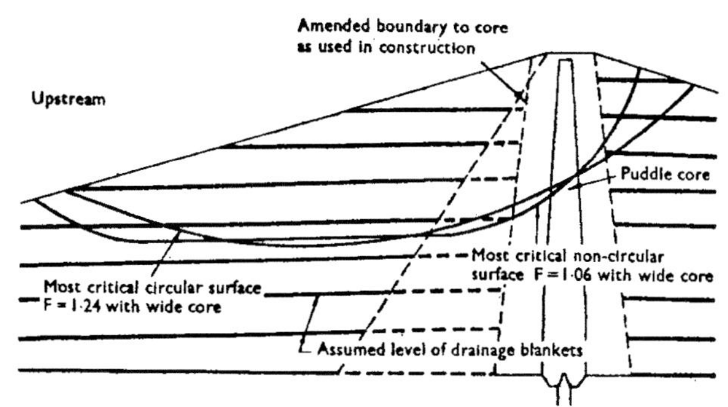

FIG. 8.—CRITICAL SLIP SURFACES IN THE EMBANKMENT FILL AT THE END OF CONSTRUCTION ON THE BASIS OF THE DESIGN ASSUMPTIONS LISTED BELOW

Fill	Puddle core
$c'=500$ lb/sq. ft	$c'=0$
$\phi'=24°$	$\phi'=24°$
$\gamma=135$ lb/ft^3	$\gamma=130$ lb/ft^3

Pore pressures were calculated assuming that the moisture content of the fill was optimum +3% and $c_v=1$ sq. ft /month.

47. The lowest value of factor of safety during construction was found to occur, as predicted, in the final banking season, but as a result of pore pressures set up in the first and wettest layers of the new season's fill. The loading condition was more severe than originally planned, since bad weather had delayed the contractor and 63 ft out of the 130 ft of fill remained to be placed in one season. Because the resulting pore pressures could be observed, permission was given for the specified rate of construction (20 ft in 3 months and 45 ft per year) to be exceeded.

48. A regular check on the factor of safety was made by means of non-circular failure surfaces, and assuming that no other section had higher pore pressures than those observed at point 71 (a piezometer placed at the same level on the upstream side of the core showed much lower pore pressures), then the factor of safety fell at the end of construction to about 1·25 based on what were considered to be the worst possible shear parameters of $c'=0$, $\phi'=24°$. With $c'=160$ lb/sq. ft and $\phi'=26\frac{1}{2}°$ the factor of safety was just in excess of 1·5. Even the latter shear parameters are probably conservative, particularly for the less plastic clay which was used for fill towards the top of the dam. A year after the completion of construction the factor of safety, based on $c'=160$ lb/sq. ft, and $\phi'=26\frac{1}{2}°$, had risen to 1·9.

49. The detailed analysis of the rate of build-up and dissipation of pore pressure to confirm the assumptions made in design is complicated by three factors. Local variations in placement water content have a very marked effect on the pore pressure distribution during the first season after placing. This is illustrated in Fig. 16 of the accompanying Paper.[1] In addition most of the piezometers used in the fill were fitted with coarse-grained porous disks.[21] These would not record the suction initially present in the rolled fill, and were erratic at low pore pressures, where they probably tended to reflect air pressure. At the higher (and therefore more dangerous) pore pressures in the wetter fill little difficulty in obtaining reliable readings occurred. The values of the pore-pressure parameters under low loads are therefore subject to some uncertainty.

50. Thirdly, at low stresses the rise in pore pressure is much more rapid than can be accounted for on the basis of laboratory tests under steadily increasing loads (for example Fig. 9). This is probably caused by additional volume change occurring several feet below the surface of the fill due to cyclical stressing by heavy construction traffic, and has been observed on other dams.[22]

51. The laboratory values of c_v and field values obtained by numerical analysis (Table 2) are compared in Fig. 10. It will be seen that the average field value of c_v for vertical flow is about 1·5 times the average laboratory value, and more than twice the conservative value assumed in the detailed analysis. Considering the limited maximum particle size of the laboratory specimens, and the differences in soil structure produced by field and laboratory compaction equipment, the agreement may be considered satisfactory. It is doubtful whether a less conservative design value of c_v could have reasonably been used.

52. The value of c_v in the most critical layer is 3·5 sq. ft/month. Most of this layer and all of the layers above came from the north borrow pit, not included in the original investigation. This material proved to have a lower clay fraction and plasticity index and this is reflected in the value of c_v.

53. The value of c_v for horizontal flow could not be calculated with equal certainty, but the piezometer readings in the first season indicated no marked increase over the value for vertical flow.

54. The rate of increase of pore-water pressure u with increase in major principal stress σ_1 *under undrained conditions* denoted[23] by $\bar{B}$ is related to the overall pore-pressure ratio r_u by an expression involving the initial pore pressure in the soil. If no drainage occurs, and u_0 is the initial pore-water pressure in the soil after compaction, then

$$u = u_0+\bar{B}.\Delta\sigma_1 \quad . \quad . \quad . \quad . \quad . \quad . \quad . \quad (3)$$

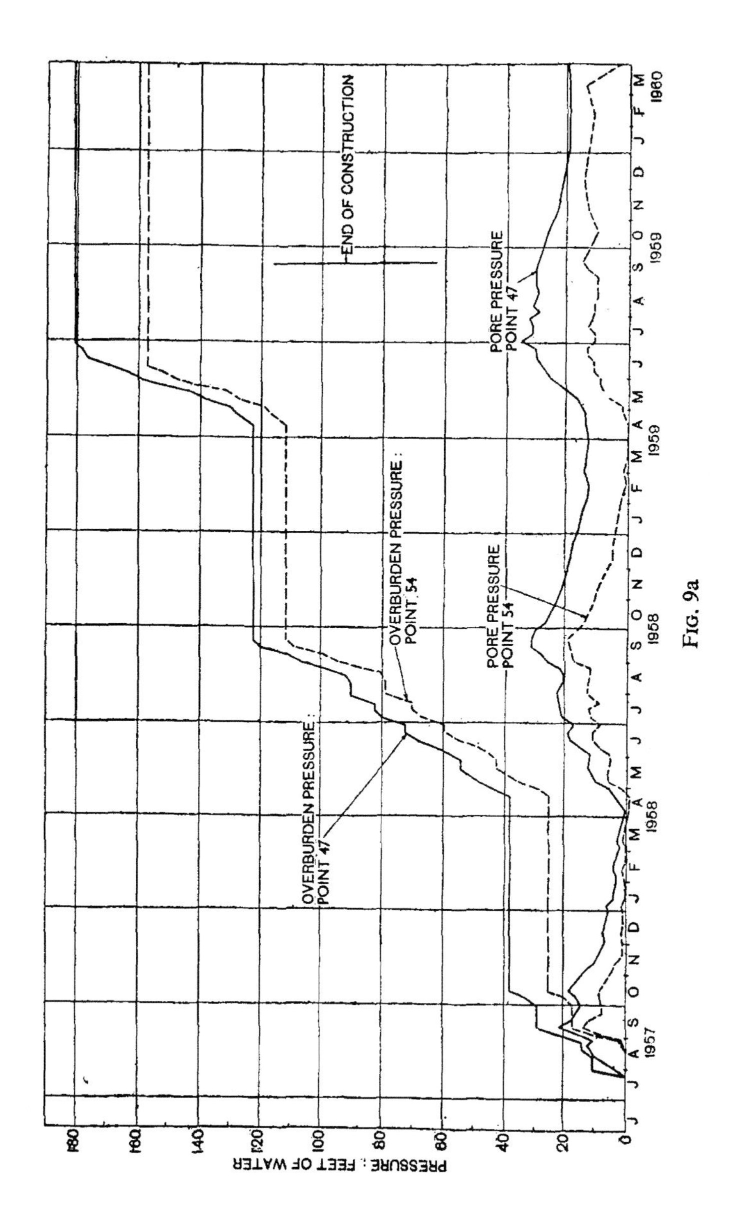

PRESSURE : FEET OF WATER
0
20
40
60
80
100
120
140
160
180
J J A S O N D J F M A M J J A S O N D J F M A M J J A S O N D J F M
1957
1958
1958
1959
1959
1960
OVERBURDEN PRESSURE : POINT 47
OVERBURDEN PRESSURE : POINT 54
PORE PRESSURE POINT 54
PORE PRESSURE POINT 47
END OF CONSTRUCTION

FIG. 9a

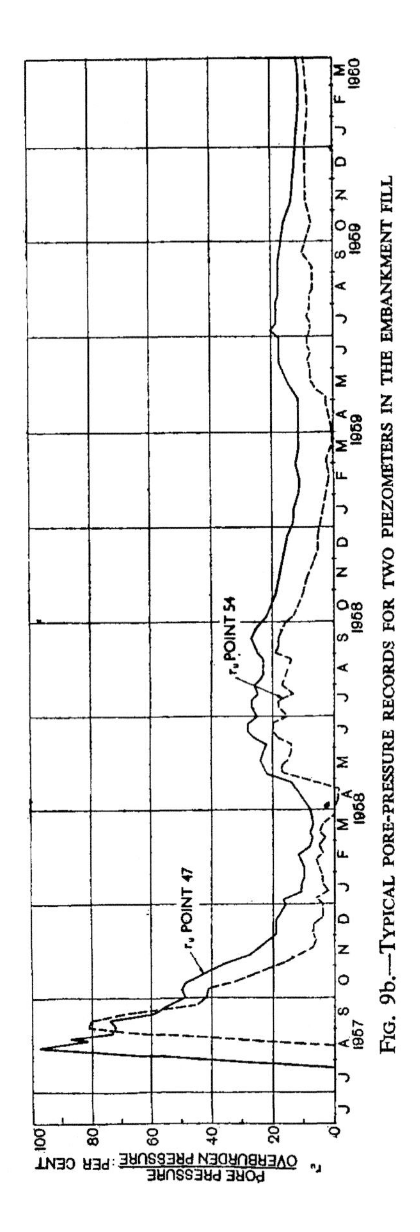

FIG. 9b.—TYPICAL PORE-PRESSURE RECORDS FOR TWO PIEZOMETERS IN THE EMBANKMENT FILL

TABLE 2.—COEFFICIENT OF CONSOLIDATION MEASURED DURING SHUTDOWN SEASONS

Layer	SEASON 1		SEASON 2		SEASON 3	
	Av. effective overburden pressure: lb/sq. in.	c_v: sq. ft/month	Av. effective overburden pressure: lb/sq. in.	c_v: sq. ft/month	Av. effective overburden pressure: lb/sq. in.	c_v: sq. ft/month
1	14	3·2	45	2·7	76	2·7
2			36	1·8	75	1·3
3			21	2·5	56	2·5
4		Piezometer readings unreliable				
5					31	3·5
6					20	4·0
7					12	4·0

In the first season $\Delta\sigma_1$ approximates to γh, the vertical head of soil, and hence:

$$r_u = \frac{u}{\gamma h} = \frac{u_0}{\gamma h} + \bar{B} \qquad (4)$$

Because the coarse porous piezometer disks could not read the value of u_0 and since the *initial* pore-pressure rise due to stressing make the early part of the pore-pressure build-up difficult to analyse, the field results are expressed in terms of r_u for the stress range represented by a complete season's work. For subsequent seasons the initial value of pore pressure is known, and $\bar{B}$ may be evaluated.

55. Values of r_u and of $\bar{B}$ from field and laboratory measurements are compared in Fig. 11, for a range of water contents. The agreement is not unsatisfactory, but it should not be forgotten that in high rainfall areas the problem in design lies not so much in estimating the pore pressure for a given

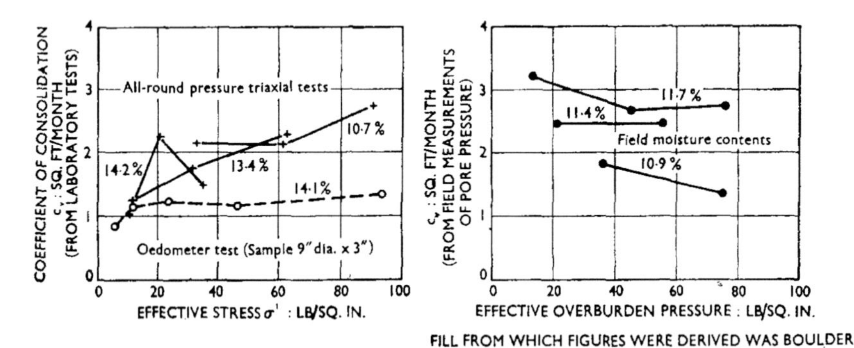

FIG. 10.—COMPACTED BOULDER CLAY: VALUES OF THE COEFFICIENT OF CONSOLIDATION c_v FROM LABORATORY TESTS AND FROM FIELD MEASUREMENTS OF PORE PRESSURE

water content as in predicting the water content range within which the contractor's plant can be made to operate economically under prevailing weather conditions. In this case the requirement in the specifications of 95% of the optimum dry density automatically excluded material more than 3·5% above the optimum water content, Fig. 11. However, site conditions and the nature of the clay prevented rubber-tired plant operating under such wet conditions, and the more extreme values of $\bar{B}$ allowed for in the design were not realized in practice.

56. The value of $\bar{B}$, and hence r_u, for any given water content and stress range is influenced by the magnitude of the shear stress (or, more precisely, by the ratios of both the minor and intermediate principal stresses to the major principal stress). This effect is illustrated by the results of the undrained triaxial compression tests plotted in Fig. 12. The influence of shear stress on the pore pressure set up within the actual fill in the absence of drainage can be predicted on this basis.

57. To enable a comparison to be made with the observed pore pressures a representative value has been calculated from the readings of piezometer 71 (Fig. 13), where a factor of safety of about 1·5 was estimated at the end of

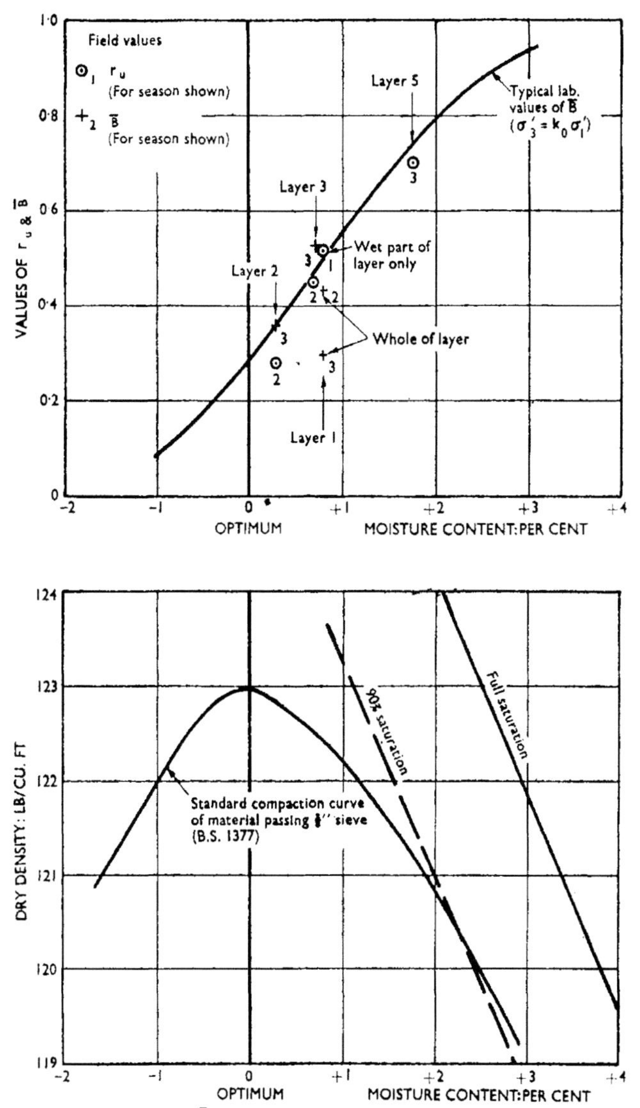

FIG. 11.—VALUES OF r_u AND $\bar{B}$: LABORATORY AND FIELD MEASUREMENTS OF PORE PRESSURE

construction. This point is plotted in Fig. 12 and is in reasonable agreement with the laboratory results.

58. The compressibility of the fill in situ can be determined from the relative settlements of the water-level indicators placed as shown in Fig. 14. The settlement measurements are also shown in Fig. 14 and compression curves derived from the first three instruments, which are vertically above one another, are given in Fig. 15. The effective stress is given by the weight of fill and the observed pore pressure. Since relative movements could be recorded only after the fill reached the level of the upper instrument in each case, the initial part of the compression curve could not be obtained. The rather erratic shape

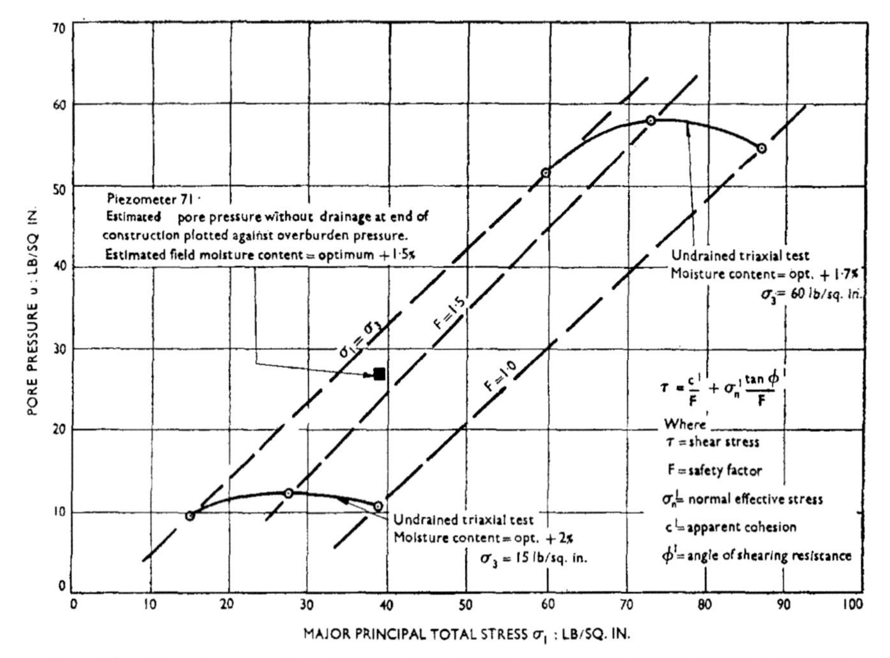

FIG. 12.—EFFECT OF SHEAR STRESS ON PORE PRESSURE IN THE TRIAXIAL TEST AND IN THE FIELD

of the early part of each plot in Fig. 15 is probably due to redistribution of water content within the layers of fill. The values are again of the same order as the laboratory values and have been used in making the final estimate of the settlement allowance.

59. The relative settlements within the puddle core were recorded in the first banking season and the following winter. The values are given in Fig. 16 and, though the limited number of observations has necessitated the use of linear extrapolations at the beginning and end of the shut-down period, they show two features of particular interest. By far the larger part of the settlement occurs, not in the shut-down period when it could be attributed to consolidation,

but during the period of rapid loading. Thus it must be attributed to lateral yield, since the puddle clay is saturated and virtually incompressible. This lateral yield* is consistent with the build-up of shear stresses within the foundation and bank fill, and owing to the ratio of height to width of the puddle core can account for large vertical displacements.

60. The large vertical displacement of the puddle core relative to the adjacent rolled fill is apparent from Fig. 16. This leads to vertical shear forces on the sides of the puddle core. These greatly reduce both the average vertical stress

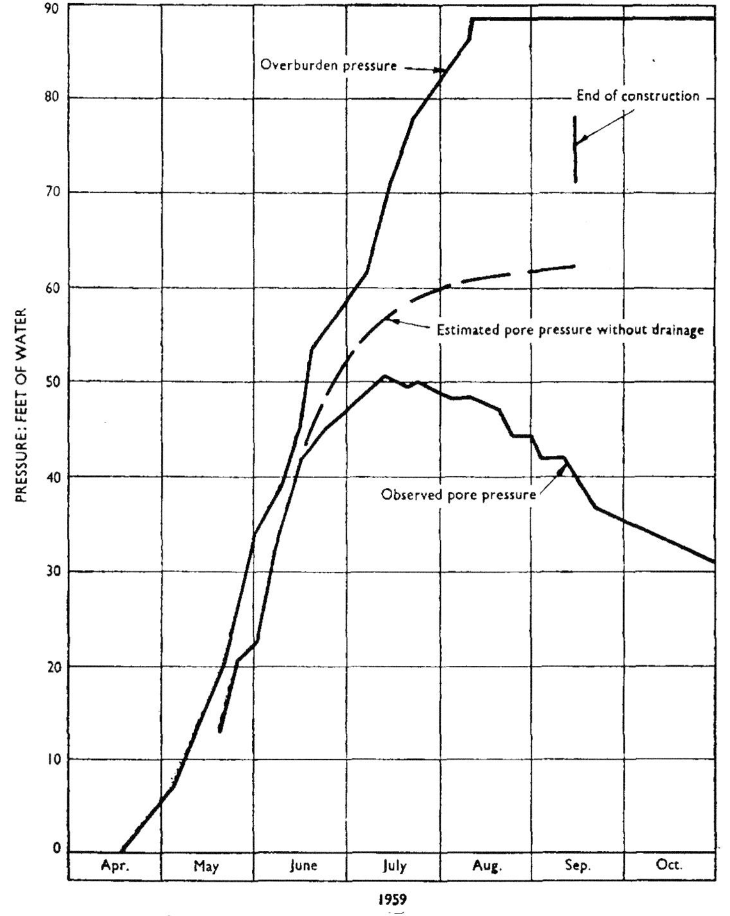

FIG. 13.—PORE-PRESSURE RECORD FOR PIEZOMETER NO. 71

* Satisfactory records of lateral yield were not obtained at Selset, but it is relevant to note that the toe of the 43 ft embankment at Chew Stoke[24] founded on a soft clay layer moved outwards by 2½ in. during construction although the factor of safety based on field measurements of pore pressure was over 1·5.

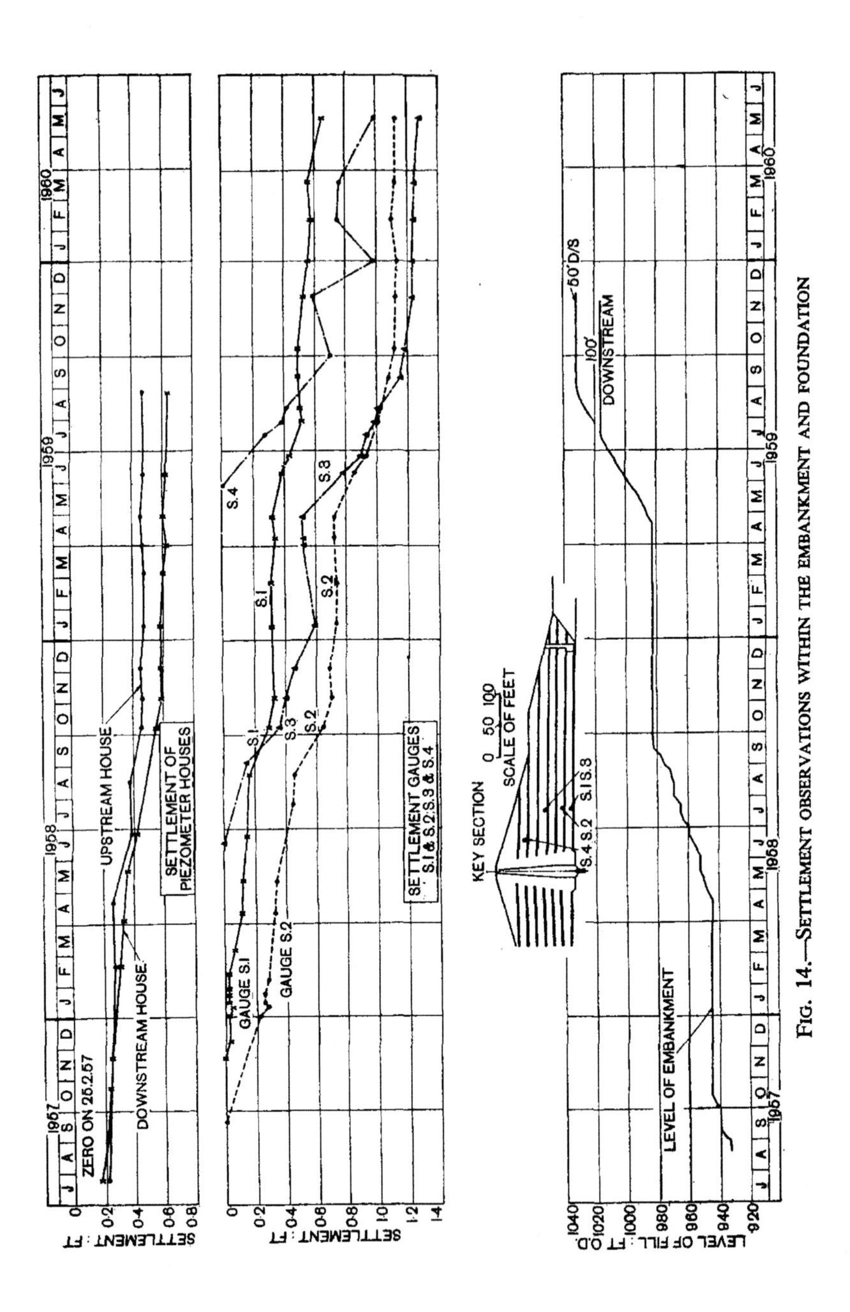

FIG. 14.—SETTLEMENT OBSERVATIONS WITHIN THE EMBANKMENT AND FOUNDATION

in the core (by silo action) and its lateral thrust on the adjacent shoulders of the embankment. An analysis of the stresses corresponding to plastic equilibrium of the core in this condition has been made[6] and leads to the expression for the major principal stress σ_1 at a depth y in the core and at a distance x from the centre line:

$$\sigma_1 = y\left(\gamma - \frac{c_u}{a}\right) + c_u\left(\frac{z}{a} - \frac{\pi}{2}\right) + c_u\left(1 + \sqrt{1 - \frac{x^2}{a^2}}\right) \quad . \quad . \quad . \quad (5)$$

where c_u is the undrained shear strength of the puddle clay
$2a$ is the average width of the core above the point considered,
γ is the bulk density of the core material
z is the depth of possible tension cracks at the upper boundary of the core.

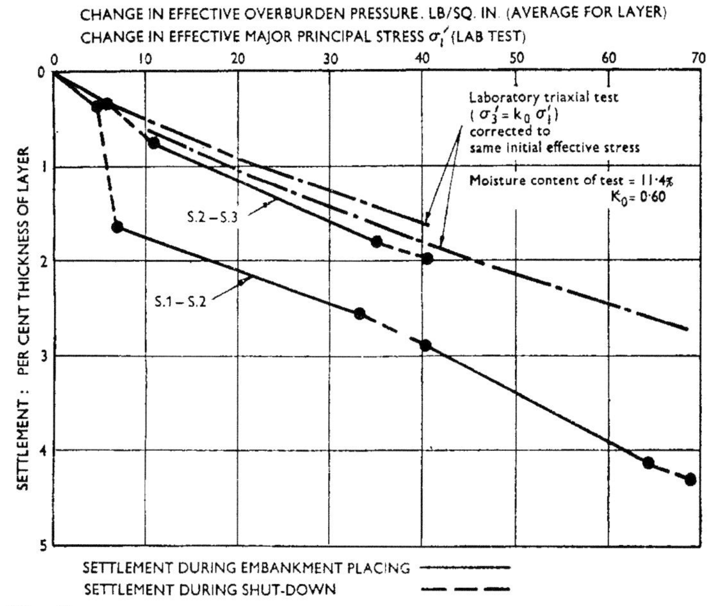

FIG. 15.—COMPRESSIBILITIES OF COMPACTED BOULDER CLAY AS MEASURED IN THE FIELD COMPARED WITH LABORATORY VALUES

Av. initial moisture content S1 – S2 = 10·8%
Av. initial moisture content S2 – S3 = 11·2%
Av. initial effective stress S1 – S2 = 10 lb/sq. in.
Av. initial effective stress S2 – S3 = 20 lb/sq. in.

61. The relief of vertical stress by side shear accounts for the much lower rate of build-up of pore pressure in the core than would be expected in almost fully saturated soft clay. This is illustrated in Fig. 17 for three piezometers located in the puddle core. Also shown are estimates of σ_1 based on eq. (5), assuming that two-thirds of the undrained strength (taken as 330 lb/sq. ft on the basis of 17 tests) is mobilized at the deformation reached in the core. The corresponding values for the full mobilization of shear strength are also given.

The agreement between the shape of the curves supports the assumption of stress relief, and also helps to explain why the long term settlement of narrow plastic clay cores is not as large as theory would otherwise suggest. Fig. 18 shows the maximum crest settlement to date, measured on a reference point set just below the surface of the puddle core.

PART 5. PROPERTIES OF THE FOUNDATION STRATA

62. Experience elsewhere[25] had indicated that a softened layer might be expected at the boundary where the stiff clay was overlain by the alluvial gravels of the River Lune. The trial pits sunk in 1955 revealed a much greater depth of softened boulder clay than had been anticipated. Undrained shear strengths as low as 2 lb/sq. in. were encountered 12 ft below the top of the clay and 20 ft below ground surface.

63. Tests on samples from boreholes supported the view that the soft zone terminated in hard boulder clay containing no further soft layers. Piezometers sealed into four of the boreholes which penetrated into the underlying rock showed artesian pressures with a maximum head of 12 ft above ground surface. These pressures could not readily account for the softened upper zone, however, and it is probable that this zone is associated with the landslides which occurred during the formation of the valley.

64. Undisturbed samples were successfully taken from the trial pits by jacking a 4-in.-dia. sampler horizontally into one side of the pit with a hydraulic jack supported by a strut from the opposite side. These samples were used to obtain the undrained strength, the shear parameters in terms of effective stress and the consolidation characteristics. The undrained strength could be more readily obtained by using the same jack to force a 4-in. or 6-in.-dia. plate into the side of the pit, and observing the load and penetration. From the correlation between load and penetration given by Skempton[26] the undrained strength was obtained. Strengths obtained in this way agreed well with those obtained from undrained triaxial tests and this method enabled a rapid and fairly detailed investigation to be made of each pit.

65. The properties of the foundation strata are summarized in Table 3 and Fig. 19. The only significant differences between the soft zones and adjacent hard clay lie in the moisture content and undrained strength, the index properties, coefficient of consolidation and other shear strength parameters being virtually the same.

66. It will also be seen from Fig. 19 that the coefficient of consolidation increases with increasing effective stress. The narrow range in which all the test values lay enabled a design dependent for its stability on consolidation during the construction period to be accepted with confidence.

PART 6. DESIGN OF SAND DRAINS TO ACCELERATE CONSOLIDATION OF SOFT CLAY FOUNDATION

67. The average shear strength in the clay foundation layer required to give a factor of safety of 1·5 against sliding under the action of the lateral thrust of the clay core was calculated to be 23·5 lb/sq. in., using the equation given in reference 6. In making this calculation the undrained strength of the puddle clay was taken as 200 lb/sq. ft on the basis of tests on other sites.[27] As the

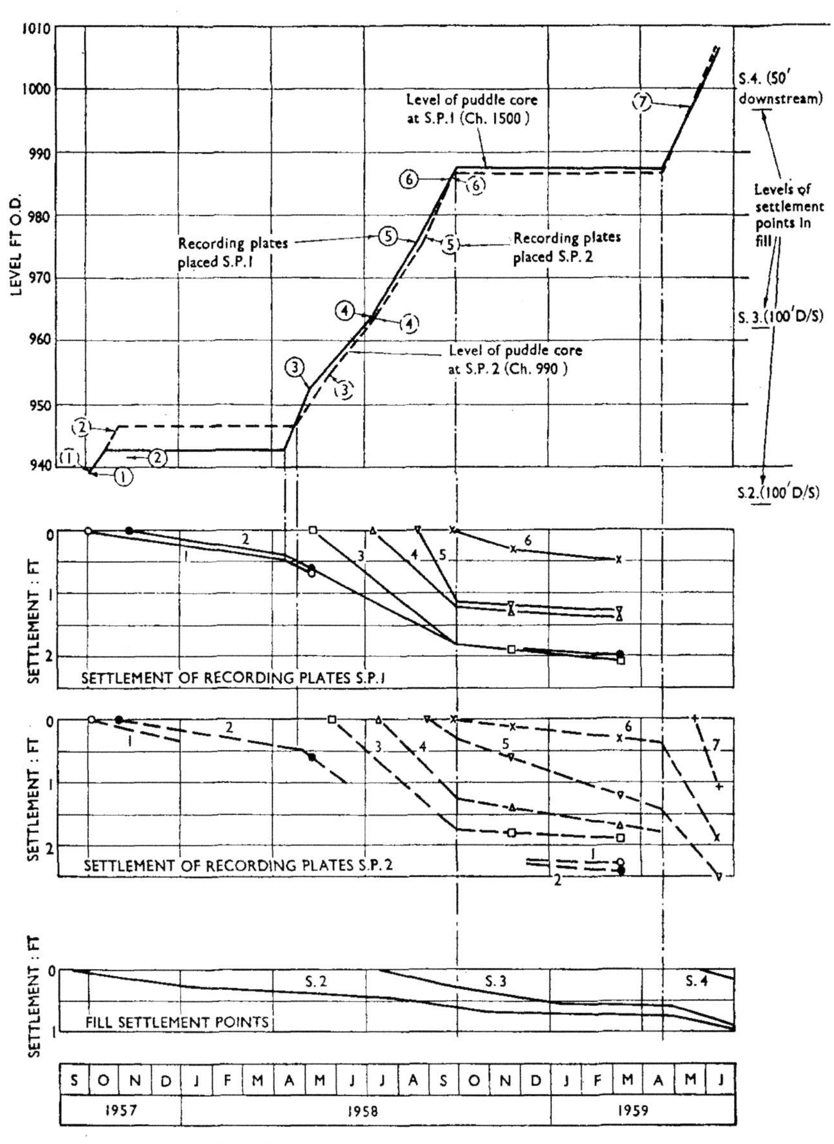

FIG. 16a.—SETTLEMENT OBSERVATIONS IN THE PUDDLE CORE

strength revealed by the trial pits, even averaged over considerable areas, was less than half the required value of 23·5 lb/sq. in., the clay either had to be removed or its strength increased by consolidation.

68. Owing to the length of the drainage paths, the distance between alluvial gravel and the relatively pervious rock being 30 to 40 ft, little consolidation

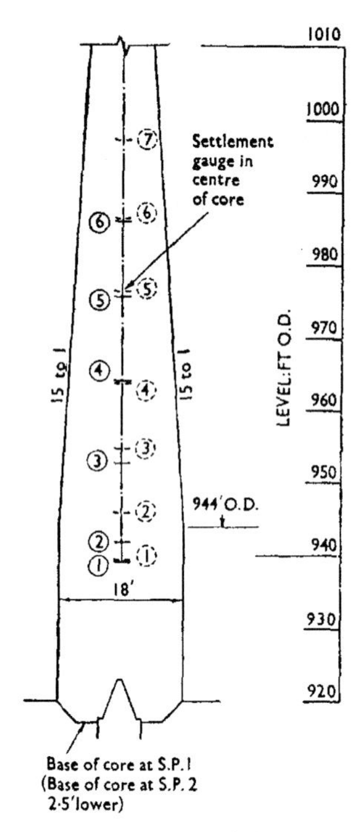

FIG. 16b.—POSITIONS OF GAUGES IN THE PUDDLE CORE

would occur in the centre of the layer even in the 3-year period of banking unless sand drains were used to accelerate the process.

69. The presence of boulders in the alluvial gravel and in the clay itself meant that a sand drain might sometimes fail to penetrate the soft zone to its full depth. In addition some small areas of soft clay might lie outside the area indicated by the site investigation. The spacing of the sand drains was therefore chosen to give a factor of safety of at least 1·5 with 75% success in a given area in locating and penetrating the soft clay; and a factor of safety in excess of 1·0 with 50% success.

70. The average values of coefficient of consolidation would have permitted a drain spacing of 12 ft on a square grid, but as it was possible that several of the lower values would occur in the same area a spacing of 10 ft was adopted, based on a value of $c_v = 0{\cdot}7$ sq. ft/month.

71. The stability analysis for the case of partial consolidation of the clay foundation was carried out in terms of effective stress using the estimated pore pressure and a series of non-circular surfaces. A typical example is given in Fig. 20. The increase in shear strength was also checked against the lateral thrust of the clay core.

72. It is interesting to note that, had the embankment slopes not been relatively flat due to the high pore pressures expected in the fill, it would have

TABLE 3.—CONSOLIDATED-UNDRAINED TRIAXIAL COMPRESSION TESTS ON FOUNDATION SAMPLES; SIZE OF SAMPLE 3 IN. × 1·5 IN. DIA. EACH TEST WAS MADE ON THREE SAMPLES FROM THE SAME 4-IN.-DIA. TUBE

Depth & type of sample	Natural moisture content: per cent	Apparent cohesion c': lb/sq. ft	Angle of friction: ϕ' degrees	Liquid limit: per cent	Plastic index: per cent
12 (U)	15·3	0	27	34	21
14·3 (R)	15·7	0	29½	34	20
15·5 (U)	14·8	0	26½	34	20
10·0 (R)	16·4	100	29	32	18
14·0 (R)	18·5	110	29	29	15
19·0 (R)	18·0	140	26	34	19
7·5 (U)	16·1	310	26½	30	17
5·8 (R)	14·7	850	24½	32	20
12·7 (U)	19·7	90	27	34	19

U. Undisturbed tube samples obtained by jacking horizontally from trial pits.
R. Samples remoulded at natural moisture content.

been much more difficult to achieve an economical solution to the foundation problem. The two aspects of the problem cannot be considered in isolation.

73. The size of the sand drains was chosen on practical grounds, it being estimated that an 18 in.-dia. mandrel would thrust aside or carry ahead of it most sizes of boulder likely to be encountered within the soft clay. To determine the set at which it could be safely assumed that stiff clay had been penetrated, a series of loading tests were run on the cast-iron shoe at the bottom of the mandrel after it had been driven to various sets. The undrained strength was estimated using a bearing capacity factor N_c of 9·6. The test results are given in Fig. 21.

74. On the basis of these results a set of 30 blows/ft for a distance of 2 ft was specified to avoid a boulder being identified as stiff clay. Alternative requirements to identify the boundaries of the soft clay areas are summarized in the accompanying Paper,[1] together with the specification for the sand used to fill the drains.

PART 7. PERFORMANCE OF THE SAND DRAINS IN THE CLAY FOUNDATIONS

75. Two means of checking the performance of the sand drains were used:—

(*a*) The field measurement of pore-water pressure at the centre of representative groups of four sand drains both upstream and downstream of the cut-off.

(*b*) The settlement of a reference point buried at the base of the fill and of the two buried instrument houses which rest on the drainage blanket immediately above the foundation.

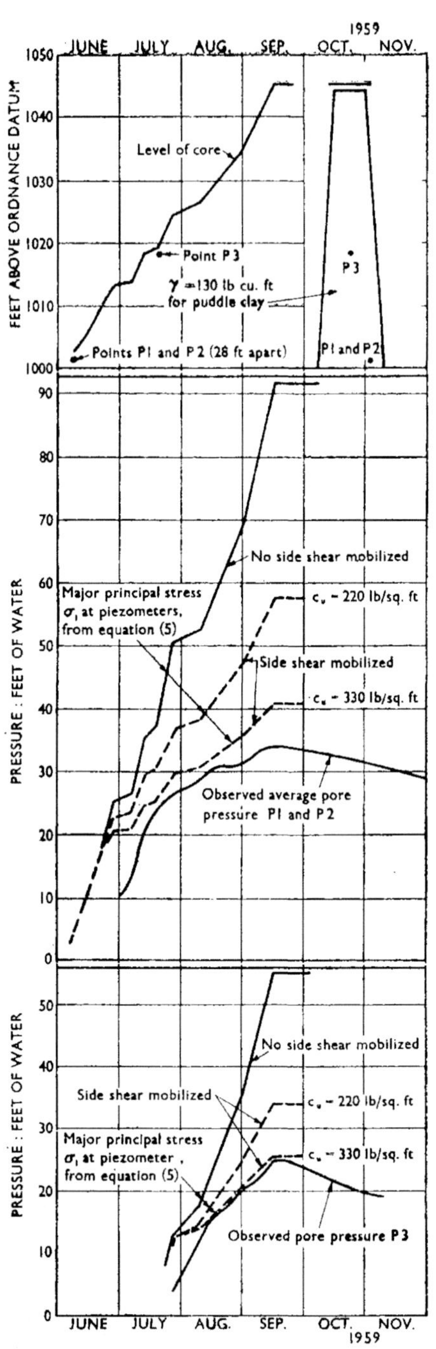

FIG. 17.—PORE-PRESSURE OBSERVATIONS IN THE PUDDLE CORE

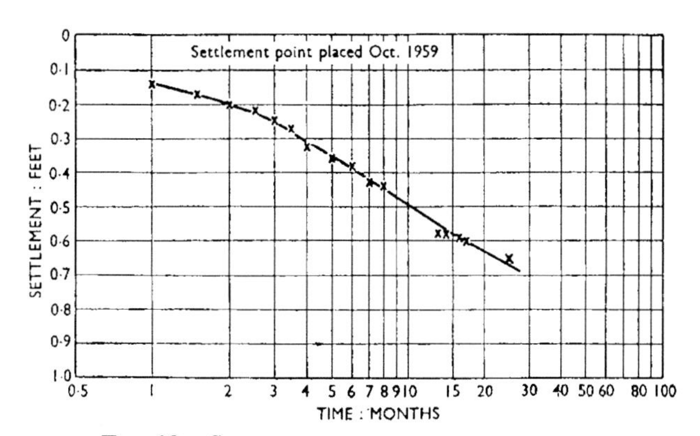

FIG. 18.—SETTLEMENT OF THE EMBANKMENT CREST

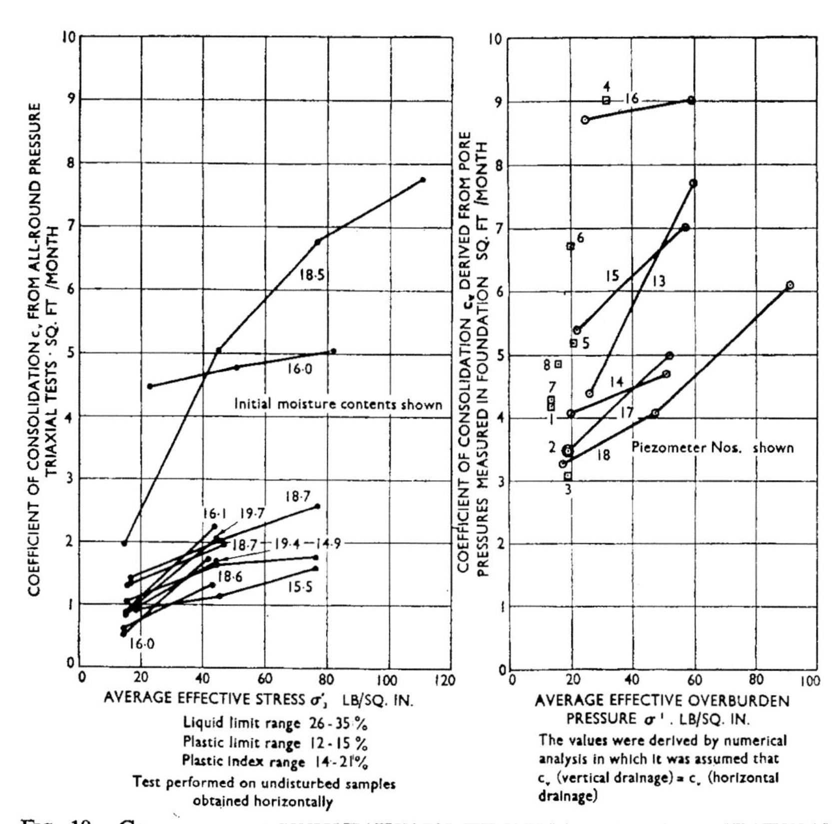

FIG. 19.—COEFFICIENTS OF CONSOLIDATION FOR THE SOFT BOULDER CLAY FOUNDATION AS MEASURED IN THE LABORATORY AND AS DEDUCED FROM FIELD MEASUREMENTS OF PORE PRESSURE

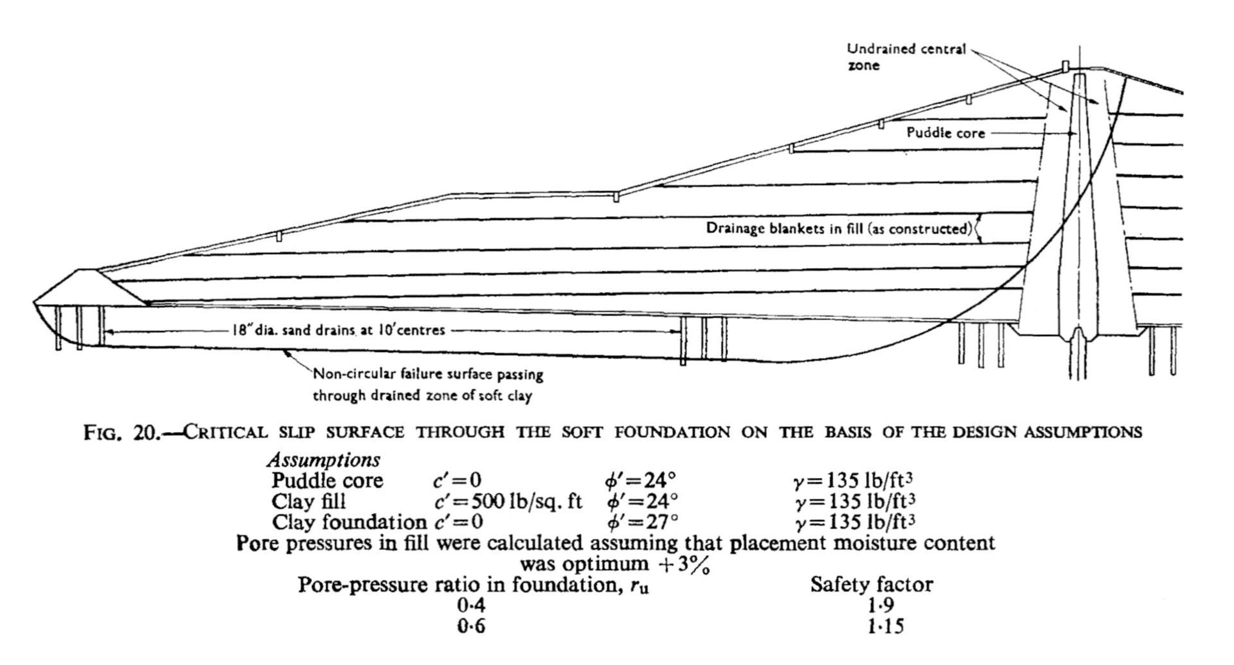

FIG. 20.—CRITICAL SLIP SURFACE THROUGH THE SOFT FOUNDATION ON THE BASIS OF THE DESIGN ASSUMPTIONS

Assumptions

Puddle core	$c'=0$	$\phi'=24°$	$\gamma=135$ lb/ft³
Clay fill	$c'=500$ lb/sq. ft	$\phi'=24°$	$\gamma=135$ lb/ft³
Clay foundation	$c'=0$	$\phi'=27°$	$\gamma=135$ lb/ft³

Pore pressures in fill were calculated assuming that placement moisture content was optimum $+3\%$

Pore-pressure ratio in foundation, r_u	Safety factor
0·4	1·9
0·6	1·15

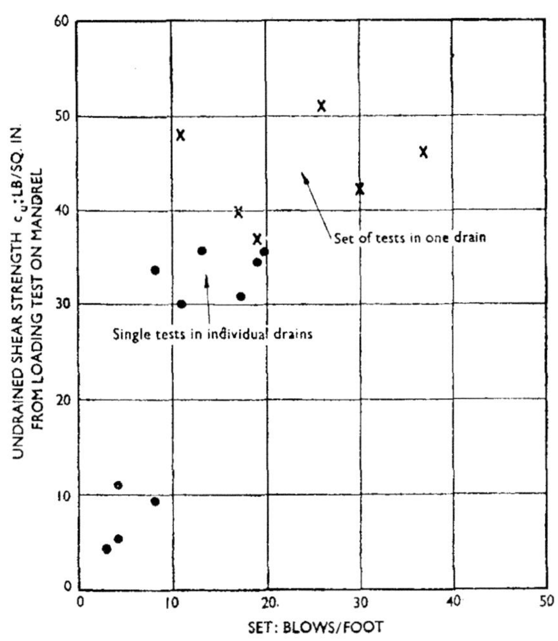

FIG. 21.—SHEAR STRENGTHS OF THE SOFT BOULDER CLAY FOUNDATION AS DERIVED FROM LOADING TESTS ON THE SHOE BENEATH THE MANDREL USED IN FORMING THE SAND DRAINS, PLOTTED AGAINST SET OF THE SHOE

76. A typical foundation piezometer record is given in Fig. 22. A continuous plot of the ratio of excess pore pressure to overburden pressure was kept for the 14 piezometers in the foundation and these gave an immediate guide to the stability of the embankment.

77. Based on the pore pressures measured in the foundation and in the fill, and with the shear parameters measured during the foundation invesitgation, the factor of safety at the end of construction was estimated to be 2·4 on the most critical non-circular surface passing through the sand drains.

78. This rather high factor of safety reflects the conservative estimates of both the coefficient of consolidation, c_v, and the percentage success in driving the sand drains made in the design calculations. The values of coefficient of consolidation deduced from a numerical analysis of the field results are shown in Table 4 and in Fig. 19. The field values are consistently higher than the laboratory values by a factor of about 3.

79. In comparing these values three points should be noted. The samples tested in the laboratory were taken from tubes driven horizontally into the sides of the trial pits so that the predominant direction of flow to the sand drain corresponds to the direction of drainage in the laboratory tests. However, owing to the difficulty of sampling boulder clay the less stony areas are more likely to be sampled, leading to values of c_v lower than the average.

80. The value of c_v depends on the ratio of permeability to compressibility, and in laboratory tests larger volume changes occur under all-round pressures

TABLE 4.—FOUNDATION PIEZOMETERS

Values of coefficient of consolidation c_v derived from numerical analysis of shutdown seasons.

Point No.	Location, distance from centre-line: feet	Average number of blows per foot*	Depth of alluvial gravel: feet	Depth of piezometer: feet	Average depth of adjacent drains: feet	Season 1		Season 2		Season 3	
						σ': lb/sq. in.†	cv: sq. ft/month	σ': lb/sq. in.	cv: sq. ft/month	σ': lb/sq. in.	cv: sq. ft/month
1	390 U/S	5	4	12	23	14	4·2				
2	190 U/S	6	9	15	27	19	3·5	After the first season a variable water level in the upstream bottom drainage blanket makes the analysis of points 1 to 8 inaccurate			
3	260 U/S	4	11·5	16	25	19	3·1				
4	250 U/S	12	4·5	15·5	22	32	9·0				
5	260 U/S	5	5·0	16·5	25	21	5·2				
6	190 U/S	5	4·5	14	25	20	6·7				
7	320 U/S	7	4·5	14	26	14	4·3				
8	310 U/S	8	7	16·5	22	16	4·9				
13	170 D/S	5	13·5	18·5	29	26	4·4	60	7·7		
14	240 D/S	6	4	13·0	24	20	4·1	51	4·7		
15	290 D/S	4	3·5	13·5	23	22	5·4	57	7·0		
16	130 D/S	14	11	17·5	26	25	8·7	59	9·0		
17	210 D/S	5	4	14·0	26	19	3·5	52	5·0		
18	40 D/S	10	0	12·5	20	17	3·3	47	4·1	91	6·1

* No. of blows to drive mandrel in adjacent sand drains at level of piezometer.
† Average effective overburden pressure.

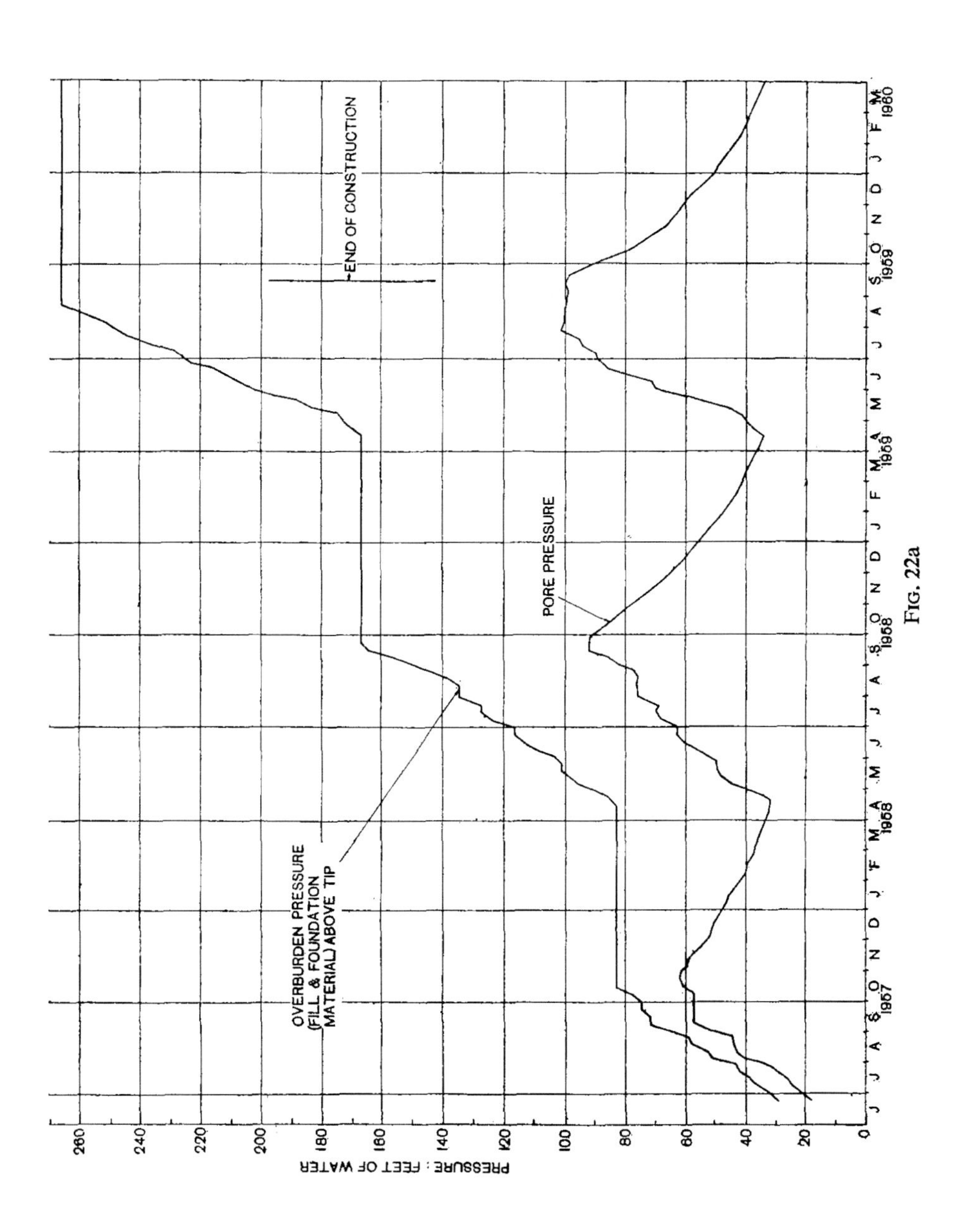
PRESSURE : FEET OF WATER
260
240
220
200
180
160
140
120
100
80
60
40
20
0
J J A S O N D J F M A M J J A S O N D J F M A M J J A S O N D J F M
1957
1958
1959
1960
END OF CONSTRUCTION
PORE PRESSURE
OVERBURDEN PRESSURE (FILL & FOUNDATION MATERIAL) ABOVE TIP

FIG. 22a

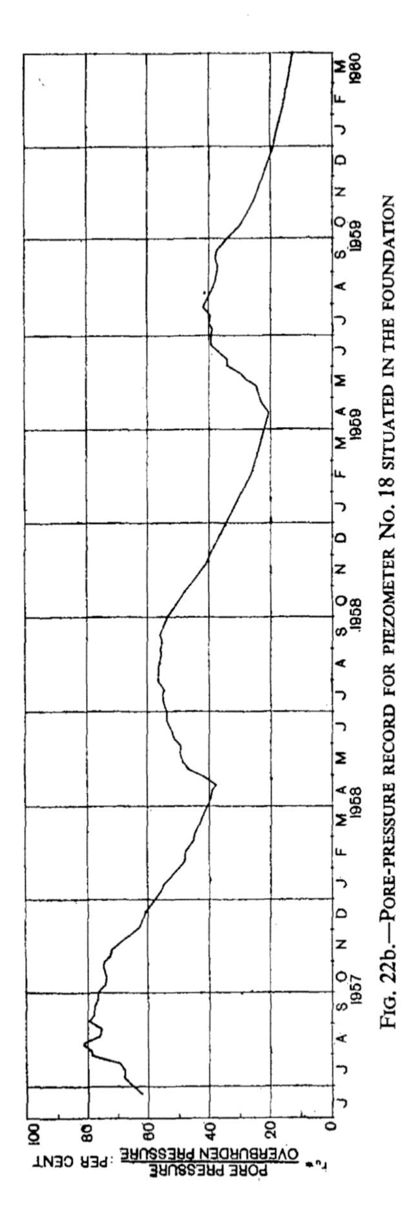

FIG. 22b.—PORE-PRESSURE RECORD FOR PIEZOMETER NO. 18 SITUATED IN THE FOUNDATION

than with no lateral strain (a ratio of about 3:2 is indicated by the data in Fig. 4). Thus c_v in the field might be underestimated by all-round pressure tests by a similar ratio. Only the upstream instrument house was located over clay of similar moisture content to the laboratory samples, and the compressibility derived from the settlement of this house is compared in Fig. 23 with the laboratory compressibilities measured under all-round pressure. Their agreement suggests that differences in compressibility only account for a small part of the discrepancy between field and laboratory values of c_v.

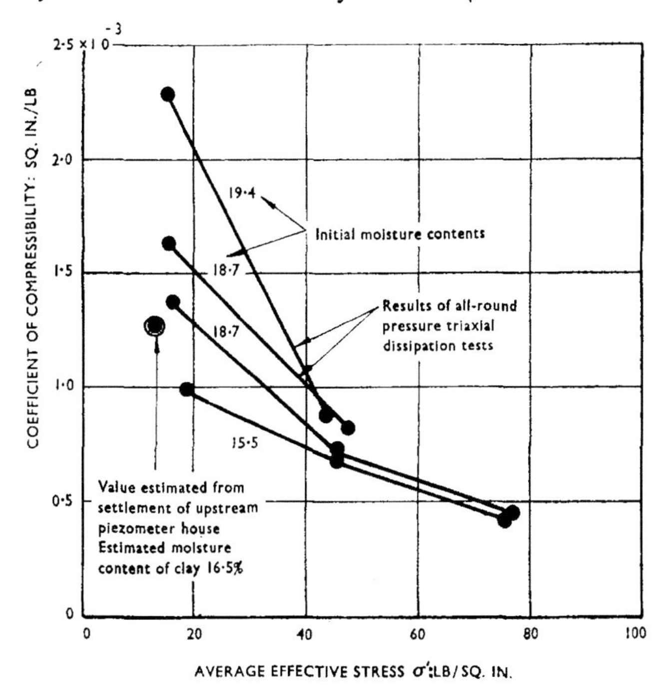

FIG. 23.—COMPRESSIBILITIES OF THE SOFT BOULDER CLAY FOUNDATION AS MEASURED IN THE LABORATORY AND IN THE FIELD

81. Leakage past the bentonite seals above the piezometers would also give an apparently higher value of c_v in the field, but leakage does not seem to have been significant as the rates of consolidation indicated by the foundation settlement records (Fig. 14) and by the piezometer records are consistent.

PART 8. PROPERTIES OF THE BOULDER CLAY IN THE LANDSLIP AREA DOWNSTREAM OF THE SOUTH ABUTMENT

82. The south side of the valley immediately downstream of the proposed embankment was an active landslide area and also showed signs of an older deep-seated rotational movement. As the construction of the embankment and impounding of water upstream were likely to raise the water table in the vicinity, measures to stabilize the area were obviously necessary.

83. To provide the necessary data for the stability analysis, four boreholes

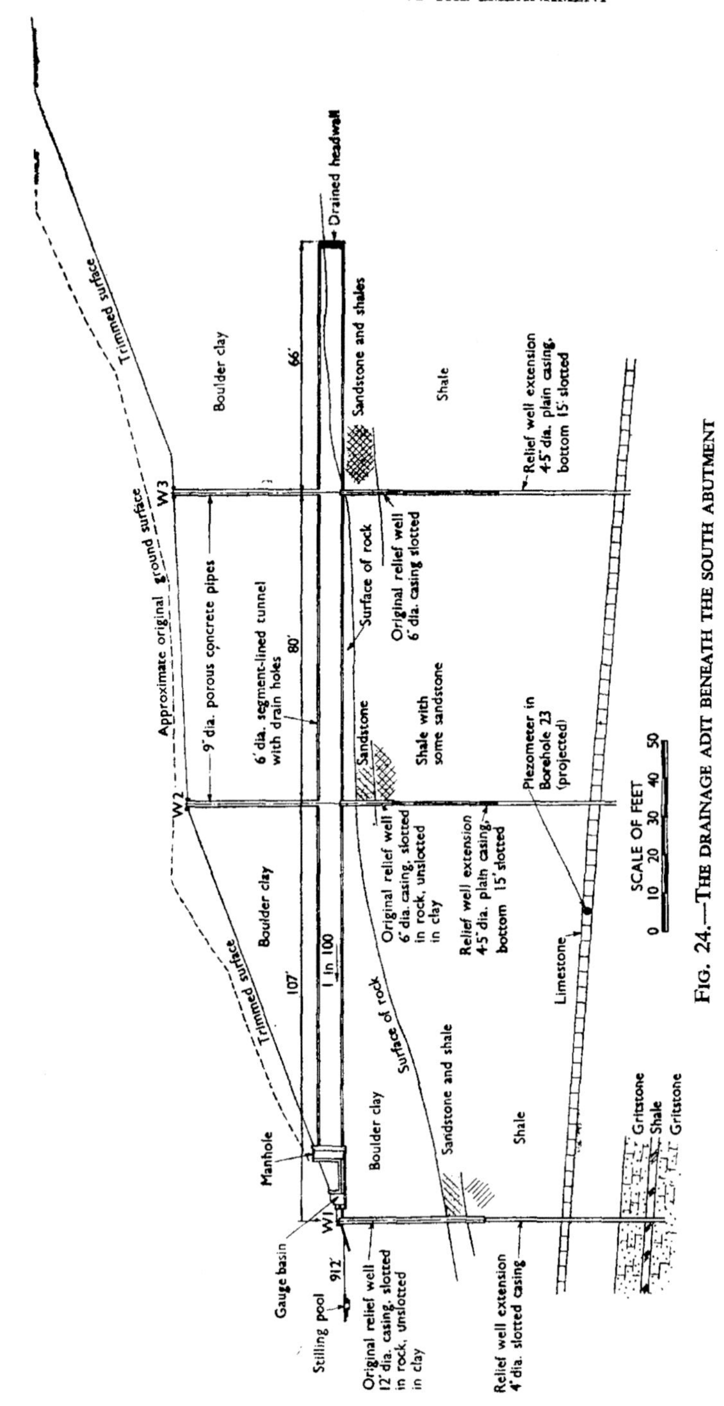

Fig. 24.—The drainage adit beneath the south abutment

were sunk and a limited number of 4-in.-dia. samples were taken, the sampling operation proving very difficult because of stones and boulders. A piezometer was sealed into the bottom of borehole 23 (Fig. 24) which penetrated into the underlying rock to measure the water pressure beneath the boulder clay, and the other boreholes were lined with open-jointed earthenware pipes. Additional data were subsequently obtained about the pore-pressure distribution within the boulder clay by sealing piezometers into shallow holes made by driving a mandrel.

84. The adit and relief wells constructed to stabilize the slope[1] provided the additional data on the underlying rock strata used in preparing Fig. 24.

85. Drained tests were run on samples prepared from six cores taken from depths between 3 ft and 42 ft and gave the average values $c' = 170$ lb/sq. ft and $\phi' = 32{\cdot}5°$. Three of the cores were sufficiently free from stones for undisturbed samples to be cut. In the other three cases remoulded samples were prepared after removing the stones, but no significant difference in the shear parameters was apparent. Two cores subsequently taken from shallow depths showed similar values, the overall average being $c' = 180$ lb/sq. ft and $\phi' = 32°$. The tests are summarized in reference 28.

86. An analysis of the existing unstable slope based on the observed pore-water pressures and the original six borehole samples gave values of factor of safety between 0·99 and 1·03. The pore-pressure data did not define a flow net very accurately, but showed no inconsistency with the full mobilization of both c' and ϕ' under long term conditions. A detailed discussion of this landslip is given elsewhere by Skempton and Brown.[28]

PART 9. STABILIZATION OF THE LANDSLIP AREA

87. It was evident that, even if no unfavourable change in the water table occurred, it was desirable to flatten the slopes in this area to avoid future slides which might damage the outlet works. The existing unstable slope was inclined at approximately 1·9:1. By flattening the slope to 2·5:1 the calculated factor of safety increased, for a given set of assumptions, from 0·99 to 1·28. This was considered an adequate increase since the laboratory data used was consistent with the limiting equilibrium of the existing slope.

88. The resulting cross-section is illustrated in Fig. 24. It was specified that the 2·5:1 slope could only be accepted in trimmed intact clay, and that hollows were not to be made up with compacted fill.

89. As the water table in the boulder clay (as indicated by the shallow piezometers) already stood almost at the ground surface in the winter, it was considered that the principal adverse change in conditions to be guarded against on impounding was an increase in uplift pressure on the base of the clay due to water flowing in the relatively more pervious rocks. The deep piezometer in borehole 23 had shown that although the water level in the rock below the crest of the lower slope stood 14 ft below ground surface, this represented an excess of 28 ft above the level of the ground at the toe. Elsewhere in the valley bottom, artesian pressures of up to 12 ft above ground level had been measured. An increase in these pressures due to seepage through or round the grouted cut-off could lead to deep-seated slips in the valley side.

90. To relieve possible increases in water pressure an adit was constructed through which vertical relief wells were sunk from the ground surface as shown in Fig. 24. The upper section of each hole above the adit was lined with a

porous concrete pipe, both to improve the drainage of the boulder clay and to provide a ready means for drilling deeper into the rock if the piezometer readings on impounding showed this to be necessary. Two additional relief wells were located at the toe of the slope downstream of the adit.

PART 10. PERFORMANCE OF THE STABILIZED AREA

91. There has been no evidence of any deep seated movement of the hillside either during impounding or subsequently. However, after a period of heavy rainfall in the winter of 1960–61 a small slump of limited area occurred in the topsoil placed on the trimmed slope. A small slip occurred in the same period at the edge of the trimmed area, where a subsequent survey shows the ground to have been left at a slope of 2 to 1. The performance so far appears to justify the choice of a 2·5:1 slope for this area.

92. The yield of the adit and of the adjacent relief wells showed little change as impounding proceeded. However, when the reservoir level reached 1,020 ft O.D. in April 1960 the level in the deep piezometer in borehole 23 had risen from 930 ft O.D. to 966 ft O.D., 10 ft above the surface of the slope vertically above it and 52 ft above the level of the valley floor at the toe of the slope. This piezometer was not recording uplift pressures immediately at the base of the boulder clay, but in a thin stratum of limestone 65 ft below the valley floor. As this implied possible uplift pressures at the toe of the slope of more than 80% of the overburden pressure (depending on the dip of the stratum), use was made of the provision for extending the relief wells. This, together with other measures taken to control uplift under the embankment itself, has led to a general reduction in uplift under the area and the piezometer now stands at 935 ft O.D. with the reservoir full.

PART 11. CONTROL OF UPLIFT PRESSURES DOWNSTREAM OF THE CUT-OFF

93. Reference has been made in the preceding parts to the special problem presented by the unstable slope on the south side of the valley. With the extensive cut-off measures employed it was not known whether the increase in uplift pressures would be of practical significance except in this area. However, two piezometers, Nos 19 and 20, were retained to measure the uplift pressure in the gritstone immediately beneath the clay stratum on which the embankment was founded, respectively 300 ft and 400 ft downstream of the cut-off in the centre of the valley (Fig. 3 of Ref. 1). Two 12-in.-dia. relief wells W7 and W8 were also drilled in the gritstone under the north slope of the valley to protect this slope and the lower sections of the overflow channel against uplift.

94. Following the initial rise in water level in the piezometer in borehole 23, three relief wells were extended into the thin limestone stratum. The yield from this area then increased from 700 gal/hour to 7,500 gal/hour, and the piezometer level dropped from 966 ft to 940 ft O.D. At this time the piezometers beneath the embankment showed pressures almost equal to those recorded before the contractor dewatered the cut-off trench area, but by August 1960 when the reservoir was full (1,037 ft O.D.) P19 and P20 showed levels of 967 ft and 963 ft O.D., respectively. Ground level at the toe of the embankment was 922 ft O.D. and this gave an estimated uplift pressure of 80% of the overburden pressure at the toe. The margin of safety against local failure was therefore

small, though the factor of safety against a slip through the core and along the base of the clay was estimated to be 1·6.

95. As the level at borehole 23 had also risen to 952 ft O.D., the existing relief well at the entrance of the adit was extended as a 4-in.-dia. hole. The gritstone was struck at a depth of 73 ft and the flow soon increased to 10,000 gal/hour and made further drilling impossible. The pressures in all three piezometers dropped at once by 4 ft and continued to fall slowly.

96. A 12-in.dia. relief well was also sunk through the gritstone beneath the boulder clay at the toe of the embankment. A steady flow of 15,000 gal/hour was obtained, and the levels at P19 and P20 dropped a further 20 ft. An observation borehole just downstream of the centre-line of the dam on the south side of the valley (Ref. 1, Fig. 3) also showed a drop in level of 15 ft. Borehole 23, however, was almost unaffected.

97. During impounding the level in W7 on the north slope rose from 937 ft to 947 ft O.D. where water overflowed at 400 gal/hour. In W8 the level rose from 946 ft to 965 ft O.D., ground level being 996 ft O.D. The additional relief wells in the valley bottom only slightly reduced these values.

98. The values given in the preceding paragraphs indicate the very large increases in uplift pressure which can occur downstream of a cut-off and the extent to which they can be controlled by relief wells. Had provision not been made to observe these pressures with piezometers they would have passed unnoticed, since the nearest outcrop of the bedrock is in Grassholme reservoir below Selset weir. Without relief wells the embankment and south abutment would have been very close to local instability.

99. The effectiveness of a single line of grout holes in controlling uplift has recently been questioned by Casagrande,[29] and the observations at Selset appear to support his conclusions.

Conclusions

100. The successful completion of the embankment at Selset demonstrates that it is possible to construct an embankment of very impervious boulder clay in a high rainfall area by accepting the control of pore pressure by drainage as the basic principle of the design. The development of reliable means of measuring pore-water pressures in the field has played a large part in making this principle acceptable as a practical engineering method for an important structure. The piezometer readings provided a continuous check on the design assumptions, and in the last season permitted a rate of construction to be used which could not have otherwise been accepted.

101. The acceleration of the consolidation of the foundation, without which the embankment would probably have become unstable at little over half the full height, could also be accepted as a basis for design with confidence as the decrease in excess pore pressure could be observed in the field.

102. The piezometers sealed into the rock beneath the boulder clay showed that, however adequate the embankment design, unexpectedly high uplift pressures could occur downstream of a grouted cut-off in a clay-blanketed valley, which if undetected might lead to local or complete failure.

103. Experience at Selset shows the importance of an adequate initial investigation of the site and borrow pits, but it also emphasizes the advantage of retaining sufficient flexibility in the design and contract to take advantage of

the additional data which become available during construction from field observations on the actual structure. Care and continuity in making these observations are amply repaid in the long run.

104. The performance of the embankment, as indicated by the field measurements of pore-pressure and settlement, is consistent with the estimates made at the design stage on the basis of laboratory tests. The results confirm the view that the principal uncertainty lies in predicting the placement conditions of the fill under the prevailing climatic conditions.

Acknowledgements

105. The Paper is published by permission of the Tees Valley and Cleveland Water Board.

106. The Authors wish to acknowledge the co-operation of Mr Julius Kennard, B.Sc.(Eng.), M.I.C.E., and Mr M. F. Kennard, B.Sc.(Eng.), A.M.I.C.E., of Edward Sandeman, Kennard and Partners throughout the work described and in the preparation of the Paper. In the period since 1954 a number of post-graduate students of Imperial College have given valuable assistance in the field, laboratory and drawing office; in particular Dr Clive Wood, B.Sc., B.E., Ph.D., Mr B. Hanson, M.Sc.(Eng.), Grad.I.C.E., Mr T. C. Kenney, B.E.. M.Sc., Mr Carl Romhild, M.S.(Harv.), M.Sc., and Mr J. D. Brown, B.E., M.Sc. Dr R. E. Gibson, B.Sc.(Eng.), Ph.D., A.M.I.C.E., has assisted with the numerical solution of consolidation problems. The continuity of the field records is largely due to Mr J. M. Jackson, B.Sc., A.M.I.C.E., assistant to the Resident Engineer.

107. Professor A. W. Skempton, D.Sc.(Eng.), M.I.C.E., F.R.S., has been associated with the work throughout as a consultant.

References

1. J. Kennard and M. F. Kennard, "Selset reservoir: design and construction". Proc. Instn civ. Engrs, vol. 21, Feb. 1962, pp. 277–304.
2. J. A. Banks, "Construction of Muirhead reservoir, Scotland". Proc. 2nd. Int. Conf. Soil Mech. and Foundation Engineering, 1948. Vol. 2, pp. 24–31.
3. J. A. Banks, "Problems in the design and construction of Knockenden Dam". Proc. Instn civ. Engrs, Pt. 1, vol. 1, July 1952, pp. 423–443.
4. G. A. R. Sheppard and L. B. Aylen, "The Usk scheme for the water suply of Swansea". Proc. Instn civ. Engrs, vol. 7, June 1957, pp. 246–274 incl. oral discussion.
5. W. W. Daehn and J. W. Hilf, "Implications of pore pressure in design and construction of rolled earth dams". Proc. 4th Congress on Large Dams, 1951, vol. 1, pp. 259–283.
6. A. W. Bishop, "The stability of earth dams". Ph.D. Thesis, University of London, 1952.
7. A. W. Bishop, "The use of the slip circle in the stability analysis of slopes". Géotechnique, vol. 5, part 1, pp. 7–17, Jan. 1955.
8. A. W. Bishop and Norbert Morgenstern, "Stability coefficients for earth slopes". Géotechnique, vol. 10, part 4, pp. 129–150, Dec. 1960.
9. B. Hanson, "Shear strength and pore pressure characteristics of compacted soils". M.Sc. Thesis, University of London, 1955.
10. A. W. Bishop and D. J. Henkel, "The measurement of soil properties in the triaxial test". Edward Arnold, London, 1957, 190 pp.
11. A. W. Bishop, "The measurement of pore pressure in the triaxial test". Proc. Conf. on Pore Pressure and Suction in Soils, pp. 38–46. Butterworths, London, 1960.

12. A. W. Bishop, I. Alpan, G. E. Blight and I. Donald, "Factors controlling the strength of partly saturated cohesive soils". Proc. Conf. on Shear Strength of Cohesive Soils, Amer. Soc. civ. Engrs, 1960.
13. J. W. Hilf, "An investigation of pore-water pressure in compacted cohesive soils". U.S. Bureau of Reclamation Tech. Memo., 654, 109 pp. 1956.
14. D. W. Taylor, "Fundamentals of soil mechanics". Wiley, also Chapman Hall, 1948.
15. B. Löfquist, "Earth pressure in a thin impervious core". Proc. 4th Congress on Large Dams, 1951, vol. 1, pp, 99–109.
16. A. W. Bishop, "The use of pore pressure coefficients in practice". Géotechnique, vol. 4, part 4, pp. 148–152, Dec. 1954.
17. K. Terzaghi and R. B. Peck, "Soil mechanics in engineering practice". Wiley, 1948.
18. U.S. Bureau of Reclamation, chapter on earth dams in "Treatise on Dams", 1951.
19. N. Janbu, "Application of composite slip surfaces for stability analysis". Proc. European Conf. on Stability of Earth Slopes, vol. 3, pp. 43–49, Stockholm, 1955.
20. T. C. Kenney, "An examination of the method of calculating the stability of slopes". M.Sc. Thesis, University of London, 1956.
21. A. W. Bishop, M. F. Kennard, and A. D. M. Penman, "Pore pressure observations at Selset dam". Proc. conf. on pore pressure and suction in soils, pp. 91–102, Butterworths, London, 1960.
22. A. D. M. Penman, "A field piezometer apparatus". Géotechnique, vol. 4, part 2, pp. 57–65, June, 1956.
23. A. W. Skempton, "The pore pressure coefficients A and B". Géotechnique, vol. 4, part 4, pp. 143–147, Dec. 1954.
24. A. W. Skempton and A. W. Bishop, "The gain in stability due to pore-pressure dissipation in a soft clay foundation". Proc. 5th Cong. Large Dams 1955, vol. 1, pp. 613–638.
25. A. W. Bishop, "Some factors involved in the design of a large earth dam in the Thames Valley". Proc. 2nd Int. Conf. Soil Mech. and Foundation Engineering, vol. 2, 1948, pp. 13–18.
26. A. W. Skempton, "The bearing capacity of clays". Building Research Cong. 1951, vol. 1, pp. 180–189.
27. A. W. Bishop, "Embankment dams: principles of design and stability analysis". Part of Chap. 9 of Hydro-Electric Engineering Practice, edited by J. Guthrie Brown, Blackie & Son, London, vol. 1, pp. 349-406, 1958.
28. A. W. Skempton and J. D. Brown, "A landslip in boulder clay at Selset, Yorkshire". Géotechnique, vol. 11, part 4, Dec. 1961, pp. 280–293.
29. A. Casagrande, "Control of seepage through foundations and abutments of dams" (The first Rankine lecture). Géotechnique, vol. 11, part 3, Sept. 1961, pp. 161–182.

The Paper, which was received on 9 August, 1961, is accompanied by twenty-seven sheets of drawings and diagrams, from which the Figures in the text have been prepared.

Written discussion on this Paper should be forwarded to reach the Institution before 15 April, 1962 and will be published in or after August 1962. Contributions should not exceed 1,200 words.—Sec.

Sixth Rankine Lecture

THE STRENGTH OF SOILS AS ENGINEERING MATERIALS

by

A. W. Bishop, M.A., D.Sc.(Eng.), Ph.D., M.I.C.E.*

INTRODUCTION

Of the Rankine Lecturers so far appointed from the United Kingdom I am the first to have spent the early years of my professional life working on the design and construction of civil engineering works. Although I became deeply involved in soil testing during this period, and spent more than a year working at the Building Research Station with Dr Cooling and Professor Skempton, the tests which I performed were carried out primarily for the solution of immediate engineering problems and only secondarily as a fundamental study of soil properties.

This period no doubt left its mark, because I find that I have retained a preference for investigating naturally occurring soils, either in their undisturbed state or in the state in which they would be used for constructing the embankments of earth or rockfill dams, or other engineering works. As a consequence, I would like to direct attention to the following four aspects of the study of the strength of soils which are not only of fundamental significance, but also of immediate practical importance to the engineer:

(1) the failure criteria which are used to express the results of strength tests and which reflect the influence, if any, of the intermediate principal stress;

(2) the behaviour of soils under the high stresses implied by the greatly increased height of earth and rockfill dams now under construction;

(3) the difficulty of determining what is the in-situ undrained strength of a soil, due to the influence both of anisotropy and of unrepresentative sampling;

(4) the influence of time on the drained strength of soils.

(1) FAILURE CRITERIA

A satisfactory failure criterion should express with reasonable accuracy the relationship between the principal stresses when the soil is in limiting equilibrium. To be of practical use it should express this relationship in terms of parameters which can be used in the solution of problems of stability, bearing capacity, active and passive pressure, etc, and which can form the currency for the exchange of information about soil properties.

If we consider soil properties in terms of effective stress (Fig. 1), the most marked feature of which the failure criterion must take account is the increase in strength as the average effective stress increases. But we will wish to apply the information obtained from testing samples in axial compression in the triaxial cell (Fig. 2) (where the intermediate principal stress σ_2' is equal to the minor principal stress σ_3') to practical problems where σ_2' is greater than

* Professor of Soil Mechanics, Imperial College, University of London

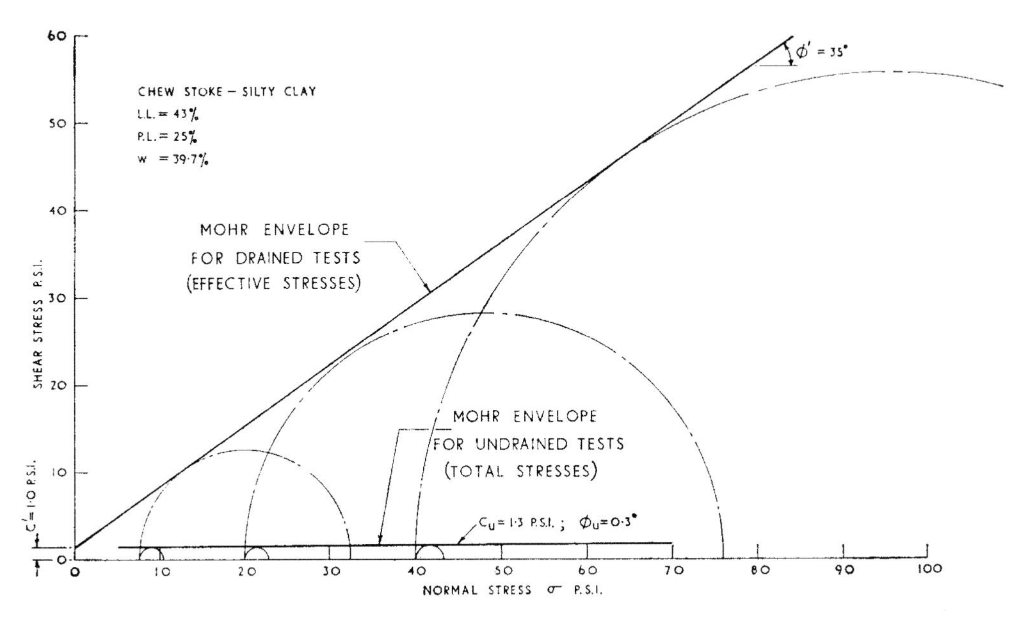

Fig. 1. Mohr envelopes for undrained and drained tests on a saturated soil, showing increase in strength with increase in effective stress

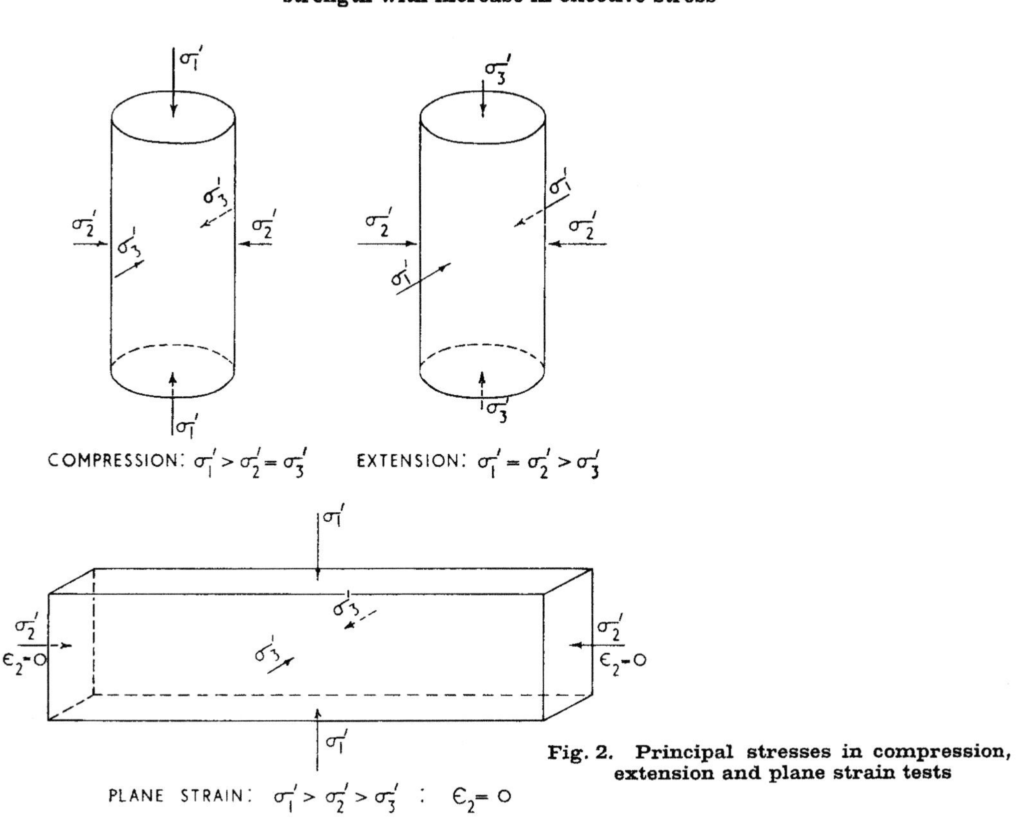

Fig. 2. Principal stresses in compression, extension and plane strain tests

σ_3' and may in the limit equal σ_1' (as in the centre of an excavation about to fail by heaving). A common special case is that of plane strain, where there is no change in length along the axis of the structure (Fig. 2). Most problems of slope stability fall into this category, and it is here that low factors of safety are most often encountered. The failure criterion should therefore also reflect the influence on the strength of the soil of the variation of σ_2' between the limiting values of σ_3' and σ_1'.

The principal failure criteria currently under discussion (see, for example, Kirkpatrick (1957), Hvorslev (1960), Scott (1963a), Roscoe *et al* (1963)) are given below. For simplicity they are given for cohesionless soils (or soils in which the cohesion intercept c' is zero). Also for simplicity σ_1', σ_2', σ_3' are chosen to denote the major, intermediate and minor effective principal stresses respectively.

The failure criteria may then be written:

Mohr–Coulomb:

$$\sigma_1' - \sigma_3' = \sin\phi' . (\sigma_1' + \sigma_3') \qquad (1)$$

Extended Tresca:[1]

$$\sigma_1' - \sigma_3' = \alpha\left(\frac{\sigma_1' + \sigma_2' + \sigma_3'}{3}\right) \qquad (2)$$

Extended von Mises:[2]

$$(\sigma_1' - \sigma_2')^2 + (\sigma_2' - \sigma_3')^2 + (\sigma_3' - \sigma_1')^2 = 2\alpha^2\left(\frac{\sigma_1' + \sigma_2' + \sigma_3'}{3}\right)^2 \qquad (3)$$

For the present discussion we are concerned with two points:

(1) that for a given stress system the strength should be proportional to the normal stress, and

(2) that the influence of the intermediate principal stress should be correctly indicated.

The first requirement is clearly satisfied by all three criteria. The second can be examined only on the basis of experimental evidence. It should be pointed out that whether or not yield takes place at constant volume is irrelevant to the present stage of the discussion, though it is relevant to any examination of the physical components of shear strength.

It will be noted that in the Mohr–Coulomb criterion the value of σ_2' has no influence on the strength, and the same principal stress ratio at failure would be expected for both compression and extension tests. This is in contrast to the extended Tresca and extended von Mises criteria, which both show an important difference between the stress ratios and angles of friction in compression and extension.

In *axial compression* $\sigma_2' = \sigma_3'$ and the extended Tresca (equation 2) and the extended von Mises (equation 3) both reduce to:

$$\sigma_1' - \sigma_3' = \alpha\left(\frac{\sigma_1' + 2\sigma_3'}{3}\right) \qquad (4)$$

In *axial extension* $\sigma_2' = \sigma_1'$ and the extended Tresca and the extended von Mises both reduce to:

$$\sigma_1' - \sigma_3' = \alpha\left(\frac{2\sigma_1' + \sigma_3'}{3}\right) \qquad (5)$$

[1] This is attributed by Johansen (1958) to Sandels. Roscoe *et al* (1958 and 1963) denote $\sigma_1' - \sigma_3' = q$ and $(\sigma_1' + \sigma_2' + \sigma_3')/3 = p$ and take $q = \alpha p$ as the failure criterion.

[2] This is attributed to Schleicher (1925, 1926).

The influence of the intermediate principal stress on strength may be more readily appreciated in terms of the variation in ϕ' which is implied by the different failure criteria as σ_2' varies between the limits σ_3' and σ_1'.

The Mohr–Coulomb criterion (equation 1) may be written:

$$\frac{\sigma_1'-\sigma_3'}{\sigma_1'+\sigma_3'} = \sin\phi' \qquad (6)$$

The relative position of σ_2' between σ_3' and σ_1' may be denoted by the parameter b, where:

$$\frac{\sigma_2'-\sigma_3'}{\sigma_1'-\sigma_3'} = b \qquad (7)$$

and b varies between 0 and 1.

The extended Tresca criterion then becomes:

$$\frac{\sigma_1'-\sigma_3'}{\sigma_1'+\sigma_3'} = \frac{1}{\frac{1}{3}+\frac{2}{\alpha}-\frac{2}{3}b} \qquad (8)$$

and the extended von Mises criterion becomes:

$$\frac{\sigma_1'-\sigma_3'}{\sigma_1'+\sigma_3'} = \frac{1}{\frac{1}{3}+\frac{2}{\alpha}\cdot\sqrt{1-b+b^2}-\frac{2}{3}b} \qquad (9)$$

The variation of ϕ' (as defined by equation 6) with b (as defined by equation 7) corresponding to the two failure criteria expressed in equations 8 and 9 is illustrated in Fig. 12 and will be compared with observed values in a later section. It will be seen that the predicted ϕ' varies between $\sin^{-1}\frac{1}{\frac{2}{\alpha}+\frac{1}{3}}$ in the compression test, and $\sin^{-1}\frac{1}{\frac{2}{\alpha}-\frac{1}{3}}$ in the extension test, which is a very marked difference since the value of α at failure is typically more than 0·8.

A great many tests have been carried out, at Imperial College and elsewhere, to examine the influence of the intermediate principal stress, and some of the principal results are illustrated below.

As the accuracy of the tests is usually called in question when they fail to fit whichever theory is in vogue, it is of interest to note several points. The error in the determination of the principal stress ratio due to the distortion of the sample *at failure* has often been considerably over-exaggerated. The strain at failure in the test series to be quoted below (Cornforth, 1961) varied in compression, from 3½% for dense sand to 6% for the middle of the range and 12% for loose sand. A sample which reached its peak stress at 6·3% axial strain is illustrated in Fig. 3. A detailed study of 4-in. diameter samples having different heights and degrees of end restraint (Fig. 4) suggests that measurement of peak strength in compression need be subject to little ambiguity (Bishop and Green, 1965).

In plane strain the failure strain ϵ_1 varied from 1·3% for dense sand to 2% in the middle of the range and 4% for loose sand. Rupture in a thin zone then occurred (Fig. 5). With these very small failure strains little uncertainty again arises in the stress calculations.

In extension the axial strain ϵ_3 at failure varied from −4% to −5% for dense and medium dense sand and rose to −9% for loose sand. In drained tests a neck begins to form at about the peak stress ratio, though it may not be very apparent to the eye (Fig. 6). If the test is stopped as soon as the peak is defined and the actual shape of the sample determined, the computed[3] value of ϕ' may be ½° to 1° higher at the dense end and about 2° higher at the loose end than the value based on average cross-sectional area. This correction has been made by Cornforth (1961) in the tests quoted.

[3] Based on the average cross section of a zone capable of containing a plane inclined at $45°-\phi'/2$.

Fig. 3. 4-in. dia. × 8-in. high compression test: Ham River sand: porous disc at each end: $n_i = 40{\cdot}2\%$, $\phi'_{max} = 37{\cdot}4°$, $\epsilon_{1f} = 6{\cdot}3\%$, test stopped at $\epsilon_1 = 7{\cdot}5\%$. (Test by Green, 1966)

Fig. 5a. Plane strain test apparatus (with cell body removed). (Wood, 1958)

Fig. 5b. Plane strain compression test sample (4 in. × 2 in. × 16 in.) of Brasted sand after failure $n_i = 35{\cdot}6\%$, $\epsilon_{1f} = 1{\cdot}7\%$, test stopped at $\epsilon_1 = 15{\cdot}0\%$. (Test run with end plattens removed) (Cornforth, 1961)

Fig. 6a. 4-in. dia. × 7-in. high extension test; Ham River sand: porous discs at each end. $n_i = 39{\cdot}2\%$, $\phi'_{max} = 40{\cdot}5°$, $\epsilon_{3f} = -7{\cdot}4\%$, test stopped at $\epsilon_3 = -8{\cdot}4\%$

Fig. 6b. 4-in. dia. × 4-in. high extension test; Ham River sand: 2/0·010-in. thick lubricated membranes at each end. $n_i = 45{\cdot}8\%$, $\phi'_{max} = 32{\cdot}6°$, $\epsilon_{3f} = -9{\cdot}5\%$, test stopped at $\epsilon_3 = -9{\cdot}8\%$

(Tests by Green, 1965)

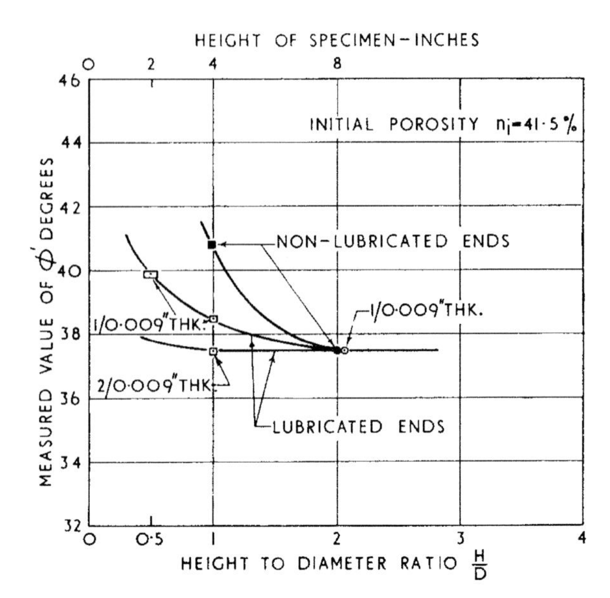

Fig. 4. Effect of sample height and degree of end restraint on measured peak ϕ' (number and thickness of lubricated membranes on end plattens indicated)

A comparison between the peak strength values, expressed in terms of ϕ' as defined above, of plane strain and axial compression tests is given in Fig. 7. This indicates values of ϕ' in plane strain higher by 4° at the dense end of the range and by $\frac{1}{2}$° in loose sand. A subsequent series of tests at Imperial College on Mol sand by Wade (1963) shows similar results. Tests on sand by Kummeneje (1957) in a vacuum triaxial apparatus and also by Leussink (1965) are in general agreement, but show less tendency to converge at higher porosities.

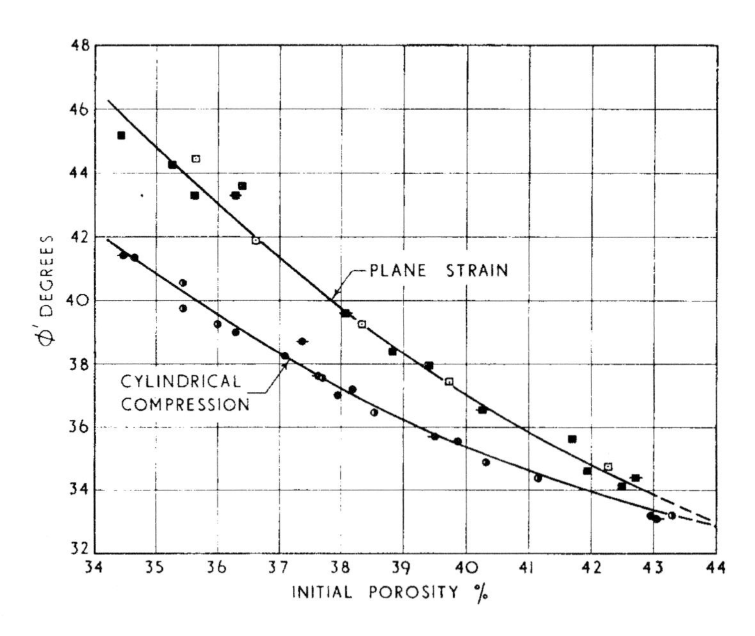

Fig. 7. Comparison of results of drained plane strain and cylindrical compression tests on Brasted sand (Cornforth, 1961)

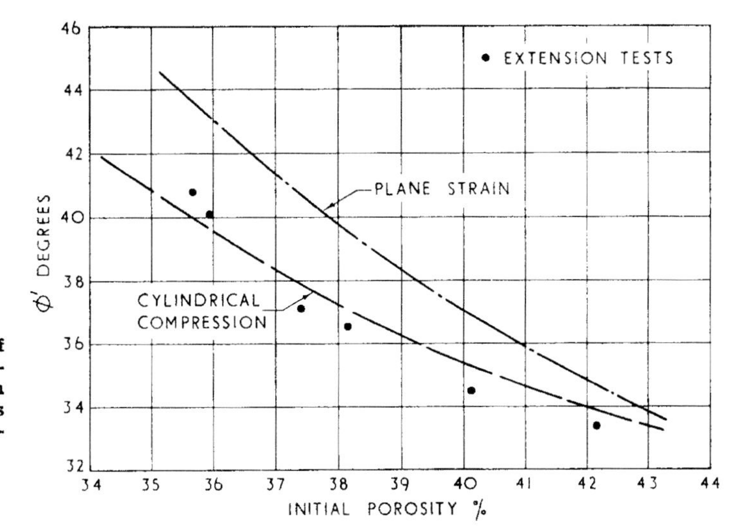

Fig. 8. Comparison of results of drained extension, compression and plane strain tests on Brasted sand (Cornforth, 1961)

The comparison between compression and extension tests is shown in Fig. 8 and it is apparent that the difference in the value of ϕ' is not significant over the range of porosities investigated. The same general conclusion is indicated by a subsequent series of tests by Green on Ham River sand,[4] using lubricated end plattens in both compression and extension tests (Fig. 9).

To examine the failure criteria a knowledge of σ_2' is required. In the compression and extension tests σ_2' is equal to the fluid pressure in the triaxial cell less the pore pressure in the sample. In the plane strain test σ_2' is determined from the load on the lubricated plattens maintaining zero strain in the σ_2' direction (Wood, 1958).

[4] Due to the variation in ϕ' with normal stress, the cell pressures in the extension tests have been selected so that the minor principal stress at failure approximates to the minor principal stress used in the compression tests.

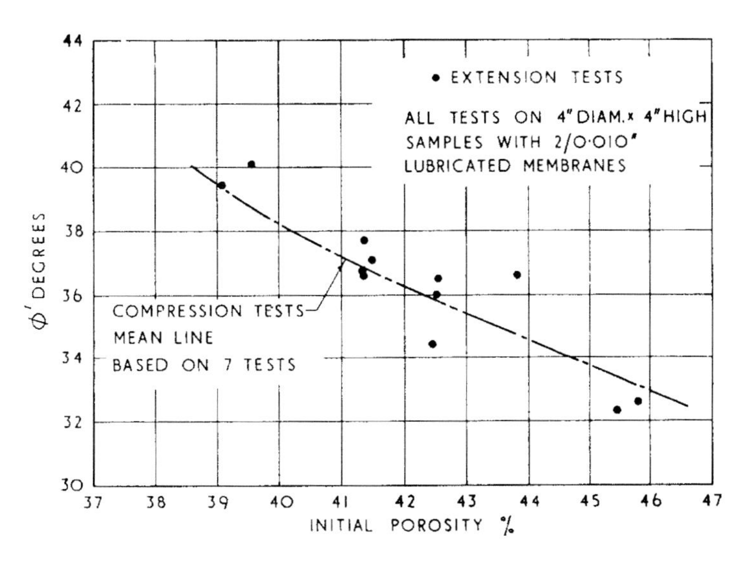

Fig. 9. Comparison of results of drained extension and compression tests on saturated Ham River sand: cylindrical samples with lubricated ends, $\frac{H}{D} = 1$ (Tests by Green, 1965)

The observed values of the ratio $\frac{\sigma_2'}{\sigma_1'+\sigma_3'}$ show a close correlation with the peak value of ϕ' (Fig. 10), over a wide range of values. The empirical expression $\frac{\sigma_2'}{\sigma_1'+\sigma_3'}=\frac{1}{2}\cos^2\phi'$ is in good agreement with the general trend, but slightly underestimates σ_2'. The expression may be derived by combining two earlier empirical relationships. Wood (1958) noted that for tests on compacted moraine carried out in the plane strain apparatus at Imperial College, the relationship between σ_2' and σ_1' at failure approximated to the expression:

$$\sigma_2' = K_0\sigma_1' \qquad (10)$$

where K_0 was the coefficient of earth pressure at rest measured with zero strain in both lateral directions (i.e. when both $\epsilon_2=0$ and $\epsilon_3=0$). Tests reported by Bishop (1958) and Simons (1958) had shown that there was an empirical relationship between K_0 and ϕ' which could be represented with reasonable accuracy by an expression due to Jaky (1944 and 1948):

$$K_0 = 1-\sin\phi' \qquad (11)$$

Combining these expressions and putting

$$\frac{\sigma_1'-\sigma_3'}{\sigma_1'+\sigma_3'} = \sin\phi' \quad \text{(from equation 6)}$$

we obtain

$$\frac{\sigma_2'}{\sigma_1'+\sigma_3'} = \frac{1}{2}\cos^2\phi' \qquad (12)$$

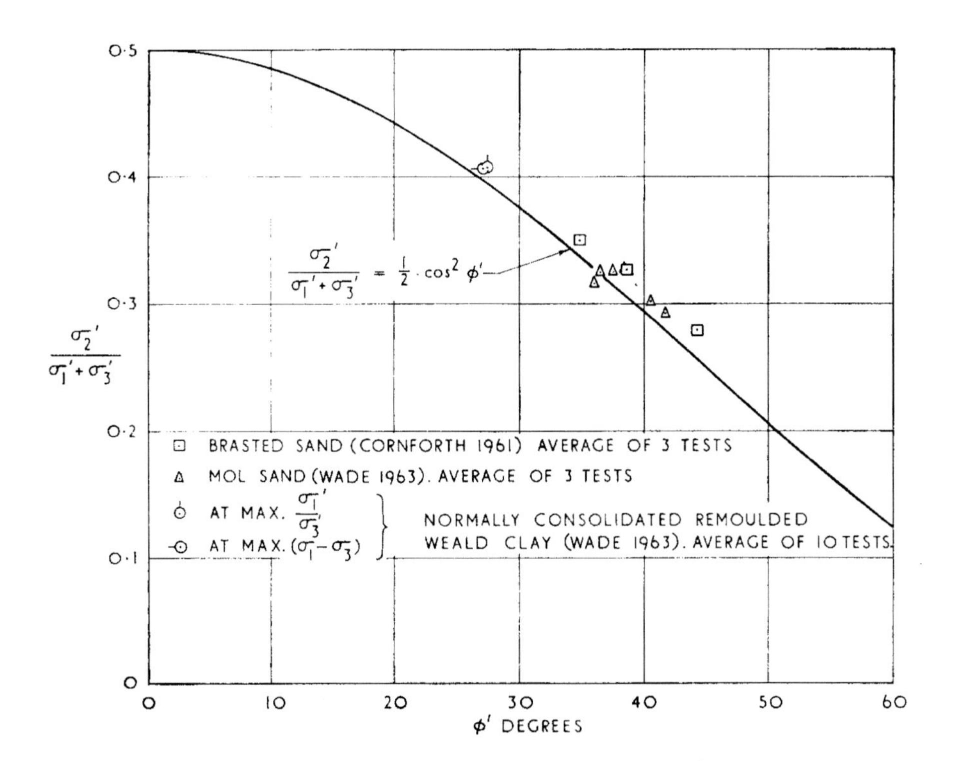

Fig. 10. Correlation with ϕ' of value of intermediate principal stress σ_2' at failure in plane strain

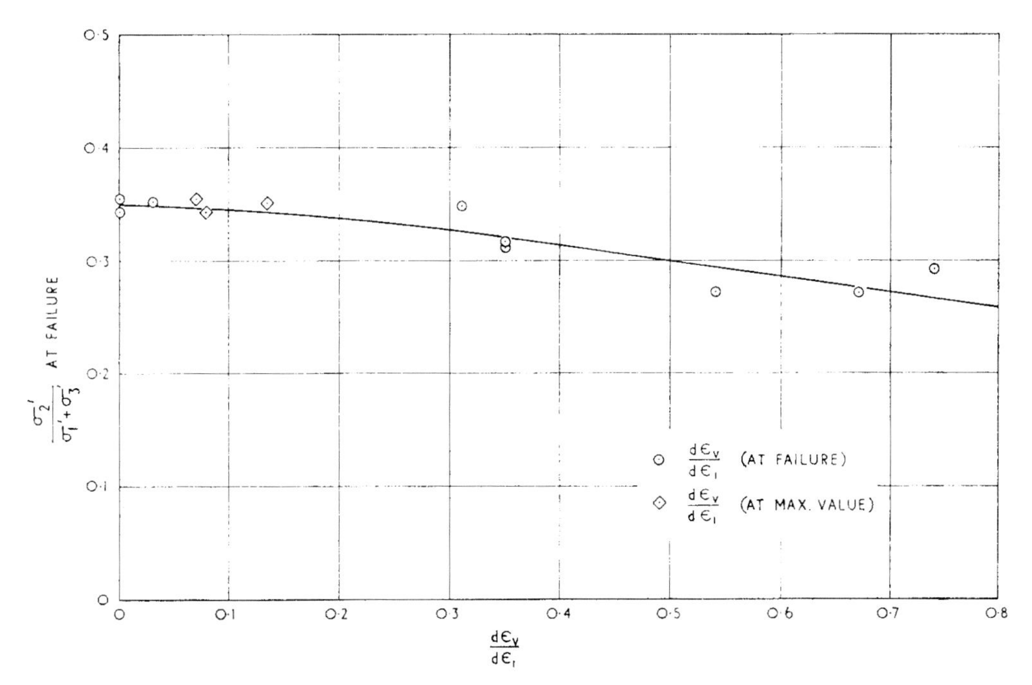

Fig. 11. Variation of $\frac{\sigma_2'}{\sigma_1'+\sigma_3'}$ with rate of volume change $\frac{d\epsilon_v}{d\epsilon_1}$ for drained tests on Brasted sand

It is of interest to note that samples which were deforming at constant volume when the peak stress ratio was reached[5] conformed to this relationship and the ratio $\frac{\sigma_2'}{\sigma_1'+\sigma_3'}$ showed no indication of being equal to $\frac{1}{2}$ as might have been expected (Fig. 11). What occurs to this ratio at strains beyond those corresponding to the maximum stress cannot be determined in the present apparatus, as zone failure always follows closely after the peak and the localized value of σ_2' cannot be measured.

With the knowledge of σ_2' we can now compare the observed values of ϕ' $\left(\text{defined by } \sin\phi' = \frac{\sigma_1'-\sigma_3'}{\sigma_1'+\sigma_3'}\right)$ with those predicted by the various failure criteria as b $\left(=\frac{\sigma_2'-\sigma_3'}{\sigma_1'+\sigma_3'}\right)$ varies between 0 and 1. Fig. 12(a) shows this comparison for loose Brasted sand, which is shearing almost at constant volume at the peak stress ratio. It will be seen that the results fit well with the Mohr–Coulomb failure criterion, whereas the extended von Mises and extended

[5] In a plane strain compression test on loose sand constant volume shear may occur firstly at the peak stress ratio, when the condition of pure shear is approximated to, and then subsequently at a lower stress ratio, when strains are largely confined to a thin slip zone in which simple shear is approximated to. In the tests on Brasted sand performed by Cornforth (1961) the peak value of ϕ' corresponding to zero rate of volume change is 34·3°. If the direction of the thin slip zone is taken to correspond to a Mohr–Coulomb slip plane a residual ϕ' of 32·3° is obtained. If, alternatively, the shear stress acting along the boundary of this zone is taken (following Hill, 1950) to be equal to the maximum shear stress within the zone, the residual value of ϕ' would be 39·2°, which is clearly unreasonable. These differences suggest that constant volume yield in pure shear may involve a failure mechanism significantly different from that associated with constant volume yield in simple shear.

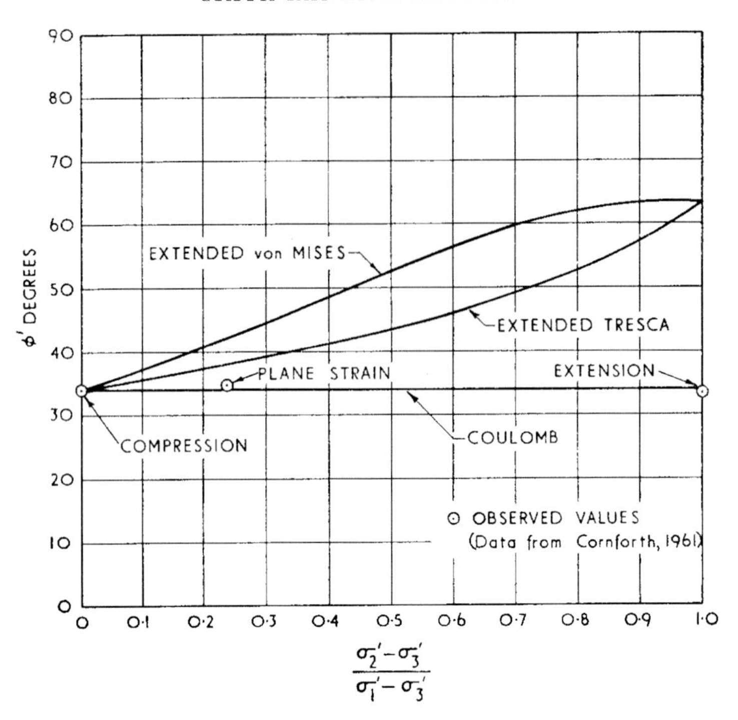

Fig. 12a. Observed and predicted values of ϕ'. Loose sand; ϕ' in compression $=34°$ and $\alpha=1{\cdot}375$

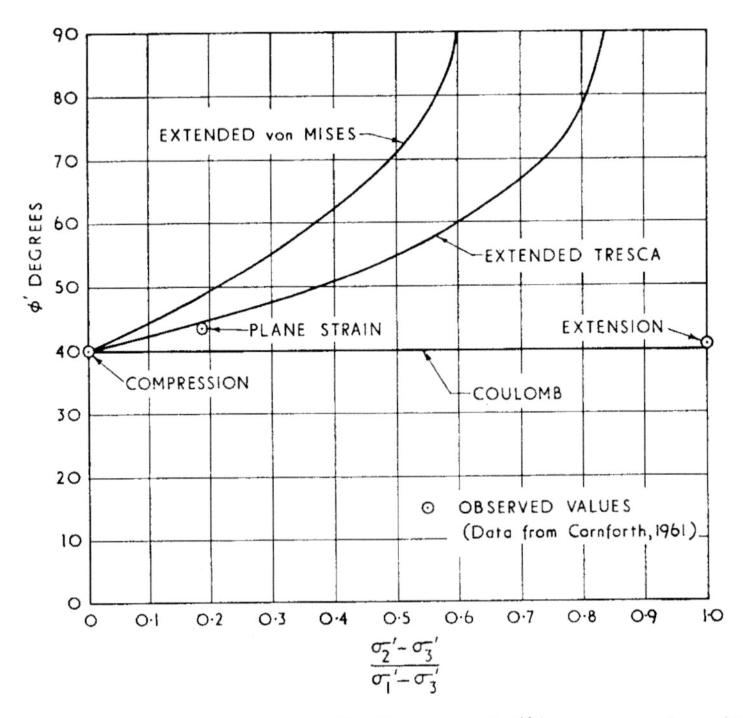

Fig. 12b. Observed and predicted values of ϕ'. Dense sand; ϕ' in compression $=40°$ and $\alpha=1{\cdot}636$

Tresca criteria predict values which differ from the observed values by more than the most pessimistic estimate of experimental error.

Fig. 12(b) shows that for dense sand, which is dilating at failure, the Mohr–Coulomb criterion again gives the best over-all fit, though for a comparison of the compression test and the plane strain case only the extended Tresca is in better agreement.[6] However, for the extension tests the extended Tresca and the extended von Mises both fail to predict meaningful results. The reason for this is of considerable interest.

In Fig. 13 the values of ϕ' in compression predicted by the two failure criteria are plotted against the value of the parameter α which is used in both expressions to indicate the increase in strength with normal stress. This α is the same as that used by Roscoe *et al* (1958). It will be seen that for the compression test the relationship between ϕ' and α is almost linear and in fact approximates to $\phi'=25\alpha$ over the range $\phi'=20°$ to $40°$. However, the value of ϕ' in extension predicted by both failure criteria rapidly diverges from that in compression as α increases, and becomes equal to $90°$ (i.e. $\sigma_1'/\sigma_3'=\infty$) when $\alpha=1{\cdot}5$. At this value of α the compression value of $\phi'=36{\cdot}9°$, which is well within the range of values encountered in dense sand.

The physical explanation can be seen from the representation of the failure criteria in a three-dimensional stress space (Fig. 14). If the axes σ_1', σ_2', σ_3' represent the magnitudes of the principal effective stresses in those three directions, we can select a plane on which $\sigma_1'+\sigma_2'+\sigma_3'=\text{constant}$, and a diagonal 00′ (normal to it) for which $\sigma_1'=\sigma_2'=\sigma_3'$ (i.e., no shear

[6] If the value of $\sigma_2'/(\sigma_1'+\sigma_3')$ in the failure zone exceeds the average value recorded in the plane strain apparatus and approaches $\frac{1}{2}$ the apparent agreement with the extended Tresca criterion no longer obtains.

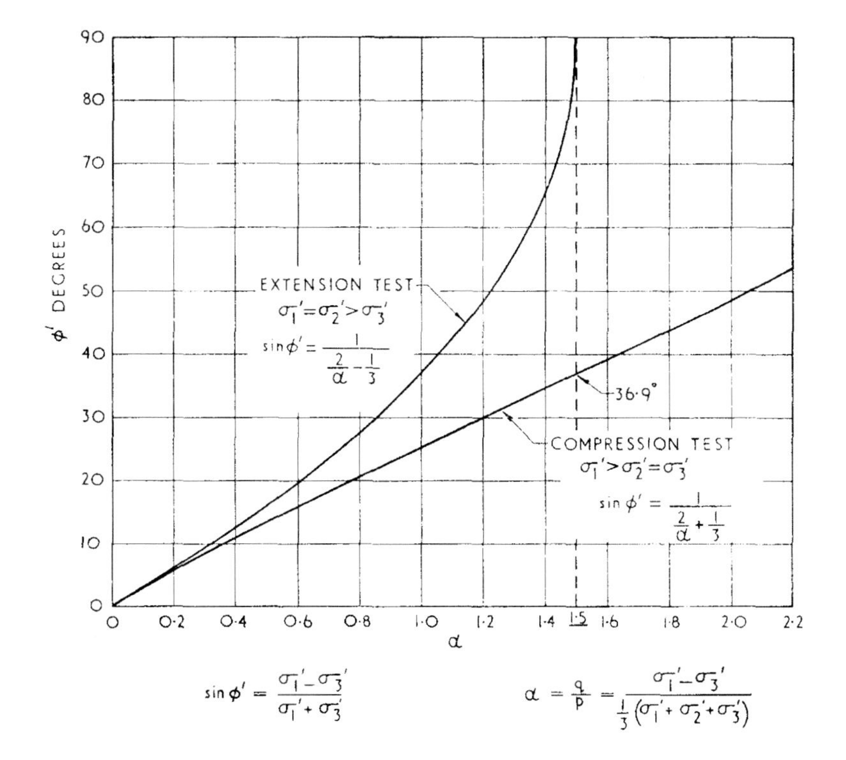

Fig. 13. Relationship between parameters used in failure criteria

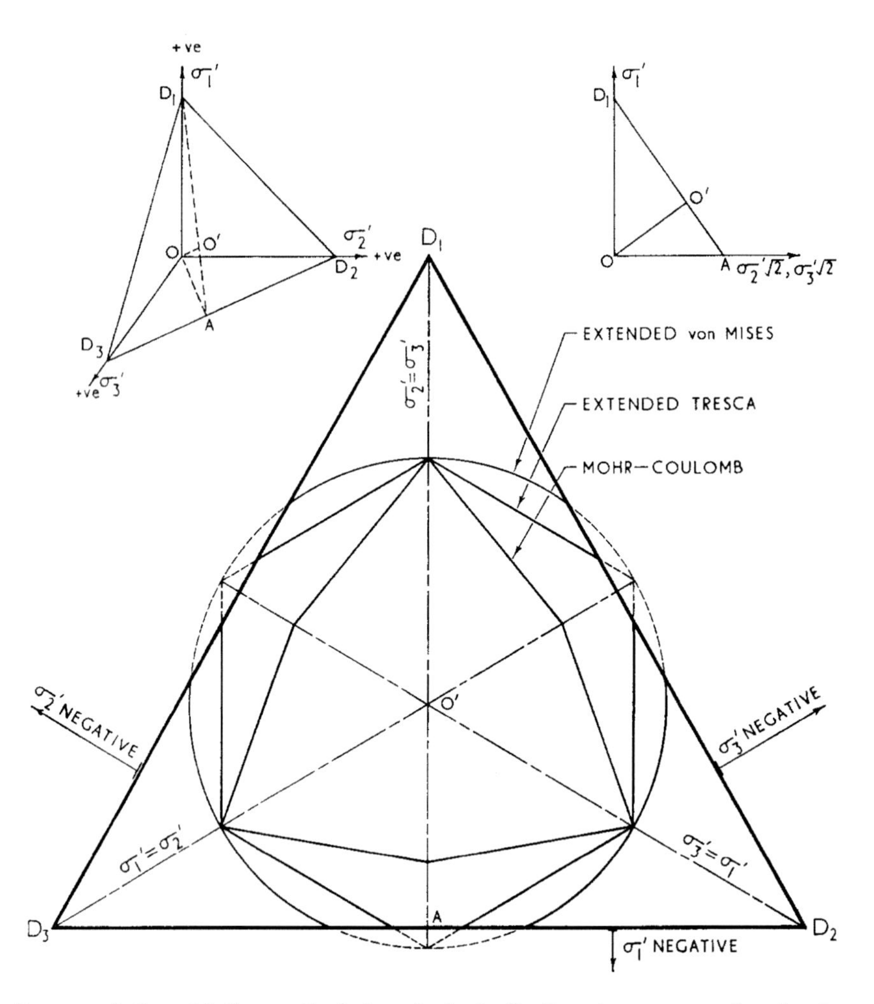

Fig. 14. Representation of failure criteria in principal effective stress space, showing boundaries of positive stress field. Sections of failure surface shown for $\phi' = 40°$ in axial compression

stress). States of limiting equilibrium expressed by the various failure criteria are then represented by pyramid shaped surfaces having their apex at 0 and showing characteristic sections on the plane $D_1D_2D_3$. These are an irregular hexagon for the Mohr–Coulomb criterion; a circle with centre 0′ for the extended von Mises criterion and the regular hexagon inscribed in this circle for the extended Tresca criterion.

In Fig. 14 these sections are plotted for $\phi' = 40°$ in axially symmetrical compression (i.e. $\alpha = 1{\cdot}636$), and it will be seen that as σ_2' moves from σ_3' to σ_1' the circle and hexagon representing the extended von Mises and Tresca criteria respectively go outside the lines D_1D_2 etc., into negative effective stress space. For a cohesionless soil (or $c' = 0$ material) this is meaningless. These failure criteria are therefore in principle unable to represent the behaviour of the denser frictional materials having a compression ϕ' of more than $36{\cdot}9°$ (i.e. $\alpha = 1{\cdot}5$) for which the extended von Mises circle is tangential to D_1D_2, D_2D_3 and D_3D_1.

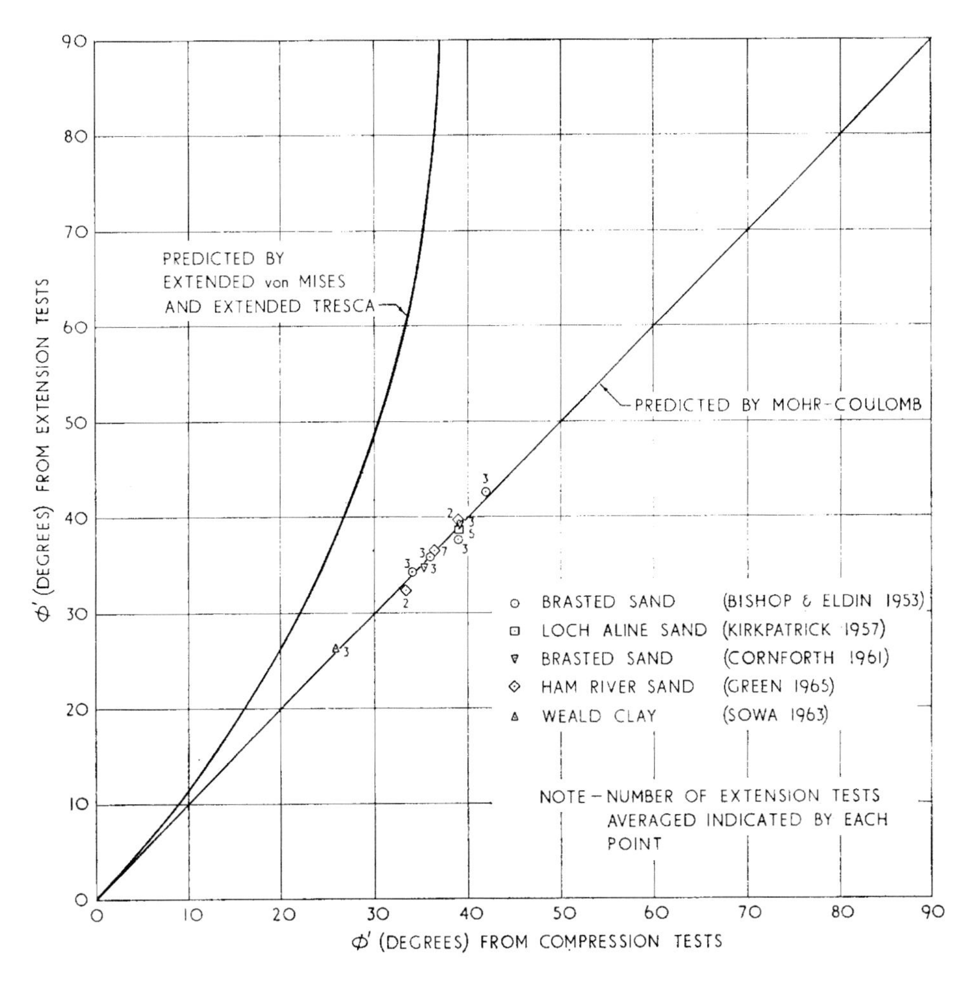

Fig. 15. Comparison of results of compression and extension tests

Even within this range the two criteria predict values of ϕ' in extension (Fig. 15), which differ from the experimental results by an amount which cannot be attributed to experimental error. The experimental results, however, strongly support the Mohr–Coulomb criterion,[7] and we must, I feel, accept the Mohr–Coulomb criterion as being the only simple criterion of reasonable generality.[8,9] It does, however, underestimate the value of ϕ' for plane strain in dense sands by up to 4° (from tests on Brasted sand by Cornforth and Mol sand by Wade).

[7] This applies to all the data illustrated in Fig. 15, to tests reported by the Norwegian Geotechnical Institute (1958), and by Wade (1963). The tests illustrated in Fig. 15 were drained tests on saturated samples consolidated under equal all round pressure, with the exception of the series by Cornforth (1961), where consolidation with zero radial strain was used. Unpublished tests at Imperial College by Walker on dry sand consolidated under equal all round pressure agree with the tests described above. In contrast early tests by Habib (1953) and Peltier (1957) show lower values of ϕ' in extension. No obvious explanation for this difference is apparent. Tests by Haythornthwaite (1960) also show a difference, but are based on a different definition of the point of failure. On the other hand undrained tests on undisturbed clays or clays initially anisotropically consolidated from a slurry (Ladd and Bailey, 1964; Smotrych, unpublished data) show higher values of ϕ' in extension. The interpretation of the test data is, however, subject to more

Tests representing the actual state of stress or strain to be encountered in practice are obviously best for precise work in this case. The present data can be expressed in terms of an extended Mohr–Coulomb criterion involving two parameters:

$$\frac{\sigma_1' - \sigma_3'}{\sigma_1' + \sigma_3'} = \frac{K_1}{1 - K_2\sqrt{b(1-b)}} \quad . \quad . \quad . \quad . \quad . \quad . \quad . \quad . \quad (13)$$

where $b = \dfrac{\sigma_2' - \sigma_3'}{\sigma_1' - \sigma_3'}$

and K_1 and K_2 are parameters determined from compression and plane strain tests (K_1 being equal to sin ϕ' in the compression and extension test).

Further studies of the variation of ϕ' with b for dense frictional materials now being carried out at Imperial College may, however, suggest a more elaborate expression.

It is clear, however, from Fig. 12 that from an engineering point of view the use of the extended von Mises or Tresca criteria would lead to a very substantial overestimate of strength for a wide range of b values.

(2) THE BEHAVIOUR OF SOILS UNDER HIGH STRESSES

There is one limitation, which is of considerable engineering significance, common to all the failure criteria so far discussed. The constant of proportionality between strength and normal stress, either ϕ' or α, is not really a constant when a wide range of stress is under consideration. This is of particular importance when drawing conclusions from tests on models on the one hand, where unpublished tests at Imperial College by Dr Ambraseys and Mr Sarma have shown that values of ϕ' may rise by up to 8° at stresses represented by a fraction of an inch of sand, and on the other hand in the design of high dams (Fig. 16) which are now reaching 1000 ft in height and imply considerably reduced values of ϕ'. The major principal stress in such a dam could approach 900 lb/sq. in.

Typical Mohr envelopes for various soils tested under a wide range of pressures are illustrated in Fig. 17. The marked curvature of many of the failure envelopes will be seen. It is of interest to note that the uppermost envelope[10] represents tests on one of the fill materials of the Oroville dam shown in the previous figure. The other materials include rockfill from the Infiernillo dam,[11] compacted glacial till, two dense sands, and one loose sand (Ham River), and undisturbed silt and two undisturbed clays, as well as one clay consolidated from a slurry. There is some indication that the curvature is most marked for soils which

(*a*) are initially dense or heavily compacted,
(*b*) are initially of relatively uniform grain size,
(*c*) if undisturbed, have been heavily over-consolidated.

In the coarser granular materials (the sands, gravels, and rockfills) the curvature is clearly associated with crushing of the particles, initially local crushing at interparticle contacts, and ultimately shattering of complete particles. This in turn is associated, in dense materials,

---

uncertainty. The influence of undrained stress path on the value of ϕ', which is of greater importance in clays, and the anisotropy of the soil structure, are likely to play a more significant part in the behaviour of clays and more experimental data is required before general conclusions can be drawn.

[8] This conclusion was reached by Kirkpatrick (1957). However, from thick cylinder tests on dense sand he obtained an increase in ϕ' of only 2° for $b = \frac{1}{2}$.

[9] The method proposed by Johansen (1958) reduces to the Mohr–Coulomb criterion for the case when the value of ϕ' in extension is equal to that in compression.

[10] The minimum ratio of sample diameter to maximum particle size which may be used without an overestimate of the strength resulting has not been fully investigated. The ratio of only 4 was used in the tests by Hall and Gordon (1963) quoted above.

[11] Details of the special triaxial cell for samples 113 cm. diameter are given by Marsal *et al* (1965).

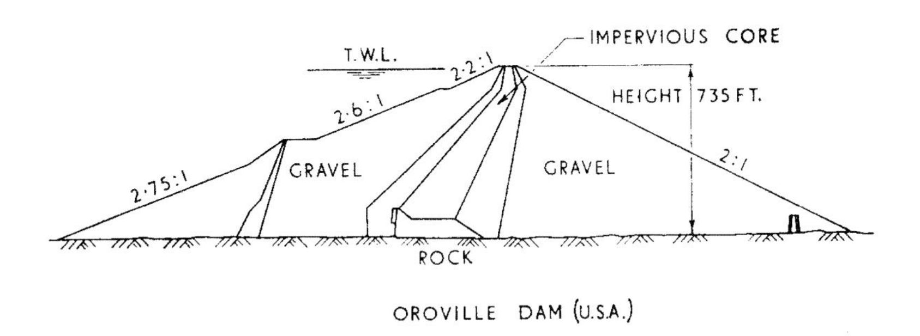

Fig. 16. Two high dams now under construction. (USCOLD Newsletter, Jan. 1963; Nitchiporovitch, 1964)

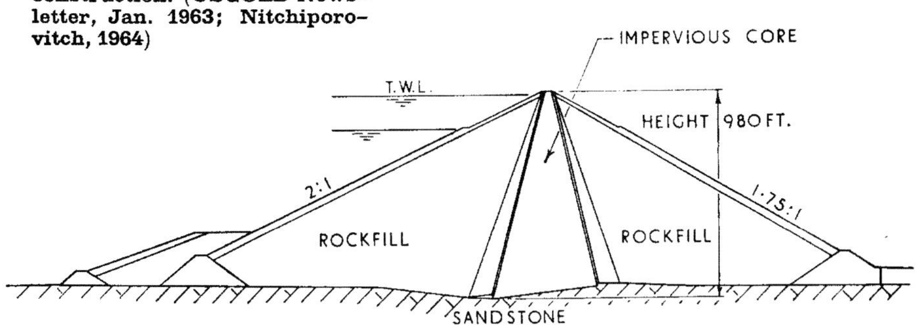

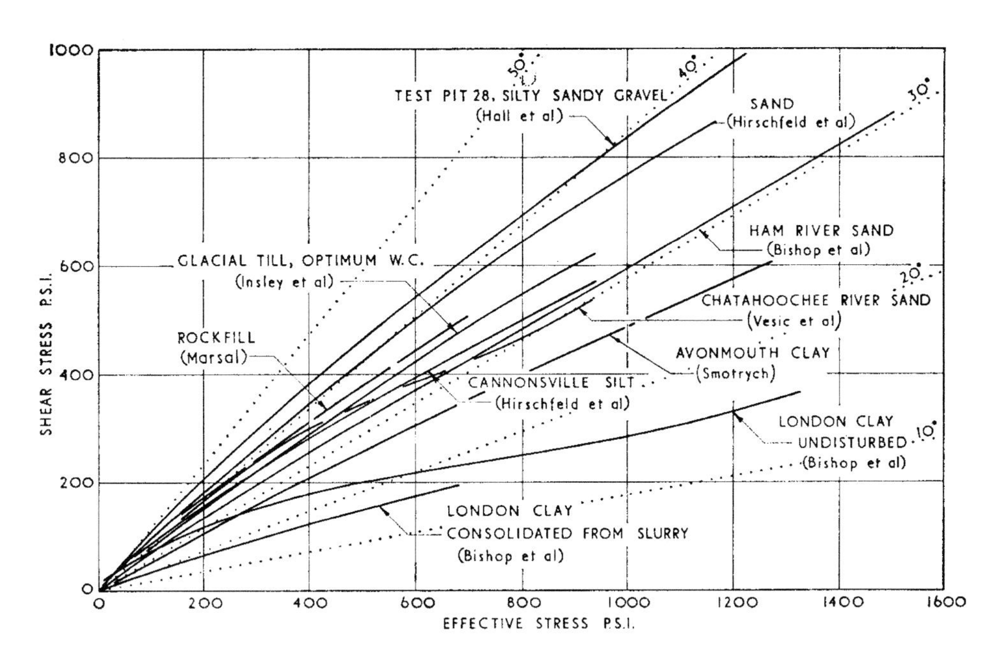

Fig. 17. Mohr envelopes for various soils under high confining pressures (all tests drained except London Clay consolidated from slurry)

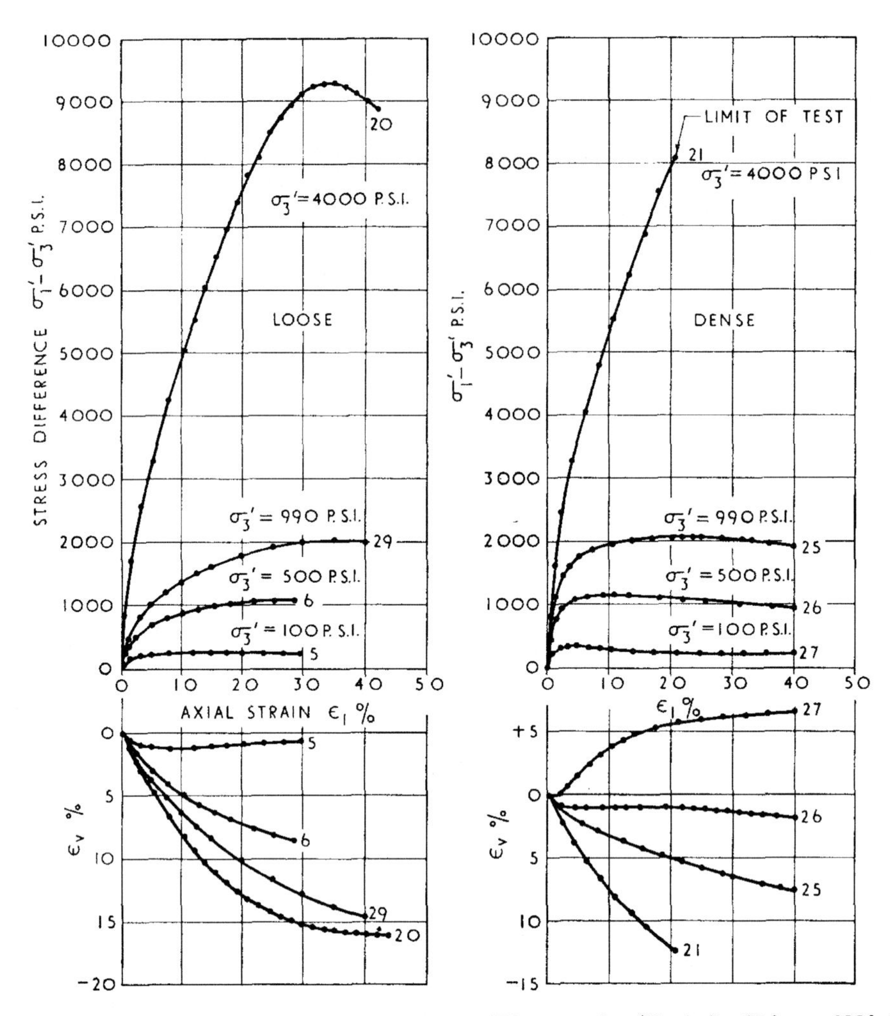

Fig. 18. Results of drained tests on saturated Ham River sand. (Tests by Skinner 1964–66)

with a greatly reduced rate of volume increase at failure, and this leads to a marked reduction in the overall value of ϕ' at failure. The dependence of ϕ' on rate of volume change was demonstrated by Taylor (1948), Bishop (1950), Hafiz (1950), and Bishop and Eldin (1953). A discussion of the various mathematical approaches to the problem is outside the scope of the present paper, particularly as the work done in crushing the particles is usually omitted from the calculations. From a practical point of view it should be noted that only London Clay and Avonmouth Clay gave ϕ' values significantly below 30°.

Typical stress, strain and volume change curves for drained tests on saturated Ham River sand are given in Fig. 18. The marked effect of pressure on the volume change during shear can be clearly seen. For loose sand the reduction in volume during shear rises rapidly with increase in σ_3' up to about 1000 lb/sq. in., and then more gradually as σ_3' is increased to 4000 lb/sq. in. Dense sand, which is strongly dilatant at low confining pressures, shows almost zero rate of volume change at failure when σ_3' reaches 500 lb/sq. in. At higher values of σ_3'

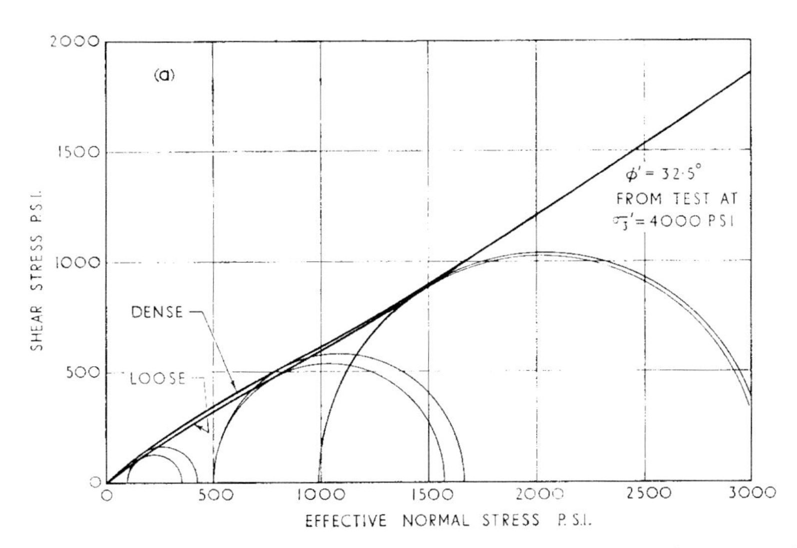

Fig. 19a. Mohr envelopes for drained tests on loose and dense Ham River sand (data from Skinner, 1964–66)

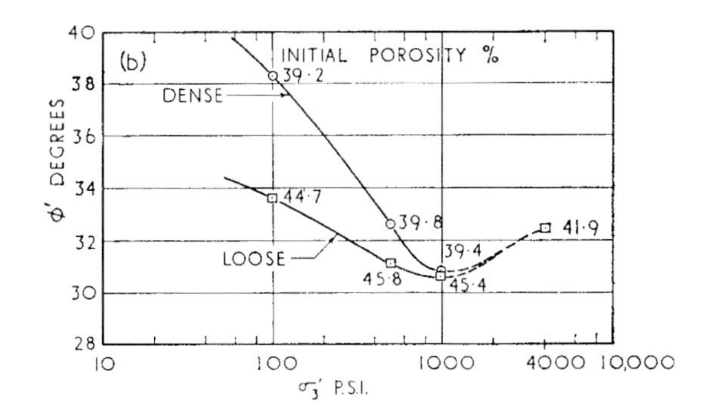

Fig. 19b. Variation of ϕ' with σ_3' (data from Skinner, 1964–66)

dense sand shows an increasingly marked reduction in volume during shear and at $\sigma_3' =$ 4000 lb/sq. in., its behaviour approximated to that of loose sand. For loose sand the stress difference at failure when $\sigma_3' = 4000$ lb/sq. in. exceeds 9000 lb/sq. in.

The Mohr envelopes for loose and dense sand are given in Fig. 19(a). The marked curvature of the envelope for the dense samples and its convergence at high stresses with that for loose samples can be seen. Fig. 19(b) shows that the difference between ϕ' for dense sand and ϕ' for loose sand drops from nearly 5° at $\sigma_3' = 100$ lb/sq. in. to only 0·2° when σ_3' is 1000 lb/sq. in.[12] It is also apparent that even for sand placed in a very loose state ϕ' is not independent of effective stress, but drops about 3° as σ_3' rises from 100 lb/sq. in. to 1000 lb/sq. in. As will be seen later, this drop is closely associated with the rate of volume change at failure.

[12] In this context ϕ' is defined by the tangent to the origin from the stress circles at the value of σ_3' under consideration.

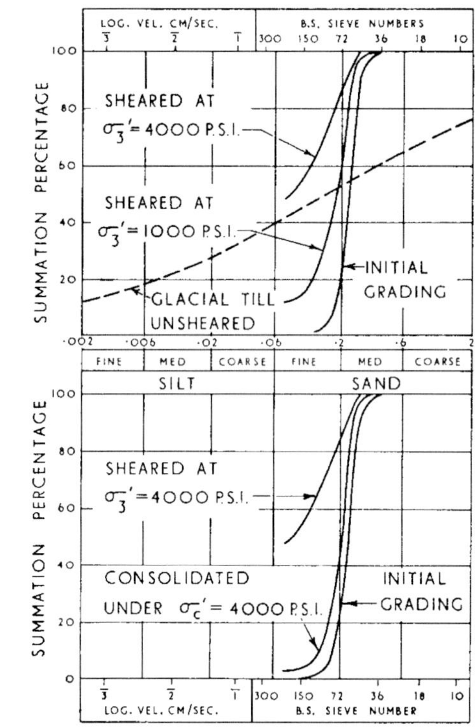

Fig. 20a. Changes in grading resulting (1) from shear at different pressures and (2) from consolidation as compared with complete shear test
Saturated Ham River sand. (Tests by Skinner, 1964–66)

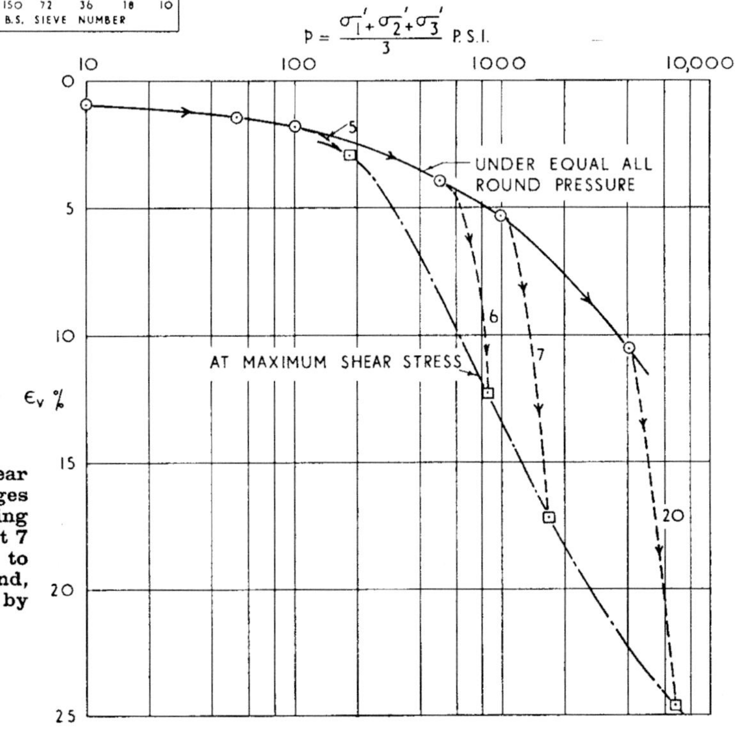

Fig. 20b. Influence of shear stress on volume changes associated with crushing of sand grains. (Test 7 terminated just prior to peak.) Ham River sand, initially loose. (Tests by Skinner, 1964–66)

In Fig. 20(a) the change in grading due to particle breakdown during compression and shear is illustrated. Several points of interest may be noted. Firstly, the combined effect of consolidation and shear leads to very marked particle breakdown even in a medium to fine sand. This effect can also be detected at much lower stresses than illustrated here. Secondly, breakdown results in a grading tending to approximate to that found in naturally occurring glacial tills, for which ϕ' has proved to be relatively insensitive to stress (Fig. 17; see also Insley *et al*, 1965).[13] This suggests the type of grading likely to be suitable for fills to withstand high stresses. Thirdly, at high stresses the particle breakdown occurs to a much greater extent during the shear stage than during the consolidation stage.

This latter point is further illustrated by plotting the associated volume changes against the average stress $\frac{\sigma_1'+\sigma_2'+\sigma_3'}{3}$ in Fig. 20(b). Where the increase in average stress is associated

[13] Further data on the breakdown of sand particles with normal and shear stress is given by Vesic and Barksdale (1963) and Borg, Friedman, Handin and Higgs (1960). The breakdown of rockfill is discussed by Marsal (1965). These points are further discussed by Bishop (1965).

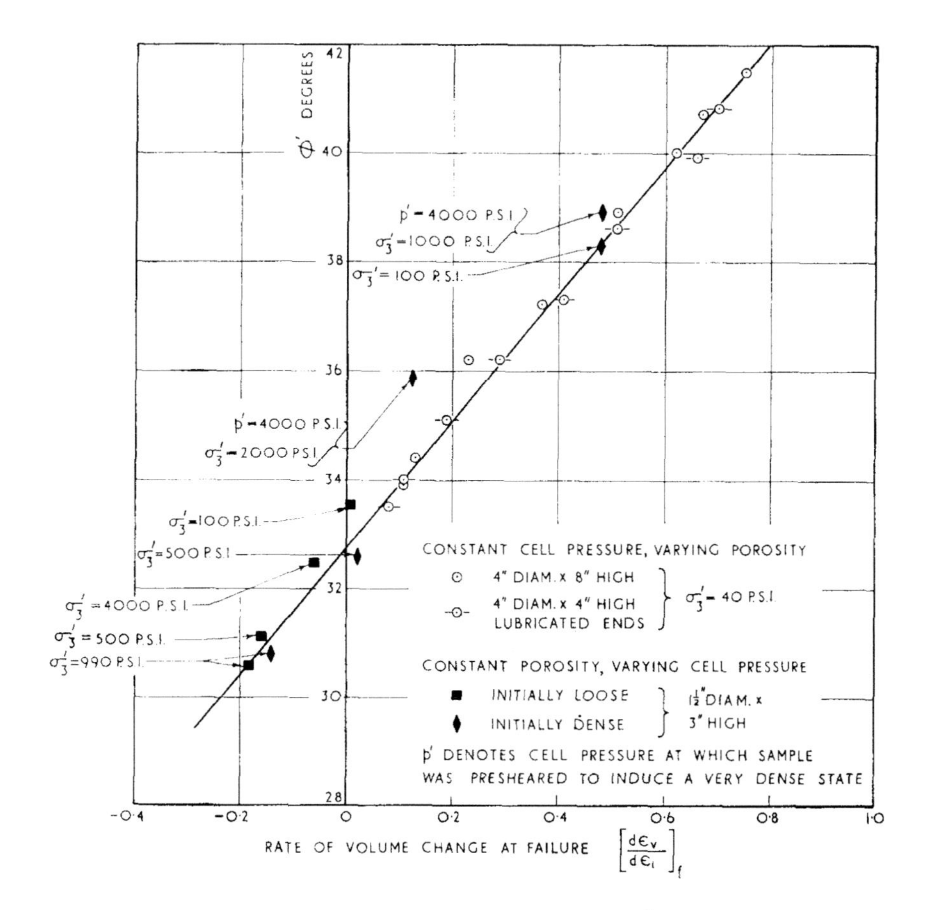

Fig. 21. The relationship between ϕ' and rate of volume change $\frac{d\epsilon_v}{d\epsilon_1}$ at failure. (Tests by Green and Skinner, 1964–66)

with a stress difference (or shear stress) the rate of volume decrease is many times greater than under all round pressure, except at low consolidation pressures.

In Fig. 21 the relationship between the value of ϕ' and the rate of volume change $\frac{d\epsilon_v}{d\epsilon_1}$ at failure (or dilatancy rate) is given. At a given pressure the dilatancy rate will vary with the initial density at which the material is placed, and a very close correlation with ϕ' will be noted. At a given initial density an increase in pressure will reduce the rate of dilatancy, and it is of interest to note how near to the previous line values obtained in this way actually lie. This is all the more remarkable when the extremely wide stress range is considered, from $\sigma_3'=40$ lb/sq. in. to $\sigma_3'=4000$ lb/sq. in. For this particular sand (Ham River), the maximum departure from the mean line is only $1\frac{1}{2}°$. The curvature of the failure envelope is thus largely accounted for by the decrease in the rate of dilatancy with increasing stress.

The variation in the rate of volume change at failure with increase in effective stress σ_3' is plotted in Fig. 22. It appears that for Ham River sand the change from positive to negative rate of dilatancy at failure occurs mainly within the stress range $\sigma_3'=0-1000$ lb/sq. in. At higher stresses the trend is reversed and failure at almost constant rate of volume change occurs after the large initial decrease in volume illustrated in Fig. 18. At failure, however, the material is no longer a fine to medium sand but a well graded silty sand with nearly 50% in the silt sizes or smaller (Fig. 20a). The greatly increased number of interparticle contacts carrying the stresses must largely counterbalance the effect of the reduction in the basic coefficient of interparticle friction with pressure (see, for example, Hafiz, 1950).

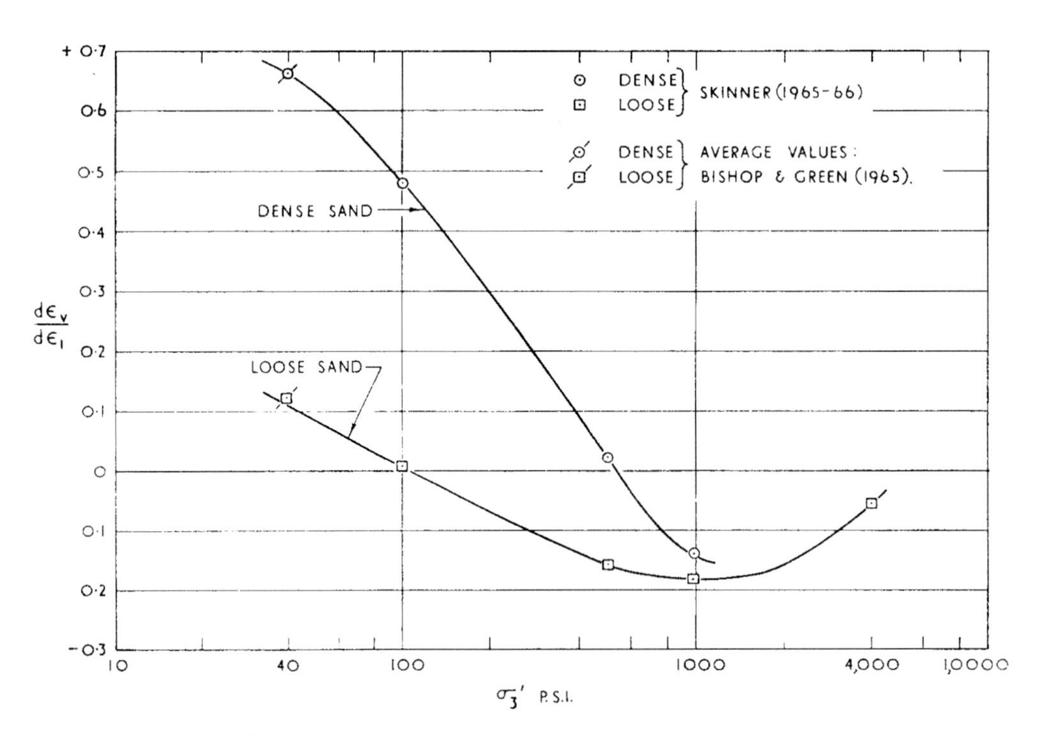

Fig. 22. Variation of rate of dilatancy at failure with increase in effective stress; saturated Ham River sand

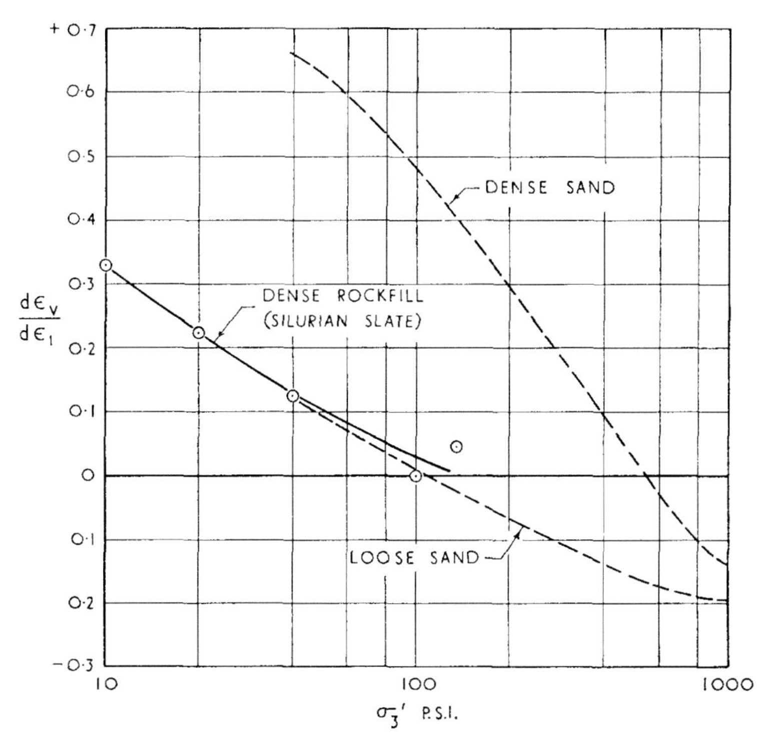

Fig. 23. Variation of rate of dilatancy at failure with increase in effective stress: dense rockfill (Silurian slate). (Tests by Tombs, 1966)

A comparison with the results of tests on 1-ft diameter samples of a compacted rockfill is made in Fig. 23. Within the working range of stress the dilatancy characteristics of the compacted rockfill (Silurian slate) approximate more closely to those of a loose sand than those of a dense sand. This is consistent with the very marked particle breakdown observed during the tests.

Provided that the influence of high pressure on ϕ' and on volume change is measured and that designs are not based on the extrapolation of low pressure tests, little difficulty is involved under drained conditions in achieving safe designs. However, fully saturated granular material under these stresses could be very dangerous under undrained or shock loading, as structural breakdown gives it an undrained stress path very similar to that of quick clay. The curves showing the relationships between stress, strain, and the build up of pore pressure are given in Fig. 24. The very small strain at failure in test 8 should be noted. The stress paths are given in Fig. 25.[14]

It will be seen that the shape of the stress path of test 8 is very similar to that observed in the low stress range for very loose sands which are susceptible to flow slides (Waterways Experiment Station, 1950; Bjerrum, 1961), and for sensitive clays (Taylor and Clough, 1951; N.G.I., unpublished data). In particular it will be noted that at the maximum value of the stress difference (represented by $\frac{1}{2}[\sigma_1' - \sigma_3'])$ the value of ϕ' mobilized is 21·3° for test 8 (Fig. 25); thereafter the rise in pore pressure more than compensates for the increase in ϕ' as the residual state is approached.

[14] Further details of these tests are given by Bishop, Webb and Skinner (1965).

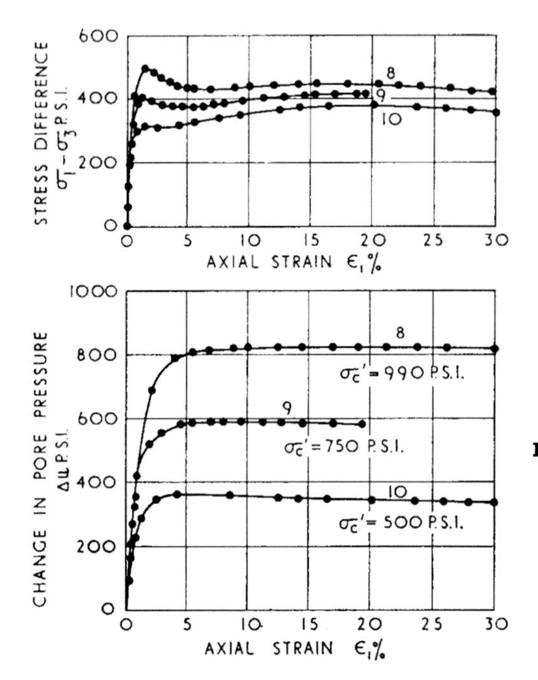

Fig. 24. Results of consolidated-undrained tests on saturated Ham River sand. (σ_c' denotes effective consolidation pressure). (Tests by Skinner, 1964)

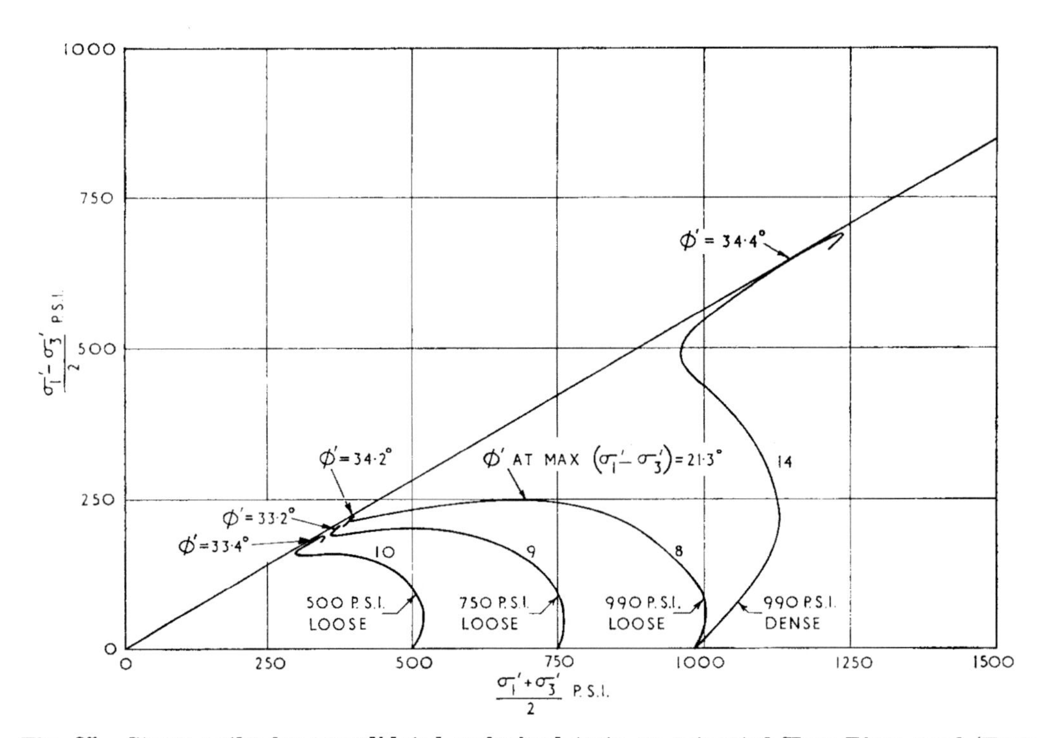

Fig. 25. Stress paths for consolidated-undrained tests on saturated Ham River sand (Tests by Skinner, 1964–66)

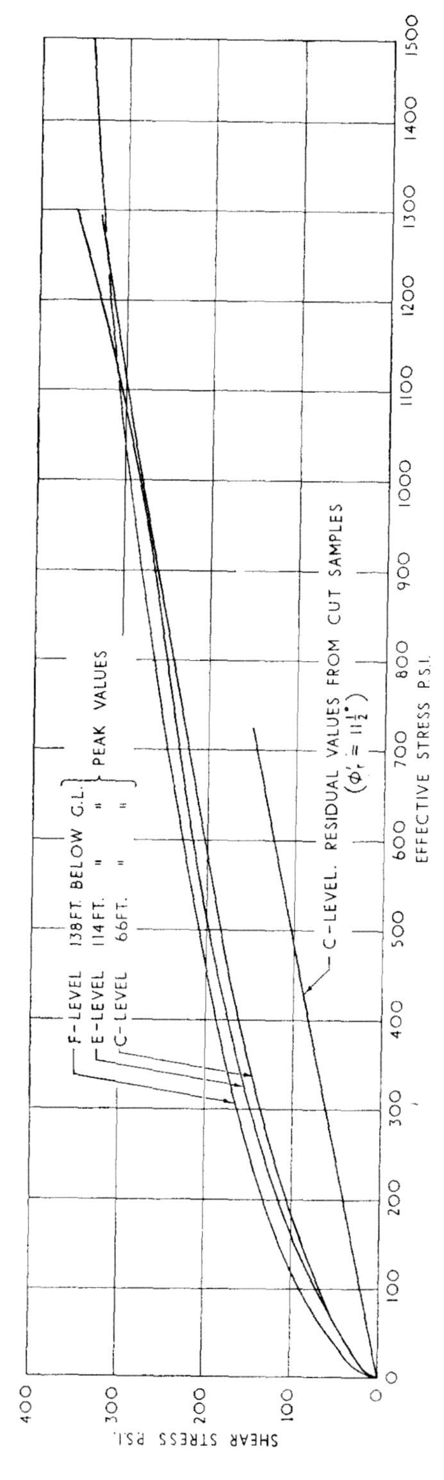

Fig. 26. Mohr envelopes for undisturbed London Clay from Ashford Common shaft (after Bishop, Webb and Lewin, 1965)

It would appear, therefore, that fill to be used fully saturated, and under high stresses, should preferably be well graded and should in any case be placed in a dense state, so that under undrained loading the pore pressure build-up is reduced and the undrained strength increased as illustrated in test 14 (Fig. 25).

There is not space to deal in detail with the behaviour under a wide range of stress of clay as an undisturbed material and as a compacted fill. One illustration must suffice. Fig. 26 shows the Mohr envelopes for undisturbed London Clay from depths of 66, 114, and 138 ft below ground level at the Ashford Common shaft.[15] The change in slope from 30° near the origin to 10° at very high stresses will be seen (the latter corresponds to an origin $\phi'=15°$). For the same material consolidated from a slurry the value of ϕ' obtained from undrained tests with pore pressure measurement (Fig. 17) drops only from 21° near the origin to about 16° (tangent from the origin). The residual angle ϕ' from pre-cut undisturbed samples from a depth of 66 ft dropped only from $11\frac{1}{2}°$ to 11°.

(3) THE IN-SITU UNDRAINED STRENGTH OF A SOIL

In using the undrained strength of a clay in the $\phi_u=0$ analysis for short-term loading the engineer is apt to imagine that, apart from sampling disturbance, a given sample of clay has a unique undrained strength, irrespective of the type of test (triaxial, vane or simple shear test) used to measure it and irrespective of the inclination or direction of the slip surface implied in the problem he is analysing.

In fact, as early as 1949, Professor J. Brinch Hansen and Dr R. E. Gibson showed that on theoretical grounds the laboratory undrained compression test should differ from the field vane test (Table 1) and that the in-situ strength on an inclined failure surface could differ from them both, being greatest in the case of active earth pressure and least in the case of passive earth pressure. This prediction was for a soil consolidated with zero lateral yield, i.e. the initial lateral effective stress was only K_0[16] times the vertical effective stress. The soil was thus subjected to an anisotropic stress history, but no anisotropy of the shear strength

Table 1

Predicted influence of orientation of shear plane on undrained strength (after Hansen and Gibson, 1949)

Value of $\frac{c_u}{p}$	Clay 1	Clay 2
Active earth pressure	0·331	0·282
Passive earth pressure	0·193	0·256
Failure on a horizontal plane . .	0·213	0·262
Vane test	0·191	0·252
Unconfined compression test:		
condition (a)	0·170	0·247
condition (b)	0·250	0·283

Clay 1. Normally consolidated, sensitive silty clay.
Clay 2. Normally consolidated, typical British post-glacial clay.
Condition (a), Condition (b): Probable limits of influence of stress release on sampling.

[15] The index properties at the 114 ft level, for example, are $W_L=70$, $W_p=27$, clay friction $=57$, activity $=0{\cdot}75$, initial water content $=24{\cdot}2$. For full details see Bishop, Webb and Lewin (1965).

[16] K_0 is termed the coefficient of earth pressure at rest and is generally within the range of 0·4 to 0·7 for normally consolidated clays.

parameters in terms of effective stress was assumed in estimating the undrained strength. Their results thus indicated the influence on the excess pore pressure at failure of stress history and, in particular, of the rotation of principal stress directions in an anisotropically consolidated soil.

Their conclusions have been rather lost sight of in recent years, due to the renewed emphasis on the unique relationship between undrained strength and water content. However, test results both in the field and in the laboratory, though still rather limited, suggest that Hansen and Gibson's conclusions are substantially correct, though complicated by two additional features. These are that

(*a*) the soil structure may in fact be significantly anisotropic with respect to its shear parameters in terms of effective stress, due to orientation and/or segregation of particles during deposition, and due to orientation arising from its subsequent history;

(*b*) the degree of mobilization of these parameters (c' and ϕ') at the peak stress difference varies with the orientation of the principal stresses at failure, as does the strain at failure.

Table 2

Some examples in which c_u depends on principal stress directions during shear (tests in situ or on undisturbed samples)

1. c_u from field vane lower than c_u from piston sampler or block samples. (Vold, 1956; Coates and McRostie, 1963).
2. c_u from field vane for vertical plane lower than for horizontal plane. (Aas, 1965.)
3. c_u from block samples with axis horizontal lower than with axis vertical in lightly over-consolidated clay. (Lo, 1965)
4. c_u from block samples with axis horizontal higher than with axis vertical in heavily over-consolidated clay. (Ward, Marsland and Samuels, 1965)
5. c_u/p for simple shear (max shear stress horizontal) lower than in triaxial compression with axis vertical. (Bjerrum and Landva, 1966)
6. c_u/p for axial extension much lower than in axial compression for samples consolidated with zero lateral yield. (Ladd and Bailey, 1964; Ladd and Varallyay, 1965)

Some of the test results which bear on this problem are listed in Table 2. Of particular interest are the field tests (Fig. 27) carried out by the Norwegian Geotechnical Institute with a series of specially proportioned vanes (Aas, 1965), as they indicate the relative magnitudes of the in-situ undrained strength on the vertical and horizontal planes respectively. The ratio varies between $\frac{1}{2}$ and $\frac{2}{3}$ for the three normally consolidated clays tested. Since the conventionally proportioned vane measures mainly the strength on a vertical cylindrical surface it may underestimate the field value relevant to some engineering problems in normally or lightly over-consolidated clays. The values are for sensitive or quick clays; values for the clays usually encountered in Britain would be of great interest.

The variation in the undrained strength of lightly and heavily over-consolidated clay with the direction of the applied major principal stress is illustrated in Fig. 28. The samples were all cut from blocks taken from vertical shafts[17] and average values are given based on a large number of tests detailed in the references. In the lightly over-consolidated Welland clay the

[17] With the exception of the tests described by Bishop (1948).

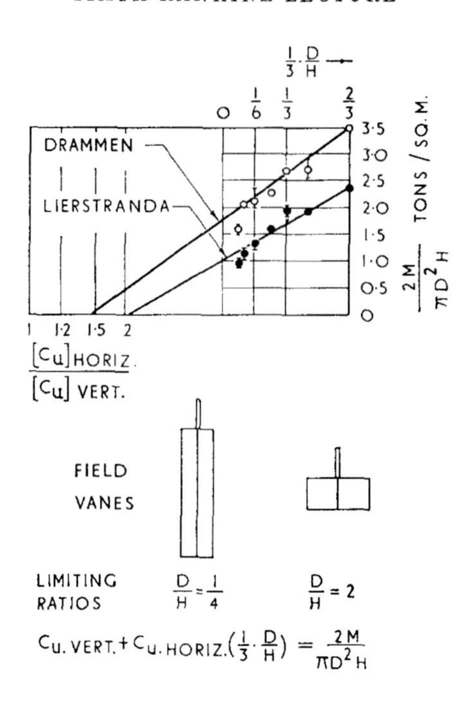

Fig. 27. Determination of anisotropy ratio $\frac{(c_u)\text{ horiz.}}{(c_u)\text{ vert.}}$ from undrained tests with vanes of different $\frac{D}{H}$ ratios. (After Aas, 1965)

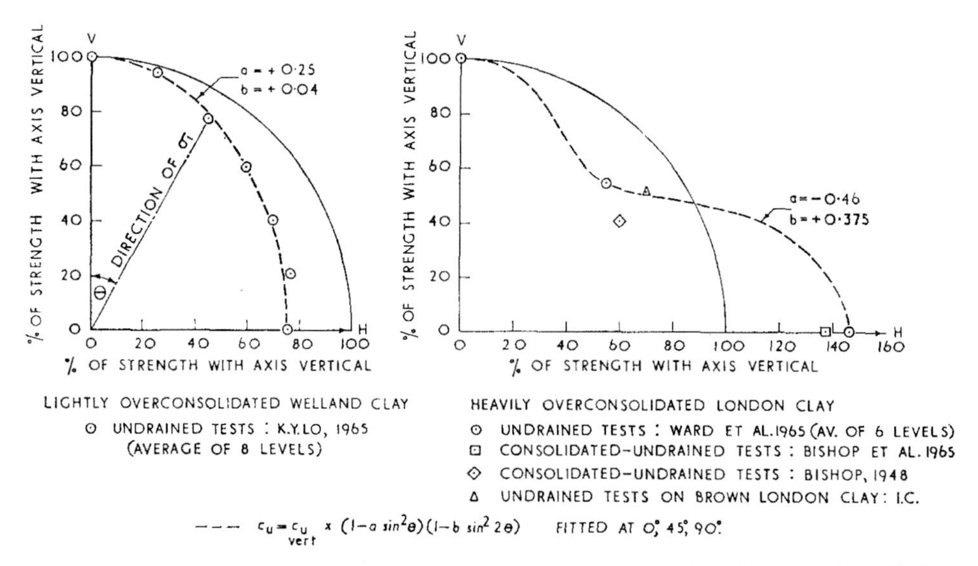

Fig. 28. Polar diagram showing variation of undrained strength with direction of applied stress: θ denotes inclination of major principal stress with respect to vertical axis

compression strength with σ_1 horizontal was about 0·75 the value with σ_1 vertical.[18] In the heavily over-consolidated London Clay the ratio of horizontal to vertical strength was 1·46 (range 1·23 to 1·63). That this difference is primarily a pore pressure phenomenon is indicated by the associated consolidated–undrained tests (Bishop, Webb and Lewin, 1965), which show ratios of the same order (e.g. 1·35), but little difference in the effective stress envelopes, the A values, however, being +0·42 for the vertical sample and +0·19 for the horizontal sample quoted.

However, inclined samples of London Clay show a great reduction in strength, presumably associated with lower shear strength parameters along the bedding planes, the undrained strength with σ_1 at 45° (i.e. the maximum shear stress parallel to the bedding planes) being 0·77 of the vertical value. A simple expression with two parameters, a and b, can be used to express these results: a reflecting the influence of pore pressures and b the directional character of c' and ϕ' as well as that of pore pressure.[19] Even in the lightly over-consolidated Welland clay the value of the second term suggests that an orientated structure is becoming manifest. Undrained tests on block samples from the brown London Clay at Maldon show a similar drop in strength in the direction of the bedding planes, as also do the earlier consolidated-undrained tests on London Clay from Walton (Bishop, 1948).

These strengths are not, of course, identical with the values of the in-situ undrained strengths in plane strain, which are of principal interest in many engineering problems, and the influence, not only of stress release on sampling,[20] but also of sample size, is of particular interest in dealing with over-consolidated clays.

Before examining these factors in more detail it should be pointed out that a drop in strength of 50% in the horizontal direction leads, in the circular arc analysis of a typical slope (Lo, 1965), to a reduction in factor of safety of only 15%–30% since the whole of the slip surface is not inclined at the least favourable angle. This, together with the reduction in measured strength which always results from sampling, has probably made the $\phi_u = 0$ analysis appear more accurate than, theoretically, it should be.

The importance of the two factors, anisotropy and the size of sample, in practical design work may be illustrated by tests carried out on the weathered London Clay foundation of a proposed embankment near Maldon in Essex. Owing to the great base width of the embankment and to the low coefficient of consolidation of the clay, undrained failure on an almost horizontal slip surface in the weathered zone represented a critical condition. I therefore asked for the normal site investigation to be augmented by a number of in-situ undrained direct shear tests on samples 2 ft × 2 ft in cross-section with their zone of maximum shear stress in the horizontal plane.

The layout of the testing equipment is shown in Fig. 29. An intact block of clay was left projecting 6 inches above the floor of the trial pit, and with the minimum of delay the shear box and loading platten were fitted over it. A load equivalent to the overburden pressure was applied through a hydraulic jack mounted beneath a strut transmitting the load to a joist

---

[18] The interpretation of cylindrical compression tests on samples cut with their axis horizontal is open to some ambiguity in a soil which may have a low undrained strength when the plane of failure is vertical, but the relative motion horizontal, as in the tests by Aas (1965). Failure may occur on such a plane in the compression specimen cut with its axis horizontal, rather than on an inclined plane forming part of a conventional plane strain failure surface, as assumed by Lo (1965).

[19] The first term agrees with that used by Lo (1965) and Casagrande and Carrillo (1944). It is simpler than that proposed by Hansen and Gibson (1949), but lack of detailed information hardly justifies a more elaborate expression. The second term is assumed to represent a satisfactory working hypothesis, though, based on work by Hill (1950) on metals in plane strain, Scott (1963b) assumes a similar term to the power $-\frac{1}{2}$.

[20] Bishop and Henkel (1953), Ladd and Lambe (1963), Skempton and Sowa (1963) and Ladd and Bailey (1964) have dealt with this problem either with no rotation of the principal stresses or (Ladd and Bailey) with the special case of axial extension.

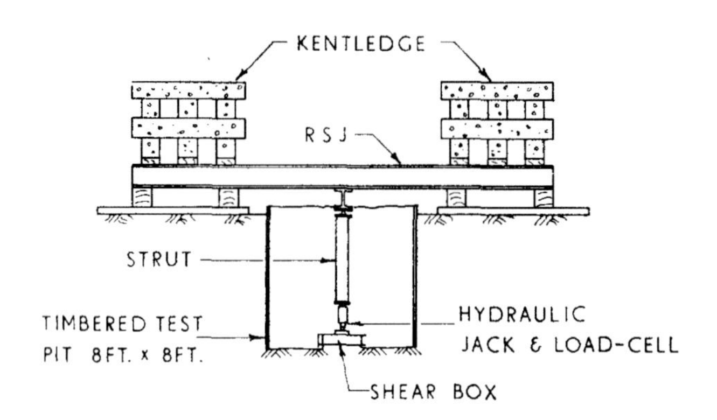

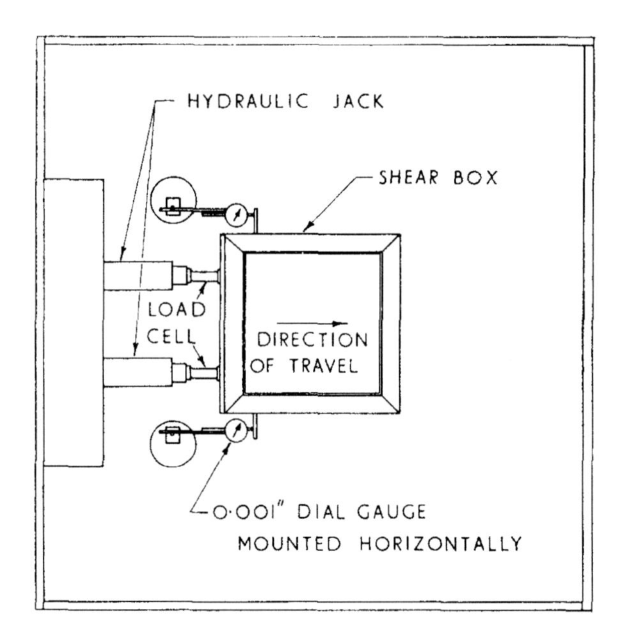

Fig. 29. Layout of direct shear test on 2 ft × 2 ft samples in the field

passing beneath kentledge on either side of the trial pit. The horizontal load was applied through two hydraulic jacks fitted with electrical load cells. Any tendency of the box to run out of the true could thus be controlled. A pair of dial micrometers recorded the horizontal displacement of the box.

The shear stress–displacement curve for the test at a depth of 11 ft in trial pit 3 is given in Fig. 30. The very small displacement (0·3 in.) at which the peak stress was reached may be

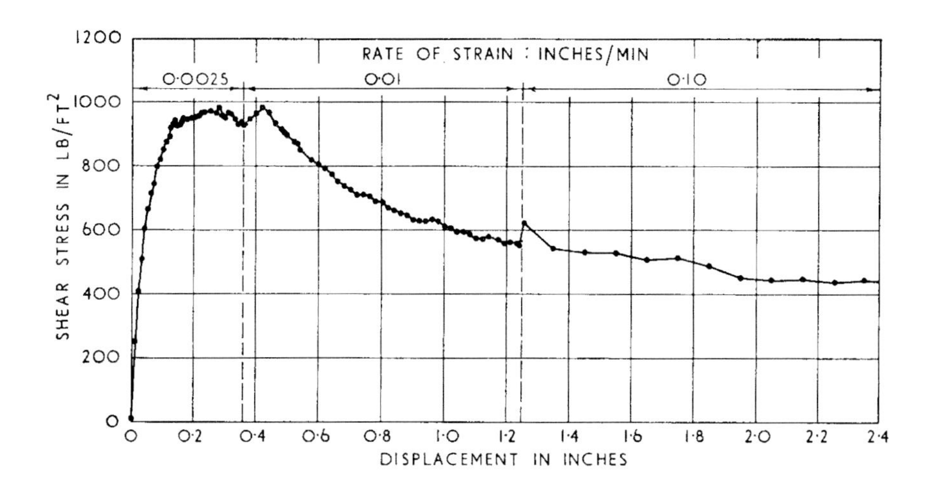

Fig. 30. Shear stress-displacement curve for undrained direct shear test on 2 ft × 2 ft sample of brown London Clay at Maldon, Essex: horizontal shear plane 11·3 ft below surface

noted. The rate of strain was controlled so that the peak stress was reached after about 1 to 2 hours. In this test, which was one of the two deep tests performed, the stress then fell off and appeared to reach an almost constant value when the limit of travel of the shear box was reached. At this stage careful sectioning of the sample revealed one or more pronounced slip surfaces running almost horizontally across the sample near the base of the shear box.

The undrained strength–depth plots are given in Fig. 31. The most notable feature is that the strengths obtained with horizontal shear on large samples are, on the average, only 55% of the strengths obtained by testing compression specimens with their axis vertical either from boreholes samples, or from tube or block samples taken in the trial pits. A reduction of this magnitude makes a conventional factor of safety of 1·5 on a conventional undrained test result on London Clay appear rather inadequate, and it is difficult to see on what grounds we can fault the in-situ tests.

Four factors may be considered in assessing the significance of the difference.

(1) *Time to failure.* The time to failure (about one hour) in the field test is greater than the duration (about 5 minutes) of the laboratory test. Other studies (La Rochelle, 1961) suggest that the effect on strength of this difference amounts to only a few per cent, but in relation to construction work the slower test in any case is the more correct.

(2) *Stress conditions in the direct shear test.* The principal stress directions in the shear box at failure are not known very precisely. An error of $(1 - \cos \phi_e) \times 100\%$ is possibly involved in estimating c_u from a shear box.[21] This is generally less than 5% for a clay of high plasticity.

(3) *Anisotropy.* Tests on block samples in the laboratory orientated so that the slip plane lay in the horizontal direction gave undrained strengths 86% of the strengths

[21] This is still controversial. Reference can be made, for example to Hill (1950), and Hansen and Gibson (1949). The interpretation of the shear box test and simple shear test will clearly be influenced by anisotropy.

with the axis vertical.[22] Similar tests on samples cut from cores taken in the trial pit with a 4-in. diameter sampler show a reduction of 87% of the strength with the axis vertical.

(4) *Sample size.* The major part (from 86%–55%) of the reduction in strength must therefore be attributed to the use in the field of a large and more representative sample.

The results of four large plate loading tests are also included on the strength–depth plot for trial pit 4. These again indicate strengths much below the values given by small laboratory samples, though rather higher than the values given by the large direct shear tests. This latter difference[23] probably reflects the fact that in the plate loading tests the slip surface is inclined to the direction of the bedding planes over much of its area, and the influence of anisotropy is reduced.

[22] Anisotropy in a larger sample may of course be more marked than in a small one due to the inclusion of a more representative structure.
[23] This difference may have been to some extent masked by the limited displacement applied in the plate loading tests.

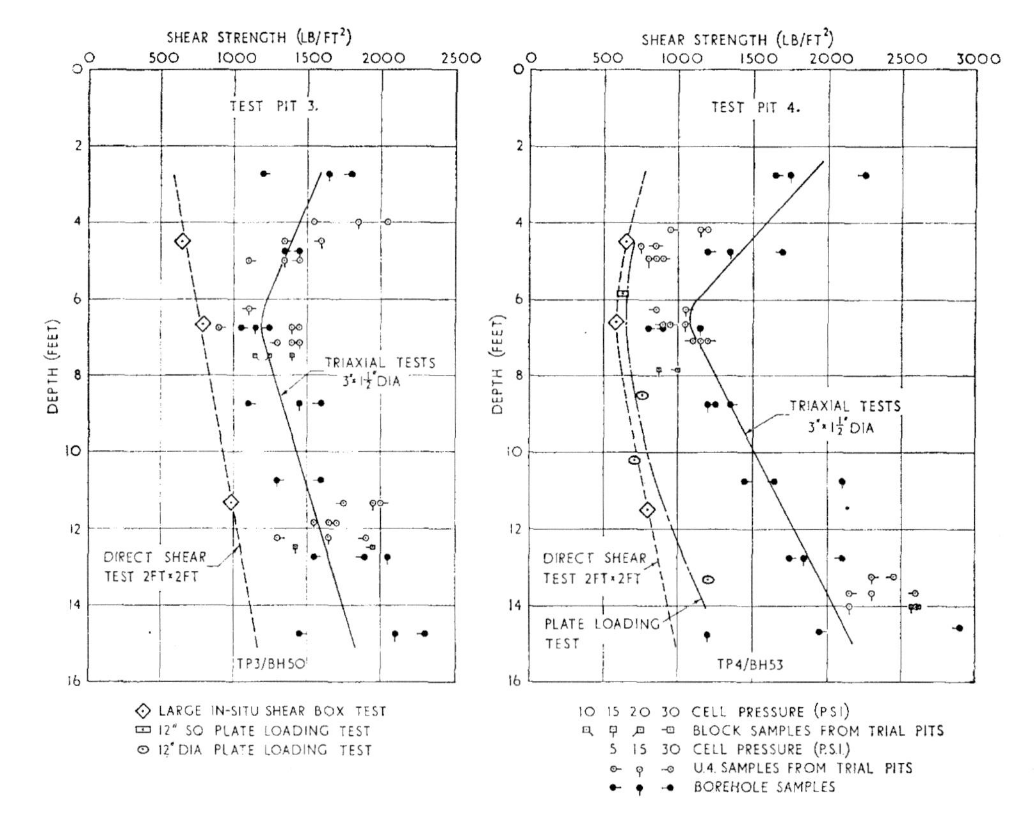

Fig. 31. Relationship between undrained strength and depth: results of 2 ft × 2 ft direct shear tests and plate loading tests compared with values obtained from 3 in. × 1½ in. dia. triaxial tests: brown London Clay from Maldon, Essex

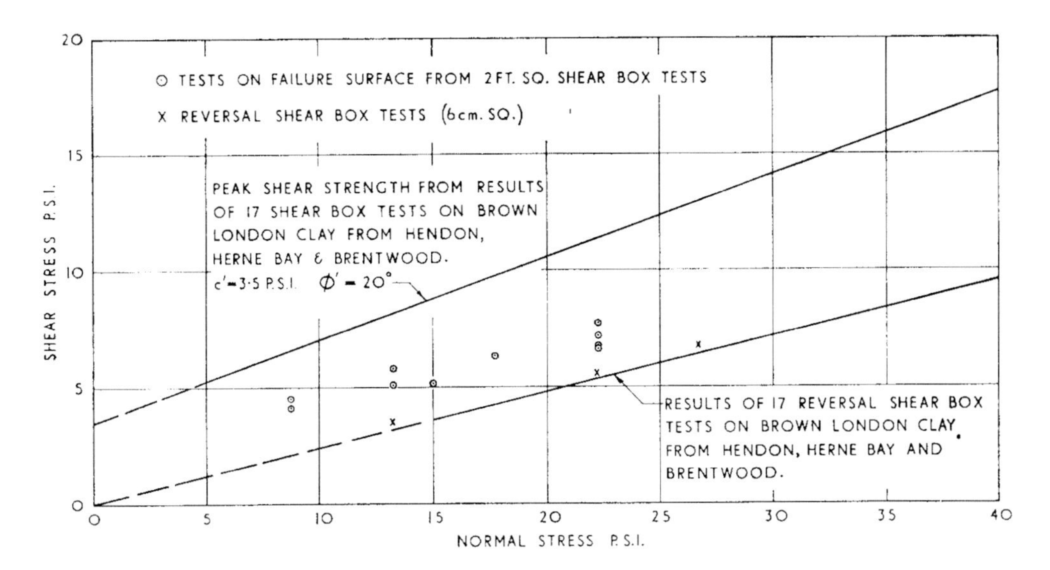

Fig. 32. Results of drained tests on samples from failure surface of 2 ft square in-situ shear box tests. (Tests by Petley, 1966)

Whitaker and Cooke (1966) have reached a similar conclusion for the deeper layers of the London Clay from the ultimate base loads measured on large bored piles. The in-situ undrained strength was, on the average, only 75% of that given by standard borehole samples tested with the axis vertical.

It is of interest to note that, in similar clay at Bradwell, Skempton and La Rochelle (1965) found that the 'undrained' strength mobilized in a slide in the side of an excavation was only about 55% of the undrained strength of 3-in. × 1½-in. diameter samples, either from boreholes or cut from blocks. In this case the slip surface was steep and not in the direction of the bedding planes, but a larger time elapsed before failure, which in itself could have accounted for a reduction to 80% of the value measured in the standard laboratory test. The further reduction of 30% due to size effect in a soil with a fissured structure at Bradwell is thus about the same as the reduction due to size effect at Maldon. This suggests that at this level in the London Clay a 2-ft square sample was adequate.

These differences overshadow refinements in our methods of analysis and suggest that we may need to rethink the means by which we determine undrained strength in engineering practice. Either samples must be large enough to include a fully representative soil structure (this could mean even larger samples in some cases than the size tested at Maldon) and they must be tested with the correct orientation, or we must apply empirical factors to the strength of our conventional laboratory specimens.

To obtain *drained* shear parameters for large samples is much more difficult. The test duration necessary for full pore pressure dissipation in the shear test described above would have been 6 months or more[24] for a clay of the type tested at Maldon. However, large scale drained tests on stiff fissured clays are clearly essential to the rational design of engineering works in or on these strata.

[24] At low stresses preferential drainage through the fissures might accelerate the consolidation of weathered samples.

It is perhaps of interest to add that drained tests run on sections cut from the Maldon in-situ tests so as to include part of the actual slip surface showed that the residual value of ϕ' had not been reached at the displacement of 2 to 3 in. applied in the tests. The test results are given in Fig. 32. The values of the residual factor R (Skempton, 1964) for 6 cm. square samples vary between 0·45 and 0·8.

(4) THE INFLUENCE OF TIME ON THE STRENGTH OF SOILS

The strength of a given soil stratum which is available to the engineer depends on time in a number of different ways.

Under a sustained load the available strength of a clay stratum changes from the undrained strength to the drained strength at a rate which depends on the coefficient of consolidation or swelling and on the length of the drainage path. The change is an increase in most foundation problems where the load increases, and a decrease in most excavation or cutting problems, where the load is decreased, and results in the first place from pore-water pressure changes in the field (for a detailed discussion see Bishop and Bjerrum, 1960).

However, as was shown by Professor Skempton in the fourth Rankine Lecture, the shear strength parameters calculated from actual slips based on a knowledge of the field pore pressure differ radically from the peak strength values measured in the laboratory in the case of over-consolidated clays of other than low plasticity.

Several factors may contribute to this discrepancy,

(*a*) the size and orientation of the test specimens will influence the values of the shear strength parameters measured in the laboratory;

(*b*) in soils showing brittle or work-softening stress–strain characteristics it is likely that failure in the field will be to some extent progressive, i.e. that the peak strength will not be mobilized simultaneously along the complete slip path. As pointed out by Webb, Lewin and myself in 1965, the release of stored energy on stress reduction under drained conditions in clay showing marked swelling characteristics is of special importance in this case. This approach has been developed by Dr Bjerrum in his recent Terzaghi Lecture to the American Society of Civil Engineers;

(*c*) the peak values of the drained shear strength parameters may be substantially time dependent in heavily over-consolidated clays and clay shales.

Little precise information exists on all three factors. Almost nothing really appears to be known about the time dependence of the drained peak strength of undisturbed clays. The technical difficulties of maintaining a known constant stress difference on a sample in an apparatus without leaks over periods of months and possibly years are considerable. The apparatus we are currently using at Imperial College is illustrated in Fig. 33. Several important features may be noted. The whole loading and strain measuring system is inside one continuous pressure vessel filled with oil, so that friction due to a seal on the loading ram is entirely avoided. The load is applied by two very long springs in tension, controlled by a screw adjustment at the base of the cell, and is transmitted to the cylindrical sample through a ram guided by a ball bushing. The load can be determined both from the length of the springs and from a proving ring mounted on the ram. Creep in the sample has little effect on the load in the springs due to their large extension, but in the early stages of the test such adjustment as is necessary due to the shortening of the sample and the change in its cross-sectional area can readily be made with the screw at the base of the cell. Deformation is measured by an oil filled dial micrometer reading to 10^{-4} in. and by a linear differential transformer reading to 10^{-5} in.

The sample is enclosed in a rubber membrane and is submerged in mercury to prevent loss of water through the rubber membrane, and to protect the membrane from contact with the

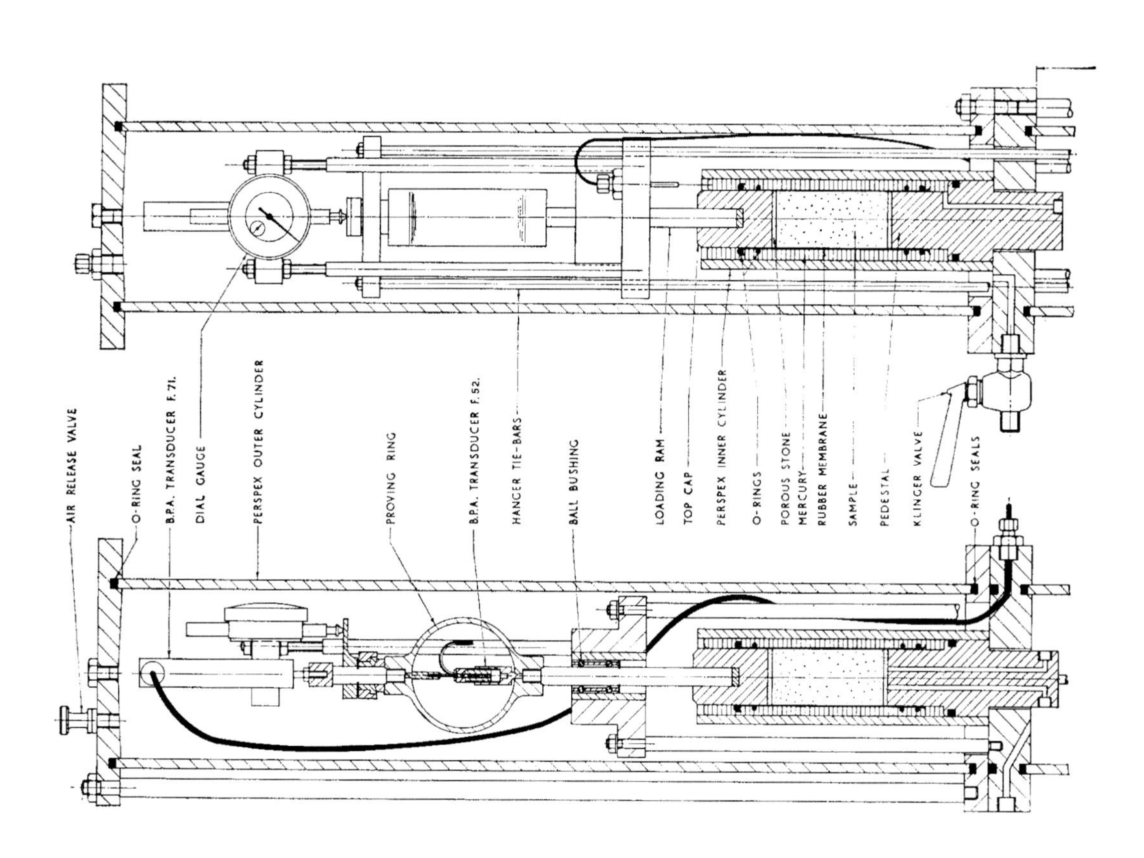
AIR RELEASE VALVE
O-RING SEAL
B.P.A. TRANSDUCER F.71.
DIAL GAUGE
PERSPEX OUTER CYLINDER
PROVING RING
B.P.A. TRANSDUCER F.52.
HANGER TIE-BARS
BALL BUSHING
LOADING RAM
TOP CAP
PERSPEX INNER CYLINDER
O-RINGS
POROUS STONE
MERCURY
RUBBER MEMBRANE
SAMPLE
PEDESTAL
KLINGER VALVE
O-RING SEALS

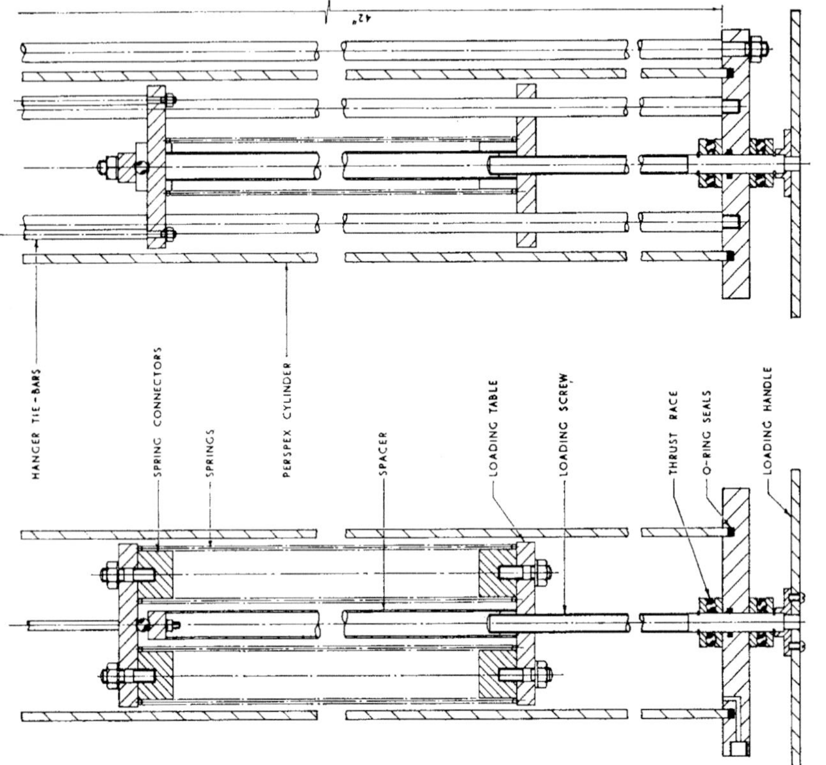

Fig. 33. Diagrammatic sections of creep cell

mineral oil. Volume change is measured with a paraffin volume gauge (Bishop and Henkel, 1962, Fig. 141) using a back pressure to ensure full saturation of the system. Volume change is also measured with a mercury-filled volume gauge, back-pressured from the cell, which is adjusted to maintain the mercury in the inner cell surrounding the sample to a constant level, determined by an electric contact.

The use of a spring-loaded system of low inertia makes the apparatus less susceptible to tremors transmitted by the structure of the building than the use of a dead load system.

The first series of tests (Fig. 34) on block samples of London Clay has been running for about seven months and has already produced some interesting information. Samples were set up and kept under sustained shear stresses, the stress levels being approximately 90%, 80%, 70%, 60%, 40%, and 16% of the peak drained strength in a test of one week's duration. These percentages are based on the results of six triaxial tests of this duration. The percentage shown first in Fig. 34 is calculated from the average of the two highest values observed. The second percentage is based on the four lower values in the series.

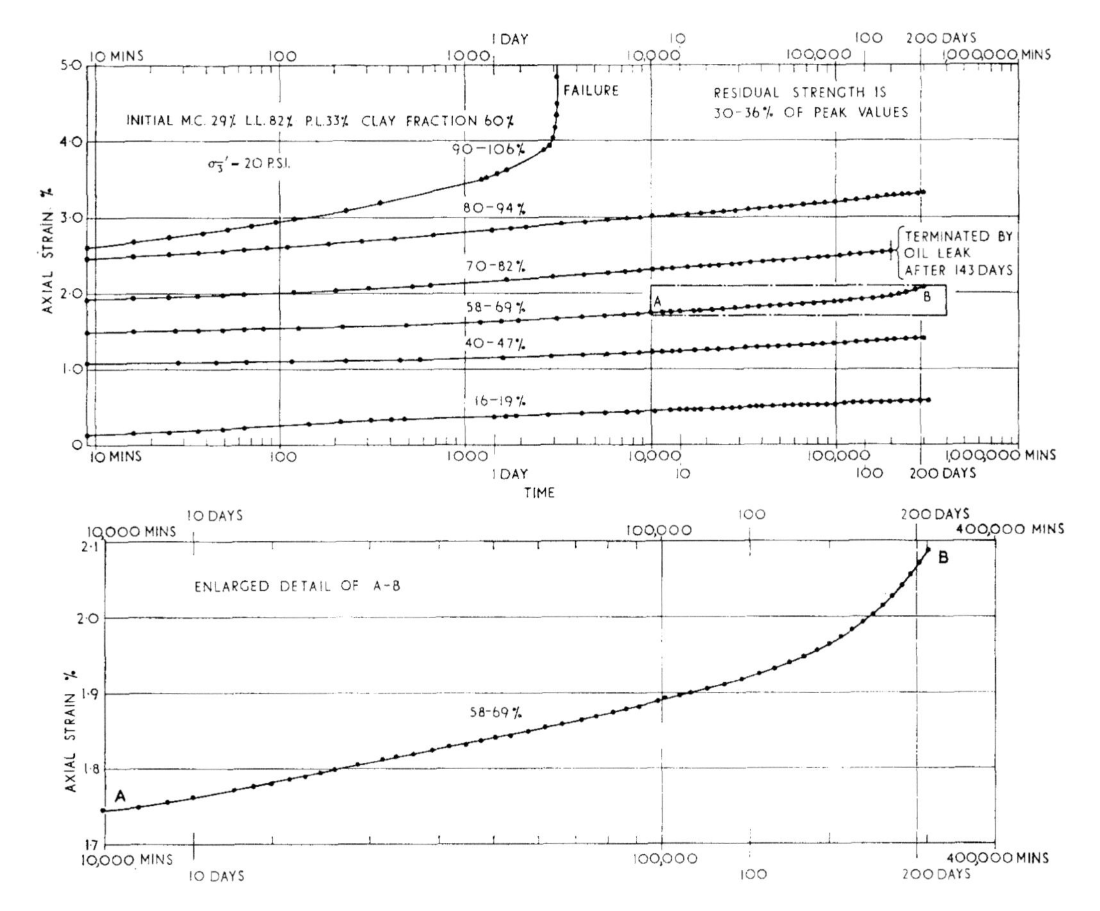

Fig. 34. Drained creep tests on undisturbed brown London Clay from Hendon. The applied principal stress differences as given above are percentages of the peak values measured in drained triaxial tests of 5 days' duration. (Tests by Lovenbury, 1965–66)

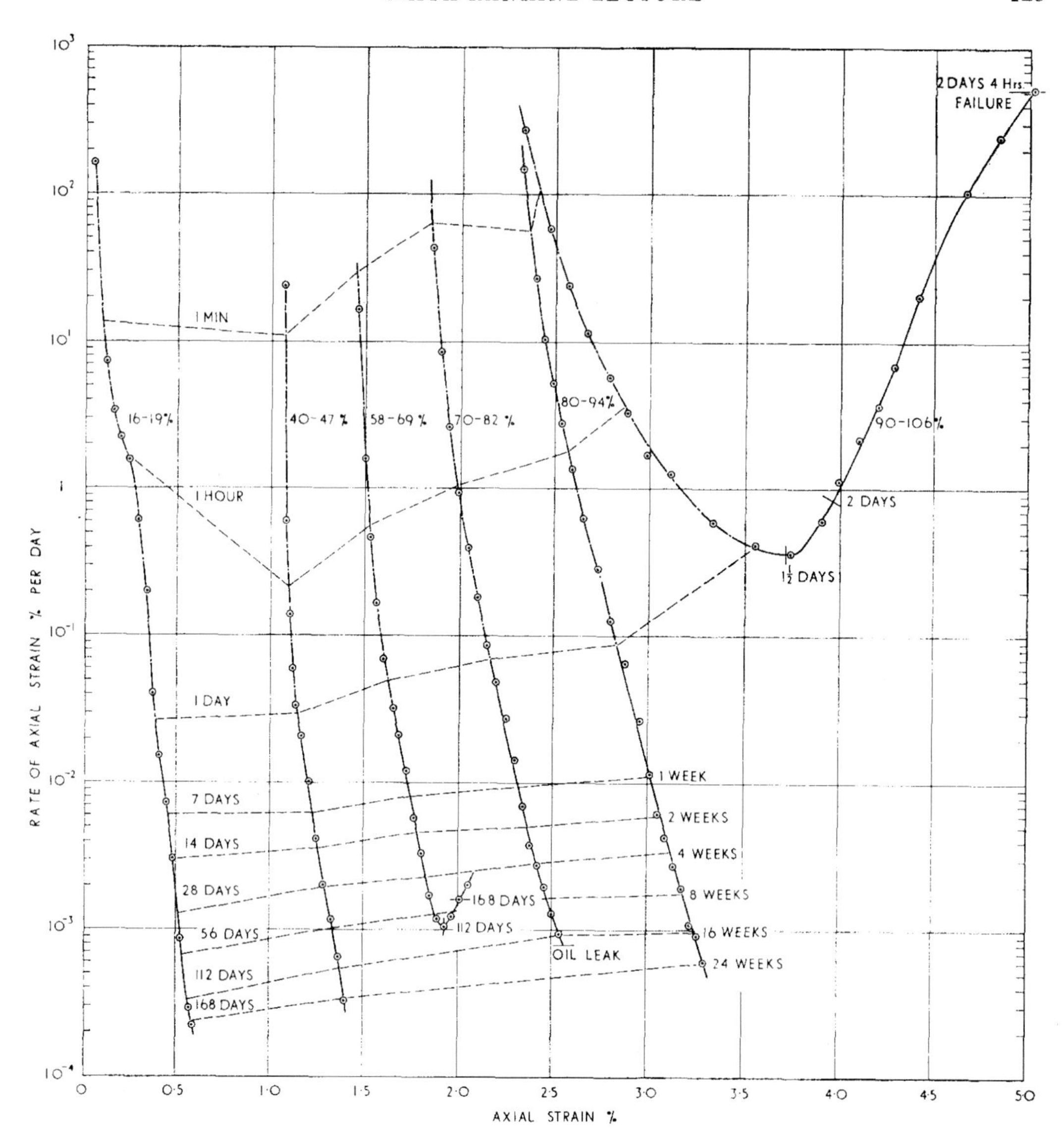

Fig. 35. Strain rate as a function of axial strain for various stress levels. (Stress levels indicated by percentages adjacent to curves). Contours represent times after application of final stress increment; drained tests on brown London Clay from Hendon. (Tests by Lovenbury 1965–66)

It will be seen that the 90% peak strength sample failed after two days, having given (at least in retrospect) sufficient warning of its intention.[25] The 80% test is still continuing with no sign of impending failure. The 70% test came to a premature end after 143 days, due, ironically, to creep in a perspex component for which we had not allowed. The 60% test showed a decreasing rate of strain for the first three and a half months (100 days), but the

[25] The period of two days is measured from the time of application of the last increment of the stress difference and full drainage under this increment would not have been achieved in the earlier part of this period.

rate has subsequently increased steadily. Whether, as the enlarged detail might suggest, this presages eventual failure is very difficult to forecast. The creep rates are given in Fig. 35, plotted against strain. Perhaps the most interesting feature of this diagram is that in undisturbed clay the creep rate does not stay constant under constant stress at any stage of the test. It either decreases fairly steadily or increases. The simpler rheological models are therefore not applicable. The 60% test, although it has speeded up, is still only proceeding at two thousandths of 1% per day and at this rate would take another 1200 days to reach the strain at which the 90% sample failed. An extrapolation of the present strain rate path suggests about 65 days more to failure.[26]

Even if the 60% sample did fail, and if we ignore the fact that the 80% sample is not yet showing a speeding-up in its strain rate, we have only accounted for part of the drop to the residual value which is approximately 30% of the higher peak drained strength. But it is a very considerable part, and may make possible a quantitative explanation in terms of the three factors listed above.

The creep tests outlined above are, of course, equally significant in relation to the performance of foundations under sustained load, where current concepts of secondary consolidation are based mainly on the results of oedometer tests with rigid lateral confinement.

CONCLUSION

In conclusion, I hope I have thrown a little light on some of the problems of soil mechanics which are both of intellectual interest and of practical importance. I hope I have also shown, in the last two sections in particular, that there is a great deal still to be found out about the actual strength of soils as engineering materials, and that not all of this can be found out in the laboratory. The investigation of full-scale failures and the carrying out of field tests of a sufficient size to be relevant and in sufficient numbers to be representative are tasks which must be shared by consulting engineers and contractors as well as by universities and research stations, and must be budgeted for.

ACKNOWLEDGEMENTS

In preparing this lecture I have been able to draw on unpublished experimental data obtained at Imperial College by G. E. Green, A. E. Skinner, H. T. Lovenbury, D. J. Petley, S. W. Smotrych, S. K. Sarma, N. N. Som, S. G. Tombs, and Dr B. P. Walker, and at the Building Research Station by P. I. Lewin. In addition I am indebted to Dr L. Bjerrum, Director of the Norwegian Geotechnical Institute for unpublished reports on field and laboratory tests.

The large shear tests at Maldon were carried out by George Wimpey and Co. Ltd under the direction of Binnie and Partners to whom I am indebted for permission to quote the results and also the results of tests on rock-fill. Acknowledgements are also due to the Science Research Council and the Civil Engineering Research Association whose grants made possible some of the tests described in this lecture.

I am particularly indebted to Gordon E. Green for assistance in assembling material for the lecture, to E. W. Harris for assistance with the illustrations, and to Dr R. E. Gibson for reading the manuscript.

[26] Since the Rankine Lecture was delivered, the speeding up of the creep rate has not increased and the time to failure cannot be predicted with any greater certainty at the present date (March 1966).

REFERENCES

AAS, G., 1965. A study of the effect of vane shape and rate of strain on the measured values of in-situ shear strength of soils. *Proc. 6th Int. Conf. Soil Mech.*, 1:141–145.

BISHOP, A. W., 1948. Some factors involved in the design of a large earth dam in the Thames Valley. *Proc. 2nd Int. Conf. Soil Mech.*, 2:13–18.

BISHOP, A. W., 1950. *Discussion.* Measurement of the shear strength of soils. *Géotechnique*, 2:1:113–116.

BISHOP, A. W., 1958. Test requirements for measuring the coefficient of earth pressure at rest. *Proc. Brussels Conf. on Earth Pressure Problems*, 1: 2–14.

BISHOP, A. W., 1965. Contribution to panel discussion, Division 2, *Proc. 6th Int. Conf. Soil Mech.*, 3.

BISHOP, A. W. *and* L. BJERRUM, 1960. The relevance of the triaxial test to the solution of stability problems. *Proc. Res. Conf. Shear Strength of Cohesive Soils*, Boulder (ASCE) 437–501.

BISHOP, A. W. *and* A. K. G. ELDIN, 1953. The effect of stress history on the relation between ϕ and porosity in sand. *Proc. 3rd Int. Conf. Soil Mech.*, 1:126–130.

BISHOP, A. W. *and* G. E. GREEN, 1965. The influence of end restraint on the compression strength of a cohesionless soil. *Géotechnique*, 15:3:243–266.

BISHOP, A. W. *and* D. J. HENKEL, 1953. The pore pressure changes during shear in two undisturbed clays. *Proc. 3rd Int. Conf. Soil Mech.*, 1:94–99.

BISHOP, A. W. *and* D. J. HENKEL, 1962. The measurement of soil properties in the triaxial test. *Edward Arnold, London*, 2nd ed.

BISHOP, A. W., D. L. WEBB *and* P. I. LEWIN, 1965. Undisturbed samples of London Clay from the Ashford Common shaft: strength–effective stress relationships. *Géotechnique*, 15:1:1–31.

BISHOP, A. W., D. L. WEBB *and* A. E. SKINNER, 1965. Triaxial tests on soil at elevated cell pressures. *Proc. 6th Int. Conf. Soil Mech.*, 1:170–174.

BJERRUM, L., 1961. The effective shear strength parameters of sensitive clays. *Proc. 5th Int. Conf. Soil Mech.*, 1:23–28.

BJERRUM, L., 1966. Mechanism of progressive failure in slopes of over-consolidated plastic clays and clay shales. 3rd Terzaghi Lecture, ASCE Conf., Miami.

BJERRUM, L. *and* A. LANDVA, 1966. Direct simple shear-tests on a Norwegian quick clay. *Géotechnique*. 16:1:1–20.

BORG, I., M. FRIEDMAN, J. HANDIN *and* D. V. HIGGS, 1960. Experimental deformation of St Peter sand: a study of cataclastic flow. *Rock Deformation*, Geological Society of America, Memoir 79, 133–192.

CASAGRANDE, A. *and* N. CARRILLO, 1944. Shear failure of anisotropic materials. *J. Boston Soc. civ. Engrs*, 31:74–87.

COATES, D. F. *and* G. C. McROSTIE, 1963. Some deficiencies in testing Leda clay. *Symposium* Laboratory shear testing of soils. Ottawa, 1963, ASTM STP 361, 459–470.

CORNFORTH, D. H., 1961. Plane strain failure characteristics of a saturated sand. Ph.D. Thesis, London.

HABIB, P., 1953. Influence de la variation de la contrainte principale moyenne sur la résistance au cisaillement des sols. *Proc. 3rd Int. Conf. Soil Mech.*, 1:131–136.

HAFIZ, M. A., 1950. Strength characteristics of sands and gravels in direct shear. Ph.D. Thesis, London.

HALL, E. B. *and* B. B. GORDON, 1963. Triaxial testing with large-scale high pressure equipment. *Symposium* Laboratory shear testing of soils. Ottawa. 1963. ASTM STP 361, 315–328.

HANSEN, J. BRINCH *and* R. E. GIBSON, 1949. Undrained shear strengths of anisotropically consolidated clays. *Géotechnique* 1:3:189–204.

HAYTHORNTHWAITE, R. M., 1960. Mechanics of the triaxial test for soils. *Proc. ASCE*, 86 SM5: 35–62

HILL, R., 1950. The mathematical theory of plasticity. Clarendon Press, Oxford.

HIRSCHFELD, R. C. *and* S. J. POULOS, 1963. High-pressure triaxial tests on a compacted sand and an undisturbed silt. *Symposium* Laboratory shear testing of soils. Ottawa, ASTM STP 361, pp. 329–339.

HVORSLEV, M. J., 1960. Physical components of the shear strength of saturated clays. *Proc. Res. Conf. Shear Strength of Cohesive Soils*, Boulder (ASCE), 437–501.

INSLEY, A. E. *and* S. F. HILLIS, 1965. Triaxial shear characteristics of a compacted glacial till under unusually high confining pressures. *Proc. 6th Int. Conf. Soil Mech.*, 1:244–248.

JAKY, J., 1944. A Nyugalmi Nyomas Tenyezoje. *Magyar Mernok-es Epitesz-Egylet Kozlonye*, Oct., 355–358.

JAKY, J., 1948. Pressure in silos, *Proc. 2nd Int. Conf. Soil Mech*, 1: 103–107.

JOHANSEN, K. W., 1958. Brudbetingelser for sten og beton. (Failure conditions for rocks and concrete.) *Byningsstatiske Meddelelser*, 19:25–44 Teknisk Forlag, Copenhagen.

KIRKPATRICK, W. M., 1957. Condition of failure for sands. *Proc. 4th Int. Conf. Soil Mech.*, 1:172–178.

KUMMENEJE, O., 1957. 'To-dimensjonale' vakuum-triaxialforsök pă törr sand. Norwegian Geotechnical Institute Internal Report F. 80 (unpublished).

LADD, C. C. *and* W. A. BAILEY, 1964. *Correspondence.* The behaviour of saturated clays during sampling and testing. *Géotechnique*, 14:1:353–358.

LADD, C. C. *and* T. W. LAMBE, 1963. Shear strength of saturated clays. *Symposium* Laboratory shear testing of soils. Ottawa, 1963, ASTM STP 361, 342–371.

LADD, C. C. *and* J. VARALLYAY, 1965. The influence of stress system on the behaviour of saturated clays during undrained shear. M.I.T. Report, July.

LA ROCHELLE, P., 1960. The short-term stability of slopes in London Clay. Ph.D. Thesis, London.

LEUSSINK, H., 1965. Contribution to panel discussion, Division 2. *Proc. 6th Int. Conf. Soil Mech.* Vol. 3.

LO, K. Y., 1965. Stability of slopes in anisotropic soils. *Proc.* ASCE, 91: SM4: 85–106.

MARSAL, R. J., 1965. Contribution to panel discussion, Division 2. *Proc. 6th Int. Conf. Soil Mech.* Vol. 3.

MARSAL, R. J., E. M. GOMEZ, A. NUNEZ G., P. CUELLAR B. *and* R. M. RAMOS. 1965. Research on the behaviour of granular materials and rockfill samples. Comision Federal De Electricidad, Mexico.
NITCHIPOROVITCH, A. A., 1964. Deformations and stability of rockfill dams. *Proc. 8th Cong. Large Dams*, 3:879–894.
NORWEGIAN GEOTECHNICAL INSTITUTE, 1958. Triaxial compression and extension tests on a saturated fine and uniform graded sand. Internal Report, F94.
PELTIER, M. R., 1957. Experimental investigations on the intrinsic rupture curve of cohesionless soils. *Proc. 4th Int. Conf. Soil Mech.*, 1:179–182.
ROSCOE, K. H., A. N. SCHOFIELD *and* C. P. WROTH, 1958. On the yielding of soils. *Géotechnique*, 8:1: 22–53.
ROSCOE, K. H., A. N. SCHOFIELD *and* A. THURAIRAJAH, 1963. An evaluation of test data for selecting a yield criterion for soils. *Symposium* Laboratory shear testing of soils, Ottawa, ASTM STP 361, 111–128.
SCHLEICHER, F., 1925. Die Energiegrenze der Elastizitat (Plastizitatsbedingung). *Zeits für ang. Math. Mech.*, 5:478–479.
SCHLEICHER, F., 1926. Der Spannungfzustand an der Fliessgrenze (Plastizitatsbedingung). *Zeits für ang. Math. Mech.*, 6:199–216.
SCOTT, R. F., 1963a. Discussion on the Mohr–Coulomb concept in shear failure. *Symposium* Laboratory shear testing of soils. Ottawa, ASTM STP 361, 75–76.
SCOTT, R. F., 1963b. Principles of soil mechanics. *Addison Wesley*, 440.
SIMONS, N., 1958. *Discussion* Test requirements for measuring the coefficient of earth pressure at rest. *Brussels Conf. on Earth Pressure problems*, 3:50–53.
SKEMPTON, A. W., 1964. Long term stability of clay slopes. 4th Rankine Lecture, *Géotechnique*, 14:2: 77–101.
SKEMPTON, A. W. *and* P. LA ROCHELLE, 1965. The Bradwell slip: a short-term failure in London Clay. *Géotechnique*, 15:3:221–242.
SKEMPTON, A. W. *and* V. A. SOWA, 1963. The behaviour of saturated clays during sampling and testing. *Géotechnique*, 13:4:269–290.
SOWA, V. A., 1963. A comparison of the effects of isotropic and anisotropic consolidation on the shear behaviour of a clay. Ph.D. Thesis, London.
TAYLOR, D. W., 1948. Fundamentals of soil mechanics. Wiley, New York.
TAYLOR, D. W. *and* R. H. CLOUGH, 1951. Report on research on shearing characteristics of clay. M.I.T.
USCOLD, 1963. Oroville Dam. Newsletter No. 10, January, 4.
VESIC, A. *and* R. D. BARKSDALE, 1963. *Discusion* Test methods and new equipment. *Symposium* Laboratory shear testing of soils. Ottawa, 1963, ASTM STP 361, 301–305.
VOLD, R. C., 1956. Undisturbed sampling of soils. *Norwegian Geotechnical Institute*, Publ. No. 17.
WADE, N. H., 1963. Plane strain failure characteristics of a saturated clay. Ph.D. Thesis, London.
WARD, W. H., A. MARSLAND *and* S. G. SAMUELS, 1965. Properties of the London Clay at the Ashford Common shaft; in-situ and undrained strength tests. *Géotechnique*, 15:4:321–344.
WATERWAYS EXPERIMENT STATION, VICKSBURG, 1950. Triaxial tests on sands—Reid Bedford Bend, Mississippi River. Report No. 5–3.
WHITAKER, T. *and* R. W. COOKE, 1966. An investigation of the shaft and base resistances of large bored piles in London Clay. *Proc. Symp. Large Bored Piles.* Instn. civ. Engrs. London.
WOOD, C. C., 1958. Shear strength and volume change characteristics of compacted soil under conditions of plane strain. Ph.D. Thesis, London.

VOTE OF THANKS

The CHAIRMAN invited Professor J. Brinch Hansen, Vice-President of the International Society for Soil Mechanics and Foundation Engineering, to propose a vote of thanks to the lecturer.

Professor Brinch Hansen said that the subject of Professor Bishop's lecture was most appropriate, as the work for which Professor Rankine became famous dealt with the behaviour of soils at failure and the application of this to practical engineering problems.

The shear strength of soils was, indeed, one of the most fundamental subjects in soil mechanics, probably the most fundamental. Different aspects of it had been studied at practically every soil mechanics laboratory in the world, but nowhere more intensively than in Great Britain. In the history of soil mechanics names, such as Skempton, Bishop, Rowe, and Roscoe would always represent milestones in the understanding of the strength properties of soils; and Imperial College came to be considered as the place where the most important work on shear strength and its practical applications had been made. In this work Professor Bishop had played a leading role.

Professor Brinch Hansen said that he did not pretend to know about all Professor Bishop's achievements but would mention a few of them. Bishop had designed and constructed a great number of different pieces of testing apparatus which were now in common use in most parts of the world. To the development and refinement of the triaxial machine, which was now generally acknowledged to be the best apparatus for measuring shear strength, Bishop had contributed more than any other person.

In the more theoretical field, mention should be made of his general method for the analysis of stability, which might well be the method most commonly used at present.

Finally, in practical engineering his new methods of designing and investigating great earth dams had proved eminently successful.

In his lecture Professor Bishop had thrown considerable light on some of the less known factors influencing shear strength, namely, intermediate principal stress, stress level, anisotropy, sampling effects, and time effects.

In the first approximation, established many years ago, the failure criterion was expressed by means of Coulomb's law, combined with the principle of effective stresses. This was simple and seemed in most cases to work well. However, in time it was found that a number of other factors, of which Coulomb's law took no account, influenced the strength to some extent. Bishop had dealt with the most important of these in his lecture.

With regard to the failure criterion, Professor Bishop had come to the conclusion that the experimental results mostly supported the Mohr–Coulomb criterion, and proposed to accept this, although it underestimated the value of ϕ' for plane strain by up to 4°.

Professor Brinch Hansen said that he was not personally very happy about this recommendation because, after all, it might mean an unnecessary extra safety factor of 2 on the bearing capacity of dense sand.

Bishop had, quite rightly, drawn attention to the effect of the stress level, which expressed itself in a curved Mohr envelope. Whatever the physical reasons for this, it was a fact which must be taken into account in their designs, especially if small scale laboratory tests were to be used as a basis for the design of full scale structures.

The often-experienced wide discrepancies between the results of bearing capacity tests in the laboratory, and theoretical values calculated on the basis of triaxial friction angles, were probably due mainly to differences in intermediate principal stress and stress level. However, the trouble was that one of these effects would usually suffice to provide an explanation. If both were admitted they might well end in the opposite ditch.

Commenting on Professor Bishop's emphasis on the effect of anisotropy, Professor Brinch Hansen said that he would, of course, be the first to agree with him. Also his point concerning the effects of sampling, or of sample size, seemed to be well established. Professor Brinch Hansen thought, therefore, that the lecturer was probably right in assuming that the apparent accuracy of the $\phi=0$ analysis in many cases might be due to the accidental cancelling of two or more considerable errors. Incidentally, the same was probably the case in other soil mechanics calculations, for instance, in settlement analysis.

Bishop's research on the effect of time on the drained strength of London Clay was extremely interesting, as it established beyond any doubt the influence of creep, not only of the deformations but also on the shear strength itself. This seemed to be an indication that the final failure criterion would probably be expressed in strains rather than in stresses; but this belonged to the future.

They were all extremely grateful to Professor Bishop for having not only drawn their attention to some important but little-known factors influencing shear strength but having also given them the experimental background of his statements and demonstrated their significance to the practising engineer.

As usual, Professor Bishop had done all this in a very clear and scientific manner, using his considerable imagination to seek new explanations, but not giving them out until he had obtained adequate experimental proof.

On behalf of all those present he was therefore happy to propose a well-deserved vote of thanks to Professor Bishop for his important lecture.

The vote of thanks was carried with acclamation.

Progressive Failure - with Special Reference to the Mechanism Causing it

Discussion by Bishop in:

Proceedings,
Geotechnical Conference, Oslo, 1967, Vol 2, 142-150

DR. BISHOP:

Progressive Failure – with Special Reference to the Mechanism Causing it

I have been asked to introduce the subject of progressive failure with special reference to the mechanism causing it.

The concept of progressive failure is not new to Soil Mechanics. It is discussed in the introduction of Terzaghi's "Theoretical Soil Mechanics" in 1943, and a very adequate definition is given in the text book "Soil Mechanics in Engineering Practice" by Terzaghi and Peck (1948). It is relevant to quote it at length.

"The term progressive failure indicates the spreading of the failure over the potential surface of sliding from a point or line towards the boundaries of the surface. While the stresses in the clay near the periphery of this surface approach the peak value, the shearing resistance of the clay at the area where the failure started is already approaching the much smaller ultimate value. As a consequence the total shearing force that acts on a surface of sliding at the instant of complete failure is considerably smaller than the shearing resistance computed on the basis of the peak values."

D. W. Taylor in his book "Fundamentals of Soil Mechanics" presented a careful consideration of the importance of progressive failure, both in laboratory testing and in the field, under the heading "Non-Uniform Stress and Strain Conditions." The latter is perhaps the operative word.

is of interest to note that, while Terzaghi and Peck were ·iting about the stability of natural clay strata in the ntext of total stress analysis and while Taylor was consider-g primarily the shearing resistance of sands in an effective ·ess context, both associated progressive failure with the distribution of shear stress along a potential failure surface d neither made any reference to time dependent properties the soil.

In this they are consistent with the Concise Oxford ictionary, which defines progress as a 'forward or onward ovement in *space*.'

Turnbull and Hvorslev, in their paper to the 1966 ASCE onference on Stability and Performance of Slopes and mbankments, also associate progressive failure with non-iiform stress and strain conditions as also did Peck in an companying paper. They illustrate the order of magnitude this non-uniformity with data drawn from my 1952 Ph. D. esis and from that of La Rochelle in 1960. I quote these gures (Figs 7, 8, 9, 10 of Turnbull and Hvorslev), since they dicate, at least qualitatively, the distribution of shear stress hich may be expected within an embankment and its foun-tion (Figs 1 and 2) and adjacent to a cut (Figs 3 and 4). r. Bjerrum, in Fig. 2 of his Terzaghi lecture illustrates the pe of non-uniformity of stress and strain to be expected the toe of a cut due to swelling consequent on high lateral resses in the ground.

The non-uniformity in terms of stress ratio, τ_{max}/σ_n', hich is more relevant to drained stability analysis than that ustrated by Turnbull and Hvorslev, is shown in Figs 5 id 6.

The values are again taken from the relaxation solution r a homogeneous elastic embankment and foundation given my 1952 thesis. It can be seen that while the shear stress a maximum towards the middle of the trajectory and decreases towards the ends, the stress ratio τ_{max}/σ_n has high values at one or both ends of the trajectory and low values in between. This suggests that in a short term analysis in terms of undrained strength we might expect failure, in this problem, to commence within the soil mass and propagate towards the surface in both directions. In a long term analysis, where effective stress parameters are used and shear strength is a function of normal stress, we might in contrast expect failure to begin at the surface, possibly at both toe and crest, and propagate inwards into the soil mass.[1])

Clearly this is a very preliminary investigation and in the case of a cut or natural slope the stresses will be influenced by the state of stress in the ground before excavation was made or erosion took place. A row of holes fitted with inclino-meters or Wilson slope indicators installed before a cut is made would give very valuable field data on the develop-ment of the slip surface from its initiation to complete failure.

These graphs have been presented in terms of the ratio of maximum shear stress to normal stress on the maximum shear stress trajectory. For any failure envelope with even a

[1]) Peck 1967 quotes an analysis due to Conlon which bears on this problem, but appears to obtain an initial shear stress distribution without reference to stress strain characteristics. He does, however, conclude that the upper part of the slope moves first towards the residual condition.

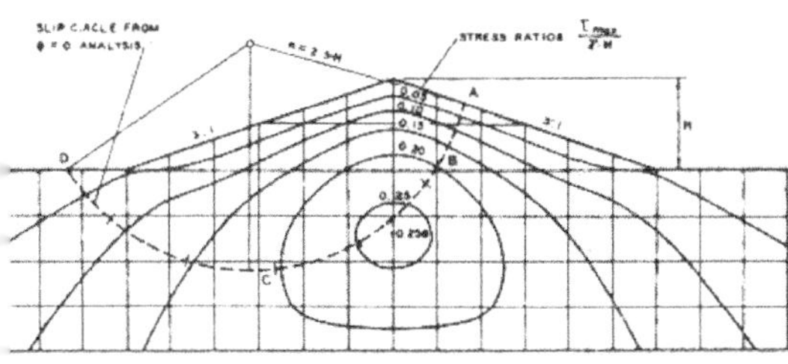

ig. 1. *Contours of maximum shear stresses in a uniform embankment and foundation. Contours determined for a uniform, cohesive and elastic material by the finite differences and relaxation method: from A. W. Bishop (1952). After Turnbull and Hvorslev (1967).*

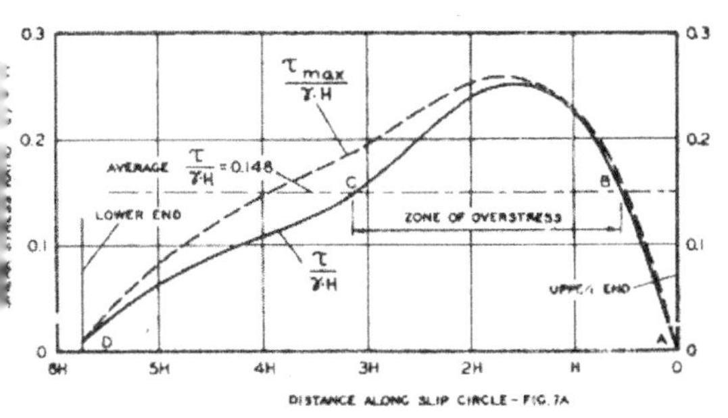

ig. 2. *Shear stresses along a slip circle in an embankment: based on data by A. W. Bishop (1952). After Turnbull and Hvorslev (1967).*

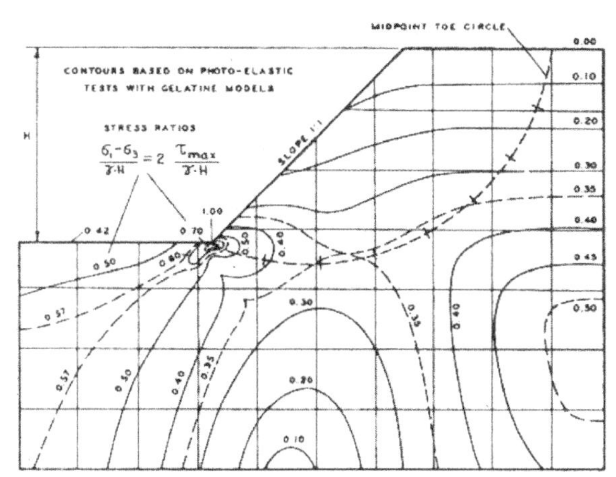

Fig. 3. *Contours of $(\sigma_1 - \sigma_3)$ in a steep excavated slope: from P. La Rochelle (1960). After Turnbull and Hvorslev (1967).*

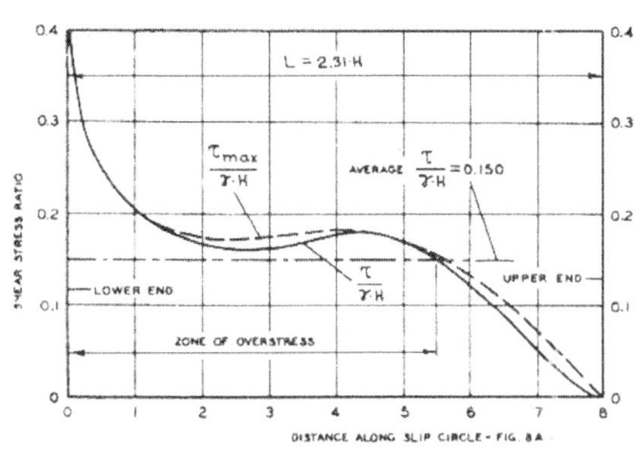

Fig. 4. *Shear stresses along a slip circle in an excavated slope: based on data by P. La Rochelle (1960). After Turnbull and Hvorslev (1967).*

small cohesion intercept the ratio of shearing resistance to normal stress rises rapidly as the normal stress decreases to very small values. Values have been included on these graphs for the peak and residual strengths of weathered London clay, the height of slope, for this purpose, being taken as 50 ft. Two cases, zero pore pressure (Figs 5 and 6) and a phreatic surface at ground level (Fig. 7) are illustrated.

Thus, as a slope is formed by excavation or erosion, or as an embankment or foundation is constructed, the shear stresses along a potential failure surface will rise in a non-uniform manner, which in principle could be determined from the stress-strain relationships for the actual soil. If locally the shear stress, or the ratio of shear stress to effective normal stress (whichever is appropriate to relevant type of analysis) reaches the limiting value, local failure will result. Local failure may also result due to a rise in pore pressure under constant loading. A further change in the loading conditions, or in the pore pressure under constant loading, will lead to a progressive extension of the zone of failure along the potential slip surface, while within this zone the shearing resistance will commence to drop from its peak to its ultimate or residual state. The state of limiting equilibrium is reached when failure has just extended to the whole of the slip surface and in this state the shearing resistance may be expected to vary from the peak value at the points failing last to the residual value, in the limit, at others.

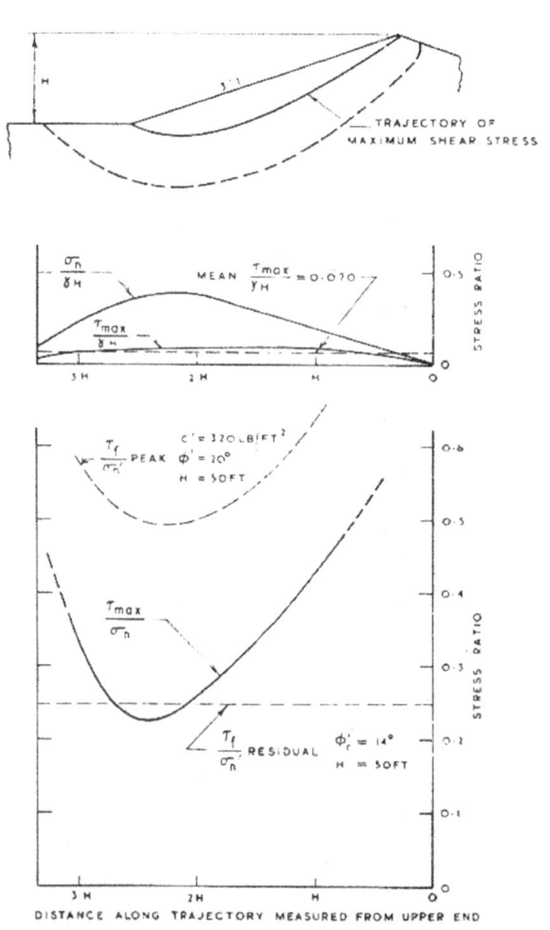

Fig. 5. *Distribution of stresses and stress ratios along a trajectory of maximum shear stress from relaxation solution by A. W. Bishop (1952).*

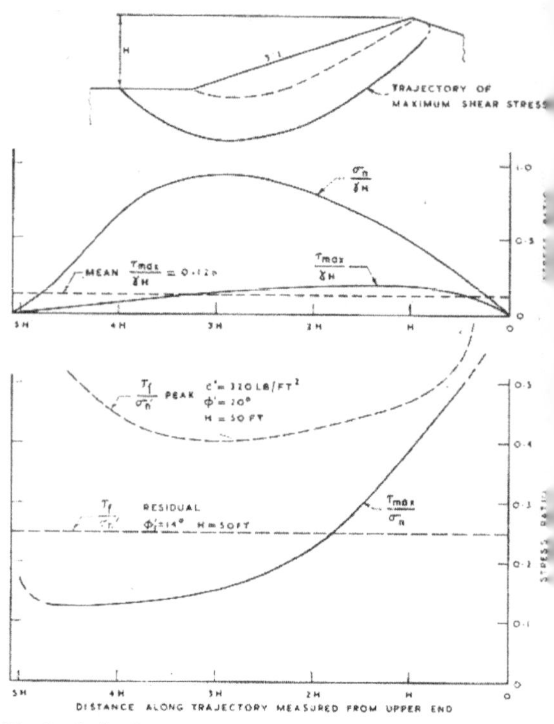

Fig. 6. *As fig. 5.*

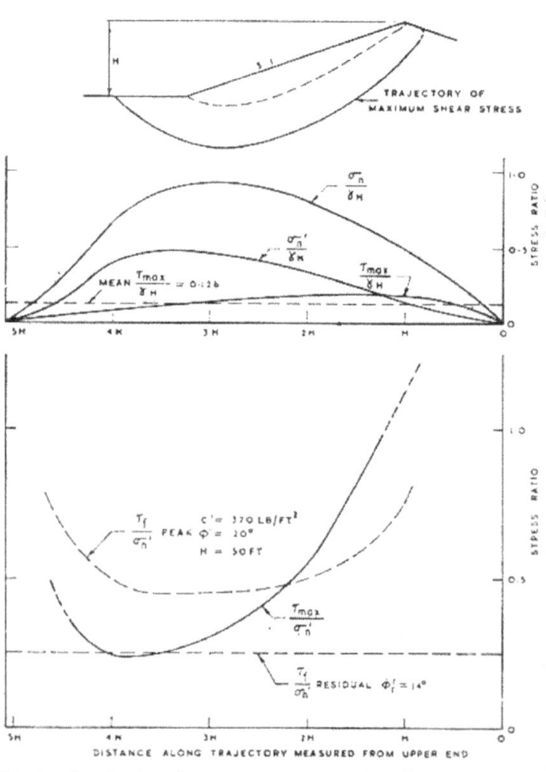

Fig. 7. *Distribution of stresses and stress ratios along a trajectory of maximum shear stress for a pore pressure ratio $r_u (= u/\gamma h)$ of 0,5: from relaxation solution by A. W. Bishop (1952).*

We must therefore conclude, with D. W. Taylor (1948), at in all problems of slope stability, bearing capacity and rth pressure, even in ideally homogeneous soils, the state limiting equilibrium is associated with non-uniform obilization of shearing resistance and thus with progressive ilure. Whether or not this fact leads to a major difficulty relating the average shear stress observed along the ilure surface in the field to the relevant laboratory shear st depends in the first place on the difference between the eak and residual strength, and secondly on the strain required for this difference to be established. In other words we must consider how brittle the soil is.

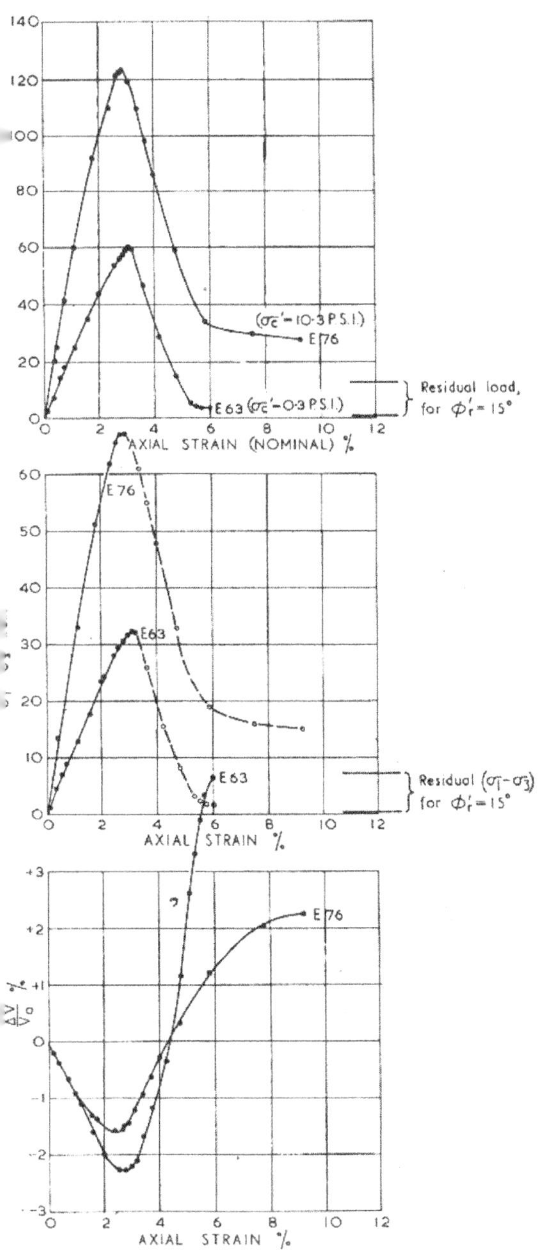

Fig. 8. *Typical results of drained triaxial compression tests on vertical undisturbed samples. (σ_c denotes effective consolidation pressure): from A. W. Bishop, D. L. Webb and P. I. Lewin (1965).*

We must in this connection discuss what we mean by a brittle soil, the term used by Dr. Bjerrum in his Terzaghi Lecture to describe stiff clays liable to progressive failure. In Fig. 8 are shown the results of drained tests on two $1\frac{1}{2}''$ dia. samples of almost intact London Clay from a considerable depth. One sample can be seen to lose 87% of its strength in passing from the peak to the residual state, the other over 95%, the percentage increasing with decreasing effective normal stress. Most of this has occurred in the first 3% of nominal axial compression after the peak stress was reached.

Without a detailed study of the factors controlling the growth of a shear surface under non-uniform stress conditions it is premature to establish parameters to define the brittleness of a clay. As a basis for discussion, however, we might consider the percentage reduction in strength in passing from the peak to the residual state (Fig. 9).

$$I_B = \frac{\tau_f - \tau_r}{\tau_f} \% \tag{1}$$

This is equal to $(1 - \lambda_R) \times 100\%$ where λ_R is Haefeli's 'residual coefficient' (Haefeli, 1965). The form which I have used has the advantage that it expresses directly the maximum percentage error which can arise due to progressive failure in a brittle soil. It is of interest to note that for typical over-consolidated soils this index varies most markedly with

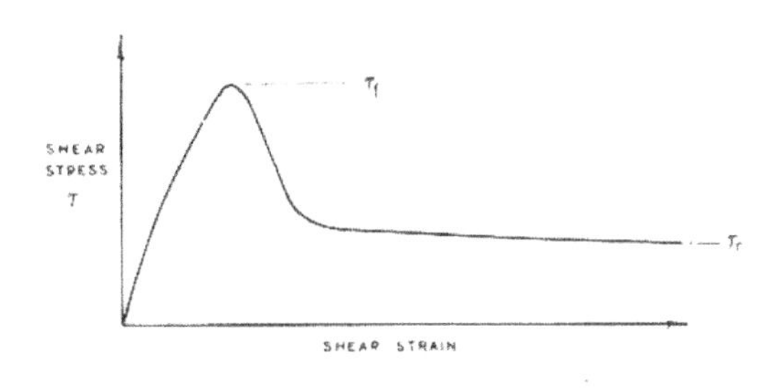

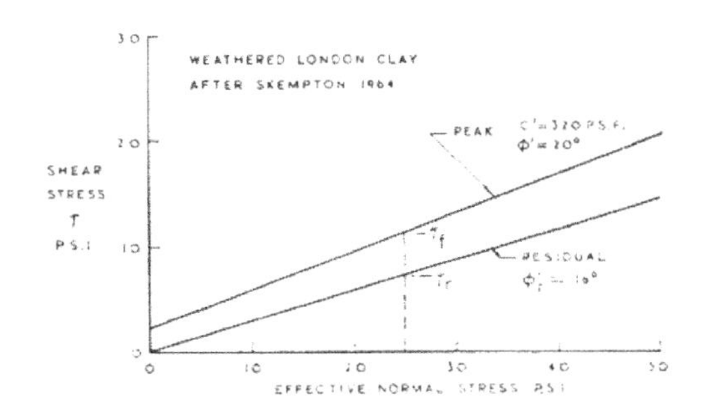

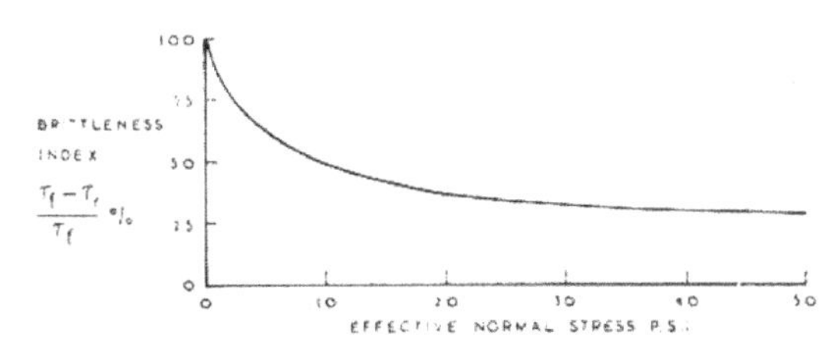

Fig. 9. *A brittleness parameter and its variation with stress.*

stress. An example is given for London Clay (Skempton, 1964) based on conventional $1^1/_2''$ dia. samples and $2.4'' \times 2.4''$ shear box results.

In the case of weathered brown London clay this index drops from 100% under very low stress to about 50% at the average depth from which these samples were taken and to about 30% at relatively high stresses. Thus for problems such as the slope I have discussed the brittleness tends to be greatest at the ends of slip surfaces where the stress ratio is highest and most critical (though we should note that as c' increases with depth, the brittleness index will be less, under a given effective stress, for shallow samples).

Brittleness may also be considered in terms of the additional energy required to make the shear failure progress from the peak to the residual state (Fig. 10). This is represented by the shaded area E_f. If the sample had shown no drop in strength after the peak the work done would have been represented by E_p. The difference between these areas, expressed as a ratio of E_p, may be a useful measure of brittleness also – I have called it 'rupture index' to distinguish it.

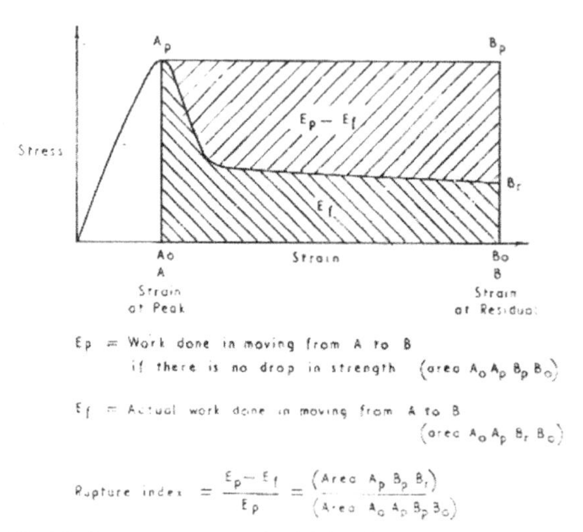

Fig. 10. *An energy parameter for brittle soils.*

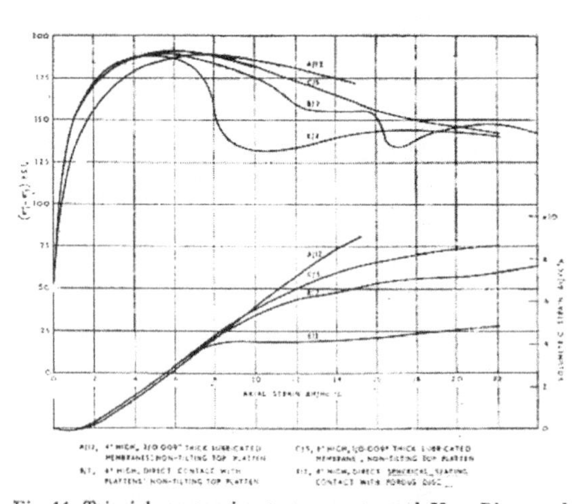

Fig. 11. *Triaxial compression tests on saturated Ham River sand: cylindrical samples 4" diameter under various test conditions: after Green (1968).*

However, great caution must be taken in evaluating ar parameter based on the shape of the stress strain curv particularly after the peak – whether measured in the triaxi compression, or in simple shear – for a number of reason The first is the tendency for failure to develop in a sing thin shear zone in many samples which show marked dilatanc or marked structural breakdown on shear. Fig. 11 gives son examples of the variation in the post peak stress-nomin strain curve for samples of dense sand having almost the sam initial density but different end conditions or sample heigh to-diameter ratios. It is of interest to note that the pea strength is hardly affected, but that a 2 : 1 sample with roug ends shows an almost brittle failure, while for a short samp with a double lubricated layer on its ends which impos uniform displacement the failure appears more plasti although at large strains the residual strengths are simila Obviously the areas under these curves would be differen The least energy is required if a single surface is permitte to develop, possibly after initiation by a local stress con centration. The implications of this fact in relation to siz of sample and method of testing are considerable, but cannc be pursued in the time at my disposal. Fundamentally, c course, one is interested in the energy per unit volume, bu energy per unit area of created slip surface may also hav to be considered.

The energy absorbed in shearing intact material as th failure surface spreads is also of great significance. Her again the type of test is of importance. The stress strai curves in Fig. 12 illustrate the difference between the result

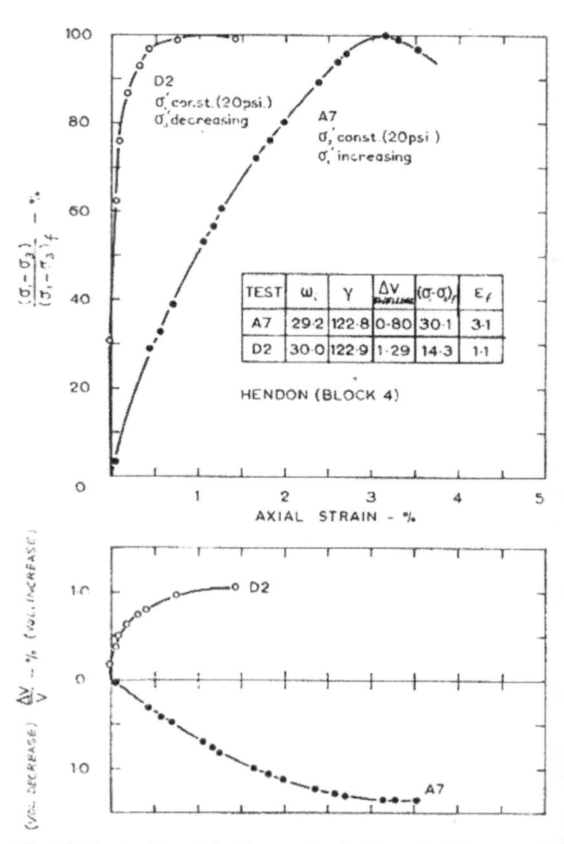

Fig. 12. *Comparison of loading and unloading triaxial test results: after Lovenbury (1968).*

btained for undisturbed samples of weathered brown London clay failed with σ_1' increasing and σ_3' constant (A. 7), nd with σ_1' constant and σ_3' decreasing (D. 2). The latter learly involves the absorption of much less external energy or a given stress difference. One might thus expect a more apid spread of progressive failure in the active than the assive zone.

It is of interest to note that the relatively intact lumps etween the more obvious fissures in the weathered zone how marked brittle failure characteristics (Fig. 13). The eak strengths in terms of effective stress are below those of $3'' \times 1^1/_2''$ dia. samples from the unweathered zone (Fig. 14) for example Bishop, Webb and Lewin, 1965), but above hose of the nearest conventional tests. These are consistent with the view that there is a continuous gradation in strength rom the intact material to the large scale test or field case where the dimensions are large compared with the fissure pacing (Fig. 15).

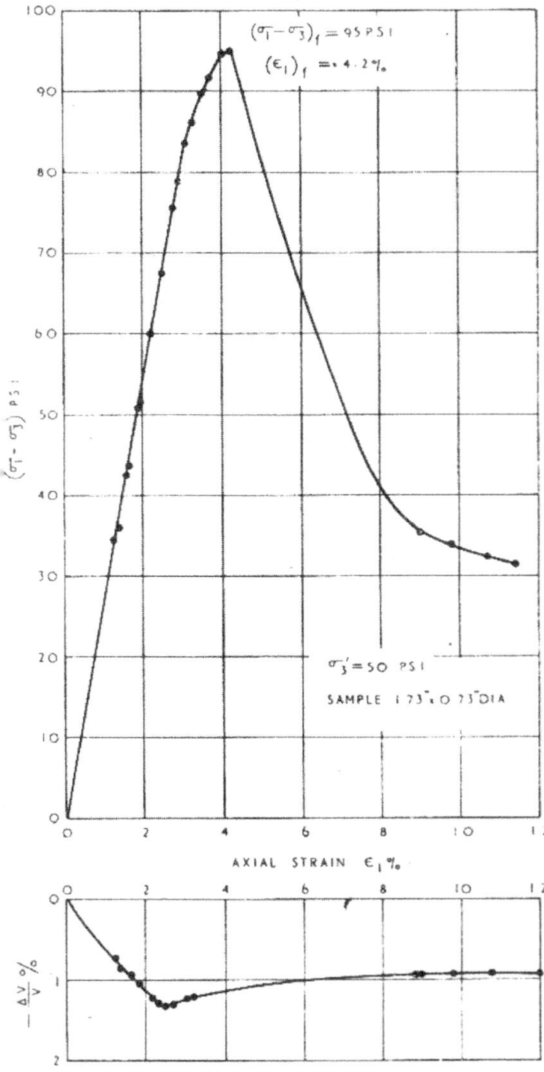

Fig. 13. *Drained triaxial compression test on small specimen cut from intact lump from 'weathered' zone of blue London Clay: axis horizontal.*

This effect may be less marked for effective stress parameters than for undrained strength, but both Marsland and Butler's tests (Paper 2/10) and our own suggest that it cannot be ignored. This means that both the brittleness index to which I referred and the Residual Factor defined in Skempton's Rankine lecture are dependent on sample size unless this is large enough to be the appropriate multiple of the fissure spacing.

However, progressive failure in a stiff fissured clay does not appear to leave the 'intact' lumps unsheared, just as in rockfill or even sand hard but brittle particles fail during shear of the mass although there is a 'fissure' surrounding each particle. The only natural failure surface I could find a photograph of is not very clear (Fig. 16), though it was in fact remarkably smooth. The lower surface of a 2 ft × 2 ft in-situ shear test (Fig. 17) and the surfaces of 2 ft × 1 ft dia. triaxial samples after failure (Figs 18, 19, 20, 21, 22) show little indication of failure going round the hard lumps (the fissure spacing is of the order of 2" in the latter tests).

The mechanism of the propagation of a shear surface in a fissured clay is clearly a subject deserving further study.

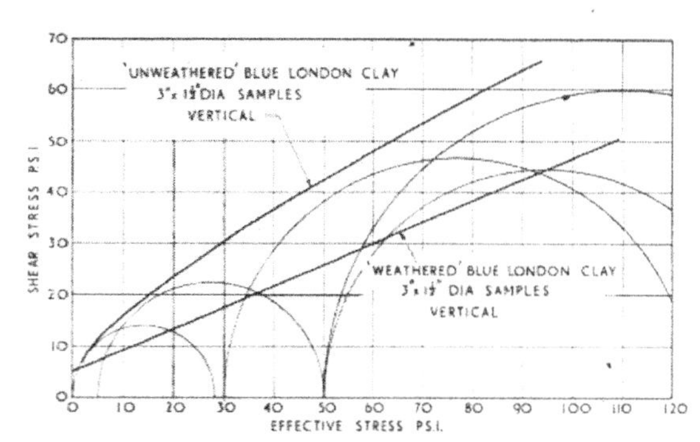

Fig. 14. *Drained tests on blue London Clay in the low stress range. Mohr circles represent tests with axis horizontal on small samples 1.7" × 0.75" dia. cut from apparently intact lumps in 'weathered' zone. (Tests by Dunlap, 1967.) Failure envelopes for comparison are for 3" × $1^1/_2$" dia. samples in 'weathered' zone 5 feet above (Agarwal, 1967), and for 3" × $1^1/_2$" dia. samples in 'unweathered' zone (Bishop, Webb and Lewin, 1965).*

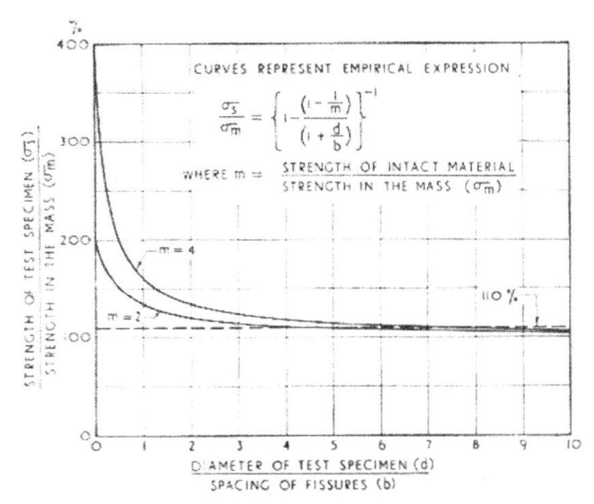

Fig. 15. *Influence of sample size on apparent strength of fissured material. After Bishop, 1956.*

This mechanism may well prove to involve concepts similar to those put forward by Griffith (1921) in his paper 'The Phenomena of Rupture and Flow in Solids' involving the small scale stress concentration around the end of the propagating surface of weakness and the interchange between different forms of energy. These include energy absorbed in forming the surface and released by relieving the stresses in the locally highly stressed material adjacent to it. Proneness to progressive failure may be directly associated with low tensile strength, in terms of effective stress, of even the intact lumps.

The importance of progressive failure in Soil Mechanics is clearly not limited to soils whose properties are markedly timedependent. Yet apart from examples of retrogressive failures such as those described by Kenney in his general report the cases in which progressive failure is considere to have played most part are problems of long term stabilit in which *delayed failure* is most notable. Examples ar given by Skempton (1964) and Bjerrum (1967) and ar discussed by Šuklje in his paper to this Conference (2/12).

I will not comment in any great detail on time effect because Professor Šuklje is going to deal with this aspect o the subject, but I will indicate briefly how time enters int the problem due to a number of factors which delay the fu development of progressive failure. As illustrated above time effects are not the primary cause of the mechanisn of progressive failure which I have described. The fou factors are:

1. The time lag, due to the low coefficient of consolidation or more usually, of swelling, in the re-adjustment o pore pressure resulting from (a) the release of stres during the formation of a slope, for example, and (b) du to the modification of ground water level resulting from the change in profile (see, for example, Bishop and Bjerrum, 1960).
2. Weathering, in relation to its effects on soil structure, with respect both to strength and to permeability.
3. The delayed release of strain energy due to time lag in the rebound curve with respect both to volume change

Fig. 16. *Part of slip surface exposed showing the smoothness of the failure surface in the 'weathered' blue London Clay.*

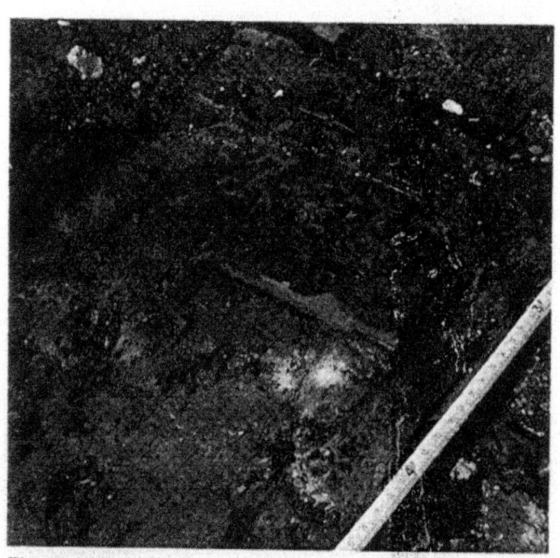

Fig. 17. *Failure surface at termination of in situ direct shear test on 2 ft. × 2 ft. dia. sample of 'weathered' blue London Clay.*

Fig. 18. *2 ft. × 1 ft. dia. sample of 'weathered' blue London Clay after undrained triaxial compression test.*

ig. 19. *Slip surface formed in London Clay in undrained triaxial compression test illustrated in Fig. 18.*

Fig. 20. *2 ft × 1 ft dia. sample of 'weathered' blue London Clay after undrained triaxial compression test.*

Fig. 21. *Slip surface formed in London Clay in undrained triaxial compression test, 2 ft × 1 ft dia. sample.*

Fig. 22. *Slip surface formed in London Clay in undrained triaxial compression test, 2 ft × 1 ft dia. sample*

and to shear deformation. This would influence, for example, the spread of a crack of the Griffith type.

4. The reduction in strength values, particularly at the peak, due to the rheological component of shear strength.

As a consequence of these factors, progressive failure, which is inherently caused by the shape of the stress strain curve, becomes time-dependent.

References

Agarwal, K. B., *Ph. D. thesis*, University of London.

Bishop, A. W., 1952. *The stability of earth dams*, Ph. D. thesis, University of London.

Bishop, A. W. and Bjerrum, L., 1960. *The relevance of the triaxial test to the solution of stability problems*. Proc. Res. Conf. Shear Strength of Cohesive Soils, Boulder, A.S.C.E., p. 437.

Bishop, A. W., Webb, D. L. and Lewin, P. I., 1965. *Undisturbed samples of London Clay from the Ashford Common shaft: strength-effective stress relationships*. Géotechnique, Vol. 15, No. 1, p. 1.

Bishop, A. W., 1966. *Soils and soft rocks as engineering materials*. Inaugural Lecture, Imperial College.

Bjerrum, L., 1967. *Progressive failure in slopes of over-consolidated plastic clay and clay shales*. 3rd Terzaghi Lecture, A.S.C.E., *93*, SM5, Part 1, p. 5.

Green, G. E., *Ph. D. thesis*, University of London.

Griffith, A. A., 1921. *The phenomena of rupture and flow in solids*. Phil. Trans. Roy. Soc., A. 221, p. 163.

Haefeli, R., 1965. *Creep and progressive failure in snow, soil, rock and ice*. Proc. 6th Int. Conf. Soil Mechanics and Foundation Engineering, Toronto, Vol. 3, p. 134.

La Rochelle, P., 1960. *The short-term stability of slopes in London Clay*. Ph. D. thesis, University of London.

Lovenbury, H., *Ph. D. thesis*, University of London.

Peck, R. B., 1967. *Stability of natural slopes*. A.S.C.E., 93, SM4, p. 403.

Skempton, A. W., 1964. *Long term stability of clay slopes*. 4th Rankine Lecture, Géotechnique, Vol. 14, No. 2, p. 77.

Taylor, D. W., 1948. *Fundamentals of Soil Mechanics*. John Wiley, New York.

Terzaghi, K., 1943. *Theoretical Soil Mechanics*. John Wiley, New York and London.

Terzaghi, K. and Peck, R. B., 1948. *Soil Mechanics in Engineering Practice*. John Wiley, New York.

Turnbull, W. J. and Hvorslev, M. J., 1967. *Special problems in slope stability*. A.S.C.E., 93, SM4, p. 499.

PHILOSOPHICAL TRANSACTIONS
OF
THE ROYAL SOCIETY
OF LONDON

A. MATHEMATICAL AND PHYSICAL SCIENCES

VOLUME 278 PAGES 511–554 NUMBER 1286

PTRMAD 278 (1286) 511–554 (1975)

3 July 1975

The influence of pore-water tension on the strength of clay

by A. W. Bishop, N. K. Kumapley and A. El-Ruwayih

PUBLISHED BY THE ROYAL SOCIETY
6 CARLTON HOUSE TERRACE LONDON SW1Y 5AG

NOTICE TO CONTRIBUTORS TO PROCEEDINGS AND PHILOSOPHICAL TRANSACTIONS OF THE ROYAL SOCIETY

The Royal Society welcomes suitable communications for publication in its scientific journals: papers estimated to occupy up to 24 printed pages are considered for the *Proceedings* and longer papers and those with numerous or large illustrations for the *Philosophical Transactions*.

Detailed advice on the preparation of papers to be submitted to the Society is given in a leaflet available from the Executive Secretary, The Royal Society, 6 Carlton House Terrace, London SW1Y 5AG. The 'Instructions to authors' are also printed in every fifth volume of the *Proceedings* A and B (volume numbers ending in 0 or 5). The basic requirements are: a paper should be as concise as its scientific content allows and grammatically correct; standard nomenclature, units and symbols should be used; the text (including the abstract, the list of references and figure descriptions) should be in double spaced typing on one side of the paper; any diagrams should be drawn in a size to permit blockmaking at a reduction to about one half linear, the lettering being inserted not on the original drawings but on a set of copies; where photographs are essential the layout should be designed to give the most effective presentation.

The initial submission of a paper should normally be through a Fellow or Foreign Member of the Society, but papers may be submitted direct to the Executive Secretary. The latest lists of Fellows and Foreign Members are to be found in the current edition of the *Year Book of the Royal Society*. In the event of any difficulty, an author is invited to send the paper direct to the Executive Secretary.

No page charge is levied, and the first 50 offprints of a paper are supplied to the author gratis.

THE INFLUENCE OF PORE-WATER TENSION ON THE STRENGTH OF CLAY

By A. W. BISHOP, N. K. KUMAPLEY and A. EL-RUWAYIH
Civil Engineering Department, Imperial College of Science and Technology, South Kensington, London, SW 7 2BU

(*Communicated by A. W. Skempton, F.R.S. – Received* 14 *August* 1974)

[Plate 17]

CONTENTS

Substantial pore-water tensions are set up by the removal, under undrained conditions, of the confining pressure from saturated clay samples from which the pore water has been free to drain during consolidation. These pore-water tensions control the mechanical behaviour of the unconfined samples, and in particular their strength and brittleness.

In the present paper the magnitude of the pore-water tensions and the change in strength on stress release are estimated theoretically for the range of pressures within which the sample remains fully saturated. Tests on samples of two clays consolidated in the triaxial apparatus under a wide range of pressure indicate that the limiting pore-water tension above which the sample ceases to remain fully saturated is related to the equivalent pore diameter. Above this limit the loss in strength on stress release is very marked.

The experimental results also show the dramatic change in brittleness resulting from high pore-water tensions. A large reduction in total stress without a change in water content is sufficient to change the failure mechanism from a plastic failure with some strain-softening to a brittle failure in which the sample shatters.

The significance of the results in relation to sampling and testing of soils for engineering purposes is briefly considered.

1. Introduction

When a pressure is applied to a saturated porous material the load is carried wholly by the solid skeleton only if the pore fluid is free to flow out of the voids. If this is prevented by the rapidity of the loading, by the low permeability of the material or by the length of the drainage path, then the load is shared between the solid skeleton and the fluid in the pore space in a ratio which is a function of their compressibilities.

An expression for the ratio of the change in pore pressure Δu to the change in the total normal stress $\Delta\sigma$ was derived by Bishop & Eldin (1950):

$$\frac{\Delta u}{\Delta\sigma} = \frac{1}{1+n(C_w/C)}, \tag{1}$$

where n denotes the porosity, C_w the compressibility of the pore fluid and C denotes the compressibility of an element of the porous material with respect to a change in ambient stress accompanied by zero change in pore pressure.

In the derivation of this expression it was assumed that the compressibility C_s of the solid material forming the skeleton of the soil or other porous material might be neglected. A more general expression which includes this term has been given by Bishop (1966, 1973)

$$\frac{\Delta u}{\Delta\sigma} = \frac{1}{1+n[(C_w-C_s)/(C-C_s)]}. \tag{2}$$

However, for a soil saturated with water, $C_w = 49\times10^{-5}\,\mathrm{m^2/MN}$ and $C_s = 2-3\times10^{-5}\,\mathrm{m^2/MN}$. The value of the compressibility of the solid skeleton is generally larger by several orders of magnitude, being equal to $3000\times10^{-5}\,\mathrm{m^2/MN}$ for a typical sample of heavily overconsolidated London Clay, for example. In these circumstances† the expression is dominated by the ratio of C to C_w and, since C is large compared with C_w, the ratio $\Delta u/\Delta\sigma$ approximates to unity in the low and medium stress range (being 0.994 for the example quoted above).

This conclusion is supported by direct measurements of pore pressure change in saturated

† For rocks and soils under high confining pressures the very low values of the compressibility C make the C_s term of relatively greater importance, and lead to low values of the ratio $\Delta u/\Delta\sigma$. For a value of $C = 9\times10^{-5}\,\mathrm{m^2/MN}$ representative of sandstone under a high confining pressure, $\Delta u/\Delta\sigma = 0.47$.

sand (Bishop & Eldin 1950) and in saturated clay (Taylor 1944; Bishop 1960). Indeed, the most reliable way of determining whether or not a soil sample is fully saturated is to measure the ratio $\Delta u/\Delta\sigma$ under undrained conditions.

Although the value of the compressibility C for a normally consolidated clay soil is very approximately proportional to the reciprocal of consolidation pressure it is only at relatively high consolidation pressures that the ratio $\Delta u/\Delta\sigma$ departs significantly from unity. From the observed compressibility of the London Clay samples discussed in later sections of this paper the calculated value of $\Delta u/\Delta\sigma$ for an increase in stress would be 0.97 and 0.80 for consolidation pressures of 20.7 MN m^{-2} (3000 lbf in^{-2}) and 62.1 MN m^{-2} (9000 lbf in^{-2}) respectively. As the soil structure is not elastic the rebound modulus is substantially different from that observed on first loading, and estimated values of $\Delta u/\Delta\sigma$ for unloading drop to 0.94 and 0.63–0.77 respectively.†

Now consider an element of saturated soil which has been subjected to a total normal stress equal in all directions of σ, the pore water being permitted to drain freely to give an initial value of the pore pressure equal to u. The effective stress, which controls the volume change and strength characteristics, is given to a very close approximation, in the case of saturated soils (Terzaghi 1936; Bishop & Eldin 1950; Bishop 1959; Skempton 1960; Bishop & Blight 1963; Bishop 1966) by the expression

$$\sigma' = \sigma - ku, \tag{3}$$

where k is a parameter which differs only marginally from unity in the case of uncemented soils at low consolidation pressures.

In principle the value of k depends on whether the effective stress equation is used with reference to changes in volume or to changes in shear strength. In the former case an expression was obtained by Bishop (1953):

$$k_v = (1 - C_s/C). \tag{4}$$

For changes in shear strength Skempton (1960) derived the expression:

$$k_s = [1 - a(\tan\psi/\tan\phi')], \tag{5}$$

where a denotes area of contact between the particles, per unit gross area of material; ψ denotes the angle of intrinsic friction of the solid material of the soil grains, and ϕ' denotes the angle of shearing resistance of the granular material. For the values of C for London Clay used in the calculation of the ratio $\Delta u/\Delta\sigma$, values of k_v of 0.999, 0.994 and 0.939 are obtained for the overconsolidated specimen, and the specimens consolidated at pressures of 20.7 MN m^{-2} (3000 lbf in^{-2}) and 62.1 MN m^{-2} (9000 lbf in^{-2}) respectively. The value of k_v is stress-path dependent, and for unloading the last two values drop to 0.989 and 0.887 respectively.

The value of k_s is less readily estimated due to uncertainty about the value of a in particular, especially for clays. A very accurate experimental examination in the low effective stress range ($\sigma - u$ = 0.36 MN m^{-2} (52.6 lbf in^{-2})) using pore pressure changes of up to 27.6 MN m^{-2} (4000 lbf in^{-2}) indicated that in the case of sand k_s did not differ from unity by more than 5×10^{-5}, which was the limit of accuracy of the test (Bishop 1966; Skinner 1975). If we assume that, for small values, a increases linearly with consolidation pressure (Skempton 1960), it follows from

† The data from which these estimates are made is given in table 4. Few direct observations are currently available of $\Delta u/\Delta\sigma$ for porous materials of low compressibility, where the departure from the truly 'undrained' condition due to the flow of water into the measuring system becomes of particular importance. Further reference is made to this problem by Bishop (1973) and examples of system compressibility are given by Wissa (1969) and Bruhn (1972).

equation (5) that the difference between the actual value of k_s and unity increases linearly with consolidation pressure. It may then be inferred that this difference would not exceed 2.8×10^{-3} at 20.7 MN m^{-2} (3000 lbf in^{-2}) and 8.6×10^{-3} at 62.1 MN m^{-2} (9000 lbf in^{-2}). This would indicate values of k_s not lower than 0.997 and 0.991 respectively. There is no evidence † that the values for clay would be lower, though clearly the physical nature of the inter-particle contact will be significantly different.

If the element of soil under consideration is now subjected to a change in total normal stress $\Delta\sigma$ under conditions of zero drainage, the pore pressure will change by Δu. If the ratio $\Delta u/\Delta\sigma$ is denoted by the parameter B (Skempton 1954) the change in effective stress is given by the expression:

$$\begin{aligned}\Delta\sigma' &= \Delta\sigma - k\,\Delta u \\ &= \Delta\sigma - \Delta u + (1-k)\,\Delta u \\ &= (1-B)\,\Delta\sigma + (1-k)\,B\,\Delta\sigma. \end{aligned} \tag{6}$$

In the low and medium range of consolidation pressures the value of B will differ from unity by less than 1 % and k is likely to differ from unity by an even smaller amount, probably less than 0.1 %. Hence the change in effective stress, and thus in volume change and strength, will lie close to the limits of direct measurement in the laboratory. This is confirmed by the negligible increase in undrained strength with increase in total normal stress reported from tests on saturated clays (Terzaghi 1932 and 1936; Jurgenson 1934; Golder & Skempton 1948; Bishop & Bjerrum 1960) and from tests on saturated sand (Bishop & Eldin 1950). Two important exceptions have, however, been reported:

(*a*) *Strongly dilatant sands tested at low confining pressures* (*Bishop & Eldin* 1950)

Here the application of the principal stress difference producing failure was associated with a substantial decrease in pore pressure. In triaxial tests at low cell pressures this decrease resulted in negative pore pressures (gauge) and the stress–strain curve terminated prematurely, due to the inability of the water in the pore space and in the measuring system to withstand the tension necessary to maintain full saturation.

(*b*) *Unconfined compression tests on some clay samples* (*Bishop* 1947; *Golder & Skempton* 1948; *Bishop & Henkel*, 1962)

It has been observed that the unconfined compression strength is often lower than the average of the strengths at higher confining pressures, although the moisture content is the same and no drainage has been permitted during shear. Since the samples are nominally saturated this difference has usually been attributed to the presence of fissures (*loc. cit*). However, it may be noted that throughout an unconfined compression test the pore-water pressure is negative (see, for example, Bishop 1960), and from the high strength of some samples substantial pore-water tensions may be inferred, though they cannot be measured directly (see, for example, Croney & Coleman 1960). In this respect clays appear to differ from the sand samples referred to in the preceding paragraph.

It is therefore of interest to examine the influence of high pore-water tensions on the strength of clay tested at constant water content in the absence of fissures.

† A series of similar tests on clay (Kumapley 1969) are consistent with tests on sand, but are not capable of the same accuracy due to the lower pressures used and to the problems of pore pressure equilization in soils of low permeability.

2. Prediction of pore-water tension and change in strength on stress release

Figure 1 illustrates the predicted pore pressure in an undrained saturated sample which has initially been consolidated under a total normal stress σ_0 with free drainage of the pore water to atmospheric pressure and then subjected to a change in total normal stress under undrained conditions. The relation between total normal stress σ and pore pressure u (gauge) is given by the expression

$$u = B(\sigma - \sigma_0). \tag{7}$$

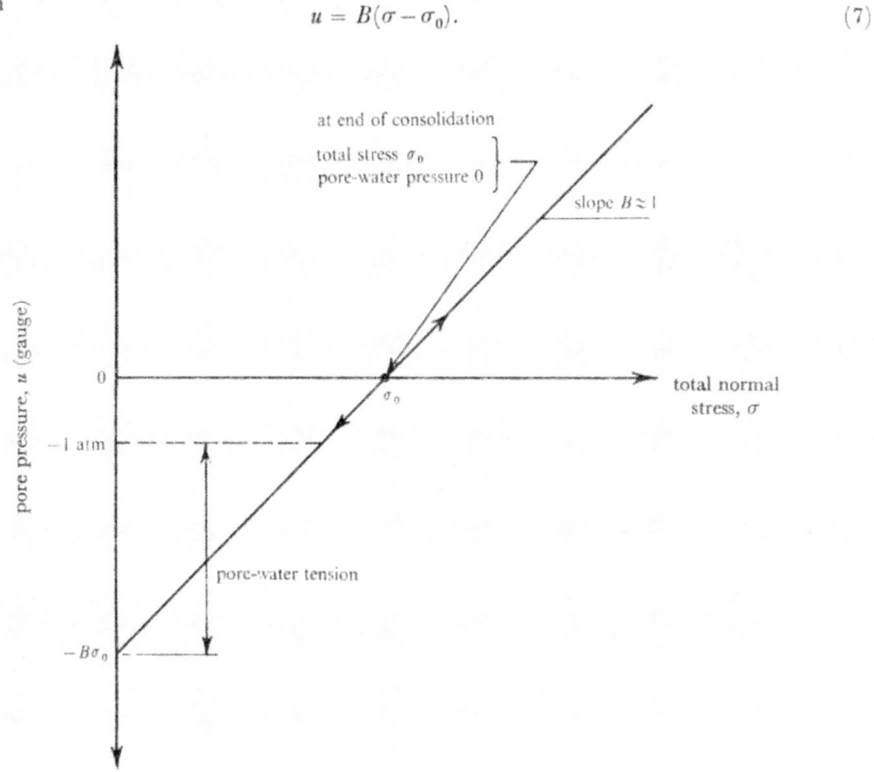

Figure 1. Relation between pore-water pressure and total normal stress for an undrained change in stress after consolidation with the total stress equal to σ_0 and the pore water free to drain to atmosphere. (Full saturation assumed.)

For an unconfined specimen the initial pore pressure before shear would thus be equal to $-B\sigma_0$, and the amount by which this negative pore pressure falls below minus one atmosphere will represent tension in the pore water. If B approximates to unity under these conditions, then the pore water of samples consolidated at a total normal stress in excess of plus one atmosphere should be in a state of tension when the sample is unconfined. In the more general case (figure 2) in which the sample drains, not to atmospheric pressure, but against a back pressure u_0, the relation between total normal stress and pore pressure is given by the expression

$$u = u_0 + B(\sigma - \sigma_0). \tag{8}$$

In the present series of tests values of σ_0 of up to 62.1 MN m^{-2} (9000 lbf in^{-2}) have been used, associated with $u_0 = 0$. A theoretical maximum pore-water tension of $B \times 62.1$ MN m^{-2} could therefore have been induced in the pore water.* Taking the value of B for unloading at this consolidation pressure as 0.63 we have a tension of 39.1 MN m^{-2} (5670 lbf in^{-2}).

As will be seen from the test results, a sample of London Clay could not sustain this pore-water tension and remain saturated.

To this negative pore pressure must be added the change in pore pressure resulting from the application of the stress difference necessary to cause undrained failure. In contrast to the case of dilatant sands, this additional pore pressure change is positive in most clays unless they have been heavily over-consolidated before the undrained test is carried out.

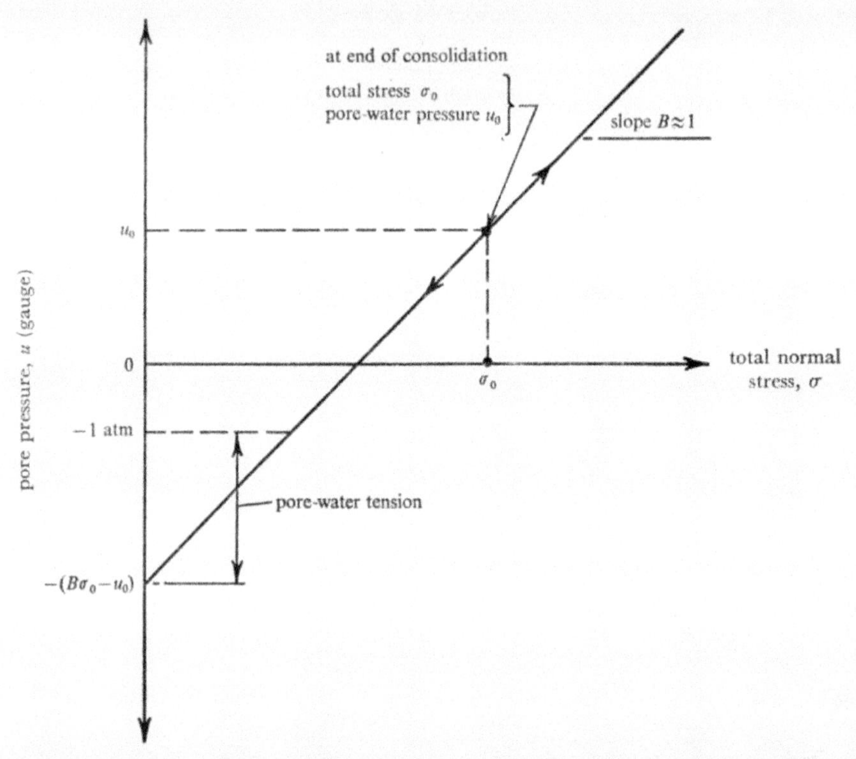

FIGURE 2. Relation between pore-water pressure and total normal stress for an undrained change in stress after consolidation with the total stress equal to σ_0 and the pore pressure equal to u_0. (Full saturation assumed.)

This additional change in pore pressure may be conveniently expressed in terms of the more general expression (Skempton 1954):

$$u = B[\Delta\sigma_3 + A(\Delta\sigma_1 - \Delta\sigma_3)], \tag{9}$$

where $\Delta\sigma_1$ denotes the change in major principal stress (i.e. the axial stress in the triaxial compression test), $\Delta\sigma_3$ the change in minor principal stress (in the conventional triaxial compression test $\Delta\sigma_3 = \Delta\sigma_2$ and equals the change in cell pressure), and A a pore pressure parameter which varies with clay type and stress history. For normally consolidated clays A is positive and generally close to unity at failure if the consolidation has been carried out under isotropic conditions (see, for example, Bishop & Henkel 1962, Table 5). For over-consolidated clays, which have been consolidated under a maximum effective consolidation pressure p_c and allowed to swell under a

* Neglecting the 0.1 MN m^{-2} difference between gauge and absolute pressure.

smaller pressure p before the undrained test stage, the value of A_f at failure becomes negative only where the over-consolidation ratio p_c/p exceeds 4 or 5 in typical cases of remoulded clay (Bishop & Henkel 1962, Fig. 82). Undisturbed samples of London Clay which have been heavily over-consolidated in nature in general do not show a negative A value at the point of maximum stress difference (Bishop, Webb & Lewin 1965; Agarwal 1967), though in undisturbed Weald Clay a value as low as -0.62 has been observed (Bishop & Henkel 1962).

The pore pressure at failure is obtained from equations (8) and (9) as

$$u = u_0 + B(\sigma - \sigma_0) + BA_f(\Delta\sigma_1 - \Delta\sigma_3)_f. \tag{10}$$

The maximum shear stress at failure under undrained conditions may be denoted c_u and is equal to $\frac{1}{2}(\Delta\sigma_1 - \Delta\sigma_3)_f$. The ratio c_u/p, where p is the effective consolidation pressure, closely approximates to a constant for normally consolidated samples of a given clay. Equation (10) may therefore be written

$$u = u_0 + B(\sigma - \sigma_0) + BA_f 2(c_u/p)(\sigma_0 - u_0). \tag{11}$$

For the present series of tests on London Clay, $c_u/p = 0.20$ for normally consolidated samples and $A_f = 1.25$ to a close approximation.† Taking the case where $u_0 = 0$ and putting $B = 1$, we have: initial pore pressure in the unconfined state

$$u = -\sigma_0, \tag{12}$$

pore pressure at failure in the unconfined state

$$\begin{aligned} u &= -\sigma_0 + 1 \times 1.25 \times 2 \times 0.20\sigma_0 \\ &= -0.50\sigma_0. \end{aligned} \tag{13}$$

The pore-water tension will thus be highest in the unconfined state before the axial load is applied and will decrease very substantially before failure.‡

Where B or k differ significantly from unity, it is apparent from equation (6) that a reduction in confining pressure will result in a decrease in effective stress. The unconfined sample will therefore behave as a slightly overconsolidated material, but the over-consolidation ratio p_c/p will be relatively small (it is equal to $1/B$ if $k = 1$) and change in the value of A_f will therefore not alter the general conclusion of the previous paragraph.

However, this decrease in effective stress will be reflected in a reduction in the undrained strength $(-\Delta c_u)$ of the sample. As will be seen from the test data reported in this paper the relation between the undrained strength c_u and the consolidation pressure p is almost linear on first loading from a slurry. However, this relation is not fully reversible (Figure 3). The results of the tests on London Clay (series 2) reported in this paper indicate that on the reduction of the effective consolidation pressure before undrained shear we have

$$\begin{aligned} [\Delta c_u/\Delta p]_{\text{unloading}} &= m(c_u/p)_{\text{first loading}} \\ &= mn, \end{aligned} \tag{14}$$

† From the results of undrained tests with pore pressure measurement at relatively high consolidation pressures reported by Bishop *et al.* (1965).

‡ This is in contrast to the position for dilatant sands, for which A_f may drop as low as -0.42, as in the tests reported by Bishop & Eldin (1950). For cohesionless soils where the angle of internal friction is ϕ', the theoretical lower limit of A_f is given by the expression (Bishop 1952):

$$A_f = -(1 - \sin\phi')/2\sin\phi'.$$

where $m = 0.5$ for London Clay consolidated from a slurry, for $-\Delta p/p_{max} = \frac{1}{2}$, and n is a constant approximating to 0.2 for this batch of London Clay. Taking $k_s = 1$, we have, from equation (6)

$$\Delta p = \Delta\sigma' = (1-B)\,\Delta\sigma. \tag{15}$$

The reduction in strength consequent on a reduction in total stress $-\Delta\sigma$ is thus

$$-\Delta c_u = (1-B)\,(-\Delta\sigma)\,mn. \tag{16}$$

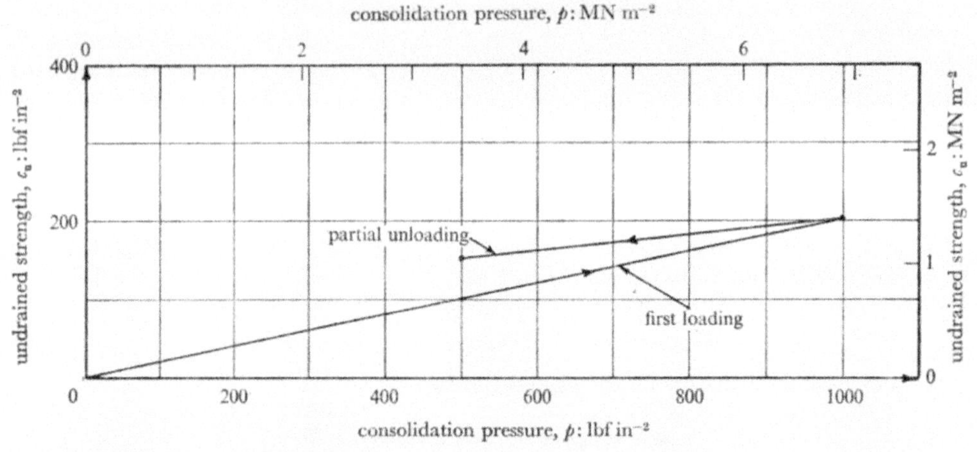

FIGURE 3. The relation between undrained strength c_u and consolidation pressure p on first loading and on partial unloading. Data from tests on London Clay, series 2.

Before the reduction in total stress, the effective consolidation pressure is $\sigma_0 - u_0$ and the undrained strength $(c_u)_0$ is equal to $n(\sigma_0 - u_0)$. The proportional reduction in strength r is

$$\frac{-\Delta c_u}{(c_u)_0}$$

and is thus given by the expression

$$r = -\frac{m(1-B)\,\Delta\sigma}{\sigma_0 - u_0}. \tag{17}$$

The accuracy of predictions made on the basis of this expression is subject to two qualifications. The stress decrement ratio $-\Delta p/p_{max}$ resulting from an undrained reduction in total stress is smaller than the values of this ratio used in obtaining representative values of m. The value of $m = 0.5$ obtained for this clay and from earlier tests on London Clay by Donald (1961) may therefore represent a slight overestimate.

Similarly the value of expansibility used in determining B is based on volume changes associated with relatively large stress changes. A preliminary examination suggests that the error involved is not important in the present context.

The theoretical predictions may conveniently be investigated experimentally in two stages:

(1) Tests to investigate the influence on undrained strength of large reductions in confining pressure which do not result in pore-water tensions. These are tests in which consolidation is carried out with a substantial back pressure u_0. It follows from equation (8) that the cell pressure

may then be reduced from σ_0 to a value equal to $\sigma_0 - u_0/B$ without inducing a negative pore pressure. Within this range of total stresses the question of any departure from full saturation does not arise.

(2) Tests to investigate the influence on undrained strength of reductions in confining pressure leading to pore water tensions of various magnitudes. In the limit these will result in departures from full saturation and thus from the assumption on which the predicted values of r are based.

3. Apparatus and testing techniques

The samples for stage 1 of the investigation were consolidated and sheared in a medium pressure range (0 – 6.9 MN m^{-2} (1000 lbf in^{-2})) triaxial cell, shown diagrammatically in figure 4. The cell consists of a stainless steel cylinder 100 mm in external diameter and having a wall thickness of 10 mm. The axial load is transmitted to the sample by a honed stainless-steel ram running in a bronze bush which is rotated at 2 rev/min by a worm drive in order to eliminate the vertical component of friction on the ram. Mechanical details are given by Bishop *et al.* (1965).

The cell pressure and back pressure were provided either by a self-compensating mercury system (Bishop & Henkel 1962) or by a hydraulic constant pressure source employing a spring loaded Amsler bleed valve. The height of the building in which the laboratory is situated permitted a pressure of 6.9 MN m^{-2} (1000 lbf in^{-2}) to be obtained from the self-compensating mercury system by the addition of only three more stages, using stainless steel cylinders connected by nylon tubing jacketed in polythene as an additional precaution against rupture.

Although the mercury system is free from hunting and from significant long-term fluctuations in pressure, the multistage arrangement is difficult to operate. In stage 2 of the investigation, where a larger pressure range was required, the system incorporating the Amsler valve was generally used. These larger pressures were found to be necessary to reach the range, in the case of London Clay, in which the unconfined tests showed a significant difference in strength from the confined tests. A high pressure triaxial cell capable of operating at 69 MN m^{-2} (10000 lbf in^{-2}) was used. This incorporates a hydraulically balanced loading ram and an internal load transducer sensitive only to the vertical component of load and unaffected by cell pressure. Some details of the apparatus are given in appendix 1.

The test specimens were 20 mm in diameter and 40 mm in height and were permitted to consolidate by drainage through the base. In the standard triaxial apparatus the sample stands on a porous ceramic disk, the lower side of which communicates through passages drilled in the pedestal to the valves controlling drainage. Undrained tests on saturated samples can be carried out satisfactorily with this system provided the cell pressure is maintained at a high enough value to ensure that the pore pressure in the sample remains positive (or in the limit does not drop below about –83 kN m^{-2} (–12 lbf in^{-2}) gauge pressure). At lower pressures, and in all tests at the stage when the cell pressure is dropped to zero to remove the sample for water content determination, the water in the drainage passages will cavitate and the sample will start to imbibe water from the porous ceramic disk.

This effect leads to an error in the final water content determination in all consolidated undrained tests, the magnitude of which depends on the dimensions of the sample, the coefficient of swelling of the clay and the time taken to strip down the triaxial cell. This effect also leads to an error, which is of more importance in the context of the present investigation, in the undrained strength of all samples in which significant pore-water tensions are created by reducing the

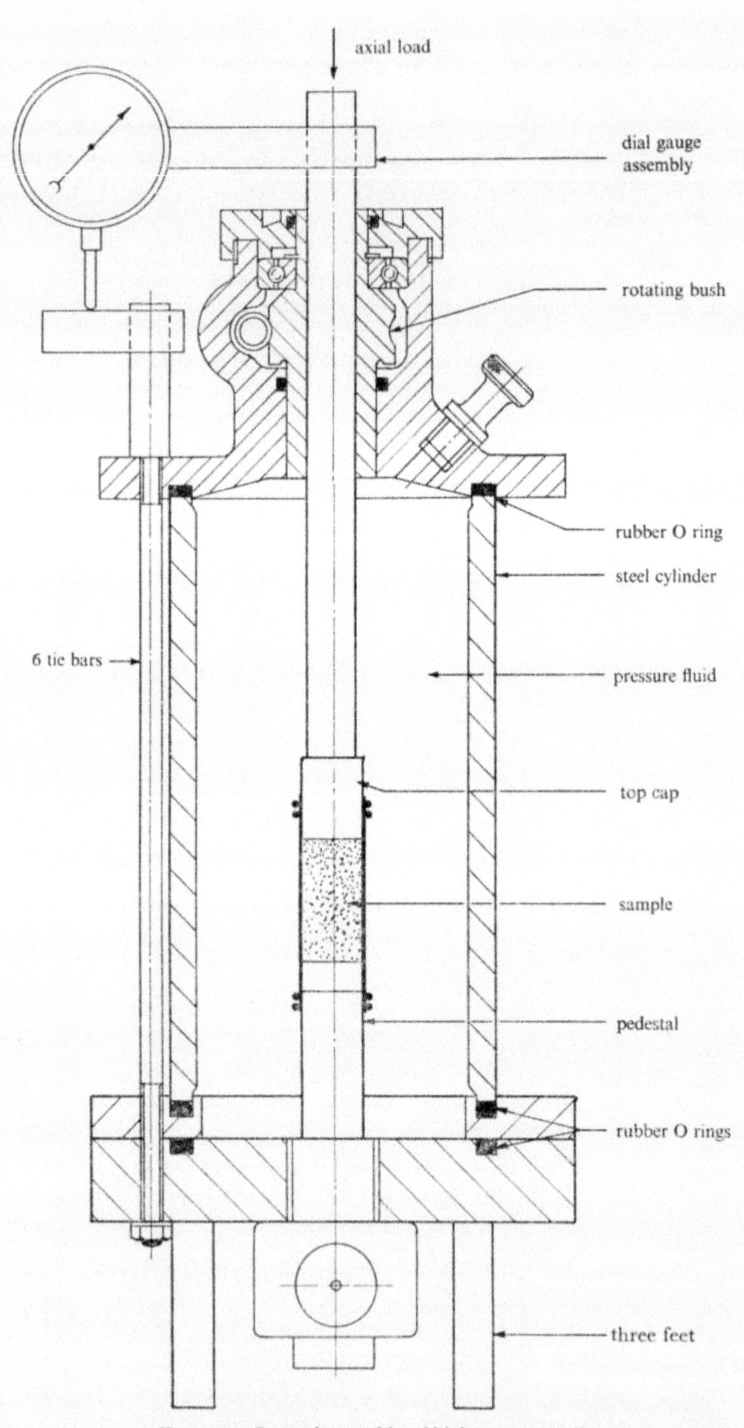

FIGURE 4. General assembly of high pressure cell.

confining pressure. Even the use of a very short duration test (about 5 min to failure from the time at which the cell pressure was reduced), together with the use of high air entry ceramic disks and a reduction in the volume of the passage between the disk and the valves, could not bring the magnitude of this effect within acceptable limits.

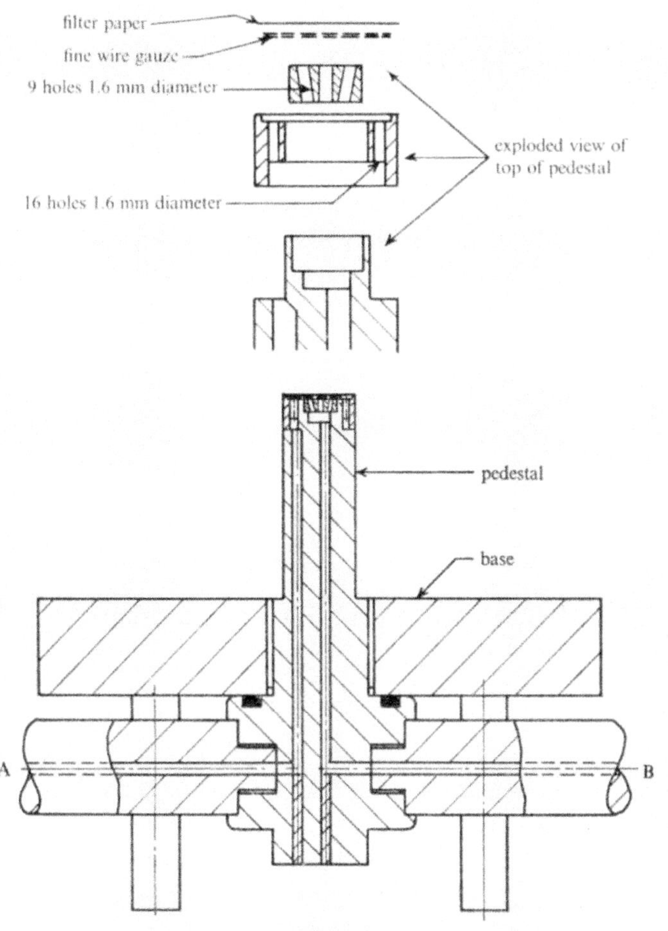

FIGURE 5. Modified triaxial cell base.

The triaxial cell base was therefore modified for stage 2 of the investigation so that, when required, all free water could be removed from below the base of the specimen by blowing dry nitrogen through the drainage system†. Details of the base are shown in figure 5. The porous ceramic disk is replaced by a disk of filter paper resting on a disk of fine wire gauze, which is recessed into the top of the pedestal so that the strands of the gauze do not puncture the rubber membrane enclosing the sample.

† A similar method was adopted by Westman (1932) in studying the effect of mechanical pressure on the moisture content of some ceramic clays. He used two porous pistons within a rigid cylinder, and removed the water from the pistons with compressed air.

Drainage is through two groups of small diameter holes (1.6 mm diameter in the intermediate pressure range and 0.8 mm diameter in the high pressure range). One group of holes is in the stainless steel collar forming the outer annulus of the pedestal and connects with the drainage passage A on one side of the cell base. The other group of holes is in the stainless steel disk forming the centre of the pedestal and connects with the drainage passage B leading to the other side of the cell base.

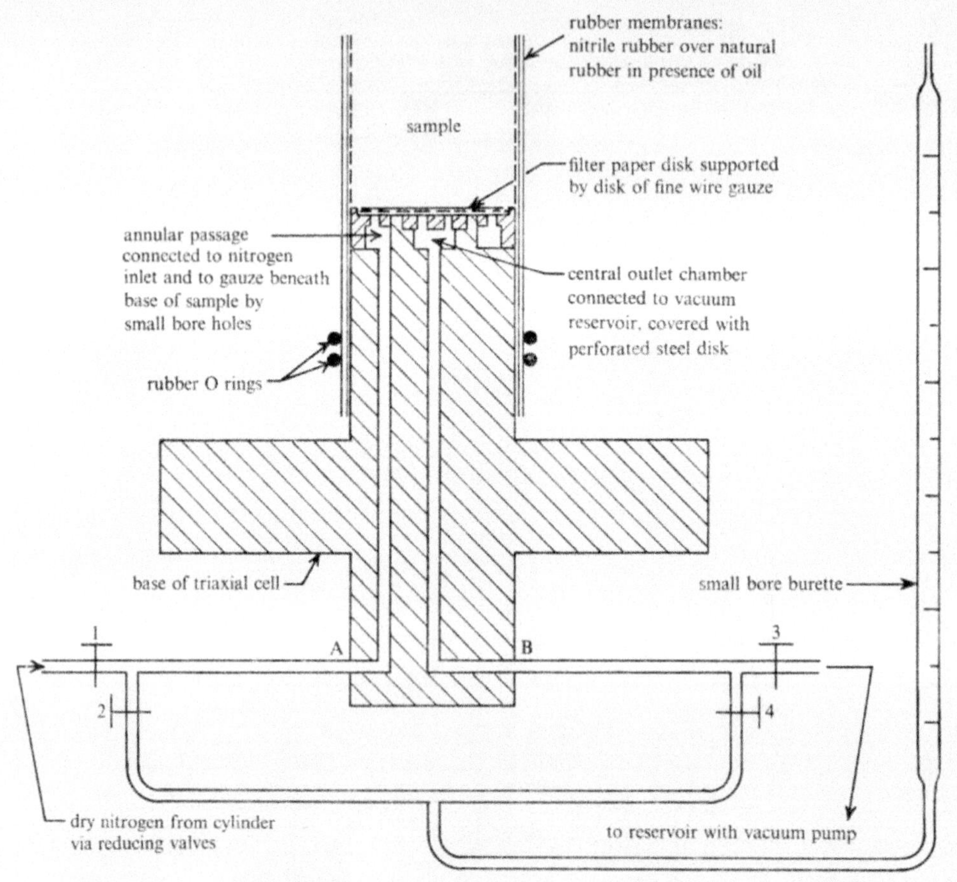

FIGURE 6. Layout of connexions to triaxial cell base for tests with high pore-water tensions (diagrammatic).

The layout of the apparatus is illustrated in figure 6. After the completion of the consolidation stage, which is carried out with a saturated pedestal and zero back pressure, the water is blown out of the connexion and burette used to measure volume change during consolidation by admitting dry nitrogen through valve 1 with valves 2 and 4 open. With valves 2 and 4 closed dry nitrogen is circulated through valve 1 to the outer annulus of the base of the sample and exhausted through the central disk to valve 3 which is opened to a reservoir connected to a vacuum pump. By admitting the nitrogen at a pressure of a little under one atmosphere (gauge) at valve 1 the base of the sample is maintained at atmospheric pressure.

The time required to remove all free water from the connecting passages and from the gauze and filter paper was determined by trial and error, and proved to be about 6 min. It was considered preferable to risk slight drying of the base of the specimen, which would not increase the measured compression strength due to the ratio of height to diameter of 2, rather than to leave free water at the base which would initiate base failure at a reduced strength.†

In order to obtain the correct moisture content values for the samples tested in the positive pore pressure range (for which undrained testing necessitated a saturated pedestal with valves 1, 2, 3 and 4 closed), an estimate of the final pore pressure was made by using equation (10) and an assumed value of A_f. The cell pressure was then dropped by an equal amount after failure, to bring the final pore pressure to zero (for $B = 1$). The system was then de-watered with dry nitrogen as above before dropping the cell pressure to zero for the removal of the sample. The close agreement apparent in table 3 between the water contents observed in the high pressure range on the unconfined samples (base de-watered before the compression test) and the confined samples (base de-watered after the compression test) indicates the success of this technique.

To ensure uniform initial conditions for all the test specimens in each series a carefully mixed clay slurry was consolidated in a 230 mm diameter oedometer to form a disk of clay about 50 mm in thickness of a strength and moisture content convenient for sample preparation. The samples were cut from the disk by means of a sharpened thin walled tube and trimmed to length before being placed in the triaxial cell.

The use of small diameter samples permitted a large series of tests to be performed on specimens cut from the same disk. To minimize handling, the disk was cut into a number of sectors, each of which was waxed, wrapped in metal foil and stored at a constant temperature until required.

The use of samples for smaller linear dimensions also brought the time required for consolidation within acceptable limits. This time depends on the square of the drainage path and owing to the necessity of de-watering the drainage system for the unconfined tests the usual expedients for shortening the drainage path of double end drainage and radial drainage could not be applied without undue complications.

4. Materials tested

The two materials used in this investigation were blue London Clay and Kaolin. The results of various classification tests are given in table 1A. The blue London Clay was from a depth of about 6.1 m in the borrow pit at Wraysbury Reservoir in Middlesex. The London Clay was deposited under marine conditions in the Eocene period. Tests reported by Bishop *et al.* (1965). suggest a preconsolidation load in this area of 4.1 MN m^{-2} (600 lbf in^{-2}). The natural water content of the clay at this level was around 29 %.

The Kaolin was a commercially prepared product mined in Cornwall and marketed as English China Clay, Spestone powder.

The initial water contents of the slurries from which the batches were prepared are given in table 1B, together with the consolidation pressures used. The batch of London Clay on which the first series of tests were performed was not allowed to swell during unloading. To reduce the pore water tensions and any consequent tendency towards partial saturation, the second batch of London Clay and the Kaolin were allowed to swell to relatively low effective stresses before storage. The higher consolidation pressures used were chosen to reduce the volume changes in the subsequent triaxial consolidation stage which might otherwise result in sample distortion when very high consolidation pressures were used.

† To determine the water content the sample was generally cut into 5 horizontal slices, and the value of the bottom slice recorded but not included in the average.

TABLE 1A. CLASSIFICATION TESTS

material	liquid limit (%)	plastic limit (%)	plasticity index (%)	clay fraction (%)	activity†	relative density of particles
London Clay (1)	74	28	46	57	0.81	2.
London Clay (2)	78	28	50	55	0.91	
Kaolin	70	40	30	68	0.44	2.64

† Plasticity index/percentage clay.
Mineralogy of London Clay: illite with kaolinite, montmorillonite and chlorite.

TABLE 1B. CONSOLIDATION OF CLAY BEFORE SAMPLING

material	maximum vertical stress		final vertical stress		water content	
	MN m^{-2}	lbf in^{-2}	MN m^{-2}	lbf in^{-2}	initial (%)	final (%)
London Clay (1)	0.28	41	0.28	41	112	43
London Clay (2)	1.10	160	0.07	10	77.5	34.2
Kaolin	1.10	160	0.02	3	71.5	42.6

5. TEST RESULTS

(a) *Stage* 1: *tests to investigate the influence on undrained strength of reductions in confining pressure which do not result in negative pore water pressures*

Samples for this series of tests were cut from batch 1 of the London Clay and were consolidated with a cell pressure $\sigma_3 = 6.9$ MN m^{-2} (1000 lbf in^{-2}) and an initial pore water pressure $u = 3.4$ MN m^{-2} (500 lbf in^{-2}). The samples were then tested, with the pedestal saturated and the drainage valves closed, at cell pressures of 6.9, 6.2, 5.5, 4.8, 4.1 and 3.4 MN m^{-2} (1000, 900, 800, 700, 600 and 500 lbf in^{-2}). The rate of axial strain approximated to $1\frac{1}{2}$ % per minute.

The results are given in table 2 and are plotted in figure 7.

TABLE 2. RESULTS OF TESTS ON LONDON CLAY (SERIES 1)

Test no.	consolidation pressures σ_3		consolidation pressures u		values of σ_3 during compression test		undrained strength c_u		c_u/p
	MN m^{-2}	lbf in^{-2}	MN m^{-2}	lbf in^{-2}	MN m^{-2}	lbf in^{-2}	MN m^{-2}	lbf in^{-2}	
1	6.9	1000	3.4	500	6.9	1000	0.673	97.6	0.195
2	6.9	1000	3.4	500	6.2	900	0.665	96.4	0.193
3	6.9	1000	3.4	500	5.5	800	0.665	96.5	0.193
4	6.9	1000	3.4	500	4.8	700	0.667	96.8	0.194
5	6.9	1000	3.4	500	4.1	600	0.657	95.3	0.191
6	6.9	1000	3.4	500	3.4	500	0.657	95.3	0.191

Although there is some scatter in the test results it will be seen that there is a small but significant decrease in strength, amounting to about 2.1 % for a decrease in total stress of 3.4 MN m^{-2} (500 lbf in^{-2}) which is equal to the effective consolidation pressure $p(= (\sigma_3)_0 - (u)_0$ to a very close approximation). This decrease corresponds to a slope of 0.2° for the Mohr envelope for undrained strength plotted against total stress.

This reduction in strength may be compared with the value predicted by equation (17) by substituting the following values:

$$m = 0.5$$

$$B = 0.983$$

$$-\Delta\sigma = \sigma_0 - u_0 = 3.4\,\mathrm{MN\,m^{-2}}\ (500\,\mathrm{lbf\,in^{-2}})$$

giving $$r = 0.86\,\%.$$

This value is smaller than the observed decrease in strength, but in view of the small values of both the predicted and observed values of r in relation to the scatter of the strength measurements plotted in figure 7, it is doubtful whether this difference is significant.

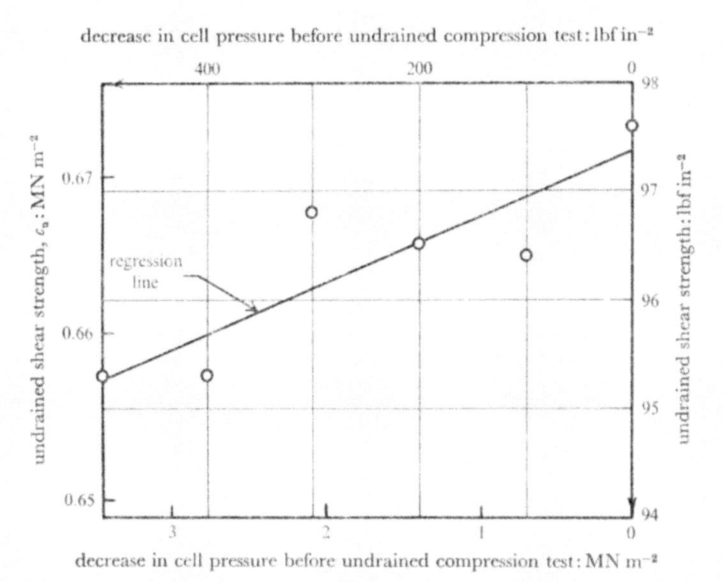

FIGURE 7. Relation between undrained shear strength and decrease in cell pressure after completion of consolidation stage: Blue London Clay, consolidated under $\sigma_3 = 6.9\,\mathrm{MN\,m^{-2}}$ (1000 lbf in^{-2}), $u = 3.4\,\mathrm{MN\,m^{-2}}$ (500 lbf in^{-2}).

(b) *Stage 2: tests to investigate the influence on undrained strength of reductions in confining pressure leading to pore water tensions*

Samples for this series of tests were cut from batch 2 of the London Clay and from the Kaolin. The samples were consolidated with a range of cell pressures 6.9–62.1 MN m^{-2} (1000–9000 lbf in^{-2}) for the London Clay and 0.86–6.9 MN m^{-2} (125–1000 lbf in^{-2}) for the Kaolin but with zero back pressure in all cases. The samples were then tested either with the cell pressure equal to the consolidation pressure ($\sigma_3 = p$) or unconfined ($\sigma_3 = 0$), the pedestal being de-watered at the appropriate stage as described in § 3. The rate of axial strain was approximately 2 % per minute.

The results are given in tables 3 and 5, but will be described separately in the next two sections.

(i) *Tests on London Clay*

Two preliminary tests on remoulded London Clay (table 3, 1 B and 2 B) had indicated that the ratio c_u/p for unconfined specimens did not differ significantly from that for confined specimens for a consolidation pressure of 6.9 MN m^{-2} (1000 lbf in^{-2}) and that much higher consolidation pressures must therefore be used for this stage of the investigation. Subsequent tests were carried out on batch 2 of the London Clay with a lower initial water content, in order to reduce the magnitude of the volume change and associated distortion, on subsequent consolidation in the triaxial cell.

The two most notable effects of high pore-water tensions on the strength and deformation

Table 3. Results of tests on London clay (series 2)

test no.	consolidation pressures σ_3 MN m^{-2}	lbf in^{-2}	u	water contents: initial w (%)	final w (%)	tests with $\sigma_3 = p$: c_u MN m^{-2}	lbf in^{-2}	c_u/p	tests with $\sigma_3 = 0$: c_u MN m^{-2}	lbf in^{-2}	c_u/p
1 B†	6.9	1000	0	54.6	19.5	—	—	—	1.44	209	0.209
2 B†	6.9	1000	0	53	19.3	—	—	—	1.45	210	0.210
3 B	6.9	1000	0	36.2	19.7	—	—	—	1.37	198	0.198
4 B	6.9	1000	0	32.4	19.9	1.39	201	0.201	—	—	—
5 B	—	—	—	—	—	—	—	—	—	—	—
6 B	6.9	1000	0	35.4	19.5	—	—	—	1.42	206	0.206
7 B	6.9	1000	0	34.8	19.8	1.52	221	0.221	—	—	—
8 B	6.9	1000	0	35.5	19.6	1.38	200	0.200	—	—	—
9 B	—	—	—	—	—	—	—	—	—	—	—
10 B	6.9	1000	0	33.4	20.0	1.40	203	0.203	—	—	—
11 B	6.9	1000	0	33.5	19.5	1.43	207	0.207	—	—	—
12 B‡	6.9	1000	0	33.6	—	—	—	—	—	—	—
	3.4	500	0	—	21.8¶	—	—	—	1.04	151	0.302
13 B	20.7	3000	0	33.6	13.8	4.40	638	0.213	—	—	—
14 B	20.7	3000	0	34.5	13.8	—	—	—	4.00	580	0.193
15 B§	6.9	1000	0	33.1	—	—	—	—	—	—	—
	0.69	100	0	—	—	—	—	—	—	—	—
	6.9	1000	0	—	17.8‡‡	1.66	241	0.241	—	—	—
16 B	20.7	3000	0	36.7	13.9	—	—	—	4.09	593	0.198
17 B	20.7	3000	0	34.7	13.4	4.07	590	0.197	—	—	—
18 B	20.7	3000	0	33.5	13.4	4.41	640	0.213	—	—	—
19 B	62.1	9000	0	33.5	10.2	11.9	1723	0.191	—	—	—
20 B‖	6.9	1000	0	33.9	19.7¶	—	—	—	1.35	196	0.196
21 B	62.1	9000	0	33.9	10.1	—	—	—	6.76	980††	0.109
22 B	62.1	9000	0	33	10.1	—	—	—	7.03	1020††	0.113
23 B	—	—	—	—	—	—	—	—	—	—	—
24 B	41.4	6900	0	33.2	11.0	—	—	—	5.69	825††	0.137
25 B	41.4	6000	—	33.5	10.8	9.14	1325	0.221	—	—	—
26 B	6.9	1000	0	35.2	—	—	—	—	—	—	—
	0.69	100	0	—	—	—	—	—	—	—	—
	6.9	1000	0	—	19.1	—	—	—	—	—	—

† Preliminary series with higher initial water content.
‡ Sample consolidated to 6.9 MN m^{-2} and allowed to swell under 3.4 MN m^{-2} before testing.
§ Sample allowed to swell and then reconsolidated before testing.
‖ Sample allowed to stand unconfined for 35 min before testing.
¶ Calculated from volume change.
†† The expansion on stress release and reduction in volume under the axial stress were not measured during these tests and are significant only for the samples which departed from full saturation. At the maximum consolidation pressure the tabulated value may represent an overestimate of the strength by about $2\frac{1}{2}$ %.
‡‡ The final water content determination appears to be in error and should approximate to that of 26 B.

characteristics of clay are illustrated very clearly by the series of stress–strain curves for progressively increasing consolidation pressure (figures 8–11). The stress–strain curves for samples consolidated with a cell pressure $\sigma_0 = 6.9\ \mathrm{MN\ m^{-2}}$ (1000 lbf in^{-2}) and a pore water pressure $u_0 = 0$ are illustrated in figure 8. It will be seen that the samples tested with the cell pressure maintained at 6.9 MN m^{-2} (and thus with the pore pressure zero at the beginning of the test and positive at failure) show a rounded stress–strain curve with a relatively gradual reduction in stress after failure. In contrast the samples tested after reducing the cell pressure to zero (and thus inducing a negative pore pressure of approximately 6.9 MN m^{-2} before the beginning of the undrained compression test) retained approximately the same peak value of strength but showed a rapid reduction in strength after failure amounting almost to brittle failure.

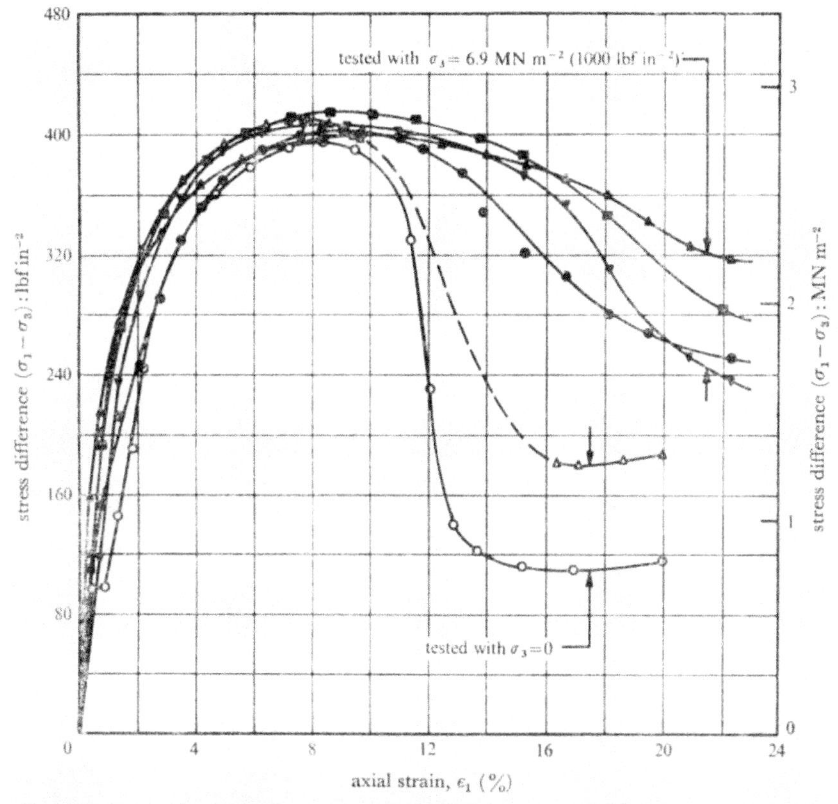

FIGURE 8. Comparison of confined and unconfined compression tests on samples of London Clay consolidated with $\sigma_3 = 6.9\ \mathrm{MN\ m^{-2}}$ (1000 lbf in^{-2}) and $u = 0$.

The fractured specimens fell apart on removal from the enclosing membrane in the case of all specimens tested under high negative pore pressures, whereas those tested under positive pore pressures could not be separated by hand along the rupture surface which was visible on the surface of the specimen. The failure mechanisms are illustrated in figure 12, plate 17, by photographs of samples subjected to the highest consolidation pressure. The corresponding curves for a consolidation pressure of 20.7 MN m^{-2} (3000 lbf in^{-2}) show the same general trends (figure 9), though there is a small but significant reduction in peak strength accompanied by a marked increase in the brittleness of the specimens tested under the condition of high negative pore pressure.

The stress–strain curves for a consolidation pressure of 41.4 MN m^{-2} (6000 lbf in^{-2}) show a very substantial reduction in peak strength in the specimen tested with a high induced negative pore pressure accompanied again by a very brittle type of failure (figure 10).

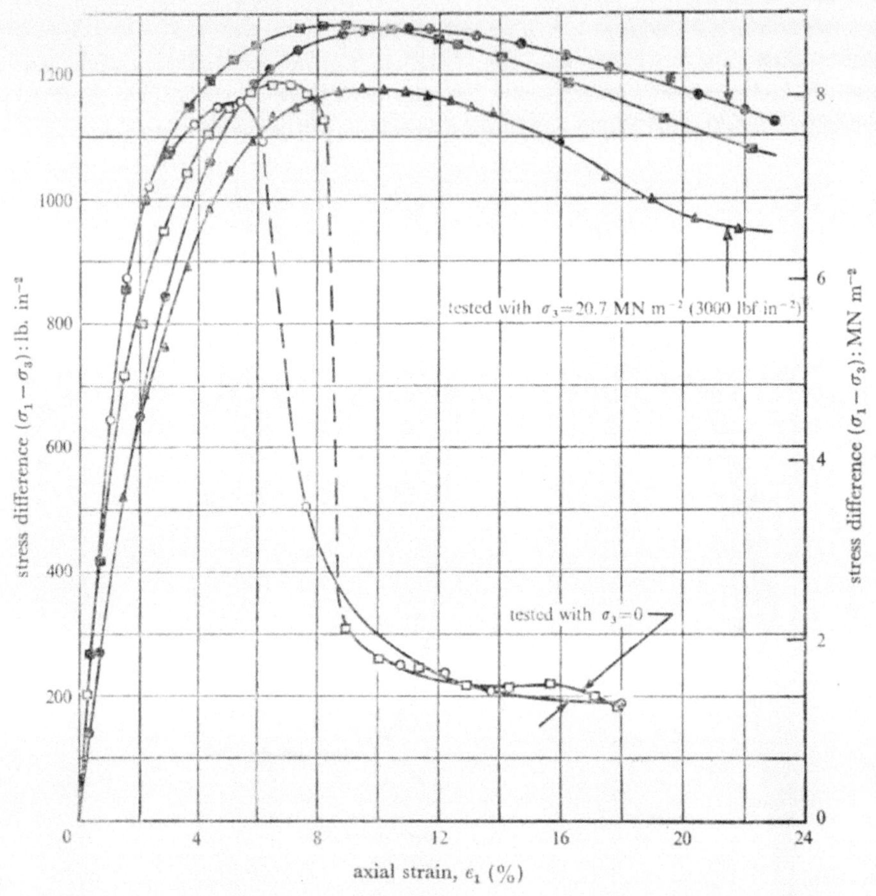

FIGURE 9. Comparison of confined and unconfined compression tests on samples of London Clay consolidated with $\sigma_3 = 20.7$ MN m^{-2} (3000 lbf in^{-2}) and $u = 0$.

The corresponding curves (figure 11) for a consolidation pressure of 62.1 MN m^{-2} (9000 lbf in^{-2}) (the maximum pressure at which the load transducer could be used) show an even larger percentage reduction in compression strength (42 %) and a very brittle failure, the fractured specimen being illustrated in figure 12.

The results are summarized on graphs relating undrained shear strength c_u and consolidation pressure p (figure 13), undrained shear strength c_u and water content as tested w (figure 14) and percentage loss in strength r against reduction in confining pressure (figure 15). It is clear from all three sets of curves that a major change in behaviour is beginning at a reduction in cell pressure of about 20.7 MN m^{-2} (3000 lbf in^{-2}). A theoretical relation between the reduction in strength r, the reduction in confining pressure $-\Delta\sigma$ and the consolidation pressure $\sigma_0 - u_0$ was obtained in §2, equation (17). In the present series of tests $u_0 = 0$, and for the samples tested unconfined the reduction in confining pressure $-\Delta\sigma = \sigma_0$. Equation (17) thus reduces to the expression:

$$r = m(1-B). \tag{17a}$$

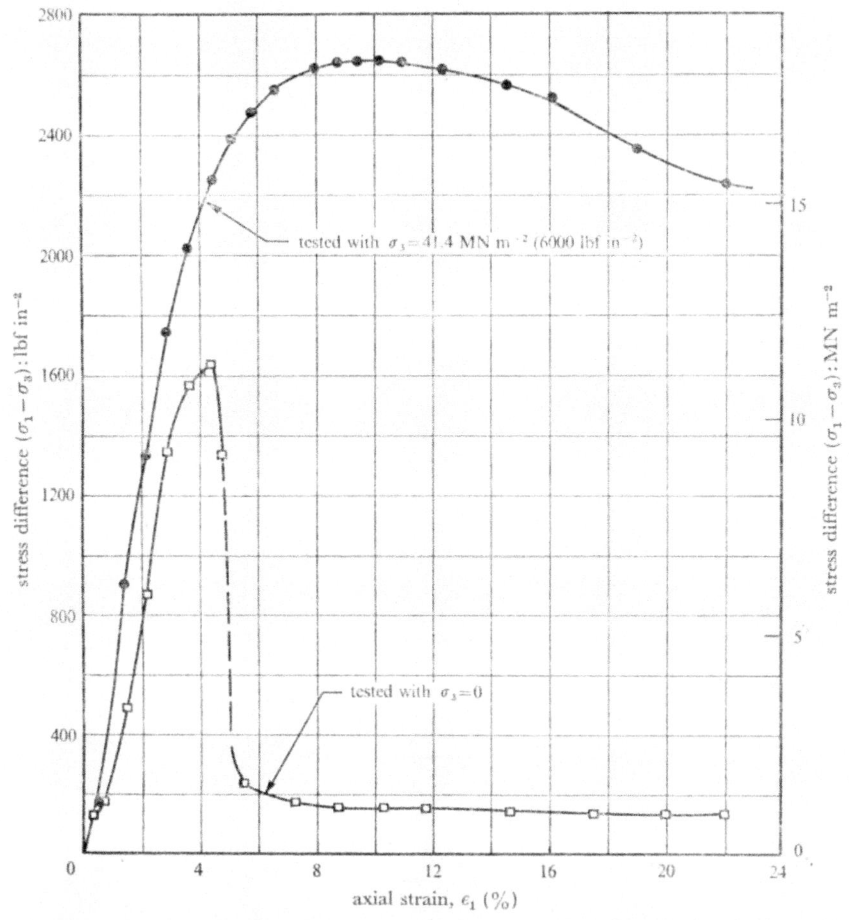

FIGURE 10. Comparison of confined and unconfined compression tests on samples of London Clay consolidated with $\sigma_3 = 41.4$ MN m^{-2} (6000 lbf in^{-2}) and $u = 0$.

For the very wide range of pressures used in the present series of tests, the relation between the void ratio e and the logarithm of consolidation pressure p departs substantially from a straight line (figure 16) and the compressibility falls to a value less than that of water. No simple relation can therefore be established between r and p, and the values of r must be obtained numerically for each stress level, using the more accurate expression for B (equation 2) for the higher consolidation pressures. In the calculation of B on unloading the expansibility of the soil structure is taken to bear a ratio λ to the compressibility C. In the lower stress range the value of λ is about

61-2

0.45 for London Clay consolidated from a slurry and allowed to swell under an equal all-round pressure. At high consolidation pressures this value might be expected to rise.

The values of r for the four values of consolidation pressure and for two assumptions about λ are given in table 4 and are plotted in figure 15. It will be seen that the theoretical relations between r and the reduction in confining pressure for samples remaining fully saturated are smooth curves which do not depart to any marked extent from straight lines except at the highest

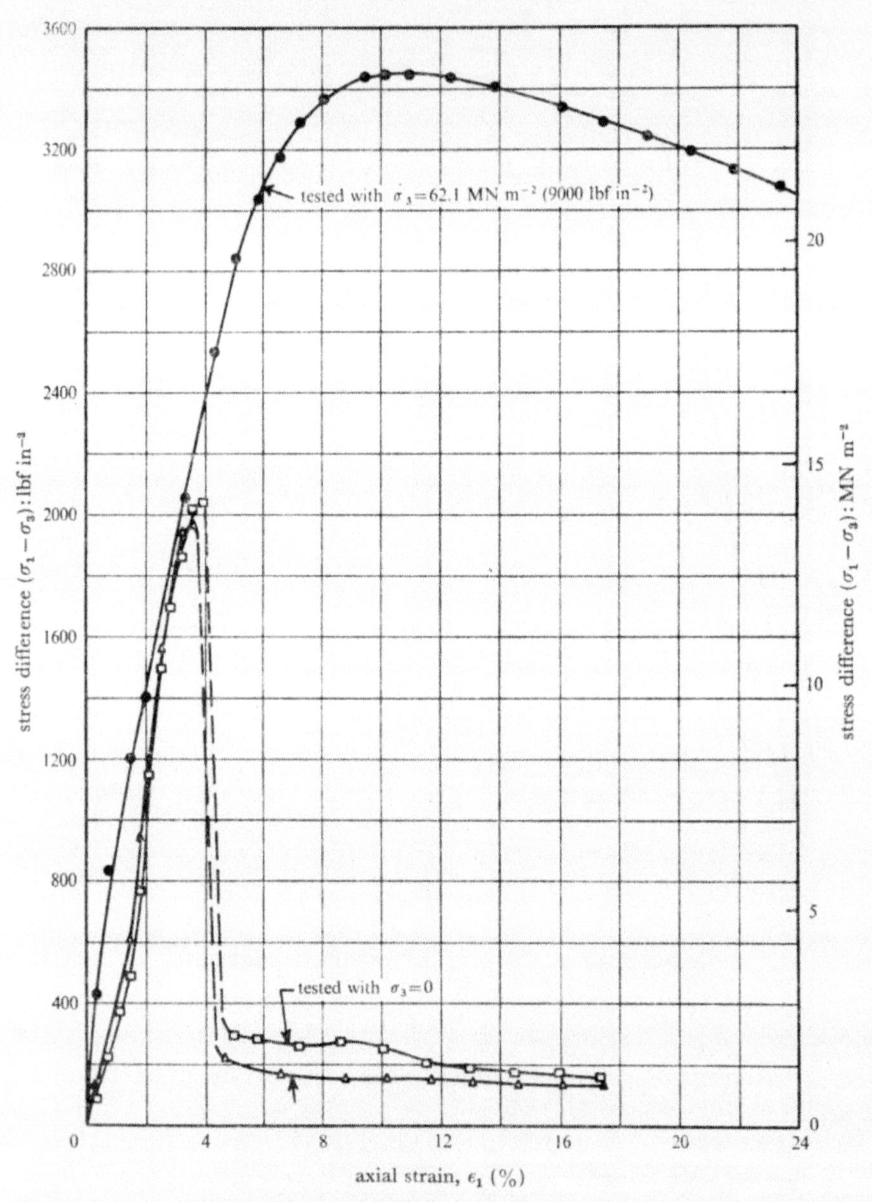

FIGURE 11. Comparison of confined and unconfined compression tests on samples of London Clay consolidated with σ_3 = 62.1 MN m^{-2} (9000 lbf in^{-2}) and u = 0.

FIGURE 12. Samples consolidated at 62.1 MN m^{-2} (9000 lbf in^{-2}) and tested (*a*) with cell pressure equal to consolidation pressure (test no. 19 B) and (*b*) unconfined (test no. 21 B). Surface texture shown in (*c*) for unconfined test.

(*Facing p.* 530)

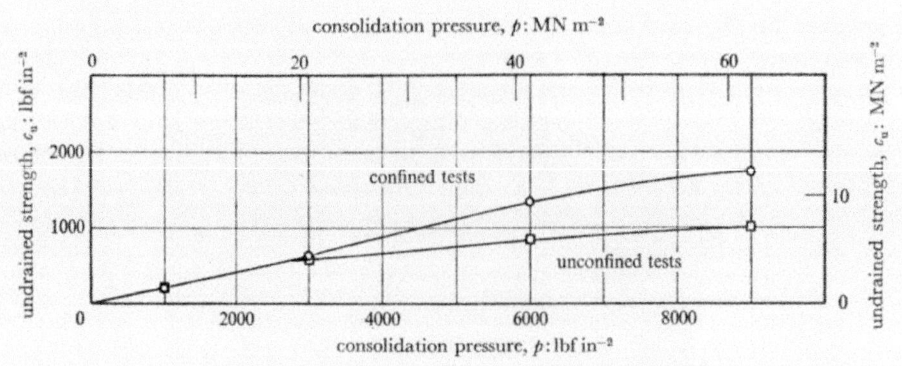

FIGURE 13. Relations between undrained strength c_u and consolidation pressure p for confined and unconfined tests on London Clay. Where a number of tests were performed at a particular pressure, the average value is plotted.

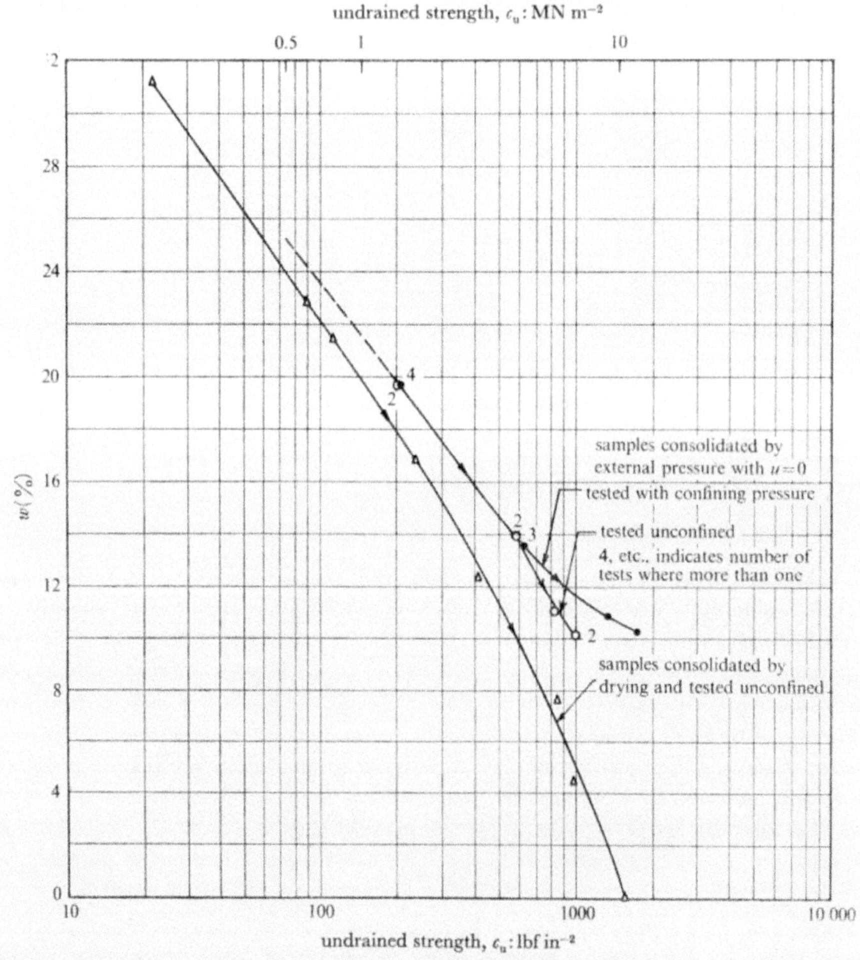

FIGURE 14. Relations between undrained strength c_u and water content w for both confined tests and unconfined tests on samples consolidated by external pressure and for unconfined tests on samples consolidated by drying: London Clay.

cell pressure. The departure of the observed line relating loss in strength to reduction in total stress from the shape of the theoretical lines† at a pressure of a little below 20.7 MN m^{-2} (3000 lbf in^{-2}) suggests that a breakdown in the condition of full saturation occurs at this point. It will be seen from table 4 that on the assumption of full saturation the predicted value of r for a reduction in total stress of 6.9 MN m^{-2} (1000 lbf in^{-2}) lies in the range 1.22–1.40 %. The observed value lies in the range 0.5–2.0 %, the latter value being obtained if one exceptionally high value is included in the average of the confined test results (see tables 3 and 7). The agreement is thus clearly within experimental error.

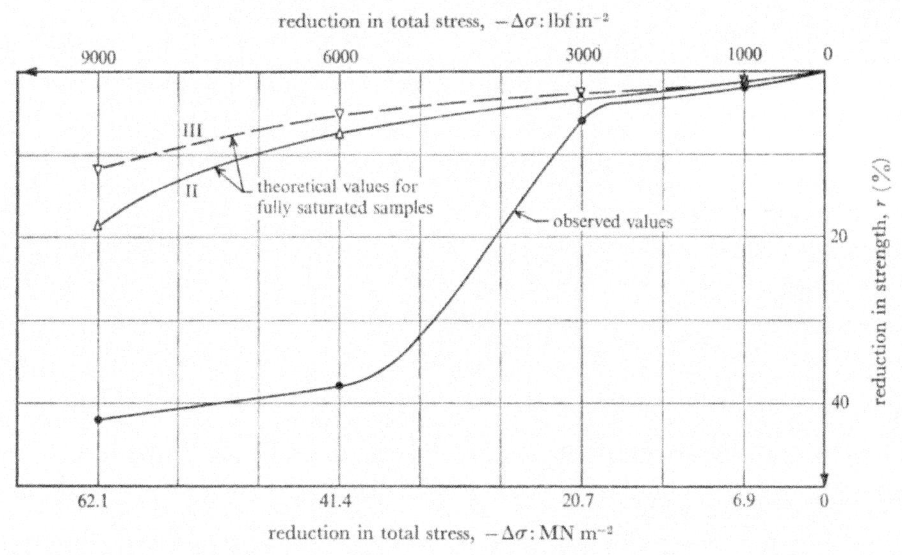

FIGURE 15. Relations between reduction in strength r and reduction in total stress $-\Delta\sigma$. A comparison between the observed values and the theoretical values based on the assumption of full saturation. Normally consolidated London Clay; initial pore-water pressure is 0.

It is also apparent from the stress–strain curves (figure 9) that a major change in the mechanism of failure is beginning to take place at this stress level. As the induced pore water tension in the unconfined samples at this consolidation pressure, if fully saturated, is $B \times p$ and thus equals approximately $0.94 \times 3000 = 19.4$ MN m^{-2} (2820 lbf in^{-2}), the breakdown might be due either to tensile failure in the pore water or due to failure of capillary menisci to prevent the invasion of the pore space by air or nitrogen in contact with the specimen or released from solution in the pore water. A series of tests was therefore carried out on Kaolin which has a radically different pore size, as is indicated by a permeability some two orders of magnitude greater than that of London Clay although the percentage of clay size particles is greater in the Kaolin (table 1A).

† If it had been assumed that a departure from unity of the value of k in equation (6) was partly responsible for the loss in strength, the shape of curves II and III would not be significantly changed. From equation (5) we would have $1-k_s = a(\tan\psi/\tan\phi')$ where a is the contact area between the particles per unit area. Since, as indicated earlier, this area is likely to be almost directly proportional to consolidation pressure (Skempton 1960), the component of r involving $1-K_s$ would likewise be almost linearly related to consolidation pressure.

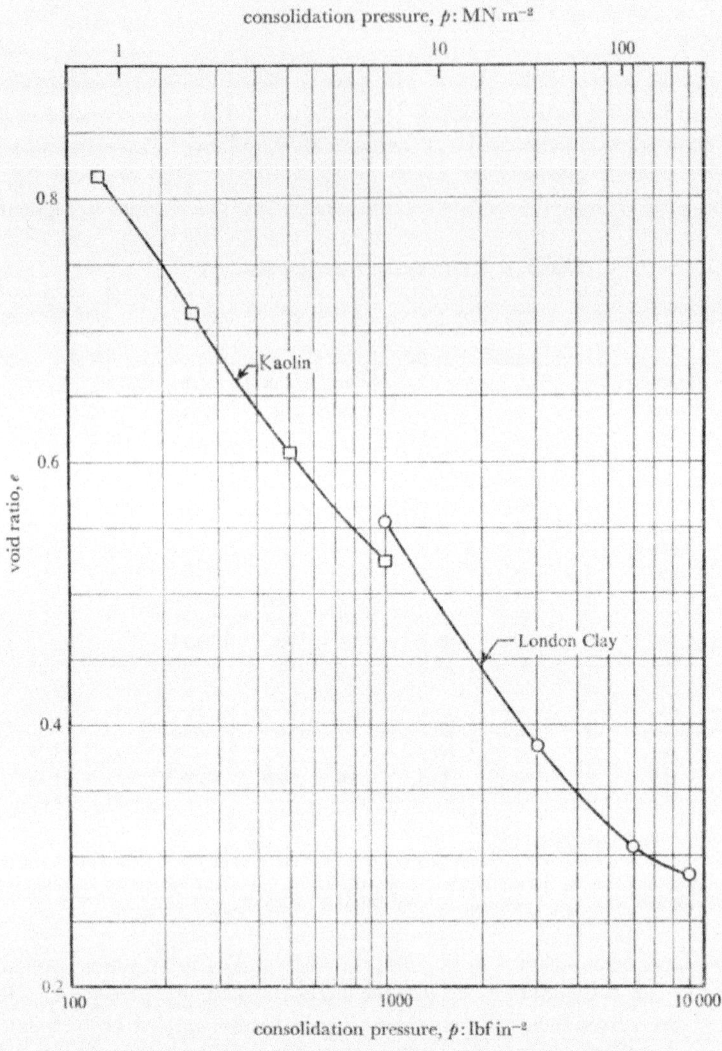

FIGURE 16. Relations between the void ratio e calculated from the final water content and the consolidation pressure p for London Clay and Kaolin.

TABLE 4. THE BASIS FOR THE DETERMINATION OF THE VALUES OF THE PARAMETER r

consolidation pressure MN m^{-2}	consolidation pressure lbf in^{-2}	compressibility, C from p–e curves m^2 MN^{-1}	expansibility, C_e from p–e curves m^2 MN^{-1}	$\lambda = C_e/C$	expansibility C_e in m^2 MN^{-1} units on basis of 3 assumptions† I	II	III	porosity n	values of r (%) for assumptions II and III: II	III
6.9	1000	15.8×10^{-3}	6.2×10^{-3}‡	0.39	6.2×10^{-3}	7.1×10^{-3}	6.2×10^{-3}	0.357	1.22	1.40
20.7	3000	45.0×10^{-4}	24.0×10^{-4}§	0.53	23.7×10^{-4}	20.2×10^{-4}	24.0×10^{-4}	0.277	3.10	2.80
41.4	6000	14.8×10^{-4}	—	—	12.4×10^{-4}	6.6×10^{-4}	10.0×10^{-4}	0.235	7.20	5.05
62.1	9000	45.0×10^{-5}	—	—	84.3×10^{-5}	20.2×10^{-5}	36.0×10^{-5}	0.222	18.5	11.7

† Assumption: I, unloading p against e curves are straight parallel lines on an e against lg p plot; II, the value of λ is constant and equal to 0.45; III, extrapolating observed C_e values using a value of λ increasing with p.

‡ Assuming a linear relation between e and lg p on unloading from 6.9 to 0.69 MN m^{-2} (1000 – 100 lbf in^{-2}).

§ Extrapolated from volume change observed over range 19.7–17.6 MN m^{-2} (2854–2554 lbf in^{-2}).

(ii) *Tests on Kaolin*

The second test on Kaolin (table 5, test 2D) gave a clear indication that the breakdown pressure for Kaolin was well below 6.9 MN m^{-2} (1000 lbf in^{-2}). The series of tests were therefore carried out for consolidation pressures in the range 0.86–6.9 MN m^{-2} (125–1000 lbf in^{-2}). The samples generally showed substantially larger strains at failure than observed for London Clay, and several samples showed a tendency to tilt associated with a reduced strength at failure.

Table 5. Results of tests on Kaolin

	consolidation pressures			water contents		tests with $\sigma_3 = p$			tests with $\sigma_3 = 0$		
test	σ_3			initial	final	c_u			c_u‡		
no.	MN m^{-2}	lbf in^{-2}	u	w (%)	w (%)	MN m^{-2}	lbf in^{-2}	c_u/p	MN m^{-2}	lbf in^{-2}	c_u/p
1D	6.9	1000	0	44.2	20.0	1.68	243	0.243	—	—	—
2D	6.9	1000	0	43.0	19.9	—	—	—	0.64	93	0.093
3D	6.9	1000	0	43.2	20.1	—	—	—	0.54	78	0.078
4D	3.4	500	0	42.2	23.2	—	—	—	0.52	75	0.150
5D	6.9	1000	0	42.7	19.6	—	—	—	0.61	88	0.088
6D†	1.7	250	0	42.9	27.2	—	—	—	0.38	55.5	0.222
7D	1.7	250	0	43.0	26.9	—	—	—	0.42	61.0	0.244
8D	1.7	250	0	42.0	27.0	0.49	71.0	0.284	—	—	—
9D	1.7	250	0	41.2	27.1	0.43	62.5	0.250	—	—	—
10D	1.7	250	0	42.0	26.8	0.49	71.0	0.284	—	—	—
11D†	3.4	500	0	42.8	22.5	0.74	108	0.216	—	—	—
12D†	3.4	500	0	42.9	23.2	0.70	101	0.202	—	—	—
13D	—	—	—	—	—	—	—	—	—	—	—
14D	—	—	—	—	—	—	—	—	—	—	—
15D	3.4	500	0	42.8	23.1	0.90	130	0.260	—	—	—
16D	0.86	125	0	42.5	30.8	—	—	—	0.25	36.3	0.290
17D	0.86	125	0	42.1	30.5	0.24	35.0	0.280	—	—	—
18D	0.86	125	0	42.6	31.3	—	—	—	0.22	32.0	0.256

† Samples tilted during compression test.

‡ The expansion on stress release and reduction in volume under the axial stress were not measured during these tests and are significant only for the samples which departed from full saturation. At the maximum consolidation pressure the tabulated value may represent an over-estimate of the strength by about 7 %.

These latter tests have been assumed to be unrepresentative. The stress–strain curves are presented in figures 17–20. It should be noted that the lowest consolidation pressure $\sigma_3 = 0.86$ MN m^{-2} (125 lbf in^{-2}) lies a little below the maximum vertical stress applied to the batch during preparation in the oedometer. Samples consolidated at this pressure are therefore lightly overconsolidated, though the increase in the value of c_u/p is small (a maximum value of 0.290 for one unconfined test as against a maximum value of 0.284 at a consolidation pressure of 1.72 MN m^{-2} (250 lbf in^{-2})). The unconfined samples at the lowest consolidation pressure do not show a significant decrease in undrained strength (or in the ratio c_u/p), but the rapid drop in strength after failure noted with London Clay is not apparent at this low stress level. At a consolidation pressure of 1.72 MN m^{-2} (250 lbf in^{-2}) a reduction in strength in the unconfined specimens is already becoming apparent (figure 18). The rapid post-peak drop in strength has also appeared.

As the consolidation pressure is raised to 3.45 MN m^{-2} (500 lbf in^{-2}), the divergence in strength becomes very marked and the brittle failure of the unconfined specimens is becoming more like that observed on the London Clay specimens with a consolidation pressure an order of magnitude higher (figure 19).

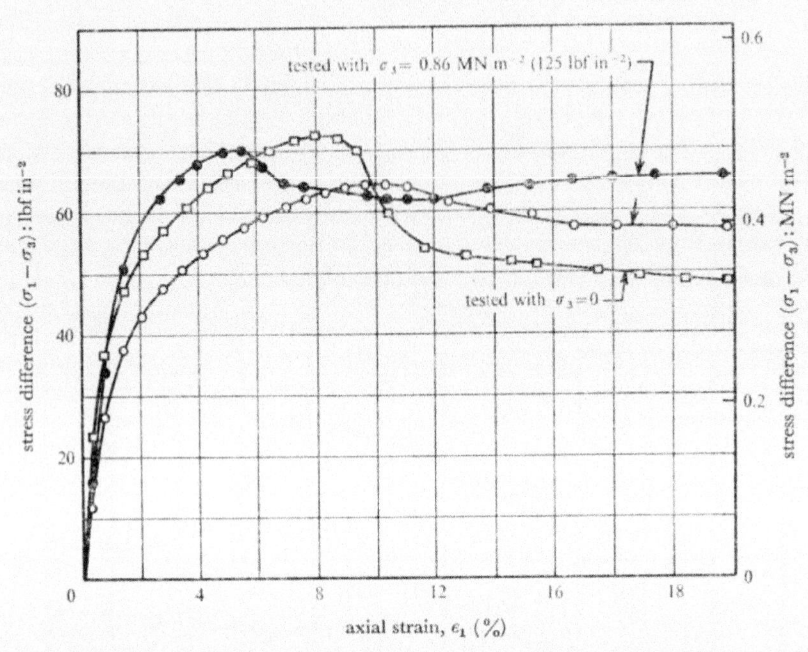

FIGURE 17. Comparison of confined and unconfined compression tests on samples of Kaolin consolidated with σ_3 = 0.86 MN m^{-2} (125 lbf in^{-2}) and u = 0.

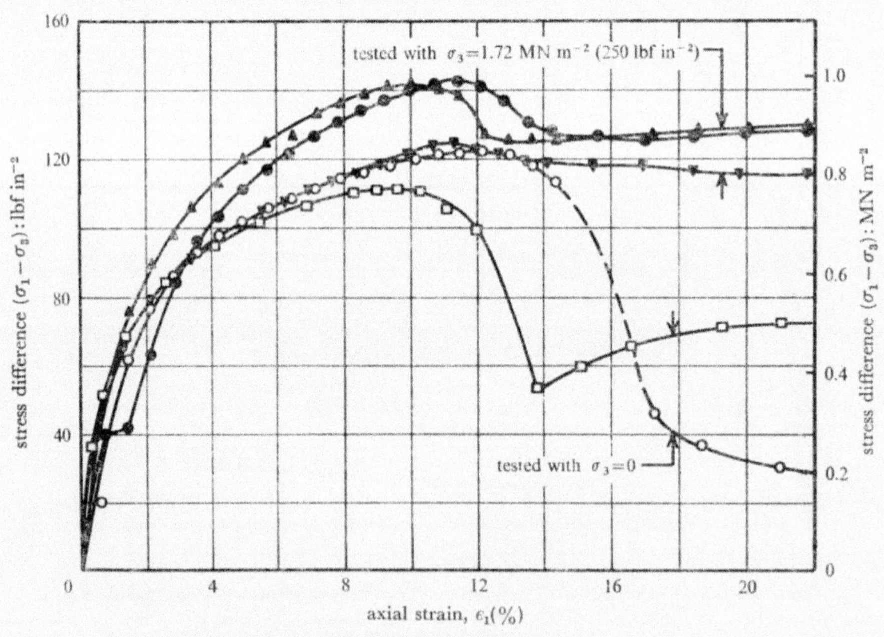

FIGURE 18. Comparison of confined and unconfined compression tests on samples of Kaolin consolidated with σ_3 = 1.72 MN m^{-2} (250 lbf in^{-2}) and u = 0.

At a consolidation pressure of 6.9 MN m^{-2} (1000 lbf in^{-2}), the divergence in strength is even more marked (figure 20), the loss in strength on the removal of the confining pressure being 65%. The brittleness of the relatively low strength unconfined samples is, however, not quite as marked as that of the unconfined samples of London Clay, which have been subjected to higher consolidation pressure (figures 10 and 11).

The results are summarized, as for the London Clay, on graphs relating c_u and p (figure 21), c_u and w (figure 22) and percentage loss in strength r and reduction in confining pressure (figure 23). It is apparent from the three sets of curves that the divergence in strength begins at a reduction in confining pressure of about 1.38 MN m^{-2} (200 lbf in^{-2}).

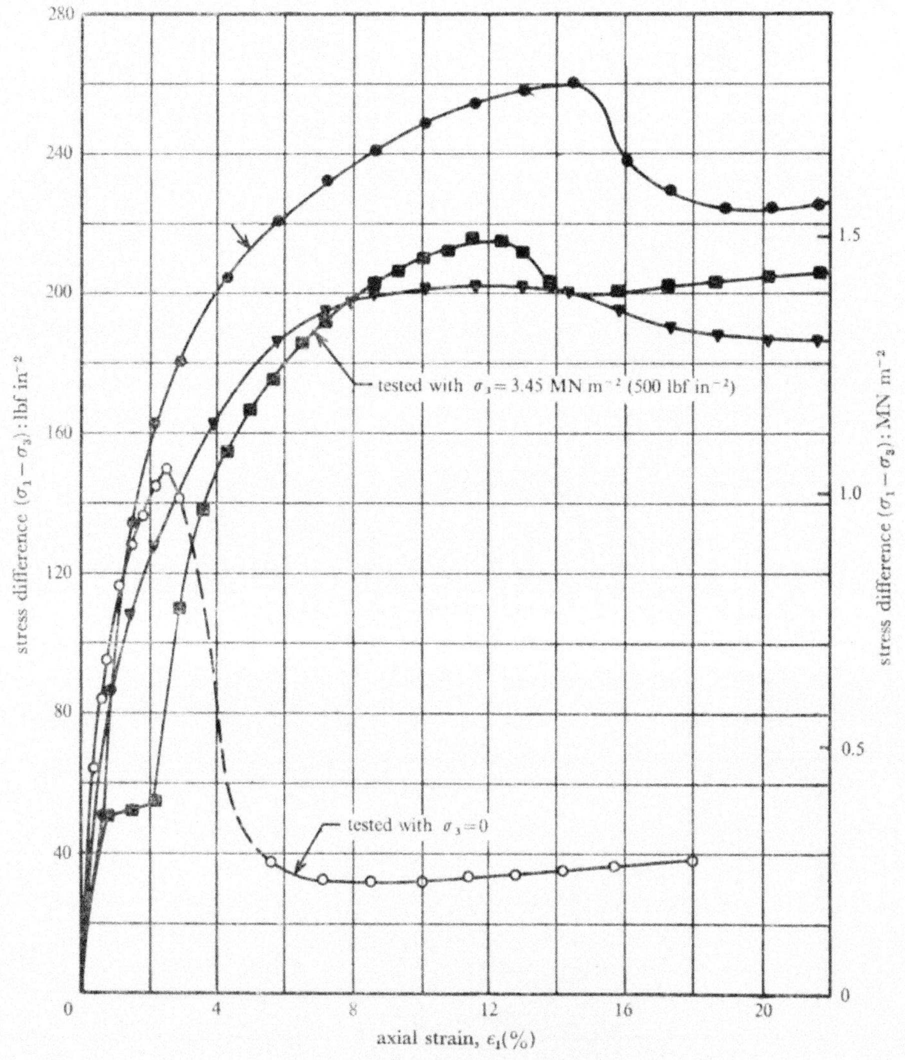

FIGURE 19. Comparison of confined and unconfined compression tests on samples of Kaolin consolidated with σ_3 = 3.45 MN m^{-2} (500 lbf in^{-2}) and u = 0.

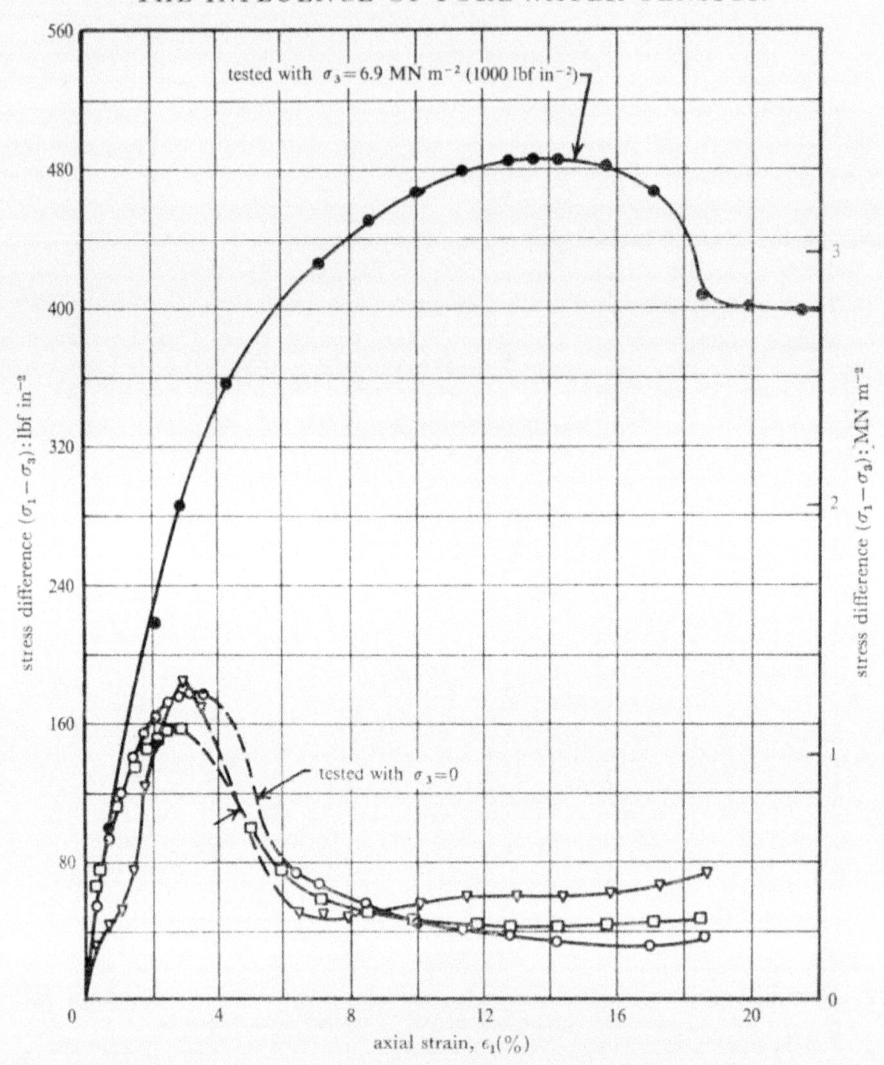

FIGURE 20. Comparison of confined and unconfined compression tests on samples of Kaolin consolidated with $\sigma_3 = 6.9$ MN m^{-2} (1000 lbf in^{-2}) and $u = 0$.

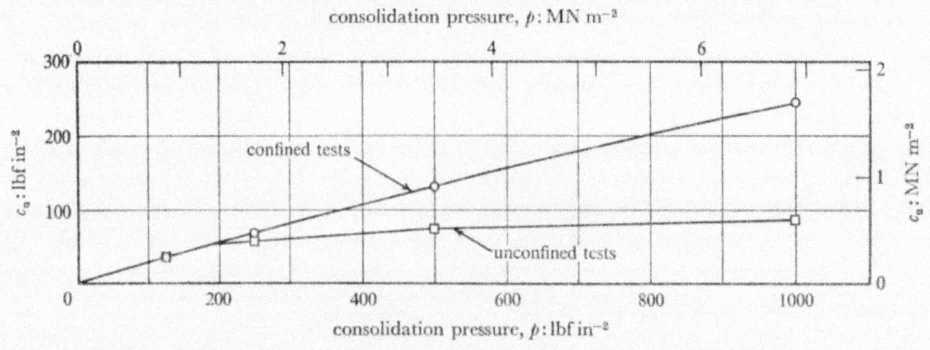

FIGURE 21. Relation between undrained strength c_u and consolidation pressure p for confined and unconfined tests on Kaolin. Where a number of tests were performed at a particular pressure, the average value is plotted.

62-2

6. Discussion of results

The results of the tests carried out in stage 1 of the investigation demonstrate very clearly that reductions in confining pressure which do not result in negative pore-water pressures have only a very minor effect on the undrained strength of saturated samples of clay. In the series of tests described this amounted to a reduction of about 2 % in strength for a reduction in confining pressure of 3.45 MN m^{-2} (500 lbf in^{-2}).

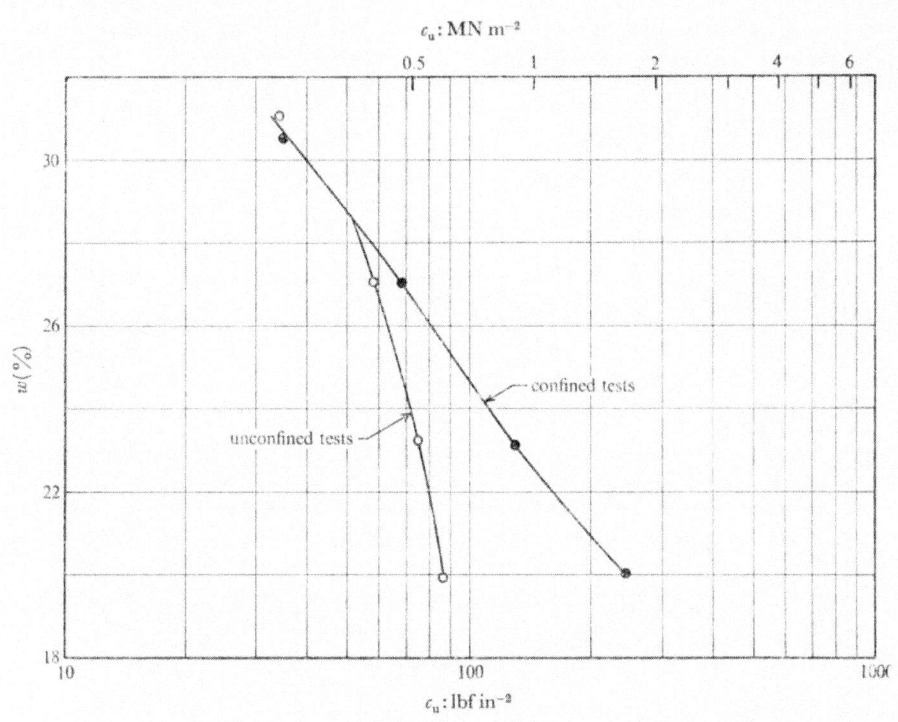

Figure 22. Relation between undrained strength c_u and water content w for both confined tests and unconfined tests on samples consolidated by external pressure: Kaolin.

The results of tests carried out under stage 2 of the investigation in which reductions in confining pressure led to pore-water tensions of various magnitudes showed equally clearly that a tensile component in the pore-water pressure is capable of replacing, with respect to its mechanical effects, an externally applied confining pressure up to a magnitude of almost 20.7 MN m^{-2} (3000 lbf in^{-2}) in the case of London Clay and about 1.38 MN m^{-2} (200 lbf in^{-2}) in the case of Kaolin.

At pore water tensions smaller than these limiting values the reduction in peak strength with confining pressure is a minor effect, of a similar order of magnitude to that noted above for saturated soils in the positive pore pressure range. However, there is a very significant change in the shape of the stress-strain curve after the peak (for example, London Clay with an inferred pore-water tension of 6.65 MN m^{-2} (965 lbf in^{-2}) before shear, figure 8). This indicates that high pore-water tensions result in greatly increased brittleness.

The loss in strength at larger cell pressure reductions indicates a failure of the pore water to sustain higher tensions without a departure from the condition of full saturation. As will be shown subsequently this is confirmed by the observed volume changes in a limited number of special tests.

In the case of the London Clay it could reasonably have been held that the pore water had failed in tension, since the inferred tensile stress in the pore water was above any value obtained by direct experimental measurement (see, for example, Temperley & Chambers 1946; Temperley 1946) although below the theoretical value of tensile strength.† However, the low value of breakdown pressure obtained for Kaolin, 1.38 MN m^{-2} (200 lbf in^{-2}), suggests that it is not a case of tensile failure in the pore water, but of the rupture of a meniscus in a capillary whose diameter is related to the effective pore size of the clay.

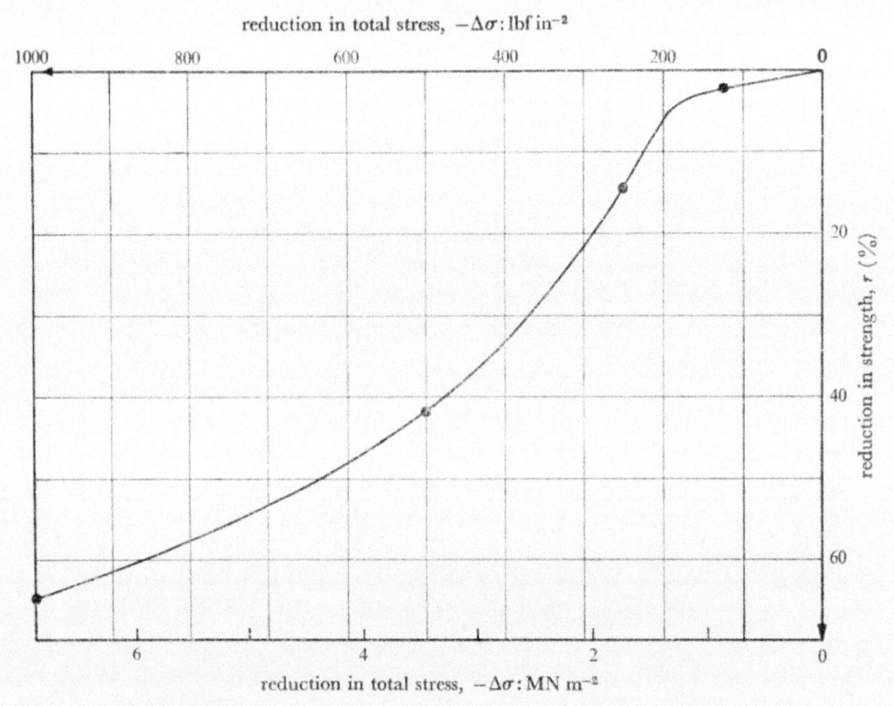

FIGURE 23. Relation between reduction in strength r and reduction in total stress $-\Delta\sigma$. Observed values from tests on Kaolin; initial pore-water pressure is 0.

The capillary rise h_c is given by the expression

$$h_c = \frac{2t_s}{R\gamma_w}\cos\alpha, \tag{18}$$

where γ_w denotes unit weight of water, t_s the surface tension, α the contact angle, and R the radius of capillary.

† Using Berthelot tubes, Temperley (1946) obtained values for the tensile strength of water ranging between 20 and 60 atm (1 atm $\approx 10^5$ N m^{-2}) and averaging 32 atm (470 lbf in^{-2}). His theoretical value is between 500 and 1000 atms. Green (1951), however, obtains a higher experimental value of 190 atm (2800 lbf in^{-2}).

The capillary pressure is therefore inversely proportional to the equivalent pore diameter, and the equivalent pore diameters for Kaolin and London Clay would therefore be expected to have the ratio of approximately 2820:200 or 14:1.

The equivalent pore diameters can be calculated independently from the permeabilities of the clays at the appropriate stress levels, the permeability K being calculated from the coefficient of consolidation c_v and the compressibility observed during the consolidation stage.†

Associating the value of permeability with the average effective stress level during consolidation we obtain values of the permeability K of 2.55–2.80×10^{-8} cm/s for Kaolin at a consolidation pressure of 1.38 MN m^{-2} (200 lbf in^{-2}) and 0.49×10^{-10} cm/s for London Clay at a consolidation pressure of 20.7 MN m^{-2} (3000 lbf in^{-2}).

Following Taylor (1948), the permeability K may be taken to be related to the equivalent pore diameter by the expression

$$K = \beta D^2 n, \tag{19}$$

where β is a constant which includes a shape factor, D denotes the equivalent pore diameter, and n denotes the porosity. Taking $n = 0.43$ for the Kaolin and 0.28 for the London Clay at the relevant pressures, and β to be the same for the two clays, we obtain the ratio of equivalent pore diameters as approximately 19:1 (in the range 18.4–19.3). This is in reasonably close agreement with the ratio of 14:1 given by the assumption that the breakdown pressure is controlled by capillarity. Indeed, very close agreement can hardly be expected, since

(1) the relation between equivalent pore diameter and permeability will be subject to the influence of surfaces forces and thus dependent on clay mineralogy and stress level, and

(2) for a given pore size distribution the 'equivalent diameter' controlling the two phenomena need not necessarily be the same.

It is of interest to note that the use of equation (18) leads to equivalent pore diameters of 2.06×10^{-1} μm for Kaolin and 1.46×10^{-2} μm for London Clay. The latter value is in the range of values obtained from shrinkage tests by Holmes (1955) who gives 2.3×10^{-2} μm and 0.9×10^{-2} μm for two clays with marked swelling characteristics and of similar clay mineralogy to London Clay (i.e. predominantly illite and kaolinite, with small amounts of montmorillonite; total clay fractions 64 % and 65 %).

In a sample subjected to a large enough reduction in total stress to cause partial saturation it is thus envisaged that passages filled with water vapour and gas (some air will have been in solution in the pore water in all tests) will propagate within the sample. In a clay with a relatively high expansibility there is no need to expect macroscopic migration of water relative to the sample boundaries and there is no significant experimental evidence of migration from the water content determinations. Whether the passages propagate from the external boundaries of the sample or from internal nuclei as well is difficult to determine, especially with small samples.

The total volume of the gas-filled voids can be calculated from the volume of the sample and its water content after the release of stress. As the samples subjected to compression tests fractured into a number of pieces accurate volume measurement was difficult. A sample subjected to the highest pressure in each series was therefore removed, without testing, for immediate volume

† Using the Terzaghi expression (Terzaghi 1943):

$$c_v = K/\gamma_w m_v,$$

where c_v denotes coefficient of consolidation and m_v denotes the compressibility and is replaced by C for equal all-round stress change.

determination, based on a series of very accurate height and diameter observations. A series of such observations was continued while the sample was allowed to dry slowly, and a final set of readings was taken after oven drying at 105 °C. A sample of each clay which had been stored after initial consolidation and swelling, but had not been heavily consolidated, was also subjected to a similar shrinkage test. The results of the four tests are presented in figures 24 and 25. The relations between dry unit weight γ_d (i.e., the weight of mineral material divided by the volume of the entire element), relative density of the particles G, void ratio e (i.e. volume of voids per unit volume of solids), water content w and degree of saturation S are given in standard textbooks on soil mechanics (see, for example, Lambe & Whitman 1969) as:

$$\gamma_d = \frac{G\gamma_w}{1+wG/S}, \tag{20}$$

$$\gamma_d = \frac{G\gamma_w}{1+e}, \tag{21}$$

$$Gw = Se, \tag{22}$$

where γ_w is the unit weight of water at the relevant temperature.

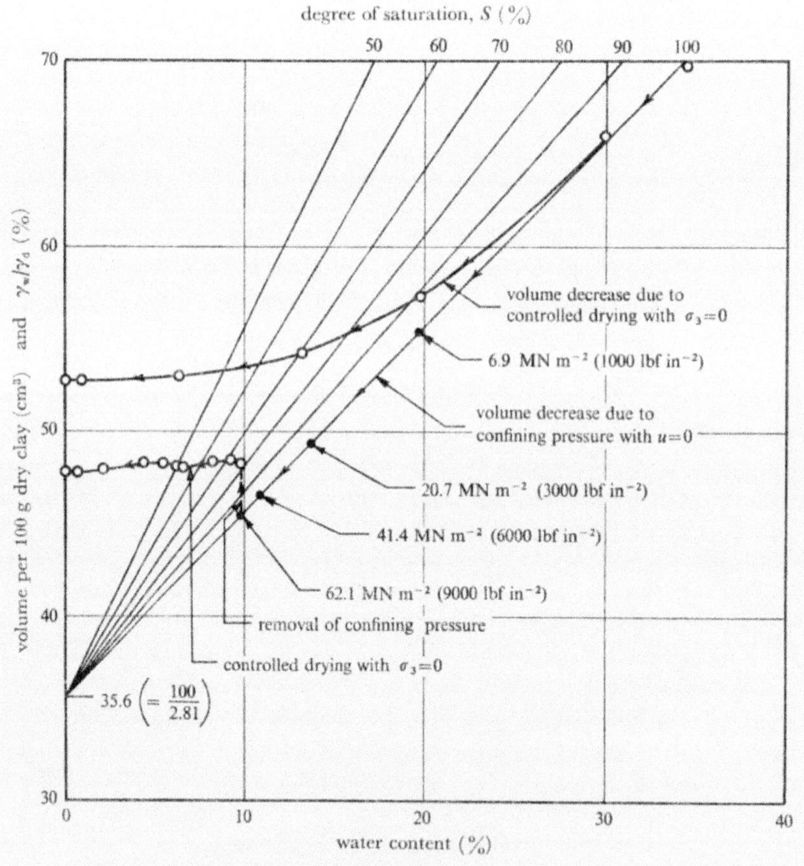

FIGURE 24. Relations between volume change and water content during consolidation, undrained stress release and drying: London Clay; relative density G assumed constant at 2.81 throughout pressure range.

Using these expressions, we can obtain the void ratio of the saturated sample at the end of consolidation from the final water content, and the void ratio after release of the cell pressure from the dry unit weight on removal from the cell.

These calculations show that on the release of cell pressure the sample of London Clay consolidated at 62.1 MN m^{-2} (9000 lbf in^{-2}), expanded 6.0 %, and that the percentage air voids was 5.7 % and the degree of saturation was 78 %. For Kaolin consolidated at 6.9 MN m^{-2} (1000 lbf in^{-2}), the corresponding values are: expansion 12.7 %, air voids 11.3 % and degree of saturation 73 %. It is of interest to note that the one dimensional consolidation tests of Westman (1932) lead to comparable values, Georgia Kaolin consolidated to 57.1 MN m^{-2} (8280 lbf in^{-2}) giving an expansion of 11.6 %, air voids of 10.4 % and a degree of saturation of 66 %.

The determination of degree of saturation and percentage air voids are subject to a small error due to the variation of γ_w with tensile stress in the water. The magnitude of this error is examined in appendix 2.

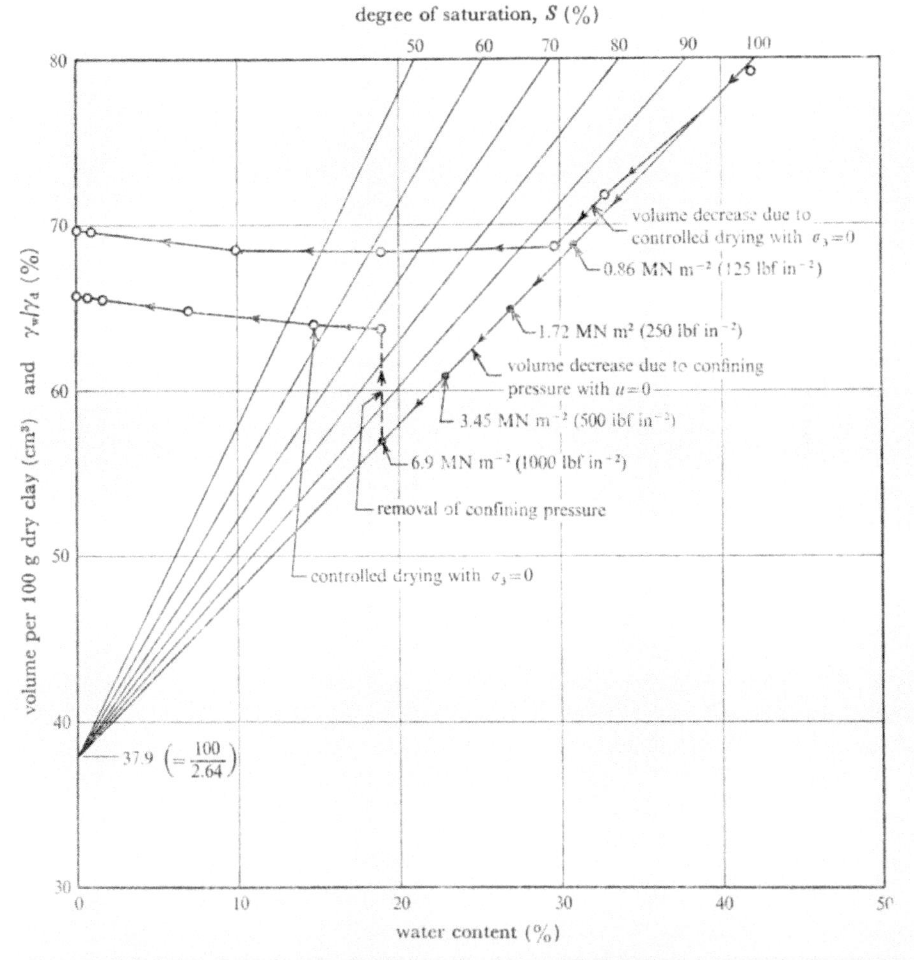

FIGURE 25. Relations between volume change and water content during consolidation, undrained stress release and drying: Kaolin; relative density G assumed constant at 2.64 throughout pressure range.

After removal from the triaxial cell the samples released from these stresses showed no significant tendency to change in volume with time (until drying took place). This suggests that, once initiated, the propagation of cavitation within the pore water is fairly rapid and complete.

On drying, the London Clay sample underwent a small decrease in volume (figure 24) of about 1 % which occurred mainly on the removal of the last 5 % of water content. In contrast the Kaolin (figure 25) showed an expansion of 3 % on drying. These results are consistent with current view of the role of water layers in soil mechanics as illustrated, for example, by the description given by Lambe & Whitman (1969) of the transmission of force through a flocculated soil. This view was forseen by Westman (1932) who considered that, at least at higher pressures, 'the data obtained could be readily explained by specifying a system of flexible, mechanically weak, solid particles in solid contact with each other'.

The primary purpose of this investigation was to study the influence of pore-water tensions on the undrained strength of clay. So far we have shown that up to a limiting pore-water tension the sign of the pore-water pressure has little influence on the strength, but a marked influence on the shape of the post-peak section of the stress–strain curve. The magnitude of the limiting pore-water tension varies radically with clay type and correlates with equivalent pore diameter in a manner similar to capillary tension. Three further aspects of the influence of pore-water tension on strength deserve a brief comment:

(*a*) the rate of increase of unconfined strength with consolidation pressure above the limiting pressure at which partial saturation occurs in the unconfined sample,

(*b*) the increased brittleness associated with pore-water tensions, and

(*c*) the consolidation of samples by the pore-water tension associated with drying.

(*a*) *The rate of increase of unconfined strength with consolidation pressure above the limiting pressure at which partial saturation occurs*

This is difficult to predict theoretically since the substantial volume increase occurring when the limiting pore-water tension is exceeded results in the sample behaving as both an over-consolidated as well as a partly saturated soil. However, several correlations of interest may be noted.

Normally consolidated and lightly over-consolidated remoulded samples of London Clay show an almost unique relationship between undrained strength and water content and hence, being fully saturated, with void ratio (see Table I–IV of Bishop (1971)). In table 6, undrained strengths are given for saturated samples tested at the void ratios observed on removal of the confining pressure for both London Clay and Kaolin (the maximum pressure in each series). For London Clay this represents an underestimate of the unconfined strength at this void ratio, though for Kaolin the agreement is close.

This difference reflects the relative importance of the contribution made by true cohesion (as defined by Hvorslev 1937) to the strength of the two clays. In table 6, values are given of Hvorslev's parameter κ (the ratio of true cohesion to consolidation pressure) taken from typical values given by Gibson (1953), together with values of the *activity* of the clay as defined by Skempton (1953).

It will be seen from table 6 that the change in the unconfined strength with consolidation pressure dc_u/dp is almost linearly related to the value of *activity* of the clay, if based on average values over the full range of partial saturation observed. The flattening of the curve for Kaolin (figure 21) suggests a better correlation with κ at higher pressures, but this is subject to confirmation by the measurement of values of this parameter in the relevant stress range.

A correlation of the unconfined strength of the samples consolidated in this range with cohesion is also consistent with the work of Schmertmann & Osterberg (1960) who showed that the cohesion term reached a peak very early in the test, while the friction term required 10–20 times the strain for its full mobilization to be approached. It will be noted that both for London Clay (figures 10 and 11) and Kaolin (figures 19 and 20) the failure strains of the unconfined samples in which full saturation is not maintained drop to 35–45 % (London Clay) and 20–25 % (Kaolin) of the failure strains of the samples tested in the positive pore pressure range. This implies that the friction contribution is reduced in the unconfined tests in which full saturation is not maintained, especially in the case of Kaolin, not only due to the reduction in effective stress but also due to the reduction in failure strain.

TABLE 6. RELATIONS BETWEEN THE INCREASE IN UNCONFINED STRENGTH ABOVE THE LIMITING PRESSURE, THE ACTIVITY OF THE CLAY, AND HVORSLEV'S COHESION PARAMETER

material	consolidation pressure		unconfined value of c_u		strength of saturated sample at same void ratio		$\frac{dc_u}{dp}$	activity†	Hvorslev κ
	$MN\ m^{-2}$	$lbf\ in^{-2}$	$MN\ m^{-2}$	$lbf\ in^{-2}$	$MN\ m^{-2}$	$lbf\ in^{-2}$			
London Clay	20.7	3000	4.04	586	—	—	0.069	0.91	0.10‡
	62.1	9000	6.89	1000	5.24	760			
Kaolin	1.72	250	0.40	58.2	—	—	0.037	0.44	0.02‡
	6.9	1000	0.60	86.3	0.59	85.0			

† Plasticity index/percentage clay.
‡ Approximate, based on typical values given by Gibson (1953) for the low pressure range.

TABLE 7. LOSS IN STRENGTH FOR 2 % POST-PEAK STRAIN FOR CONFINED AND UNCONFINED TESTS

material	consolidation pressure		loss in peak strength when tested unconfined (%)	confined test – loss in strength for 2 % post peak strain (%)	unconfined test – loss in strength for 2 % post peak strain (%)
	$MN\ m^{-2}$	$lbf\ in^{-2}$			
London Clay	6.9	1000	0.5(2)†	1.3	3.2
	20.7	3000	6	0.9	63
	41.4	6000	38	0.4	88
	62.1	9000	42	0.6	87
Kaolin	0.86	125	3	8.6	7.5
	1.72	250	11	6.0	6.6
	3.45	500	42	7.1	63
	6.89	1000	65	1.4	42

† The value of 2 % is obtained if the fifth test, which differs substantially from the average, is included.

(b) *The increased brittleness associated with pore-water tensions*

One of the most noteworthy features of the tests on both London Clay and Kaolin is the change in the shape of the post-peak section of the stress–strain curve with increasing pore-water tension, in particular when this exceeds the limit at which full saturation can be maintained.

Even in the range of positive pore-water pressure, saturated samples of both London Clay and Kaolin are strain softening, the brittleness index I_B† being estimated to be about 70 % for

† The brittleness index I_B for undrained tests is defined by (Bishop 1971) as

$$I_B = [(c_u)_f - (c_u)_r]/(c_u)_f,$$

where the suffices f and r denote failure (peak) and residual states respectively.

normally consolidated and lightly overconsolidated remoulded London Clay (Bishop 1971, Table I–VII). However, the rate of decrease in strength observed at the strains achieved in the triaxial test is relatively small (figures 8–11 and 17–20). In contrast, the unconfined samples having the higher initial consolidation pressures show a sudden loss of strength occurring mainly within an additional 1–2 % of axial strain (figures 10, 11, 19 and 20, and table 7). It is apparent from table 7 that this abrupt drop in strength is more marked in the London Clay than in the Kaolin. In the London Clay it amounts to 63 % for a cell pressure reduction of 20.7 MN m^{-2} (3000 lbf in^{-2}). In the Kaolin the drop in strength does not rise to 63 % until the limiting pressure (*ca.* 1.38 MN m^{-2} (200 lbf in^{-2})) has been exceeded by more than 100 %. The actual value of the brittleness index I_B is not readily ascertained from triaxial tests due to the limited strains imposed. In addition the top of the sample in the present apparatus is constrained by friction from moving freely sideways to accommodate the kinematics of a large displacement on a single rupture surface. The highest value of I_B observed was for London Clay at the maximum consolidation pressure (figure 11), which gave a value of 92.5 %. The absence of cohesion across the rupture surface when the sample was removed from the rubber membrane suggests that the true value of I_B was very close to 100 % for this sample and for all samples with high pore-water tensions.

(*c*) *The consolidation of samples by the pore-water tension associated with drying*

A series of unconfined compression tests was carried out on samples of the London Clay consolidated by slow shrinkage from the same initial water content as those consolidated by a confining pressure. The strength–water content relations are plotted in figure 14.

It will be noted that the strengths are considerably lower, for a given water content, than those of samples consolidated by a confining pressure, and tested unconfined. From the curve representing the strength–water content relation it can be seen that at a water content of 19.7 % (corresponding to a consolidation pressure of 6.9 MN m^{-2} (1000 lbf in^{-2})), the strength of a sample consolidated by shrinkage is 75 % of that consolidated by a confining pressure.

At 13.6 % water content (corresponding to 20.7 MN m^{-2} (3000 lbf in^{-2})) the strength is 61 % of that of the sample consolidated by a confining pressure. At a water content of 10.2 % (corresponding to a consolidation pressure of 62.1 MN m^{-2} (9000 lbf in^{-2})), the strength is 59 % of that of a sample consolidated by a confining pressure and tested unconfined, and only 33 % of that of a sample tested under a confining pressure equal to the consolidation pressure.

It is thus apparent that there is no unique relation between strength and water content, even for the unconfined samples at water contents corresponding to consolidation pressures less than the value leading to partial saturation in unconfined specimens.

An explanation for this difference may be found in the volume-change water-content relation for the shrinkage test on the London Clay plotted in figure 24. The equation for the saturation lines on this plot may be obtained by re-arranging equation (20)

$$\frac{\gamma_w}{\gamma_d} = \frac{1}{G} + \frac{w}{S}. \qquad (23)$$

It will be seen that the degree of saturation for a sample following the shrinkage curve is 92 % at 19.7 % water content, 73 % at 13.6 % water content and 57 % at 10.2 % water content. These values may be compared with 100, 98 and 78 % respectively for the samples consolidated under a confining pressure and released (the first and second values being estimated from strength changes, the third being a measured value). The void ratio of the sample following the shrinkage curve is thus higher, at each reference water content, than that of a sample consolidated by an applied pressure and then tested unconfined.

It is of interest to note that a plot of strength against void ratio at the beginning of the compression test (figure 26) indicates that for London Clay no unique relation exists on this basis either and that uniqueness is unlikely to be achieved even after correcting for the volume change

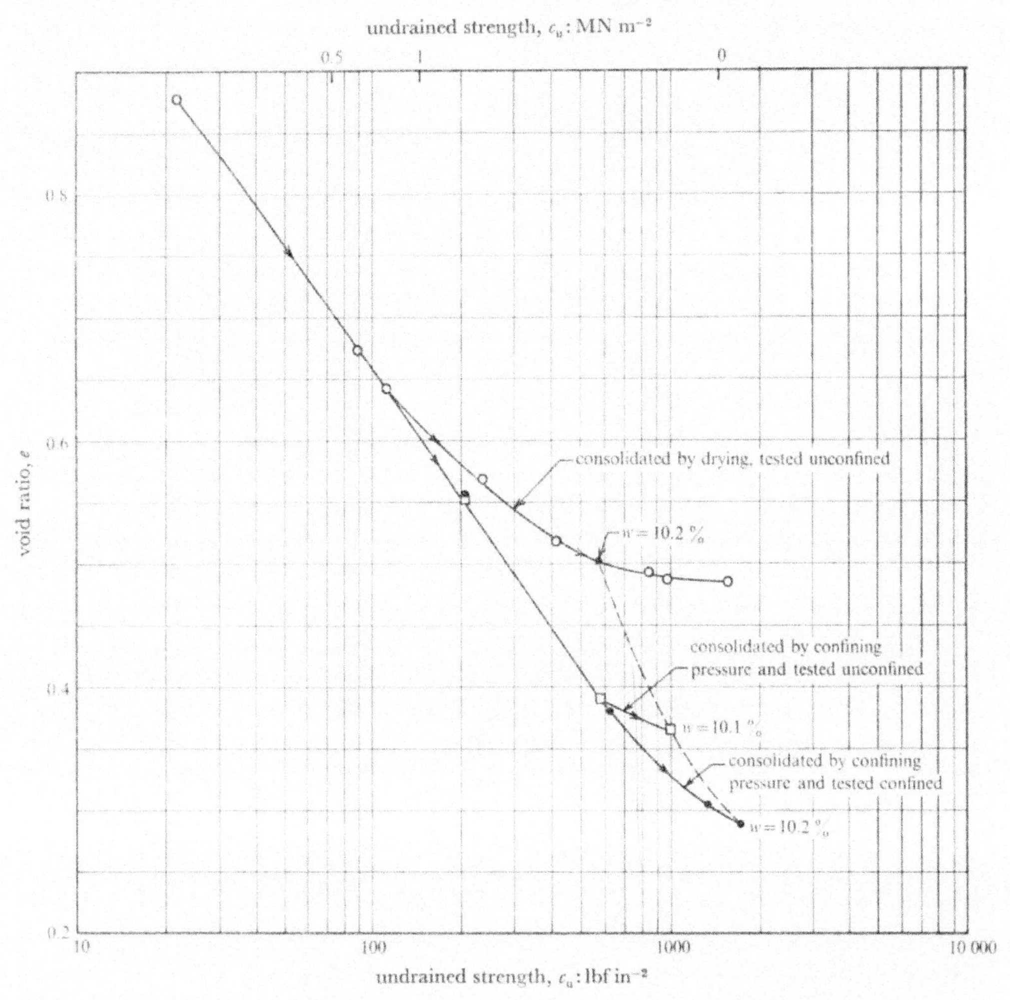

FIGURE 26. Relations between undrained strength c_u and void ratio e at the beginning of the compression test for confined tests, unconfined tests on samples consolidated by external pressure and unconfined tests on samples consolidated by drying: London Clay.

preceding failure in compression. In the limit a sample on the drying curve has a strength of 10.7 MN m^{-2} (1557 lbf in^{-2}) for an initial void ratio of 0.484 and zero water content, while the confined test at the highest consolidation pressure gives almost the same strength (11.9 MN m^{-2} (1723 lbf in^{-2})) for an initial void ratio of 0.287 and a water content of 10.2 % (figure 26). Even with the assumption of a Poisson ratio of 0.2 for the dry sample the void ratio at failure drops only to 0.407.

A discussion of the high pore-water tensions associated with drying in the range of water contents under consideration and of the modified effective stress equation necessary to relate these pore-water tensions to the strength of partly saturated soils is outside the scope of this paper. Reference may be made to Holmes (1955), Bishop (1959), Croney & Coleman (1960), Aitchison (1960), Bishop *et al.* (1960) and Bishop & Blight (1963).

The only directly comparable test data are given by Croney & Coleman (1960) who carried out standard unconfined compression tests on samples of undisturbed London Clay dried to known initial suctions in the pressure-membrane and vacuum-desiccator apparatus. The maximum value of c_u observed was about 3.79 MN m^{-2} (550 lbf in^{-2}) at an inferred suction of 689 MN m^{-2} (100 000 lbf in^{-2}) and almost zero water content. This value of c_u is only 35 % of the value obtained in the present series of tests on remoulded clay of similar index properties, and reflects the non-uniformity of particle size distribution in a natural soil, together with the effects of specimen size on strength in an undisturbed soil with a fissured structure. This latter effect could account for a reduction of almost 50 % in undrained strength in the relevant range of sample sizes in the weathered zone of the London Clay (Bishop 1971, Fig 1–29).

7. Conclusions

(1) For saturated samples of clay the influence on undrained strength of large reductions in confining pressure which do not result in negative pore-water pressures is very small. This is in general agreement with predictions based on the relative expansibilities of the soil structure and of the pore water.

(2) The influence on undrained strength of large reductions in confining pressure which do result in pore-water tensions is likewise very small until a limiting tension is reached above which a departure from full saturation occurs. The value of this tension is about 19.4 MN m^{-2} (2820 lbf in^{-2}) (i.e., 0.94 × 3000) for the London Clay and about 1.38 MN m^{-2} (200 lbf in^{-2}) for the Kaolin. This value correlates with equivalent pore diameter in the same way as the limiting capillary tension.

Larger reductions in confining pressure lead initially to rapidly increasing reductions in undrained strength which reach 42 % for the London Clay and 65 % for the Kaolin with the ranges of pressure investigated. It should be observed that the limiting pore-water tension and the percentage reduction in strength for a given reduction in total stress are not unique soil characteristics, but will depend on stress history. For example, reductions in confining pressure of 20.7 and 41.4 MN m^{-2} (3000 and 6000 lbf in^{-2}) will not lead to the same loss in strength when applied to a sample consolidated to 62.1 MN m^{-2} (9000 lbf in^{-2}) as when applied (as in the present tests) to samples consolidated only to 20.7 and 41.4 MN m^{-2} respectively. This is due to initial differences in expansibility and in equivalent pore diameter.

(3) The gain in unconfined strength with consolidation pressure above the limiting pressure correlates both with the activity of the clay and with the Hvorslev cohesion component.

(4) Large pore-water tensions are associated with a very marked change in shape of the post-peak section of the stress–strain curve, the unconfined samples of London Clay in particular showing a very brittle behaviour and negligible cohesion after rupture. This clearly indicates that undrained residual strength is a function of total normal stress, as suggested by Bishop (1971).

(5) Samples consolidated by drying do not show the same strength–water content or strength-void ratio relations as those consolidated by applied pressure at the relatively high pore-water

tensions under consideration. This appears to be due, in part at least, to a departure from full saturation at a much higher water content and thus at a smaller pore-water tension than that inferred for samples consolidated by an applied pressure and tested unconfined. It may to a smaller extent reflect the influence of residual stresses resulting from pore-pressure gradients during drying.

(6) The test results have important implications with regard to the stress release which inevitably occurs on the taking of samples from deep boreholes or shafts, to the choice of laboratory test conditions, and to the field behaviour of clay strata subject to undrained stress reduction associated with deep cuts and excavations, seismic disturbance or shock-waves.

Since the state of stress in the ground does not in general approximate to an equal all-round pressure, the pore-water tension in an ideal undisturbed and fully saturated sample on stress release is not equal to the effective vertical stress in the ground before sampling even for the case of $B = 1$. In normally consolidated clay strata the coefficient of earth pressure at rest K_0 (the ratio of the horizontal effective stress to the vertical effective stress) lies in the range 0.4 to 0.7 depending on the plasticity index, on the basis both of laboratory tests (Bishop 1958; Simons 1958) and field data (Bjerrum & Anderson 1972). On the basis of elastic theory the ratio of the pore-water tension $-u_s$ to the vertical effective stress σ'_v would be $\frac{1}{3}(1+2K_0)$ for $B = 1$. Values of $-u_s/\sigma'_v$ in the range 0.6–0.8 would thus be expected for perfect sampling.

However, normally consolidated soil does not behave as an ideal elastic material, and pore pressure changes measured in the laboratory on samples subject to the stress changes occurring during perfect sampling (Bishop & Henkel 1953; Skempton & Sowa 1963; Ladd & Lambe 1963) give lower values particularly in soils of low plasticity. On the basis of these results values of $-u_s/\sigma'_v$ in the range 0.35–0.75 might be expected in real soils.

However, mechanical disturbance during the sampling operation can account for a far greater reduction in the ratio $-u_s/\sigma'_v$ in the case of normally consolidated soils. Laboratory tests (Bishop & Henkel 1953) show that a single application and removal of a shear stress equal to the strength may reduce this value by more than 50 %, while complete remoulding may reduce it by more than an order of magnitude (Croney & Coleman 1960; Bishop 1960; Skempton & Sowa 1963). Field data presented by Ladd & Lambe (1963) suggests that a drop of 80 % below the value for 'perfect sampling' is typical for current sampling procedures in normally and lightly overconsolidated clays.

In contrast heavily overconsolidated soils have values of K_0 in excess of unity, show a behaviour which approximates more closely to the ideal elastic assumption and do not show a marked drop in pore-water tension on remoulding. In fact in some heavily over consolidated clays a single application and removal of a shear stress may increase the residual pore-water tension (Bishop & Henkel 1953) as also will complete remoulding (Croney & Coleman 1960). The cutting of a smaller sample from a large block sample consolidated under a known equal all-round stress resulted in a small but not significant reduction in the residual pore-water tension in the case of London Clay (Skinner, unpublished data).

Field data for over-consolidated clay thus show very high ratios of $-u_s/\sigma'_v$, varying with depth from 2.3 to 1.3 for the London Clay east of London (Skempton 1961) and from 2.7 to 1.7 for the London Clay west of London (Bishop *et al.* 1965). Values of K_0 calculated from these ratios lie in the ranges 2.8–1.5 and 3.4–1.7 respectively. Higher values of $-u_s/\sigma'_v$ have been reported by Blight (1967) for an expansive clay.

The largest value of residual pore-water tension reported from any of these sites is -0.76 MN

m^{-2} (110 lbf in^{-2})† for west London at a depth of 42 m (138 ft). For this value to be achieved in a *normally consolidated* sample of medium plasticity it would have to be recovered from a depth of some 1000 m on the assumption that the initial pore pressure distribution is hydrostatic.

The maintenance of full saturation under the residual pore-water tensions encountered in normal sampling operations should thus present no difficulty in homogeneous clays of pore size similar to the London Clay or even the Kaolin used in the present series of tests. However, the pore water in any macro-voids or segregated zones of coarser particles such as varves and silty partings is likely to cavitate and drain into the adjacent clay. Likewise continuity of the pore water will be lost across fissures, joints and bedding planes. This may result in a complete loss in strength, or in a substantial reduction in the unconfined strength (see, for example Bishop & Henkel 1962, Fig. 66).

The reduction in pore-water tension in the more disturbed zones of a borehole sample will result in migration of pore water, at constant overall volume, to the least disturbed zones and thus to a general reduction in strength.

The mitigation of these effects in the laboratory by the re-application of the estimated *in situ* total stresses without drainage or with drainage to an appropriate back-pressure is outside the scope of the present paper, as are such factors in the sampling procedure as the choice of sample size to minimize the proportion of very disturbed material and the limitation of access to free water particularly in overconsolidated clays.

The two factors of most significance in the field behaviour of clay masses subject to undrained stress release are the increased brittleness and the loss in strength when the pore-water tension results in loss of continuity of the pore water in silty partings, fissures and joints.

The authors are indebted in particular to Dr A. E. Skinner, who has been associated with the senior author for a number of years in the development of high pressure triaxial equipment. Dr Skinner has been largely responsible for putting the idea of a hydraulically balanced ram into practice and designed the internal load transducer used in this investigation.

Mr L. D. Wesley carried out a very careful series of relative density determinations on the two clays.

Mr D. Evans has given valuable assistance in the laboratory and Mr E. Harris has prepared the illustrations.

Appendix 1

Two novel features incorporated in the high pressure triaxial cell are the hydraulically balanced loading ram and the internal load transducer:

(a) *The hydraulically balanced ram*

The central section of the loading ram (figure A 1.1) is enlarged to form a piston having a diameter $\sqrt{2}$ times that of the ram, sliding with an appropriate oil seal in a cylinder formed in the head of the triaxial cell. Oil at cell pressure communicates through passages drilled in the ram with the annular upper surface of the piston. The cylinder below the piston is bled to atmospheric pressure.

If σ_3 is the fluid pressure in the cell, and d the diameter of the ram, then the upthrust on the

† Pore-water tensions of this magnitude cannot be measured directly and are obtained by a number of indirect methods discussed in the papers referred to.

ram due to the cell pressure is $\frac{1}{4}\pi d^2\sigma_3$. The downthrust on the piston is $\frac{1}{4}\pi[(\sqrt{2}d)^2 - d^2]\,\sigma_3$ and thus also equals $\frac{1}{4}\pi d^2\sigma_3$. However high the cell pressure, the axial load applied to the ram is equal only to the compression strength of the sample together with the ram friction.

A further advantage is that there is no change in the total volume of oil in the triaxial cell and loading head as the ram enters the cell. Thus the use of a high strain rate does not lead to a build-up of pressure in the cell due to hydraulic resistance in the small bore pressure tubing or to the characteristics of the pressure control system.

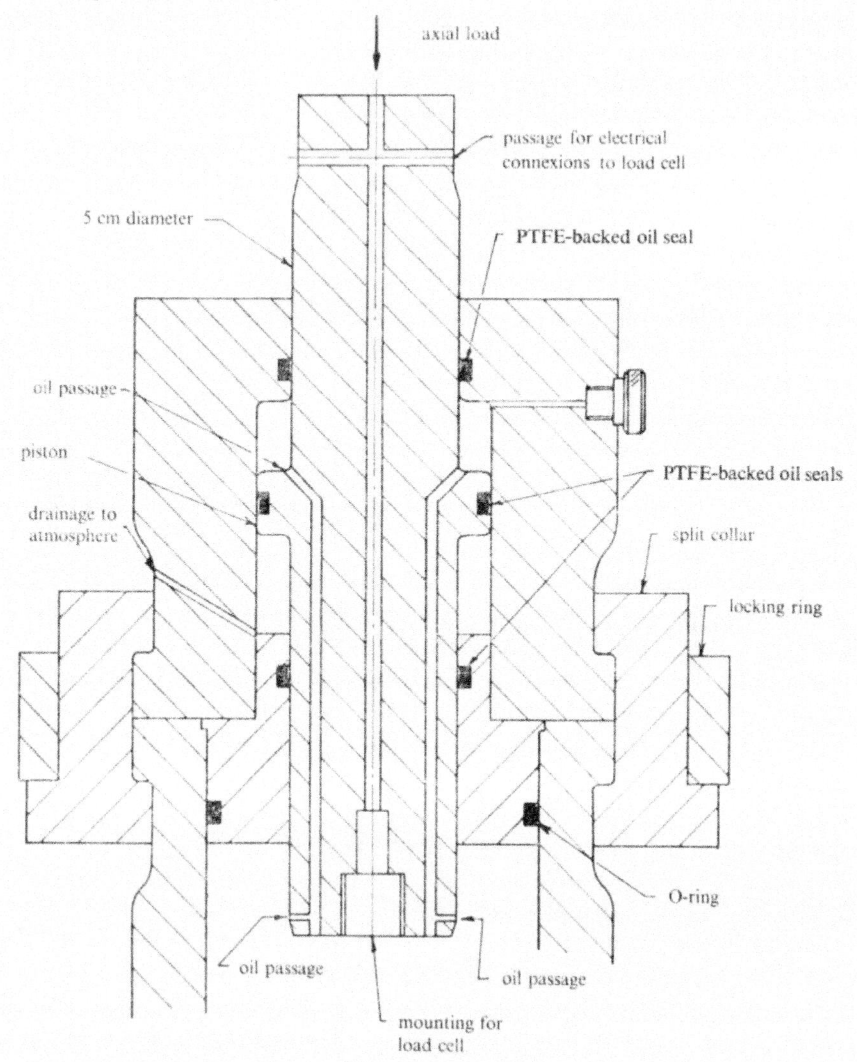

FIGURE A 1.1. Head of triaxial cell with hydraulically balanced ram.

(b) The internal load transducer

The load transducer consists, in effect, of three triangular cantilevers of uniform thickness radiating from a common boss and bearing on the edge of a groove in the cylindrical loading cap (figure A 1.2). As the bending moment and moment of resistance both increase linearly from the apex of each triangular cantilever the surface strains are almost uniform, and the calibration is therefore insensitive to the location of the electrical resistance strain gauges.

The cylindrical loading cap is filled with oil and sealed with a flexible nitrile rubber cover. A change in cell pressure thus has no significant effect either on the zero or on the calibration. The resistance gauges are wired to give complete insensitivity to eccentricity of loading and to the horizontal component of load.

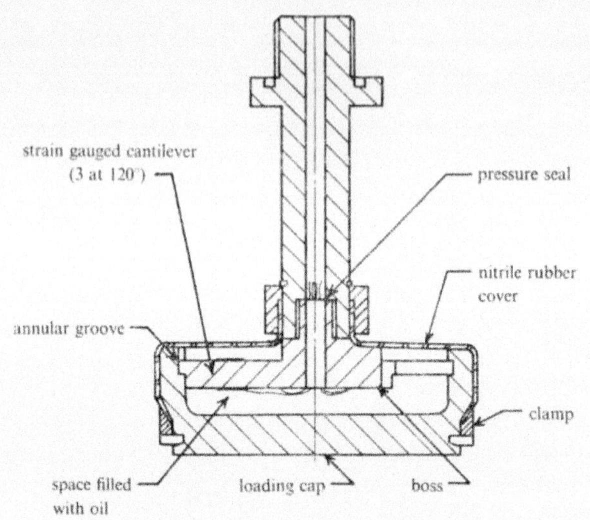

FIGURE A1.2. Internal load transducer.

APPENDIX 2

Equations (20)–(22) are based on the assumption that the unit weight of water in the voids of the soil is equal to the value of the unit weight of water used as the reference value for the determination of the relative density G. This should be the unit weight of water at 4 °C. This point was appreciated by Taylor (1948), but is not referred to in the description of the standard method of measurement of specific gravity in B.S. 1377 (1967). The difference due to temperature is small but just significant at the usual standard laboratory temperature of 20 °C, when the unit weight of water is 0.9982 gf/ml.

The difference due to the compressibility (or expansibility) of water is substantially greater at high pore-water pressures (or tensions), and is significant at the tensions induced in the tests on London Clay by stress release or drying. On the assumption that the coefficient C_w has the same value in tension as in compression (an assumption made by Temperley & Chambers 1946) the unit weight of water would decrease by 1.01 % at a tension of 20.7 MN m^{-2} (3000 lbf in^{-2}) and by 3.03 % at 62.1 MN m^{-2} (9000 lbf in^{-2}).

The pore-water tension in the samples in which partial saturation has occurred cannot be estimated with accuracy, but will clearly be in excess of the breakdown value (19.4 MN m^{-2} (2820 lbf in^{-2}) for London Clay) and below the value calculated for the particular consolidation pressure on the assumption of full saturation. This gives values of the unit weight of water ranging from 0.9888 to 0.9750 gf/ml for the highest consolidation pressure. These values imply possible errors in the calculated degree of saturation in the range 1–2½ %, and in the calculated percentage air voids of 4–9 % for the sample at the highest consolidation pressure.

The expansion which would have occurred on stress release if there had been no departure from full saturation can be calculated from the expression for undrained compressibility $\bar{C}$ given by Bishop (1966, 1973)

$$\bar{C} = \frac{nC_{\mathrm{w}} + (1-n)\,C_{\mathrm{s}} - C_{\mathrm{s}}^2/C}{1 + nC_{\mathrm{w}}/C - (1+n)\,C_{\mathrm{s}}/C}. \tag{A 2.1}$$

For the sample consolidated under 62.1 MN m^{-2} (9000 lbf in^{-2}) this gives a value for the expansion of 0.60 %.

As this expression for undrained compressibility is based on the same physical assumptions as the pore pressure equation (equation (2)), in particular that all the pore water has the compressibility of bulk water, it is of interest to check its accuracy. The present triaxial cell was not designed to measure undrained compressibility, but on the assumption of isotropy under cyclical applications of equal all-round pressure the value of $\bar{C}$ can be calculated from the changes in height of the specimen. The observations were complicated by a time-dependent interchange of water between the sample and the rubber membrane enclosing it as the pore-water tension is changed. Corrected 'instantaneous' values for two different porosities do not suggest any substantial error in equation (A 2.1).

A further assumption in the use of equations (20)–(22) to determine final void ratio under stress and degree of saturation on undrained stress release is that the value of the relative density G of the clay particles remains unchanged for the stress range under consideration. On the basis of elastic behaviour a maximum increase in the unit weight of the material comprising the particles would be approximately 0.2–0.3 % at an effective stress of 62.1 MN m^{-2} and with $u = 0$ (from equation (9) of Bishop 1973). This increase is not significant in the present context. However, Rieke & Chilingarian (1974) have presented data showing substantial decreases in the unit weight of clay particles with increase in consolidation pressure, the values for illite, for example, being 2.68, 2.53 and 2.38 gf/cm^3 at consolidation pressures of 13.8, 98.5 and 243 MN m^{-2} respectively.

The description of the method by which these results were obtained is given by Cebell & Chilingarian (1972), who carried out a series of high pressure consolidation tests in an oedometer somewhat similar to that of Westman (1932). However, Cebell & Chilingarian assume that on undrained removal of the total stress the pore water continues to fill the void space whatever the magnitude of the consolidation pressure used. In the light of the test results described in the present paper this appears to the authors to be an untenable assumption.

If equation (23) is expressed as a relation between the dry unit weight of the unconfined soil sample and $G\gamma_{\mathrm{w}}$, the unit weight of the clay particles, we have:

$$\frac{1}{\gamma_{\mathrm{d}}} = \frac{1}{G\gamma_{\mathrm{w}}} + \frac{w}{S\gamma_{\mathrm{w}}}. \tag{A 2.2}$$

The value of $G\gamma_{\mathrm{w}}$ can only be determined from measurements of γ_{d} and w if assumptions are made about the values of S and γ_{w}. The assumption that $S = 1.000$ (full saturation) and $\gamma_{\mathrm{w}} = 1.000$ gf/ml leads to values of $G\gamma_{\mathrm{w}}$ of 2.59 for London Clay and 2.22 for Kaolin consolidated at pressures of 62.1 and 6.9 MN m^{-2} respectively. Alternatively if the values of $G\gamma_{\mathrm{w}}$ measured at low pressure of 2.81 and 2.64 gf/cm^3 are assumed to apply, the term $S\gamma_{\mathrm{w}}$ has the value of 0.78 for London Clay and 0.73 for Kaolin. While the value of γ_{w} for the 9.8 % of water in the London Clay and the

19.0 % of water in the Kaolin is a matter of controversy, it is unlikely to differ from that of free water by as much as 22 and 27 % respectively. The assumption of partial saturation is more readily acceptable, though clearly a closer investigation of the factors influencing the values γ_w and G in clays subject to large changes in effective stress, pore pressure and void ratio would be of interest.

References

Aitchison, G. D. 1960 Relationships of moisture stress and effective stress functions in unsaturated soils. *Proc. Conf. Pore Pressure and Suction in Soils*, 47–52. London: Butterworth.

Agarwal, K. B. 1967 The influence of size and orientation of samples on the undrained strength of London Clay. Ph.D. Thesis, University of London.

Bishop, A. W. 1947 Strength variations in London Clay. *Silicates Industriels* **13**, 109–113.

Bishop, A. W. 1952 The stability of earth dams. Ph.D. Thesis, University of London.

Bishop, A. W. 1953 Private communication to Dr A. S. Laughton (see Skempton 1960).

Bishop A. W. 1958 Test requirements for measuring the coefficient of earth pressure at rest. *Proc. Brussels Conf. on Earth Pressure Problems* **1**, 2–14.

Bishop, A. W. 1959 The principle of effective stress. *Teknisk Ukeblad* no. 39, 859–863.

Bishop, A. W. 1960 The measurement of pore pressure in the triaxial test. *Proc. Conf. Pore Pressure and Suction Soils*, 38–46. London: Butterworth.

Bishop, A. W. 1966 Soils and soft rocks as engineering materials. *Inaug. Lect. Imp. Coll. Sci. Technol.* **6**, 289–313.

Bishop, A. W. 1971 Shear strength parameters for undisturbed and remoulded soil specimens. In *Stress–strain behaviour of soils*, 3–58, 134–139. Cambridge: Proc. Roscoe Memorial Symp.

Bishop, A. W. 1973 The influence of an undrained change in stress on the pore pressure in porous media of low compressibility. *Geotechnique* **23**, 435–442.

Bishop, A. W., Alpan, I., Blight, G. E. & Donald, I. B. 1960 Factors controlling the strength of partly saturated cohesive soils. *Proc. Res. Conf. on Shear Strength of Cohesive Soils*, Boulder, 503–532.

Bishop, A. W. & Bjerrum, L. 1960 The relevance of the triaxial test to the solution of stability problem *Proc. Res. Conf. Shear Strength of Cohesive Soils*, Boulder, 437–501.

Bishop, A. W. & Blight, G. E. 1963 Some aspects of effective stress in saturated and partly saturated soils. *Geotechnique* **13**, 177–197.

Bishop, A. W. & Eldin, G. 1950 Undrained triaxial tests on saturated sands and their significance in the general theory of shear strength. *Geotechnique* **2**, 13–32.

Bishop, A. W. & Henkel, D. J. 1953 Pore pressure changes during shear in two undisturbed clays. *Proc. 3rd Int. Conf. Soil. Mech., Zurich* **1**, 94–99.

Bishop, A. W. & Henkel, D. J. 1962 *The measurement of soil properties in the triaxial test*, 2nd ed. London: Edward Arnold.

Bishop, A. W., Webb, D. L. & Lewin, P. I. 1965 Undisturbed samples of London Clay from the Ashford Common Shaft: Strength–effective stress relation. *Geotechnique* **15**, 1–31.

Bishop, A. W., Webb, D. L. & Skinner, A. E. 1965 Triaxial tests on soil at elevated cell pressures. *Proc. 6th Int. Conf. Soil Mech., Montreal* **1**, 170–174.

Bjerrum, L. & Anderson, K. H. 1972 In-situ measurement of lateral pressures in clay. *Proc. 5th Europ. Conf. Soil Mech., Madrid* **1**, 11–20.

Blight, G. E. 1967 Horizontal stresses in stiff and fissured lacustrine clays. *4th Regional Conf. Soil. Mech., Cape Town.*

British Standards Institution 1967 *Methods of testing soils for civil engineering purposes*. B.S. 1377.

Bruhn, R. W. 1972 A study of the effects of pore pressure on the strength and deformability of Berea Sandstone in triaxial compression. *U.S. Department of the Army Technical Report – Engineering Study*, no. 52.

Cebell, W. A. & Chilingarian, G. V. 1972 Some data on compressibility and density anomalies in halloysite, hectorite and illite clays. *Am. Assoc. Petroleum Geol. Bull.* **56**, 796–821.

Croney, D. & Coleman, J. D. 1960 Pore pressure and suction in soil. *Proc. Conf. Pore Pressure and Suction in Soils*, 31–37. London: Butterworth.

Donald, I. B. 1961 The mechanical properties of saturated and partly saturated soils with special reference to the influence of negative pore water pressures. Ph.D. Thesis, University of London.

Gibson, R. E. 1953 Experimental determination of the true cohesion and true angle of internal friction in clays. *Proc. 3rd Int. Conf. Soil Mech., Zurich* **1**, 126–130.

Golder, H. Q. & Skempton, A. W. 1948 The angle of shearing resistance in cohesive soils for tests at constant water content. *Proc. 2nd Int. Conf. Soil Mech., Rotterdam* **1**, 185–192.

Green, R. B. 1951 Ordinary liquid water-substance, its thermodynamic properties, dynamic behaviour and tensile strength. D.Sc. thesis, Massachusetts Institute of Technology.

Holmes, J. W. 1955 Water sorption and swelling of clay blocks. *J. Soil Sci.* **6**, 200–208.
Hvorslev, M. J. 1937 Über die Festigkeitseigenschaften gestörter bindiger Böden. *Ingvidensk. Skr. A*, No 45. (English translation no. 69–5, Waterways Experiment Station, Vicksburg, Miss., 1969.)
Jurgenson, L. 1934 The shearing resistance of soils. *J. Boston Soc. civ. Engrs.* **21**, 242–275.
Kumapley, N. K. 1969 Triaxial tests on clays and silts at elevated cell pressures. Ph.D. Thesis, University of London.
Ladd, C. C. & Lambe, T. W. 1963 Shear strength of saturated clays. *Symp. on Laboratory Shear Testing of Soils, Ottawa, ASTM STP* 361, 342–371.
Lambe, T. W. & Whitman, R. V. 1969 *Soil mechanics*. New York: John Wiley.
Rieke, H. H. & Chilingarian, G. V. 1974 *Developments in sedimentology*. London: Elsevier.
Schmertmann, J. H. & Osterberg, J. O. 1960 An experimental study of the development of cohesion and friction with axial strain in saturated cohesive soils. *Proc. Res. Conf. Shear Strength of Cohesive Soils*, Boulder, 643–694.
Simons, N. 1958 Discussion on 'Test requirements for measuring the coefficient of earth pressure at rest'. *Proc. Brussels Conf. on Earth Pressure Problems* **3**, 50–53.
Skempton, A. W. 1953 Soil mechanics in relation to geology. *Proc. Yorkshire geol. Soc.* **29**, no. 3.
Skempton, A. W. 1954 The pore pressure coefficients A and B. *Geotechnique* **4**, 143–147.
Skempton, A. W. 1960 Effective stress in soils, concrete and rocks. *Proc. Conf. Pore Pressure and Suction in Soils*, pp. 4–16. London: Butterworth.
Skempton, A. W. 1961 Horizontal stresses in an overconsolidated Eocene clay. *Proc. 5th Int. Conf. Soil Mech., Paris*, **1**, 351–357.
Skempton, A. W. & Sowa, V. A. 1963 The behaviour of saturated clays during sampling and testing. *Geotechnique* **13**, 269–290.
Skinner, A. W. 1975 The effect of high pore water pressures on the mechanical behaviour of sediments. Ph.D. Thesis, University of London.
Taylor, D. W. 1944 Cylindrical compression research program on stress-deformation and strength characteristics of soils. *M.I.T. 10th Progress Report to US Engineers Department*.
Taylor, D. W. 1948 *Fundamentals of soil mechanics*. New York: John Wiley.
Temperley, H. N. V. 1946 The behaviour of water under hydrostatic tension. II. *Proc. phys. Soc.* **58**, 436.
Temperley, R. N. V. & Chambers, LL.G. 1946 The behaviour of water under hydrostatic tension. I. *Proc. phys. Soc.* **58**, 420.
Terzaghi, K. 1932 Trägfahigkeit der Flachgründungen. *Int. Assoc. Bridge Struct. Eng., Prelim. Publ.*, 659–683.
Terzaghi, K. 1936 The shearing resistance of saturated soils and the angle between the planes of shear. *Proc. 1st Int. Conf. Soil Mech., Harvard*, **1**, 54–56.
Terzaghi, K. 1943 *Theoretical soil mechanics*. New York: John Wiley.
Westman, A. E. R. 1932 The effect of mechanical pressure on the imbibitional and drying properties of some ceramic clays, I, II. *J. Am. Ceramic Soc.* **15**, 552–563, and **16**, 256–264.
Wissa, A. E. Z. 1969 Pore pressure measurement in saturated stiff soils. *J. Soil Mech. Fdn. Div. Am. Soc. civ. Engrs.* **95**, SM4, 1063–1073.

PHILOSOPHICAL TRANSACTIONS
OF
THE ROYAL SOCIETY
OF LONDON

A. MATHEMATICAL AND PHYSICAL SCIENCES

VOLUME 284 PAGES 91–130 NUMBER 1318

PTRMAD 284 (1318) 91–130 (1977)

4 January 1977

The influence of high pore-water pressure on the strength of cohesionless soils

by A. W. Bishop and A. E. Skinner

PUBLISHED BY THE ROYAL SOCIETY
6 CARLTON HOUSE TERRACE LONDON SW1Y 5AG

NOTICE TO CONTRIBUTORS TO PROCEEDINGS AND PHILOSOPHICAL TRANSACTIONS OF THE ROYAL SOCIETY

The Royal Society welcomes suitable communications for publication in its scientific journals: papers estimated to occupy up to 24 printed pages are considered for the *Proceedings* and longer papers and those with numerous or large illustrations for the *Philosophical Transactions*.

Detailed advice on the preparation of papers to be submitted to the Society is given in a leaflet available from the Executive Secretary, The Royal Society, 6 Carlton House Terrace, London SW1Y 5AG. The 'Instructions to authors' are also printed in every fifth volume of the *Proceedings* A and B (volume numbers ending in 0 or 5). The basic requirements are: a paper should be as concise as its scientific content allows and grammatically correct; standard nomenclature, units and symbols should be used; the text (including the abstract, the list of references and figure descriptions) should be in double spaced typing on one side of the paper; any diagrams should be drawn in a size to permit blockmaking at a reduction to about one half linear, the lettering being inserted not on the original drawings but on a set of copies; where photographs are essential the layout should be designed to give the most effective presentation.

The initial submission of a paper should normally be through a Fellow or Foreign Member of the Society, but papers may be submitted direct to the Executive Secretary. The latest lists of Fellows and Foreign Members are to be found in the current edition of the *Year Book of the Royal Society*. In the event of any difficulty, an author is invited to send the paper direct to the Executive Secretary.

No page charge is levied, and the first 50 offprints of a paper are supplied to the author gratis.

THE INFLUENCE OF HIGH PORE-WATER PRESSURE ON THE STRENGTH OF COHESIONLESS SOILS

By A. W. BISHOP and A. E. SKINNER
Civil Engineering Department, Imperial College of Science and Technology, South Kensington, London SW7 2BU

(*Communicated by A. W. Skempton, F.R.S. – Received* 12 *August* 1975 – *Diagrams for reproduction received* 8 *March* 1976)

[Plate 1]

CONTENTS

The influence on the mechanical properties of saturated particulate materials of the component of stress carried by the water filling the pore space is fundamental to both theoretical and experimental studies in soil mechanics.

The rôle of pore pressure in controlling compressibility and shear strength is expressed in Terzaghi's *principle of effective stress* to a degree of accuracy which is sufficient for most engineering purposes. However, the precise significance of the small but finite area of interparticle contact has remained uncertain in the application of this equation to shearing resistance.

In the present paper the possible errors associated with the use of current expressions for intergranular stress and effective stress are examined. These errors are of significant magnitude at high values of pore pressure and low values of the yield stress of the solid forming the particles. A very accurate experimental investigation has been carried out into the sensitivity of shearing resistance to large changes in pore pressure (up to 41.4 MN/m^2), using particulate materials ranging in strength from Quartz sand to lead shot.

The results indicate that the simple Terzaghi effective stress equation $\sigma' = \sigma - u$ is consistent with all the observations, though for Quartz sand a range of pore pressure changes an order of magnitude higher is desirable for additional confirmatory evidence.

1. Introduction

During the five decades which have elapsed since Terzaghi (1923) first stated the principle of effective stress the physical basis of the principle and the equations used to express it have been the subject of periodic review and occasionally of lively controversy. Throughout this period there has been a tendency to ignore the intention underlying Terzaghi's definition of effective stress and to identify effective stress with intergranular stress (for example Bruggeman, Zangar & Brahtz 1939; Taylor 1944, 1948; Scott 1963). While the consequent error is small for the relatively low values of pore pressure and interparticle contact area usually encountered in engineering practice, significant errors arise at the high values of pore pressure found in ocean bottom sediments, in oil and gas reservoirs and in various geophysical studies. Likewise in concrete and rocks, where the area of contact is large even at low stresses, the use of a correct effective stress equation is particularly important.

Terzaghi (1936) restated the principle of effective stress in the following terms (in current terminology):

'The stresses at any point of a section through a mass of soil can be computed from the *total principal stresses* σ_1, σ_2, σ_3 which act in this point. If the voids of the soil are filled with water under a stress u, the total principal stresses consist of two parts. One part, u, acts in the water *and* in the solid in every direction with equal intensity. It is called the *neutral stress* (or pore-water pressure). The balance $\sigma_1' = \sigma_1 - u$, $\sigma_2' = \sigma_2 - u$ and $\sigma_3' = \sigma_3 - u$ represents an excess over the neutral stress u and has its seat exclusively in the solid phase of the soil.'

'This fraction of the total principal stresses will be called the *effective principal stresses*.... A change in the neutral stress u produces practically no volume change and has practically no influence on the stress conditions for failure.... Porous materials (such as sand, clay and concrete) react to a change in u as if they were incompressible and as if their internal friction were equal to zero. All the measurable effects of a change in stress, such as compression, distortion and a change in shearing resistance are exclusively due to changes in the effective stresses σ_1', σ_2', σ_3'. Hence every investigation of the stability of a saturated body of soil requires the knowledge of both the total and the neutral stress'.

It will be noted that in this statement the emphasis is on the exclusive relation between all the measurable effects of a change in stress (compression, distortion and a change in shearing resistance) and changes in the effective stresses σ_1', σ_2', σ_3'. Terzaghi's simple expression for effective stress

$$\sigma' = \sigma - u \tag{1}$$

has been shown (Bishop & Eldin 1950; Bishop 1955; Skempton 1960) to hold rigorously for the case of volume change if the two conditions in the statement quoted are put in the form

(1) the soil grains are incompressible,

(2) the yield stress of the grain material, which controls the contact area and intergranular shearing resistance, is independent of the confining pressure (as in the theories of friction due to Terzaghi (1925) and Bowden & Tabor (1942)).

It was also inferred by Bishop (1955) and demonstrated more rigorously by Skempton (1960) that the simple expression for effective stress $\sigma' = \sigma - u$ should hold for changes in shear strength if condition (2) were satisfied.

Actual soils do not fully satisfy either condition. Skempton (1960) has examined the significance of the departures from both conditions (1) and (2) and has derived modified expressions for effective stress. In the next section the various expressions for effective stress and intergranular stress will be briefly presented and discussed.

2. Expressions for intergranular stress and effective stress

(a) *Intergranular stress*

It has been considered by a number of investigators that, *a priori*, the mechanical behaviour of a particulate mass is controlled by the forces acting at the particle-to-particle contacts. Taylor (1948, p. 126) states: 'In concepts of stresses... the surface that must be considered is the one containing the points of grain-to-grain contact, in order that it may include the points of action of the forces which make up intergranular stress. Thus the unit area should be visualized as a wavy surface which is tangent to but does not cut through soil grains, and which at all points is as close as possible to a flat surface.'

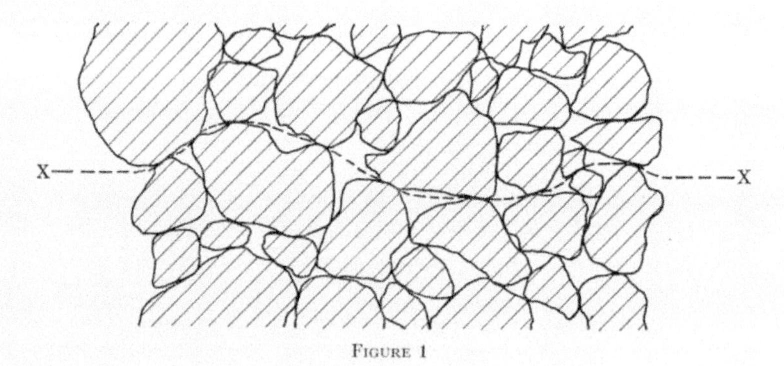

Figure 1

Consideration of the equilibrium of the forces on such a flat surface (figure 1) leads to the relation between total stress σ, the intergranular stress σ_i (defined as the average intergranular force per unit area normal to the surface) and the pore-water pressure u

$$\sigma = \sigma_i + (1-a)\,u$$

or

$$\sigma_i = \sigma - (1-a)\,u = (\sigma - u) + au, \qquad (2)$$

where a denotes the contact area of the soil particles per unit area of the surface (and projected onto the surface).

Taylor (1944) uses this expression in a discussion of the significances of pore pressure changes in undrained triaxial tests on Boston Clay and estimates the value of a on this basis, obtaining a value of about 3 %. Taylor suggested that the term 'neutral stress' should be applied to the product $(1-a)\,u$.

Most investigators have, however, agreed that a is small in cohesionless soils, and probably in clays, at the stress levels commonly encountered in engineering practice. Thus, provided u is of the same order of magnitude as $(\sigma - u)$ the term au may be neglected with little loss in accuracy. However, if u is very large compared with $(\sigma - u)$, the significance of the term au is radically

12-2

changed. For $u = 100(\sigma - u)$ a value of a of 1 % would lead to a product au equal in magnitude to $(\sigma - u)$ and a doubling of the magnitude of σ_i.

Since the magnitude of a cannot be measured directly and is a matter of some uncertainty in many soils, the validity or otherwise of the intergranular stress concept is clearly a matter of considerable importance.

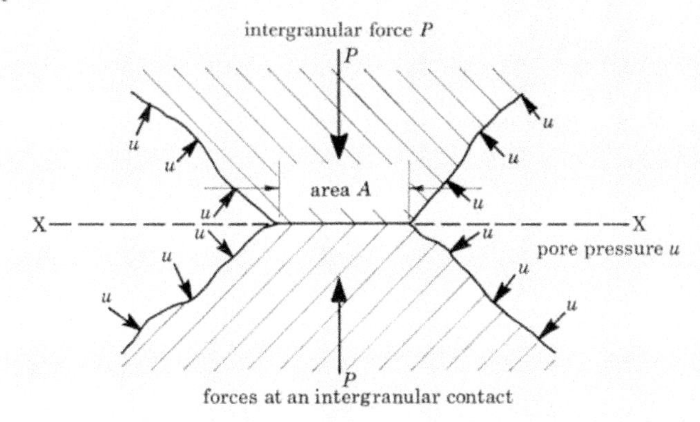

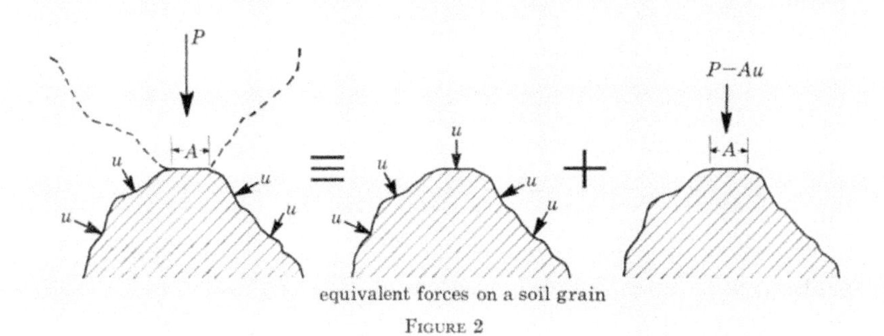

FIGURE 2

(b) *Effective stress for changes in volume*

Bishop & Eldin (1950) consequently examined the forces acting at an interparticle contact in the presence of pore-water under pressure (figure 2) and concluded that the distortion of the soil particle was not a function of the intergranular contact pressure, but of the excess of this pressure over the pore pressure acting on the adjacent surface of the particle. If the compressibility of the material comprising the soil particle was neglected as small compared with the bulk compressibility of the porous granular mass, then, irrespective of the magnitude of the interparticle contact area a, the equation for effective stress reduced to

$$\sigma' = \sigma - u.$$

Effective stress in this context† is defined as the function of total stress and pore pressure which controls volume change.

† The terminology used in the original paper by Bishop & Eldin (1950) reflected the current identification of the term effective stress with intergranular stress.

For this demonstration that the Terzaghi effective stress equation was independent of the magnitude of a, the two conditions stated in section (1) were necessary and sufficient.

The effect on the effective stress equation of a departure from condition (1), the incompressibility of the soil grains, was examined by Bishop (1953), who derived the expression

$$\sigma' = \sigma - (1 - C_s/C)\, u \tag{3a}$$

or

$$\sigma' = (\sigma - u) + (C_s/C)\, u, \tag{3b}$$

where C_s denotes the compressibility of the solid material forming the soil grains and C denotes the bulk compressibility of the porous mass for the relevant stress range.

For soils in the low stress range the bulk compressibility C is very large compared with the value of C_s. Skempton (1960) has tabulated data showing that for soils ranging from normally-consolidated clay to dense sand the ratio C_s/C lies in the range 0.00003–0.0015 for a consolidation pressure $(\sigma - u)$ of approximately $100\,\text{kN/m}^2$ (table 1). For a consolidation pressure of $20\,\text{MN/m}^2$ Skempton suggested that the ratio C_s/C is unlikely to exceed 0.01 in the case of clays. More recent test data from high pressure triaxial tests on London Clay (Bishop, Kumapley & El-Ruwayih 1975) indicate that at a consolidation pressure of $62.1\,\text{MN/m}^2$, C_s/C may rise to 0.04.

TABLE 1. (AFTER SKEMPTON 1960.)

(Compressibilities at $p = 1\,\text{kg/cm}^2$; water $C_w = 48 \times 10^{-6}$ per kg/cm^2)

material	compressibility $\text{cm}^2\,\text{kg}^{-1}$		
	C	C_s	$\frac{C_s}{C}$
quartzitic sandstone	5.8×10^{-6}	2.7×10^{-6}	0.46
Quincy granite (100 ft deep)	7.5×10^{-6}	1.9×10^{-6}	0.25
Vermont marble	17.5×10^{-6}	1.4×10^{-6}	0.08
concrete (approx. values)	20×10^{-6}	2.5×10^{-6}	0.12
dense sand	$1\,800 \times 10^{-6}$	2.7×10^{-6}	0.0015
loose sand	$9\,000 \times 10^{-6}$	2.7×10^{-6}	0.0003
London clay (over-cons.)	$7\,500 \times 10^{-6}$	2.0×10^{-6}	0.00025
Gosport Clay (normally cons.)	$60\,000 \times 10^{-6}$	2.0×10^{-6}	0.00003

Tests to examine the validity of the effective stress equation for volume change have been carried out by Laughton (1955) using high pore-water pressures in a sealed oedemeter. Laughton tested both lead shot, to ensure large values of interparticle contact area a which could be measured on unloading the sample, and Globigerina ooze from the bed of the eastern Atlantic ocean.

In both cases the test data demonstrate the validity (to within experimental error) of the effective stress equation $\sigma' = \sigma - u$, although in the case of the lead shot the observed value of a rose to 95% at the highest effective stress ($100\,\text{MN/m}^2$). In a re-examination of the test data Skempton (1960) has shown that the effective stress equation (3) gives a marginally better fit in the case of lead shot when the value of C_s/C rises to 0.05 at the highest consolidation pressure.

Skempton (1960) also examined the consequence of a departure from condition (2) on the effective stress equation for volume change and found it to be numerically unimportant in most cases.

It is of interest to note that, since the bulk properties of a granular mass subject to a change in stress are not those of an ideal elastic material, the value of C is not a unique parameter for a particular soil (as is C_s) but depends on the stress level, the previous stress path, the sign and magnitude of the stress change, and the rate of loading (Bishop & Blight 1963). Furthermore, although effective stress as defined by equation (3) determines the overall volume change, it is the component $\sigma - u$ which determines the change in compressibility C (Bishop 1973).

While it is apparent from the preceding discussion that the more rigorous expression for effective stress with respect to volume change need only be used for soils at very high consolidation pressures, for concrete and for porous rocks the term C_s/C is of much greater significance even in the low stress range. Data presented by Skempton (1960) indicate values of C_s/C of around 0.12 for concrete and from 0.08 to 0.46 for various rocks (table 1).

(c) *Effective stress for changes in shear strength*

Terzaghi (1936) considered that the same effective stress equation $\sigma' = \sigma - u$ applied equally to both changes in volume and to changes in shear strength.

Bishop (1955) considered that, if an analogy could be drawn between interparticle friction and metallic friction, then the contact area would control the frictional forces and this area, like the deformation of the soil particles, would be controlled by σ' $(= \sigma - u)$. Hence it might be inferred that the expression for effective stress $\sigma' = \sigma - u$ would be valid for shear strength irrespective of contact area, provided condition (2) was satisfied.

Skempton (1960) presented a formal analysis of interparticle friction in the presence of pore water under pressure, arriving at the same conclusion for the special case represented by condition (2), but also extending the analysis to include the more general case of grain materials whose strength is a function of confining pressure of the form:

$$\tau_i = k + \sigma \tan \psi, \tag{4}$$

where τ_i denotes shear stress at failure, σ denotes normal stress, k denotes intrinsic cohesion, and ψ denotes the angle of intrinsic friction of the solid.

For these materials Skempton obtained an expression for effective stress with respect to change in shear strength:

$$\sigma' = \sigma - \left(1 - \frac{a \tan \psi}{\tan \phi'}\right) u \tag{5a}$$

$$= (\sigma - u) + au \frac{\tan \psi}{\tan \phi'}, \tag{5b}$$

where a denotes interparticle contact area, as before, and ϕ' denotes the angle of shearing resistance of the granular mass (in terms of effective stress).

Skempton (1960) examined published test data from jacketed and unjacketed triaxial tests on Marble and Solenhofen Limestone and concluded that the areas of contact given by equation (5) (0.15 and 0.45 respectively) were consistent with the changes in strength with confining pressure obtained with the jacketed samples. He also concluded that area of contact deduced from concrete was about 0.2, and, though not subject to any independent check, was not unreasonable. For soils no critical test data was available, and Skempton concluded that Terzaghi's equation would be a valid approximation due to the small value of a.

In deriving equation (5) Skempton assumed that there is a direct relation between ϕ', the angle of shearing resistance of a non-cohesive granular material, and μ, the coefficient of friction at an interparticle contact, quoting the expression due to Caquot (1934) for constant volume shear:

$$\tan \phi' = \tfrac{1}{2}\pi\mu. \tag{6}$$

Similar though not identical relations have been obtained by Bishop (1954) and Horne (1969), but involve simplifying physical assumptions and, in the case of the earlier expression, a mathematical approximation.

This form of relation is, however, not supported by a series of very careful tests carried out by Skinner (1969, 1975) on particles of almost identical shape but of widely differing values of the coefficient μ (figure 3). Skinner's results are supported by independent tests carried out on rock fragments and gravel by Tombs (1969) and discussed by Bishop (1969).†

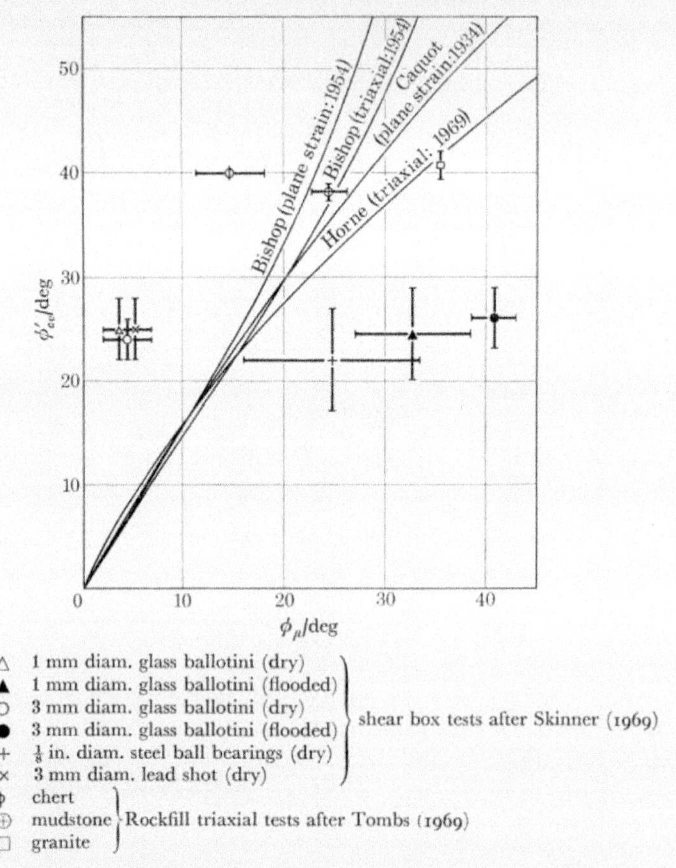

FIGURE 3. Theoretical and experimental relations between ϕ_μ and ϕ_{cv} (Skinner 1969).

Lack of direct relation between ϕ' and μ is associated with the complexity of particle movement in a particulate mass subject to a shear strain, which involves not only interparticle slip but particle rotation and out-of-plane displacements even under an overall plane strain displacement.‡ It suggests, furthermore, that one of the steps in the derivation of the effective stress

† Experimental support for the relation comes primarily from the tests in which μ has not in fact been measured directly and where its physical significance is open to serious criticism (see Horne 1969; Skinner 1969; Bishop 1969; Procter & Barton 1974).

‡ A detailed study using marked particles and X-ray stereo techniques has been carried out by Y. Sharma and is in course of preparation for publication.

equation (5) may be based on an assumption of doubtful validity, at least for cohesionless soils. This places added emphasis on the need for direct experimental verification of the effective stress equation for shear strength.

3. Testing programme, apparatus and testing techniques

The testing programme was designed generally to explore the influence of high pore-water pressures on the strength of cohesionless soils and in particular to discriminate between the intergranular and effective stress equations.

The difference between the three equations lies in the term involving the product *au* where the Terzaghi effective stress equation is

$$\sigma' = \sigma - u, \tag{1}$$

the intergranular stress equation is

$$\sigma_i = (\sigma - u) + au, \tag{2}$$

and Skempton's effective stress equation is

$$\sigma' = (\sigma - u) + mau, \tag{5}$$

where m is $\tan\psi/\tan\phi'$ and is of the order 0.3 for sand consisting of quartz particles and 0.03 for lead shot (Skempton 1960). The significance of the au term in influencing the strength, which, by definition, is controlled by σ', will depend on the magnitude of au relative to $(\sigma - u)$. In seeking, for reasons of experimental accuracy, to maximize the ratio r, where $r = au/(\sigma - u)$, we may note that, for a given particulate material, the magnitude of a depends almost linearly on $(\sigma - u)$ (Bishop & Eldin 1950; Skempton 1960). Hence we have

$$\begin{aligned} r &= \frac{au}{(\sigma - u)} \\ &= \frac{n(\sigma - u)\,u}{(\sigma - u)} \\ &= nu, \end{aligned} \tag{7}$$

where n is a constant depending on the strength parameters of the material. Thus the percentage difference in strength between the predictions of the three expressions will depend on the magnitude of the change in pore pressure (for $(\sigma - u)$ = constant), but will be independent of the actual magnitude of the consolidation pressure $(\sigma - u)$ and of the associated value of contact area a, depending instead on their ratio as represented by the parameter n. This means that the testing programme must involve the highest pore pressures (and consequently the highest cell pressures in the triaxial apparatus) consistent with the accurate measurement of small changes in strength, and must include materials in which the yield stress of the particles is relatively low.

The accurate measurement of small changes in strength at high cell pressures in the conventional triaxial apparatus is rendered almost impossible by (*a*) the friction on the loading ram and (*b*) the magnitude of the load due to the cell pressure acting on the inner end of the ram, this load being very large compared with that due to the strength change to be detected.

The error due to friction on the loading ram can be avoided by measuring the load inside the cell with an electric load transducer. However, the problem then arises of the sensitivity of the transducer itself to large changes in cell pressure. As the purpose of the present series of tests was not merely to measure the strength changes correctly, but to demonstrate incontrovertibly that they had been measured correctly, the alternative method of rotating the bushing enclosing the loading ram was adopted (figure 4). Since the frictional force opposes the relative motion between the ram and the bushing, a rotary motion of about 2 rev/min is sufficient to reduce the vertical component of friction to negligible proportions at normal rates of axial displacement. The problems of excessive oil leakage and 'wobble' due to the loss of a common axis to the inner and outer cylindrical surfaces of the bushing during machining and honing call for a very high standard of workmanship in manufacture.

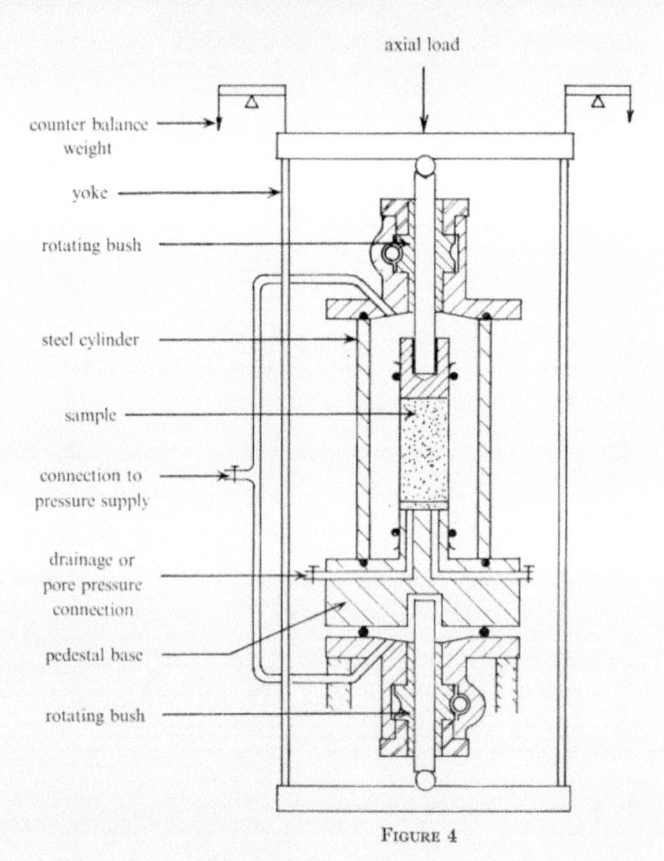

FIGURE 4

The relative magnitudes of the uplift on the end of the loading ram and the axial load required to shear the sample can be readily estimated for a maximum ratio of cell pressure σ_3 to consolidation pressure $\sigma_3 - u$ of *ca.* 100 and a ram to sample diameter ratio of $\frac{2}{3}$ (as used on the high pressure cell). From the geometry of the Mohr circle (figure 5) it follows that, for a cohesionless soil, the stress difference at failure (on the basis of the Terzaghi expression for effective stress) is given by the expression

$$(\sigma_1 - \sigma_3) = (\sigma_3 - u)\frac{2\sin\phi'}{1-\sin\phi'}. \tag{8}$$

Hence, for a value of $\phi' = 35°$, the ratio of the axial load due to uplift to the axial load on the sample at failure (neglecting the area change during compression) is

$$\frac{\sigma_3 A_r}{(\sigma_1-\sigma_3)\,A_s} = \frac{100\times(\sigma_3-u)\times 1.0^2\times\pi}{(\sigma_3-u)\times 2.69\times 1.5^2\times\pi},$$
$$= 16.5, \tag{9}$$

where A_r and A_s denote the cross-sectional areas of the ram and sample respectively.

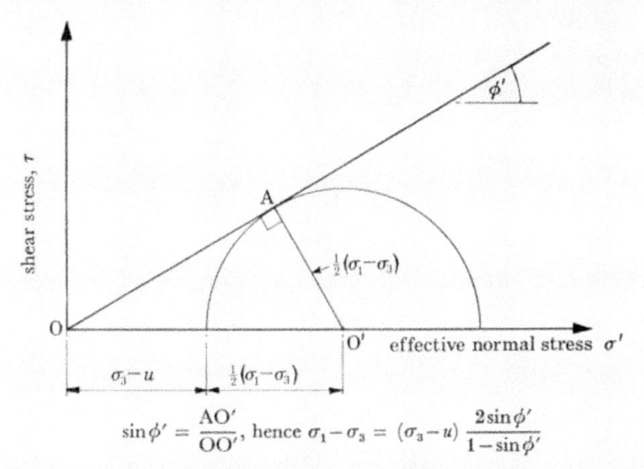

$$\sin\phi' = \frac{AO'}{OO'}, \text{ hence } \sigma_1-\sigma_3 = (\sigma_3-u)\,\frac{2\sin\phi'}{1-\sin\phi'}$$

FIGURE 5. The geometry of the Mohr diagram.

Thus to detect a change of 0.5 % in shear strength it would be necessary to measure the total axial load to an accuracy of 0.03 %, which is beyond the confidence limit of most load measuring devices.

The difficulty was resolved by balancing out hydraulically the load due to uplift with an opposed ram of identical diameter in a similar rotating bush (figure 4). A prototype triaxial cell was built (figure 6*a*, plate 1) mainly from components currently in use at Imperial College (Bishop, Webb & Skinner 1965) having a maximum cell pressure capacity of 6.9 MN/m² (1000 lbf/in²). With carefully matched rams (15.88 mm or 0.625 in nominal diameter) the maximum change in axial load for a change in cell pressure of 6.9 MN/m² was found to be 0.27 N (0.06 lbf) and was thus negligible relative to the load changes to be measured.

The success of the prototype led to the adoption of the same principle for a high pressure cell with a capacity of 69 MN/m² (10 000 lbf/in²). The technical problems to be overcome in actually manufacturing a rotating bush cell to operate in this pressure range were found to be formidable for three reasons in particular.

Firstly, the principle of rotating the bushing to eliminate the vertical component of friction was found to be ineffective if conventional oil seals were used. Hence control of the loss of the pressure fluid between the ram and the bushing depended solely on the length of the leakage path and the fineness of the fit. The length also necessitated a relatively stiff ram to avoid buckling in the higher load range. The dimensions chosen were a ram diameter of 25.4 mm (1.0 in) and an external bush diameter of 381 mm (1.5 in) with minimum and maximum radial clearances of 0.0038 and 0.0051 mm (0.00015 and 0.0002 in) internally and 0.0051 and 0.0064 mm (0.0002 and 0.00025 in) externally.

FIGURE 6. (*a*) 6.90 MN m^{-2} triaxial apparatus. (*b*) 68.95 MN m^{-2} triaxial cell.

(*Facing p.* 100)

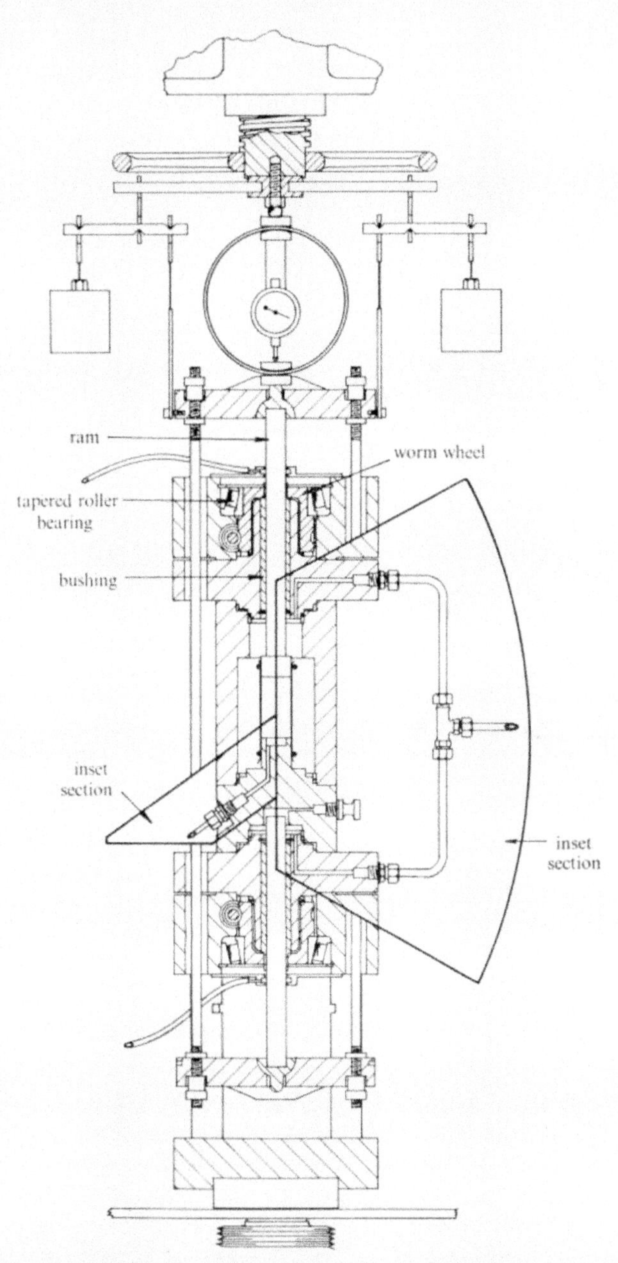

FIGURE 7. Triaxial cell with hydraulically balanced rams suitable for sample confining pressures up to 69 MN m^{-2}.

Honing to these small tolerances and the avoidance of seizing were made possible by the use of a special (spheroidal) cast iron bushing in conjunction with a stainless steel ram. The internal and external drainage paths along the bushing were 165 mm (6.500 in) and 143 mm (5.625 in) respectively. The oil loss from the pair of bushings was only about 0.75 l/h using Germ Dynobear L oil (relative density 0.892 at 15.5 °C; viscosity Redwood No. 1 at 21.1 °C 342 s; light machine tool lubricant with increased oiliness characteristic which prevents stick–slip behaviour) at the maximum cell pressure of 69 MN/m² (10 000 lbf/in²).

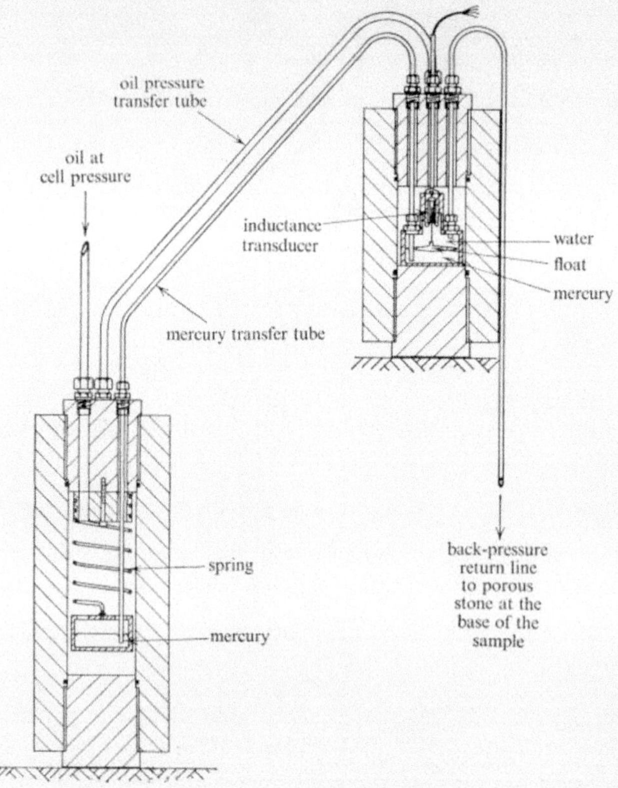

FIGURE 8. Apparatus used for $\sigma_3 - u$ = const. tests at high confining pressures, incorporating a volume gauge for tests involving high back pressures.

The second problem was the uplift force due to the cell pressure acting on the inner end of the bushing itself. Even using the minimum wall thickness convenient for manufacture and for transmitting the torque under operating conditions (6.35 mm or 0.25 in), the axial load at maximum cell pressure is 43.6 KN (9812 lbf). This load had to be carried by a thrust bearing with the minimum of vibration and friction. A tapered roller bearing (figure 7) was used for this purpose and was mounted on the boss of the bronze worm wheel which served to transmit both the torque to the rotating bush and the axial load from the bush to the inner race of the bearing. The friction between the bush and the worm wheel due to the axial load was sufficient to carry the torque necessary under most operating conditions without any form of key.

The third problem was the distortion of the loading head itself, due to the high cell pressure, which could result in unacceptable changes in diameter of the bore within which the bushing had to run. Because of the difficulty of determining these changes in bore diameter analytically, a model of the head was made in brass, strain gauged and tested. As a result of these observations the loading head was redesigned so that the major thrust on the head passed directly across to the clamping rings via the cell wall or pedestal base (figure 7).

The success of the mechanical system and the accuracy of the workmanship involved is indicated by the observation that the difference in axial load for a change in confining pressure of 69 MN/m^2 (10 000 lbf/in^2) was less than 0.27 N (0.06 lbf).

One other aspect of the apparatus is of special interest. In the main series of tests the difference between the cell pressure and the pore-water pressure $(\sigma_3 - u)$ had to be maintained very accurately at a constant value while the cell pressure and pore-water pressure were varied through a range some 100 times the magnitude of $(\sigma - u)$. This was achieved by building a system equivalent to the constant pressure self-compensating mercury control system described by Bishop & Henkel (1962) and immersing it in fluid at the operating pressure of the cell, contained in two pressure vessels separated by an appropriate vertical distance (figure 8).

As in the widely used low pressure version, the difference between the pressure (σ_3) in the fluid above the mercury surface in the suspended cylinder in the lower vessel and in the return line (u) from the fluid above the mercury in the fixed cylinder in the upper vessel is the difference in level multiplied by the difference in the unit weights of mercury and the operating fluid in the return line R (figure 9). As the return line is connected to the porous element at the base of the sample, the change in the mercury level in the upper cylinder provides a measure of the volume change in the test specimen. This measurement is based on the volume of pore water moving in or out of the specimen and is subject to corrections for membrane penetration when $(\sigma_3 - u)$ is changed, and for the compressibility of the pore water in the sample, of the soil particles, of the water in the cell base and return line and for the expansion of the thick-walled tubing when u is changed with $(\sigma_3 - u)$ held constant.

The mercury level was sensed with an inductance transducer, the core of which was attached to a flat conical stainless steel float (figure 8). The capacity of the system as a volume gauge was 20 cm^3. The overall discrimination of the system was 1×10^{-4} cm^3, but the repeatability was only of the order of ± 0.01 cm^3. As the initial volume of the specimens was of the order of 90 cm^3 this degree of accuracy was adequate for studying the shear strength and dilatancy characteristics of the soil as a granular mass, but not for investigating in detail the influence of grain compressibility on overall volume change.

The change in the pre-set value of $\sigma_3 - u$, which would result from a change in the mercury levels consequent on a volume change in the sample, is automatically compensated for by the calibrated spring in the lower pressure vessel. As the weight of mercury in the cylinder suspended from this spring changes due to the change in mercury level in the upper cylinder a spring of suitable characteristics adjusts the lower mercury level to maintain constant pressure to an accuracy of better than ± 1 kN/m^2 (*ca.* ± 0.1 lbf/in^2).

The high pressure cell is illustrated in figure 6*b*, plate 1.

The test programme itself was in principle very simple and consisted of the observation of the strength changes resulting from large changes in σ_3 and u, the difference $(\sigma_3 - u)$ being held constant to a high degree of accuracy. Since the natural scatter of a series of separate tests on individual samples might mask small strength changes, advantage was taken of the relatively

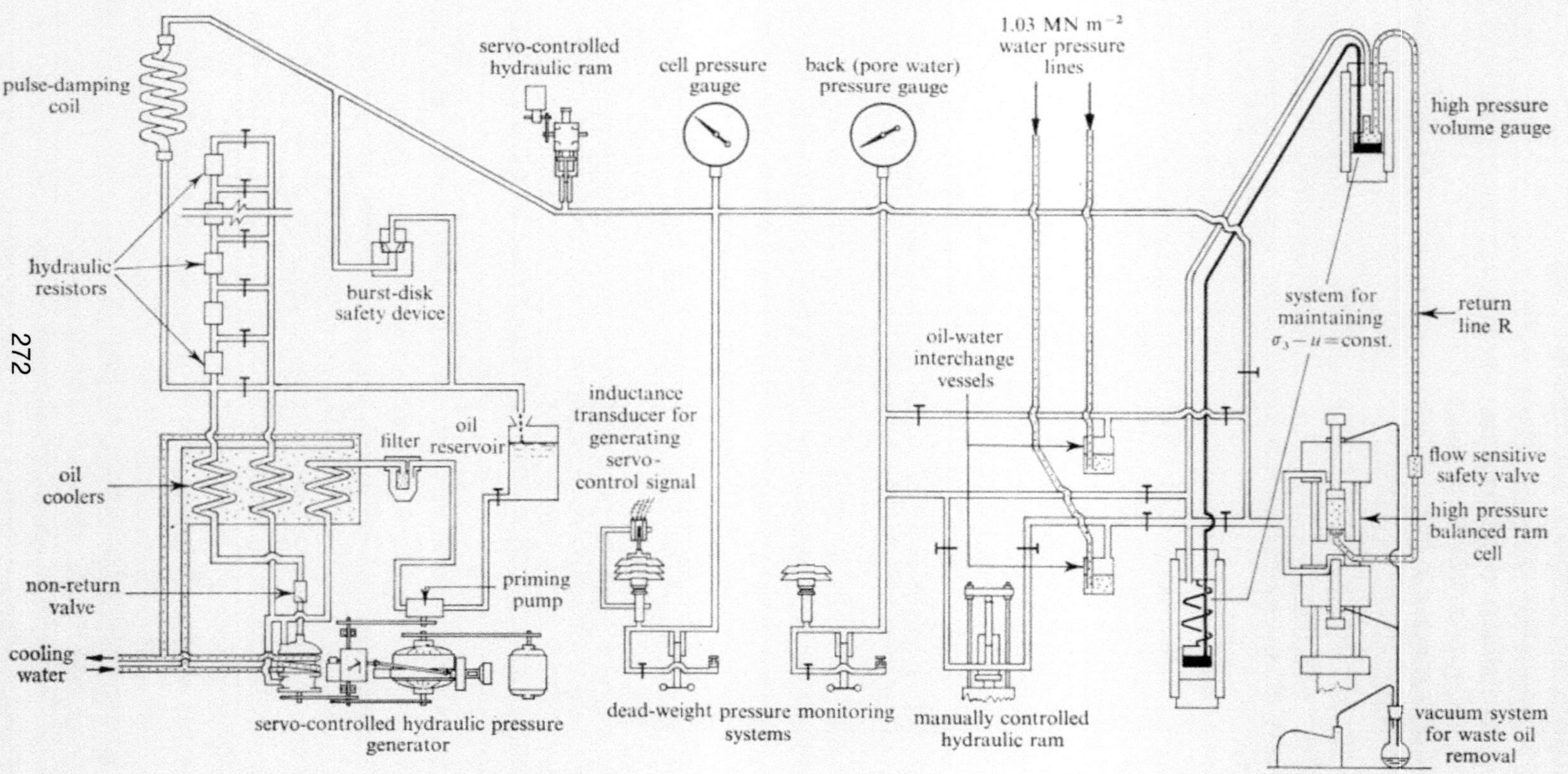

Figure 9. Diagrammatic layout of triaxial apparatus used for tests with cell pressures up to 69 MN m^{-2}.

small rate of change of mobilized strength with strain in loose granular materials to perform multistage tests on a limited number of specimens. If the au term were found to have an effect on strength, this could then be detected with an accuracy of about $\pm 0.5\%$ from the discontinuities in the stress–strain curve (as indicated diagrammatically in figure 10).

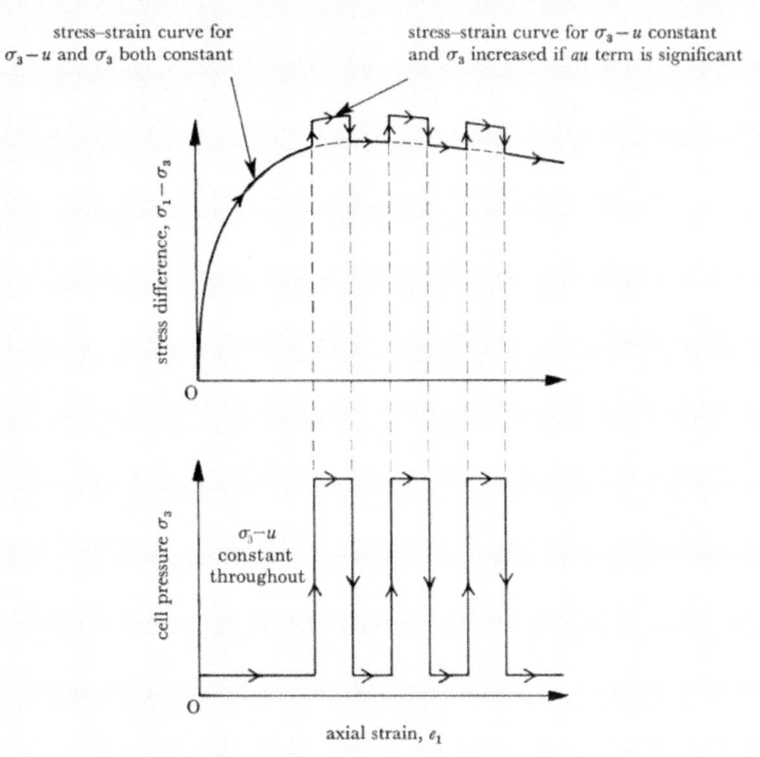

FIGURE 10. Effect on shearing resistance, in a drained triaxial compression test with $\sigma_3 - u$ constant, of the same large change in both confining pressure σ_3 and pore pressure u if the term au is significant.

All tests were performed on saturated material and the samples were sedimented under water within a rubber membrane enclosed by a split former as described by Bishop & Henkel (1962). After the sample cap had been placed in position and the membrane sealed to it a small negative pressure was applied to the pore water. This consolidated the sample by drainage through the porous ceramic disk set in the pedestal, and gave it sufficient strength for the former to be removed and the initial dimensions measured. The remaining components of the cell were then assembled and the balance of the pressure difference $(\sigma_3 - u)$ applied, the volume change being measured as described above.

During the application of the stress difference $(\sigma_1 - \sigma_3)$ the tests were run as controlled rate of strain drained tests with constant $(\sigma_3 - u)$, the axial load, axial displacement and volume change being recorded. The test procedure in general follows Bishop & Henkel (1962) and is described in detail by Skinner (1975).

4. Materials tested

Four materials were tested and their particle size distribution curves are given in figure 11. They are

(1) *Ham River sand.* This is a sieved fraction of a naturally occurring gravel and is composed largely of quartz. The detailed mineral composition is given in table 2.

Ham River sand was selected as its mechanical properties have been extensively studied both at Imperial College and at the Building Research Establishment over the past 25 years in relation to strength and deformation, and to the performance of model foundations.

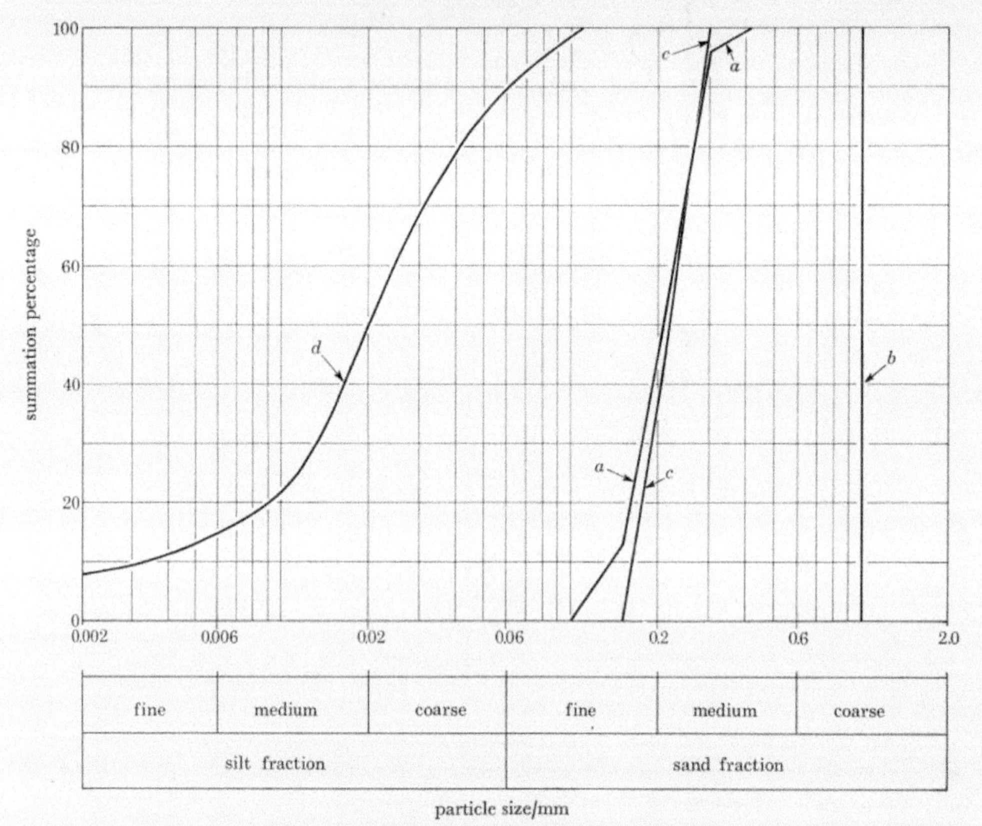

Figure 11. Particle size distribution curves for (*a*) Ham river sand; (*b*) lead shot; (*c*) crushed marble and (*d*) Braehead silt.

(2) *Lead shot.* This material had a uniform particle size of 1 mm, and was selected as a convenient and readily available material with a very low yield stress. The interparticle contact area was thus almost the maximum obtainable for a given stress level, in contrast to that of quartz sand, which is almost the minimum.

(3) *Crushed marble.* Since lead shot has a very low value of the intrinsic angle of friction ψ (*ca.* $\frac{3}{4}°$), its behaviour does not provide a very critical test of Skempton's effective stress equation

(equation (5)), since even when the product au is large for a given stress level, the factor $\tan\psi/\tan\phi'$ is very small. Calcite has the advantage of a relatively low yield stress k associated with a value of ψ of 8° (i.e. about one half of that of quartz) and can therefore provide more critical test data.

(4) *Braehead silt.* This material is a naturally occurring silt, composed mainly of quartz, but having an average particle size more than an order of magnitude smaller than the Ham River sand. The detailed mineral composition is given in table 2. Since the force per contact, for a given geometrical arrangement of particles, is proportional to d^2, where d is the equivalent diameter of the particle, this force is likely to be at least two orders of magnitude smaller for the silt than for the Ham River sand. Furthermore, about 7 % of the particles are smaller than 2 µm and the material has a significant plasticity index (6 %). It might be expected to indicate the beginning of any trend in behaviour as the grading moves towards that of materials classified as clays.

Table 2

mineral composition of Ham River sand		mineral composition of Braehead silt	
quartz	96.22 %	quartz	95.20 %
limonite	3.66 %	illite	2.80 %
zircon	0.10 %	kaolinite	1.60 %
staurolite	0.02 %	chlorite	0.40 %

analysis by Midgley (Building Research Station)

The time scale of fully drained tests on clay is so very much longer (involving weeks rather than hours) that a different pressure control and data logging system is required. Clays were therefore not included in the present programme.

5. Test results and their implications

The most extensive series of tests was run on the Ham River sand, involving substantial variations in the value of $(\sigma - u)$ as well as in the coupled values of σ and u. Individual tests were then run on each of the selected materials.

(a) *Ham River sand*

Two tests run in the prototype apparatus with $(\sigma_3 - u)$ equal to 69.0 kN/m² (10.0 lbf/in²) and a variation of σ_3 between 345 kN/m² (50 lbf/in²) and 6.9 MN/m² (1000 lbf/in²) are illustrated in figures 12 and 13. It is immediately apparent that for a change in the value of u of 6.55 MN/m² (950 lbf/in²) there is no discernible change in shearing resistance resulting from a change in pore pressure with $(\sigma_3 - u)$ held constant.

The results of three tests run in the high pressure apparatus with $(\sigma_3 - u)$ equal to 363 kN/m² (52.6 lbf/in²) and a variation of σ_3 between 1.03 MN/m² (150 lbf/in²) and 27.58 MN/m² (4000 lbf/in²) are shown in figures 14, 15 and 16. Substantially the same conclusion can be drawn as before, although slight irregularities are observed while the changes in σ_3 and u are actually being made. This is attributed to small pressure gradients resulting from the flow of the relatively compressible pore fluid under the large pressure changes, in particular in the small bore high pressure lines connecting the pressure vessels maintaining value of u at a constant difference from σ_3.

These three tests are of particular interest since the value of $(\sigma_3 - u)$ is typical of the range encountered in engineering practice and the change in u of 26.5 MN/m² (3850 lbf/in²) exceeds this value by a factor of 73 times. Had the value of the contact area a equalled the maximum value envisaged for sand by Bishop & Eldin (1950), its value on the relevant shear surface in the present tests would have been *ca.* 0.4 %. The value of the term Δr (equal to $a\Delta u/(\sigma_n - u)$)† would have been of the order 0.004×46.8 i.e. 0.19. Both of the theories involving an au term would thus have predicted substantial discontinuities in the stress–strain curve when the value of u was varied with $\sigma_3 - u$ constant.

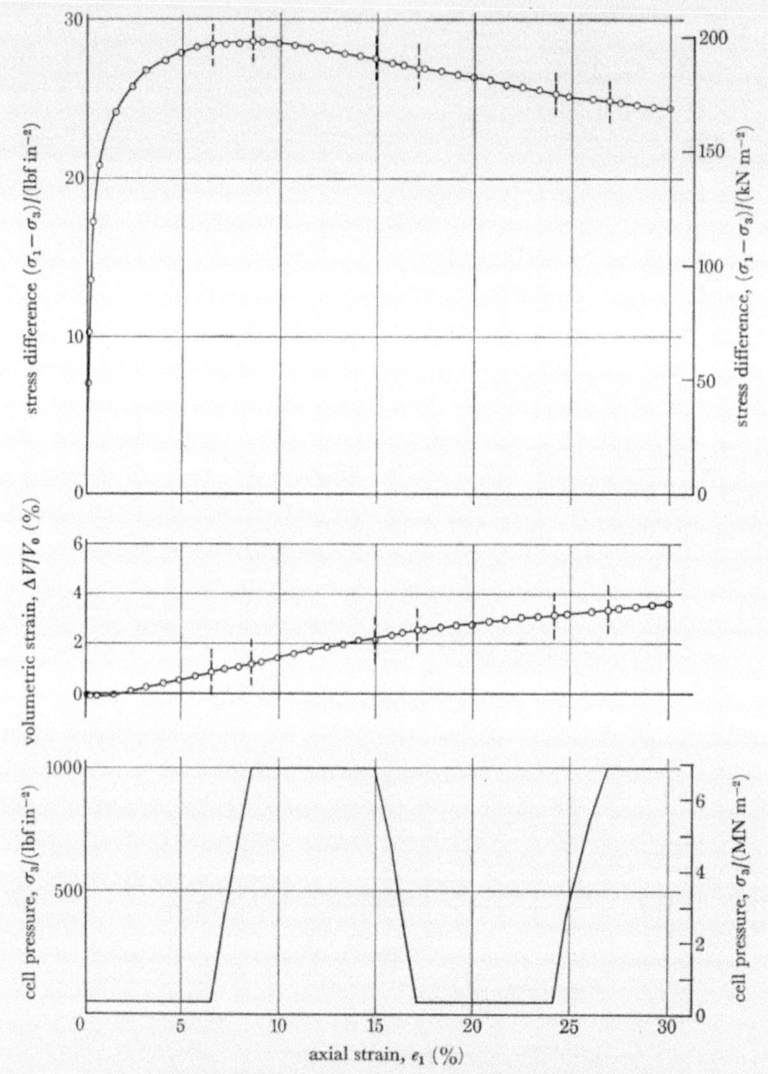

FIGURE 12. Ham river sand tested with $(\sigma_3 - u) = 69.0$ kN m^{-2} (10 lbf in^{-2}) (sample porosity at the start of the shear stage = 43.4 %).

† In any state of stress other than that of isotropic stress the value of a in a particulate material and the value of the ratio $\Delta u/(\sigma - u)$ will have directional properties. Skempton (1960) has presented his analysis in terms of the Mohr–Coulomb failure criterion, in which the normal effective stress on the shear surface (*ca.* $(\sigma_n - u)$) is $(1 + \sin\phi')$ times the minor effective stress (*ca.* $(\sigma_3 - u)$). This point is discussed in more detail in subsequent paragraphs.

It is therefore pertinent to re-examine the possible range of values of the contact area a, in particular in the light of the micro-indentation hardness data published, for example, by Brace (1963).

Bishop & Eldin (1950) related σ_p, the value of the average interparticle contact pressure, to S, the value of the crushing strength of the grains, by the expression

$$\sigma_p - u = bS, \tag{10}$$

where the constant b will depend on the type of surface failure produced, but with a minimum value probably not much lower than unity.

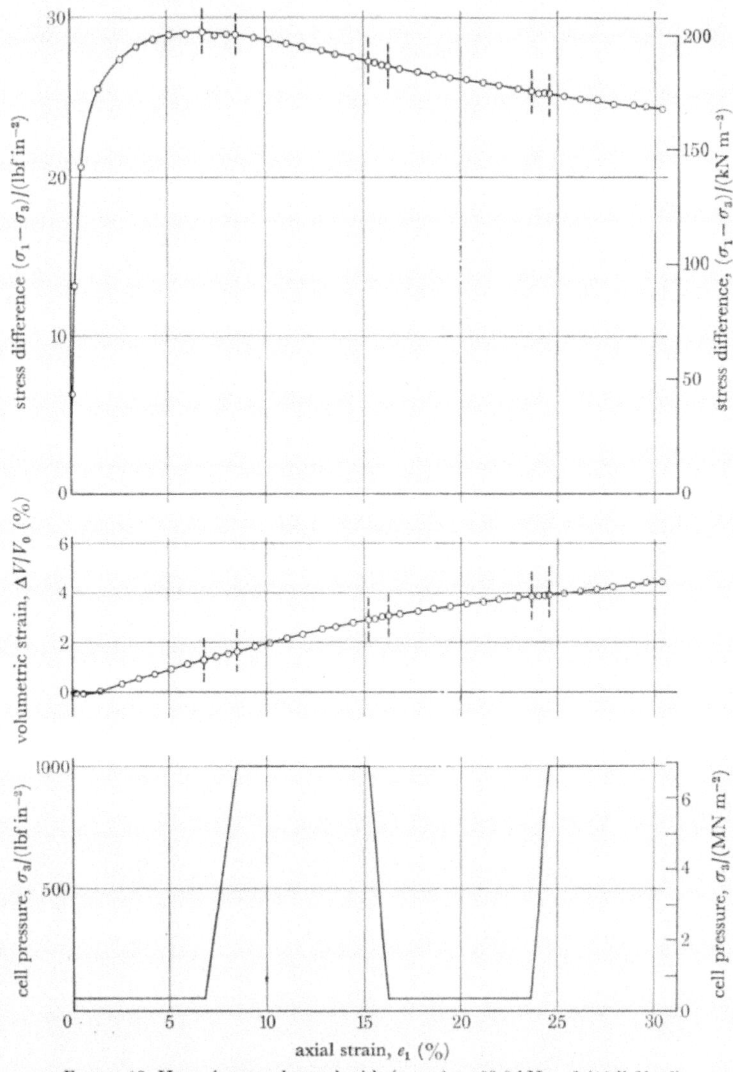

FIGURE 13. Ham river sand tested with $(\sigma_3 - u) = 69.0$ kN m^{-2} (10 lbf in^{-2}) (sample porosity at the start of the shear stage = 40.9%).

14-2

The intergranular stress σ_i (defined as the intergranular force per unit area) was related to σ_p by the expression

$$\sigma_i = a\sigma_p. \tag{11}$$

Since σ_i is related to σ and u by equation (2):

$$\sigma_i = (\sigma - u) + au,$$

we have, from equations (10) and (11),

$$(\sigma - u) + au = abS + au$$

or

$$a = \frac{\sigma - u}{bS}. \tag{12}$$

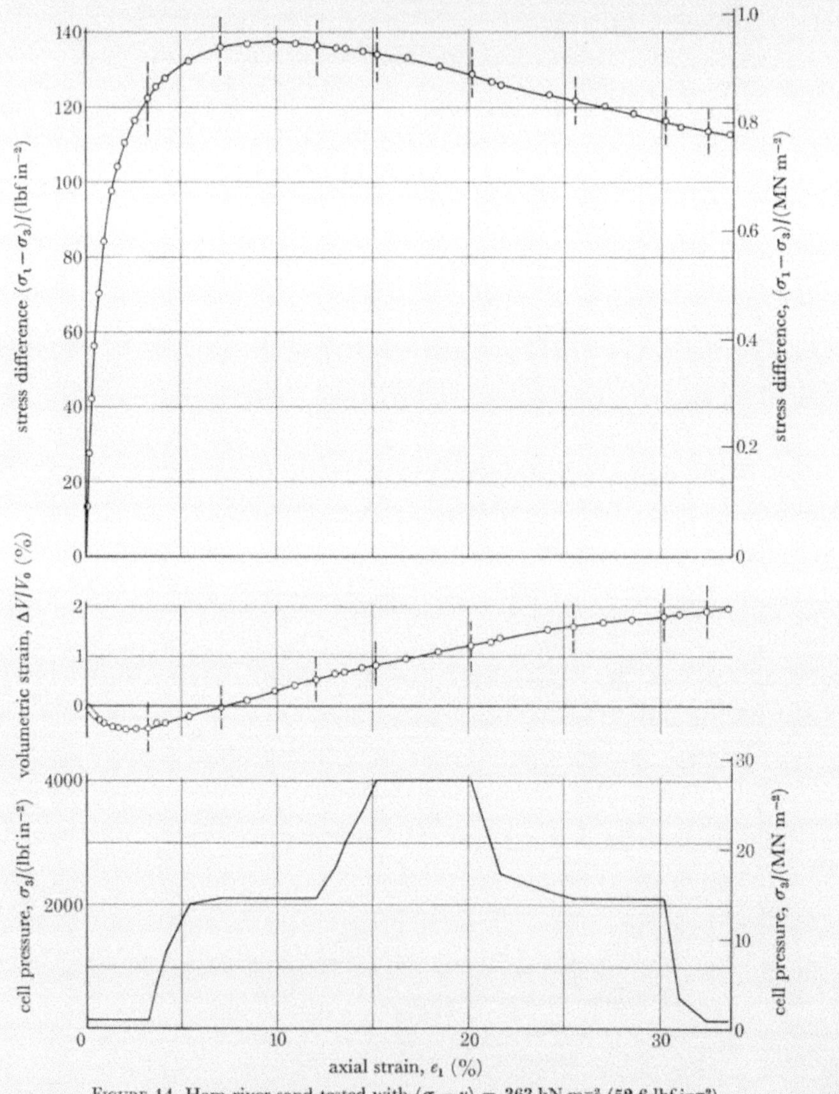

FIGURE 14. Ham river sand tested with $(\sigma_3 - u) = 363$ kN m^{-2} (52.6 lbf in^{-2}).

Skempton (1960) gives an expression for the general case of a cohesive particulate material with an initial contact area a_0 at zero stress, with the assumption that junction growth during shear is negligible:

$$a - a_0 = \frac{\sigma - u}{Mk}, \tag{13}$$

where k is the intrinsic cohesion defined in equation (4) and

$$M = \frac{2\cos\psi}{1-\sin\psi}.$$

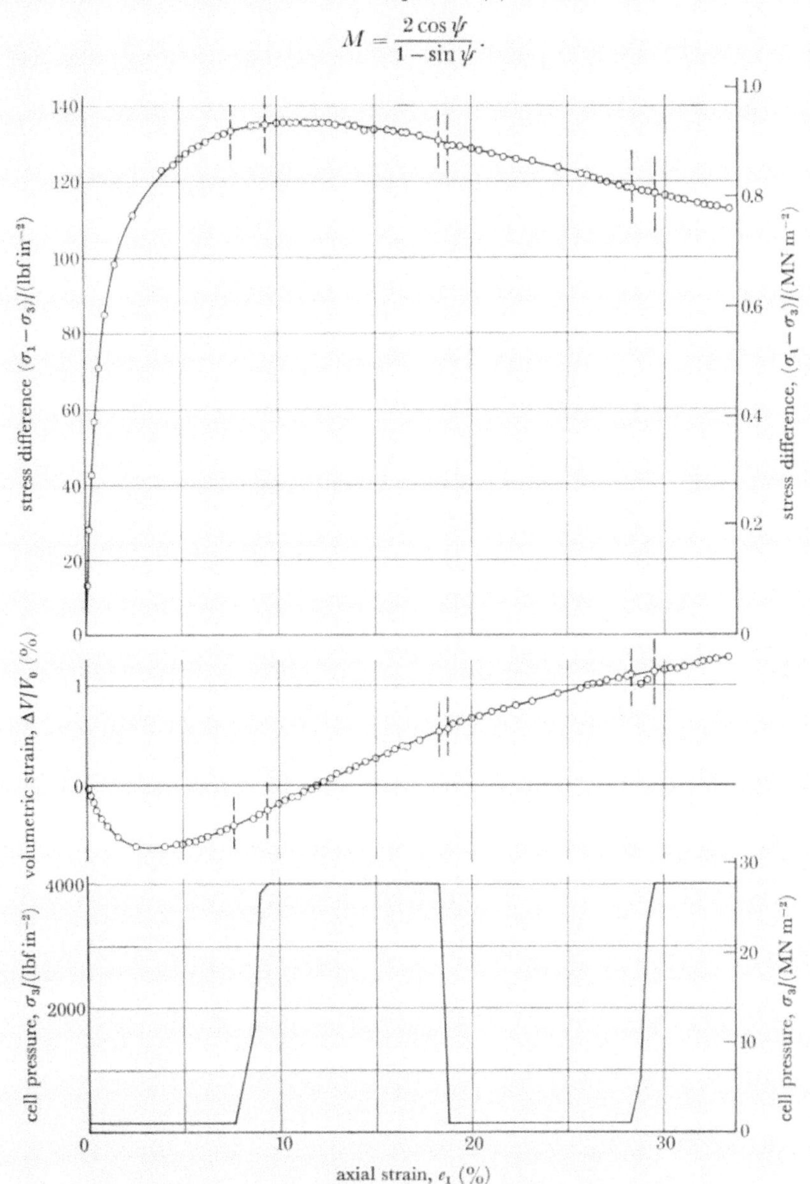

FIGURE 15. Ham river sand tested with $(\sigma_3 - u)$ = 363 kN m^{-2} (52.6 lbf in^{-2}).

Since it follows from the geometry of the Mohr diagram (figure 17) that,

$$\sigma_1 - \sigma_3 = k\frac{2\cos\psi}{1-\sin\psi} + \sigma_3\frac{2\sin\psi}{1-\sin\psi}, \tag{14}$$

then S, which is equal to $\sigma_1 - \sigma_3$ when σ_3 is zero, is given by the expression

$$S = k\frac{2\cos\psi}{1-\sin\psi}, \tag{15}$$

i.e.

$$S = Mk. \tag{16}$$

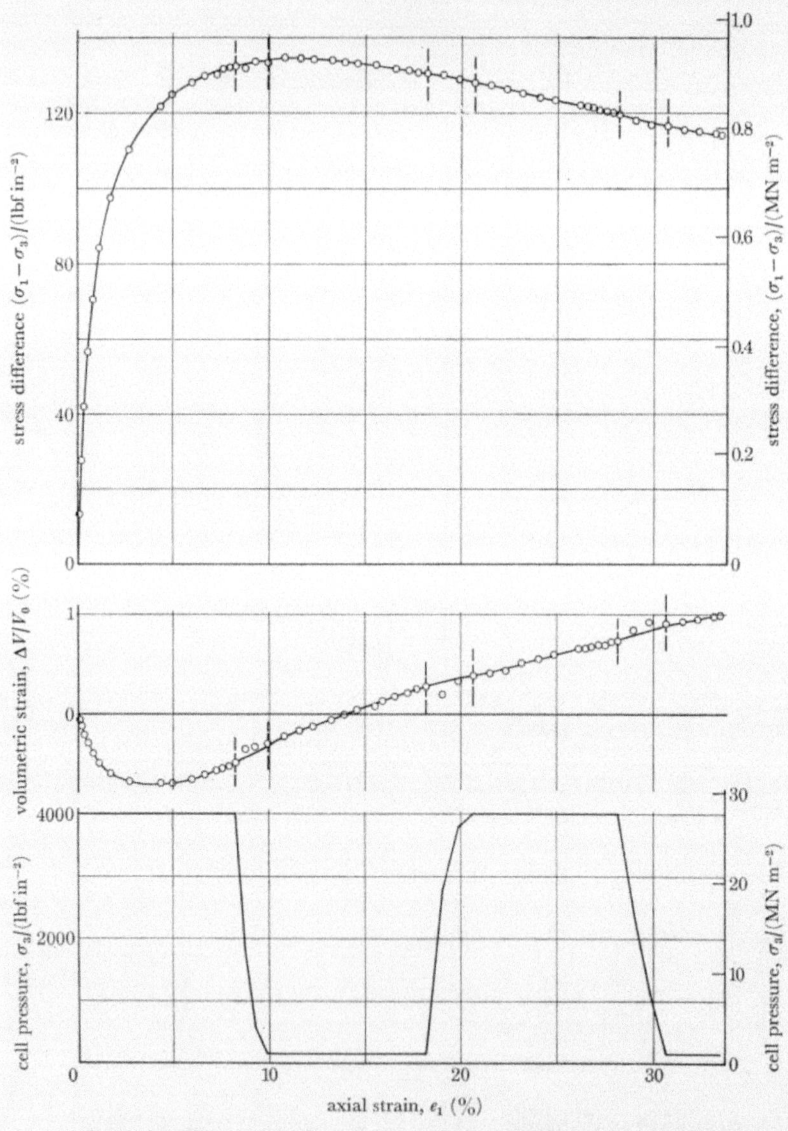

FIGURE 16. Ham river sand tested with $(\sigma_3 - u) = 363$ kN m^{-2} (52.6 lbf in^{-2}).

For a sand recently sedimented in water (as in the present case) a_0 may be taken as zero. Equations (12) and (13) then give the same value of contact area a if the coefficient b in equation (12) is assumed to be equal to unity.†

Taking the values of k and ψ for intact quartz from Skempton (1960), 932 MN/m^2 (9500 kgf/cm^2) and $13\frac{1}{4}°$ respectively, we have from equation (15) a value of S equal to 2255 MN/m^2. This is more than an order of magnitude greater than the value of 138 MN/m^2 used by Bishop & Eldin (1950) on the basis of the strength of granites and quartzites under low hydrostatic pressures. The calculated contact area on the shear surface corresponding to the state of stress in the tests run with $(\sigma_3 - u)$ = 363 kN/m^2 drops from 0.4 % to *ca.* 0.025 %.

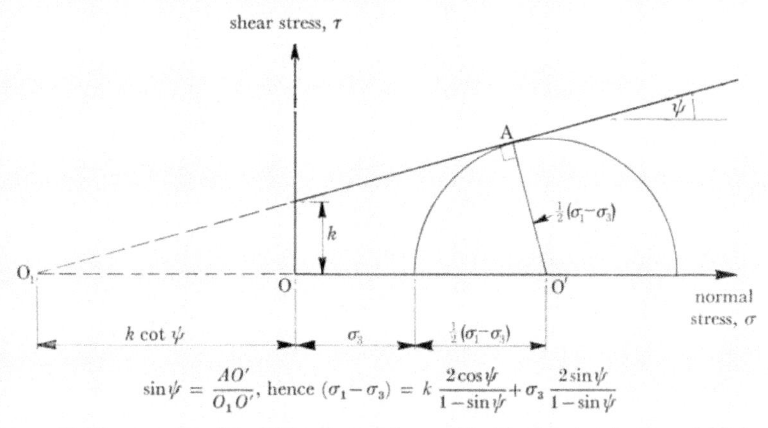

$$\sin\psi = \frac{AO'}{O_1O'}, \text{ hence } (\sigma_1 - \sigma_3) = k\,\frac{2\cos\psi}{1-\sin\psi} + \sigma_3\,\frac{2\sin\psi}{1-\sin\psi}$$

FIGURE 17. The geometry of the Mohr diagram for a solid with intrinsic cohesion and friction.

Even with this lower value of a the value of $a\Delta u/(\sigma_n - u)$ is 0.012. This would imply discontinuities of 1.2 % in the resistance to shear as the value of u was varied between the upper and lower limits if the intergranular stress controlled the shear strength. These would have been clearly discernible. In the case of Skempton's effective stress equation the factor $\tan\psi/\tan\phi'$ reduces the implied percentage discontinuity to 0.4 % (for $\psi = 13\frac{1}{4}°$ and $\phi' = 34.2°$ as in the test illustrated in figure 15.)This value is at the limit of discrimination of the test. On this basis, and for pore pressure changes of this magnitude, it is apparent that the predictions of the Terzaghi effective stress equation and Skempton's effective stress equation do not differ significantly from each other or from the observed behaviour of quartz sand in a multistage test.

However, the prediction of the value of a, which forms the basis both of the intergranular stress equation and of Skempton's equation, has so far been made without attempting to evaluate the influence either of junction growth or of the actual magnitude of the interparticle contact forces. Before examining these two factors it is necessary to look more closely at the conditions under which the Mohr-Coulomb failure criterion can be used to relate the shear strength τ_f on a plane on which the normal effective stress is σ_n' (to which Skempton's analysis applies) to the principal stresses in an axially symmetrical compression test (as used in the present experimental program).

† There is evidence that b may rise to about 3 for metals (Bowden & Tabor 1950). The predicted contact area would be correspondingly reduced, a point which is discussed further in relation to the tests on lead shot where there is direct evidence of the magnitude of the area of contact.

In the absence of pore pressure the effective stresses equal the total stresses and the Mohr–Coulomb criterion is given by the expression

$$\tau_f = c + \sigma \tan \phi. \tag{17}$$

The theory can be applied rigorously only if the parameters c and ϕ are invariant with respect to changes in the inclination θ of the reference plane (figure 18), changes in the stresses on this plane being given by the expressions:

$$\tau = \tfrac{1}{2}(\sigma_1 - \sigma_3) \sin 2\theta, \tag{18}$$

$$\sigma_n = \tfrac{1}{2}(\sigma_1 + \sigma_3) + \tfrac{1}{2}(\sigma_1 - \sigma_3) \cos 2\theta, \tag{19}$$

where σ_1 and σ_3 denote the major and minor principal stresses.

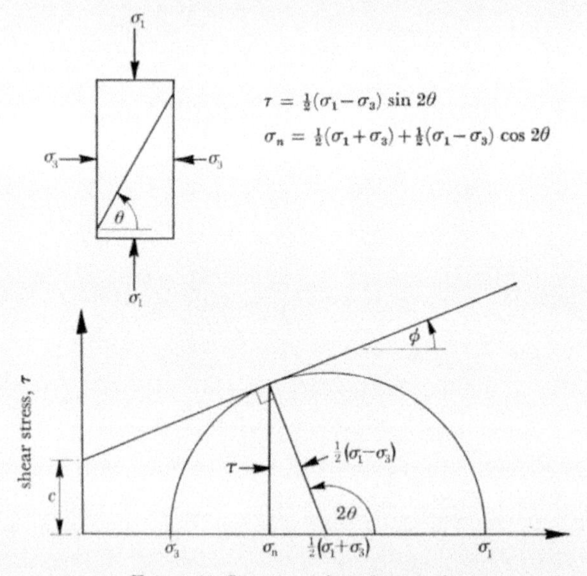

FIGURE 18. Stresses on the reference plane.

It follows from the conventional analysis for isotropic materials that the shear surface is inclined at an angle θ, where $\theta = 45° + \frac{1}{2}\phi'$, and that the principal stress difference is related to the minor principal stress by the expression

$$(\sigma_1 - \sigma_3)_f = c\frac{2\cos\phi}{1-\sin\phi} + \sigma_3\frac{2\sin\phi}{1-\sin\phi}. \tag{20}$$

This application of the Mohr–Coulomb failure criterion to effective stresses presents no difficulty if the Terzaghi expression $\sigma' = \sigma - u$ is used, since u is invariant with respect to changes in the inclination θ of the reference plane. However, in terms of intergranular stress, we have the expression

$$\tau_f = c' + \{(\sigma - u) + au\} \tan \phi'. \tag{21}$$

Skempton's expression for effective stress gives

$$\tau_f = c' + \left\{(\sigma - u) + au\frac{\tan\psi}{\tan\phi'}\right\} \tan \phi'. \tag{22}$$

In both cases a is *not* invariant with respect to changes in θ. The expression for a (without junction growth) given in equation (13) is

$$a = a_0 + \frac{\sigma - u}{Mk}.$$

Substituting in equation (22) we have

$$\tau_f = c' + \left\{(\sigma - u) + u\left(a_0 + \frac{\sigma - u}{Mk}\right)\frac{\tan\psi}{\tan\phi'}\right\}\tan\phi'.$$

Assuming a_0 to be an isotropic property, and collecting the terms which vary with σ we obtain:

$$\tau_f = c' + ua_0\frac{\tan\psi}{\tan\phi'}\tan\phi' + (\sigma - u)\left\{1 + \frac{u}{Mk}\frac{\tan\psi}{\tan\phi'}\right\}\tan\phi'. \tag{23}$$

In terms of the Mohr–Coulomb failure criterion we now have a new material in which the parameters c' and ϕ' are replaced by $(c')_u$ and $(\phi')_u$ where

$$(c')_u = c' + u\left(a_0\frac{\tan\psi}{\tan\phi'}\right)\tan\phi', \tag{24}$$

and

$$(\phi')_u = \arctan\left\{\left(1 + \frac{u}{Mk}\frac{\tan\psi}{\tan\phi'}\right)\tan\phi'\right\}. \tag{25}$$

The term $(\sigma - u)$ then correctly reflects the variation in normal stress with change in the value of θ, since u is invariant for each particular stage of a multistage test.

We may therefore replace c and ϕ in equation (20) by the expressions given in equations (24) and (25) and replace σ_3 by $(\sigma_3 - u)$. For small differences between the parameters $(c')_u$ and c', and $(\phi')_u$ and ϕ', we may obtain the change in the stress difference at failure $(\sigma_1 - \sigma_3)_f$ with a change in u by differentiating this expression with respect to u†, $(\sigma_3 - u)$ being constant as in the tests. This leads to the expression

$$\frac{d(\sigma_1 - \sigma_3)_f}{du} = a_0\frac{\tan\psi}{\tan\phi'}\frac{2\sin\phi'}{(1 - \sin\phi')} + \frac{\tan\psi}{\tan\phi'}\frac{2\sin\phi'}{(1 - \sin\phi')}\left\{\frac{1}{Mk}(c'\cos\phi' + (\sigma_3 - u)(1 + \sin\phi'))\right\}. \tag{26}$$

Now it follows from the geometry of the Mohr diagram that the component of $(\sigma - u)$ normal to the shear surface is given by the expression

$$(\sigma_n - u) = c'\cos\phi' + (\sigma_3 - u)(1 + \sin\phi'). \tag{27}$$

The component of area of contact which is a function of normal stress is given by the equation

$$a_f = \frac{(\sigma_n - u)}{Mk}. \tag{28}$$

Thus the expression for the change in strength with change in pore pressure, $(\sigma_3 - u)$ being held constant, reduces to

$$\frac{d(\sigma_1 - \sigma_3)_f}{du} = a_\tau\frac{\tan\psi}{\tan\phi'}\frac{2\sin\phi'}{(1 - \sin\phi')}, \tag{29}$$

where a_τ is the total contact area on the shear surface and $a_\tau = a_0 + a_f$.

This expression is the same as that arrived at by rather different reasoning by Skempton (1960). The validity of adding components of contact area may be questioned, especially in a complex

† Provided $\sin\phi'$ differs substantially from unity, a condition which is satisfied in all particulate materials.

material such as concrete, but this is not relevant to a discussion of cohesionless particulate materials in which $a_0 = 0$ and $c' = 0$. However, two other points which have to be considered are the influence of junction growth and the effect of the magnitude of the interparticle forces.

It should be noted that if the contact area ceases to be a linear function of the normal stress component $(\sigma - u)$ and varies with the magnitude of the shear stress τ, which is zero on the major and minor principal planes and a maximum when $\theta = 45°$, then the analysis presented in the preceding paragraphs is no longer strictly applicable. However, the quantification of the phenomenon of junction growth is itself subject to some uncertainty (see, for example, Bowden & Tabor 1954, 1964). It is clear that the stress path of the junction material would be of the form indicated in figure 19, and, with clean surfaces, could in the limit reach the intrinsic failure envelope of the material. Some bounds can be put to the probable values of interparticle contact pressure under normal pressure alone (σ_p) and at slip (σ_{nj}) from experimental observations of the coefficient of friction μ and the results of the micro-indentation hardness test.

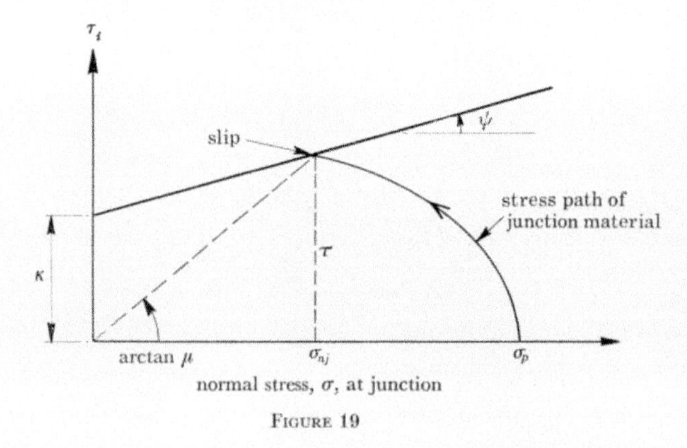

FIGURE 19

It follows from figure 19 that the lower limit of the stress σ_{nj} corresponds to the upper limit of the observed coefficient of friction. In a series of test conducted so as to ensure a clean contact between a slider and plate a value of $\mu = 0.8$ was obtained for quartz by Skinner (1975). Taking values of $\kappa = 941$ MN/m^2 and $\psi = 16\frac{1}{4}°$ for quartz (based on a re-assessment of published data), we obtain from the geometry of figure 19 a lower limit for σ_{nj} of 1851 MN/m^2.

As a simplification we may use the expression:

$$a = \frac{(\sigma_n - u)}{\sigma_{nj}}. \tag{30}$$

This gives the maximum value of the contact area a in the test run with $(\sigma_3 - u) = 363$ kN/m^2 as

$$a = \frac{567 \times 10^{-3}}{1851}$$
$$= 3.06 \times 10^{-4},$$

the value of $(\sigma_n - u)$ being 567 kN/m^2.

The corresponding minimum values of σ_p implicit in equation (12) (with $b = 1$) and in equation (13) are both 2510 MN/m^2. However, depending on the geometry of the interparticle contact and on the yield pattern, the value of b may be as high as 3, leading to a value of

$\sigma_p = 7530\,\mathrm{MN/m^2}$. Without junction growth the value of a, from equation (12), would lie in the range 2.26×10^{-4} to 5.53×10^{-5}.

These values represent the largest probable values of a. The micro-indentation hardness tests on quartz published by Brace (1963), and illustrated in figure 20, suggest that at small values of the interparticle contact force the value of σ_p required to cause yield may be substantially greater than at the larger loads used on a typical slider. Based on an analysis of the stress path the corresponding value of the coefficient of friction is lower and is associated with a higher value of σ_{nl}. For an interparticle force of 2 gf (the minimum used by Brace (1963)) the value of μ drops to 0.4 and the value of σ_{nj} rises to $5688\,\mathrm{MN/m^2}$ (Skinner 1975). For the stress level under consideration the value of a (from equation (30)) drops to a value of 6.53×10^{-5}.

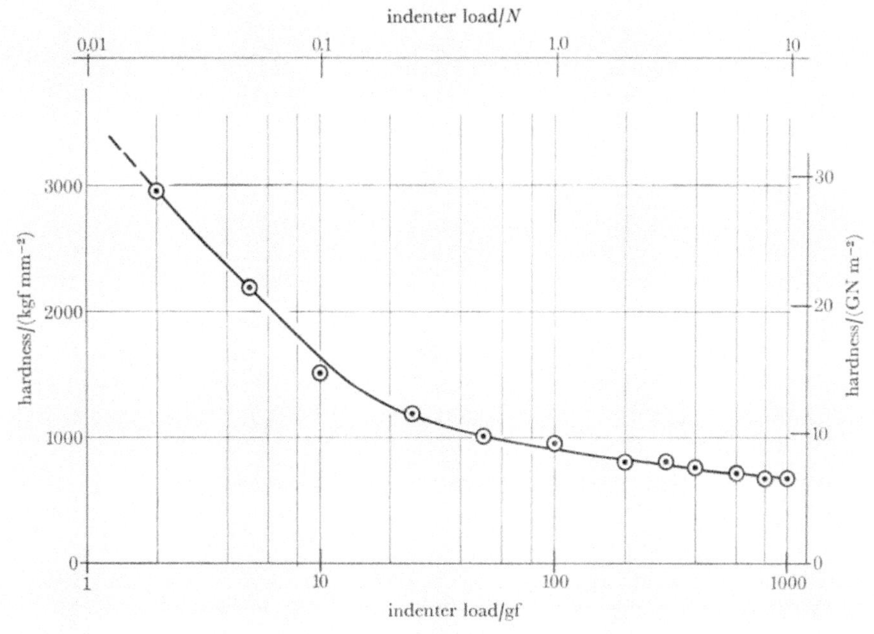

FIGURE 20. Hardness against load applied to indenter: data from Brace (1963).

It is of interest to note that for a regular cubical packing of uniform spherical particles the interparticle force P is given by the expression (Bishop 1965):

$$P = \sigma d^2. \tag{31}$$

An equivalent uniform material having the mean particle size of Ham River sand (0.2 mm diameter) and subjected to the present stress ($567\,\mathrm{kN/m^2}$) would thus have an average interparticle force P equal to 2.3 gf. This force almost certainly represents the upper limit for particles of this mean size and stress level. It may thus be inferred that the value of the contact area a tends towards the lower limit in the present tests.

The corresponding values of $\mathrm{d}(\sigma_1 - \sigma_3)_f/\mathrm{d}u$ and of the percentage change in compression strength are given in table 3, on the assumption that the error in applying equation (29) to the cases involving junction growth is acceptable in view of the other uncertainties involved. As convincing evidence of a discontinuity in the stress–strain curve requires a jump of about 0.5 % in strength it will be seen that the interpretation to be placed on the absence of jumps depends on the significance attached to the influence of particle contact force. For a pore pressure change of this magnitude (26.5 MN/m²) the use of the Terzaghi equation is fully justified. The evidence against the intergranular stress equation is just significant, but the test is not definitive in terms of Skempton's effective stress equation.

TABLE 3. PREDICTED CHANGES IN STRENGTH FOR A CHANGE IN PORE PRESSURE $\Delta u = 26.5$ MN/m², THE VALUE OF $(\sigma_3 - u)$ BEING CONSTANT AND EQUAL TO 0.363 MN/m²

		intergranular stress equation (2)		Skempton's effective stress equation (5)		Terzaghi effective stress equation (1)	
assumption about contact area a	value of a	$\frac{\Delta(\sigma_1-\sigma_3)_f}{\text{kN/m}^2}$	$\frac{\Delta(\sigma_1-\sigma_3)_f}{(\sigma_1-\sigma_3)_f}$ (%)	$\frac{\Delta(\sigma_1-\sigma_3)_f}{\text{kN/m}^2}$	$\frac{\Delta(\sigma_1-\sigma_3)_f}{(\sigma_1-\sigma_3)_f}$ (%)	$\frac{\Delta(\sigma_1-\sigma_3)_f}{\text{kN/m}^2}$	$\frac{\Delta(\sigma_1-\sigma_3)_f}{(\sigma_1-\sigma_3)}$ (%)
without junction growth, and with $b = 1$	2.26×10^{-4}	15.4	1.65	6.61	0.71	0	0
without junction growth, and with $b = 3$	5.53×10^{-5}	3.78	0.41	1.62	0.18	0	0
with junction growth, max. probable value	3.06×10^{-4}	20.9	2.24	8.97	0.96	0	0
with junction growth, value corresponding to load of 2 gf per contact	6.53×10^{-5}	4.46	0.48	1.91	0.21	0	0

A test with a greatly increased value of $\sigma_3 - u$ is the most obvious way of increasing the interparticle contact forces and thus reducing the yield stress at the contacts. Such a test is illustrated in figure 21, the breaks in the curves at the pressure change zones indicating the difficulty of maintaining $(\sigma_3 - u)$ constant while adjusting both σ_3 and u continuously once the range of the self-compensating mercury control was exceeded. Here the value of $(\sigma_3 - u)$ is 6.91 MN/m² (1002.6 lbf/in²) and $\Delta u = 41.4$ MN/m² (6000 lbf/in²).

If we consider the point on the stress–strain curve at which the axial strain is 7.52 %, and the stress difference $(\sigma_1 - \sigma_3) = 12.8$ MN/m² (1852 lbf/in²), then the mobilized value of $\phi' = 28.7°$. Proceeding as before, and considering only the cases with junction growth, we obtain the values given in table 4. Since the value of $(\sigma_n - u)$ on the failure surface is now 10.2 MN/m² and is thus 18 times larger than in the test discussed above, the interparticle contact force might be expected to correspond to a contact area mid-way between the limiting values. The absence of a step in the stress–strain curve would then constitute strong evidence against the validity of the intergranular stress equation (2), and significant evidence against Skempton's effective stress equation (5).

However, samples of Ham River sand sheared at this stress level (i.e. 6.9 MN/m²) show a substantial degree of particle crushing as the shear strain increases. The change in particle size distribution observed in an earlier series of tests and described by Bishop (1966) is illustrated in figure 22. While it is difficult to quantify the effect of particle fracture on the average interparticle force, it will clearly involve a substantial reduction. The prediction of the influence of pore pressure change using Skempton's equation thus returns to the range of values where discrimination is difficult, unless the magnitude of the pore pressure change is increased to a value well beyond the range of the present equipment.

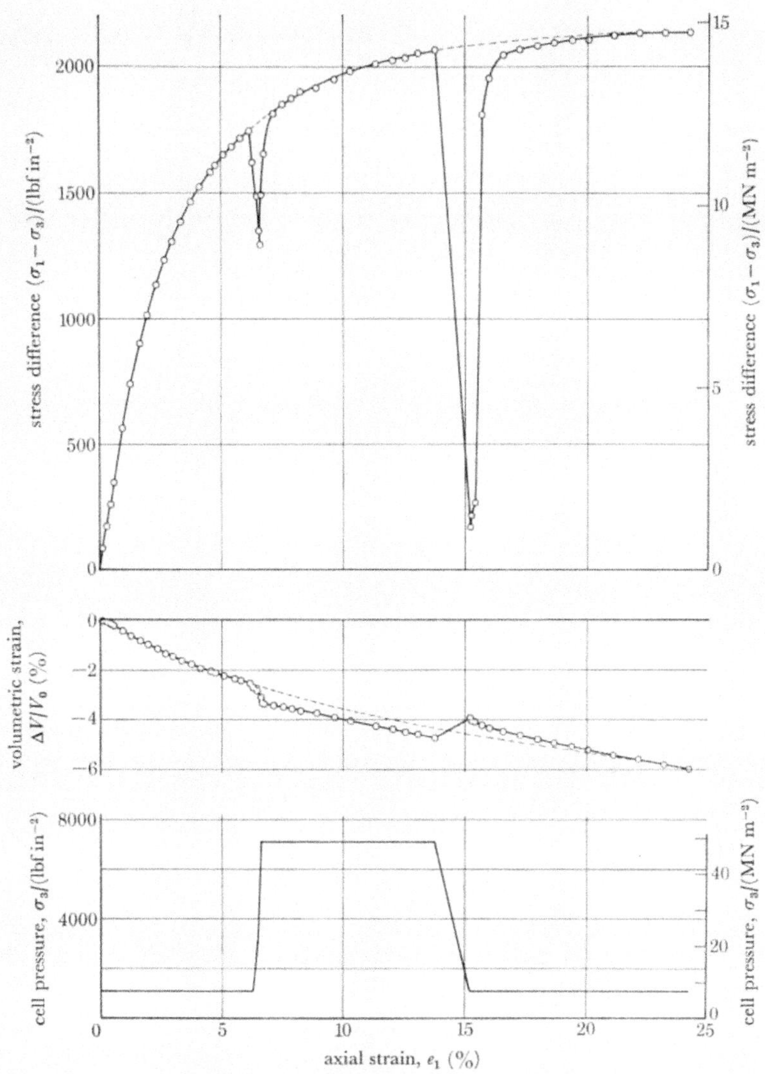

FIGURE 21. Ham River sand tested with $(\sigma_3 - u) = 6.91$ MN m^{-2} (1002.6 lbf in^{-2}).

TABLE 4. PREDICTED CHANGES IN STRENGTH FOR A CHANGE IN PORE PRESSURE $\Delta u = 41.4$ MN/m^2, THE VALUE OF $(\sigma_3 - u)$ BEING CONSTANT AND EQUAL TO 6.91 MN/m^2; HAM RIVER SAND

		intergranular stress equation (2)		Skempton's effective stress equation (5)		Terzaghi effective stress equation (1)	
assumption about contact area a	value of a	$\dfrac{\Delta(\sigma_1-\sigma_3)_f}{\text{kN/m}^2}$	$\dfrac{\Delta(\sigma_1-\sigma_3)_f}{(\sigma_1-\sigma_3)_f}$ (%)	$\dfrac{\Delta(\sigma_1-\sigma_3)}{\text{kN/m}^2}$	$\dfrac{\Delta(\sigma_1-\sigma_3)_f}{(\sigma_1-\sigma_3)_f}$ (%)	$\dfrac{\Delta(\sigma_1-\sigma_3)}{\text{kN/m}^2}$	$\dfrac{\Delta(\sigma_1-\sigma_3)_f}{(\sigma_1-\sigma_3)_f}$ (%)
with junction growth, max. probable value	5.53×10^{-3}	423	3.31	225	1.76	0	0
with junction growth, value corresponding to load of 2 gf per contact	1.18×10^{-3}	90	0.71	48	0.38	0	0

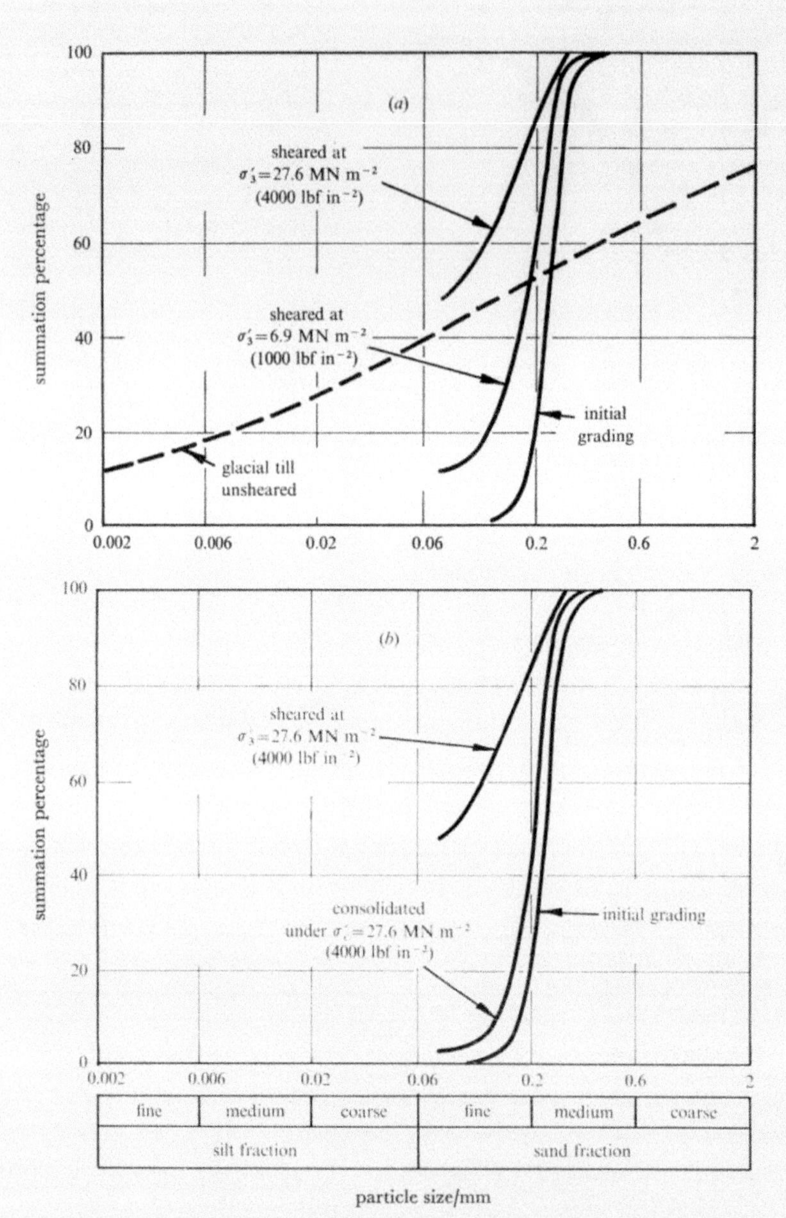

FIGURE 22. Changes in grading of saturated samples of Ham River sand resulting (*a*) from shear at different pressures and (*b*) from consolidation as compared with complete shear tests (tests by Skinner, 1964–66).

(b) Lead shot

Owing to the low yield stress of lead, this material offered the prospect of definitive results at relatively modest stress levels. The results of a test run in the high pressure apparatus with $(\sigma_3 - u)$ equal to 363 kN/m² (52.6 lbf/in²) and a variation of σ_3 between 1.03 MN/m² (150 lbf/in²) and 27.58 MN/m² (4000 lbf/in²) are shown in figure 23. Once again, there is no discernible change in shearing resistance resulting from a change in pore pressure of 26.5 MN/m² (3850 lbf/in²) with $(\sigma_3 - u)$ held constant.

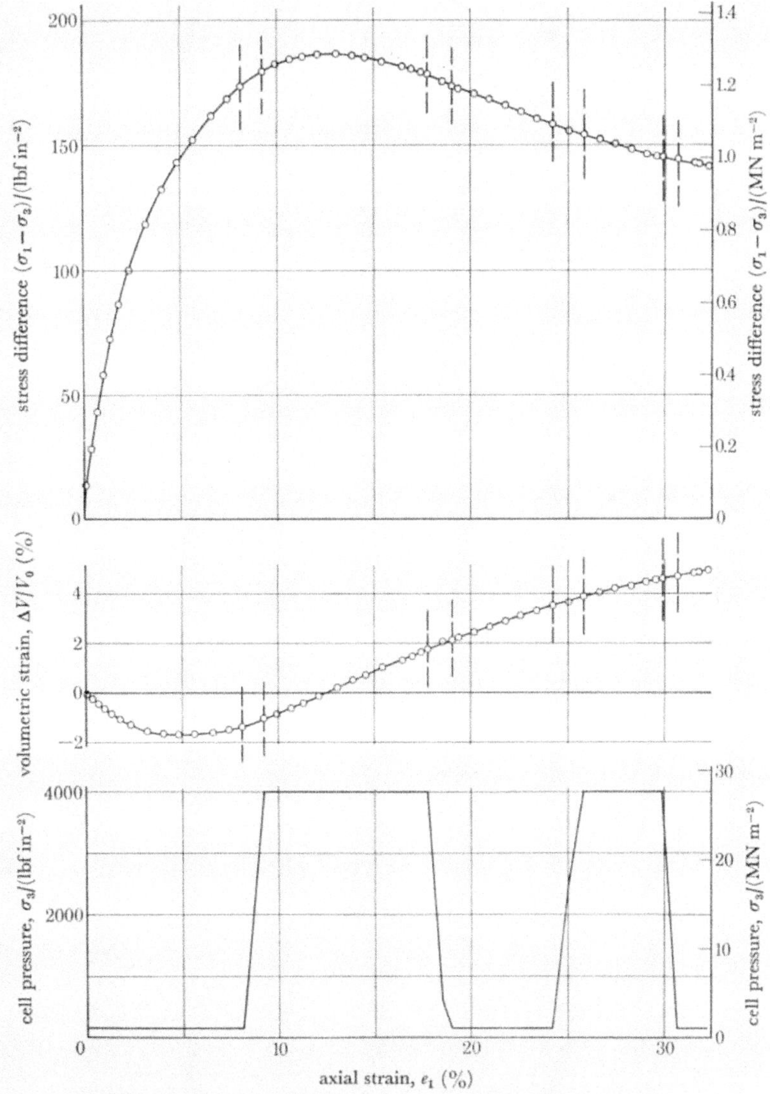

FIGURE 23. Lead shot tested with $(\sigma_3 - u) = 363$ kN m⁻² (52.6 lbf in⁻²).

The area of contact may be derived by three independent methods. Skempton (1960) gives the values of the intrinsic parameters k and ψ for lead as $9.81\ \mathrm{MN/m^2}$ ($100\ \mathrm{kgf/cm^2}$) and $\frac{3}{4}°$ respectively. Taking the value of $(\sigma_n - u)$ at an axial strain of 9.24 % as $591\ \mathrm{kN/m^2}$ we obtain a value of contact area a of 2.97 % using Skempton's expression (equation (13)). For $b = 1$ the same value is obtained from equation (12), but with $b = 3$, which seems more appropriate in the light of values obtained on other metals by Bowden & Tabor (1954), the value of a drops to 0.991 %.

Micro-indentation hardness tests on representative lead shot from this series of tests gave a mean normal stress H_v of $93.1\ \mathrm{MN/m^2}$ ($950\ \mathrm{kgf/cm^2}$). On the assumption that $H_v = 3, S = 6\ \mathrm{Mk}$, this value corresponds to $k = 15.3\ \mathrm{MN/m^2}$ and suggests a slightly stronger material than the lead quoted by Skempton. Taking the contact pressure as approximating to the indentation hardness value we obtain a value of $a = 0.635\ \%$. The low value of coefficient of friction (0.1) measured in direct observations of interparticle friction (Skinner 1975) suggests that friction is controlled by surface films on the lead and that junction growth is not very significant in this case.

TABLE 5. VALUES OF CONTACT AREA a FROM OEDOMETER TESTS ON LEAD SHOT†

vertical pressure p in oedometer		value of contact area a on release of stress
$\mathrm{MN/m^2}$	$\mathrm{kgf/cm^2}$	
2.65	27	0.03
12.55	128	0.11
100.42	1024	0.95

† Data from Laughton (1955).

Thirdly, Laughton (1955) measured with a microscope the facets caused by subjecting lead shot to a series of normal stresses in an oedometer. Since these facets represent the inelastic component of deformation, they indicate the minimum values of a. Laughton's observations, which are also quoted by Skempton (1960), are presented in table 5. It will be seen that up to a contact area of 11 % the relation between a and the vertical pressure p is almost linear and is given by the expression

$$a = 0.0120p(1 - 0.0216p), \tag{32}$$

where p is expressed in $\mathrm{MN/m^2}$.

In the oedometer the vertical pressure p is applied through a piston and the sample is confined in a rigid-walled cylinder. The radial stress, neglecting side friction, is equal to $K_0 p$, where K_0 is the coefficient of earth pressure at rest and approximates numerically to $(1 - \sin\phi')$ on first loading (Jaky 1948; Bishop 1958). Laughton did not examine the variation of contact area with the direction of the plane of reference and the average contact area measured will therefore correspond to a normal stress of $\frac{1}{3}(1 + 2K_0)\,p$.

In the present test the peak value of $\phi' = 39.7°$ and the equivalent normal stress σ is related to p by the expression

$$\sigma = 0.573p. \tag{33}$$

For the value of $(\sigma_n - u) = 591\ \mathrm{kN/m^2}$ we thus obtain $a = 1.21\ \%$. It is of interest to note the relatively close agreement of this direct experimental value with the values deduced from equation (12) with $b = 3$ and from the micro-indentation hardness tests, having regard to the fact that they refer to different samples of lead.

The magnitudes of the predicted changes in strength for this particular test are given in table 6. It will be seen that the intergranular stress equation predicts an increase in strength, for the

TABLE 6. PREDICTED CHANGES IN STRENGTH FOR A CHANGE IN PORE PRESSURE $\Delta u = 26.5$ MN/m², THE VALUE OF $(\sigma_3 - u)$ BEING CONSTANT AND EQUAL TO 0.363 MN/m²; LEAD SHOT

assumption about contact area a	value of a	intergranular stress equation (2)		Skempton's effective stress equation (5)		Terzaghi effective stress equation (1)	
		$\frac{\Delta(\sigma_1-\sigma_3)_f}{\text{kN/m}^2}$	$\frac{\Delta(\sigma_1-\sigma_3)_f}{(\sigma_1-\sigma_3)_f}$ (%)	$\frac{\Delta(\sigma_1-\sigma_3)_f}{\text{kN/m}^2}$	$\frac{\Delta(\sigma_1-\sigma_3)_f}{(\sigma_1-\sigma_3)_f}$ (%)	$\frac{\Delta(\sigma_1-\sigma_3)_f}{\text{kN/m}^2}$	$\frac{\Delta(\sigma_1-\sigma_3)_f}{(\sigma_1-\sigma_3)_f}$ (%)
without junction growth with $b = 3$, using Skempton (1960) values of k and ψ	0.991×10^{-2}	894	72.3	14.4	1.17	0	0
without junction growth, using results of micro-indentation test	0.635×10^{-2}	573	46.4	9.24	0.75	0	0
using interpretation from Laughton (1955) experimental values	1.21×10^{-2}	1091	88.3	17.6	1.42	0	0

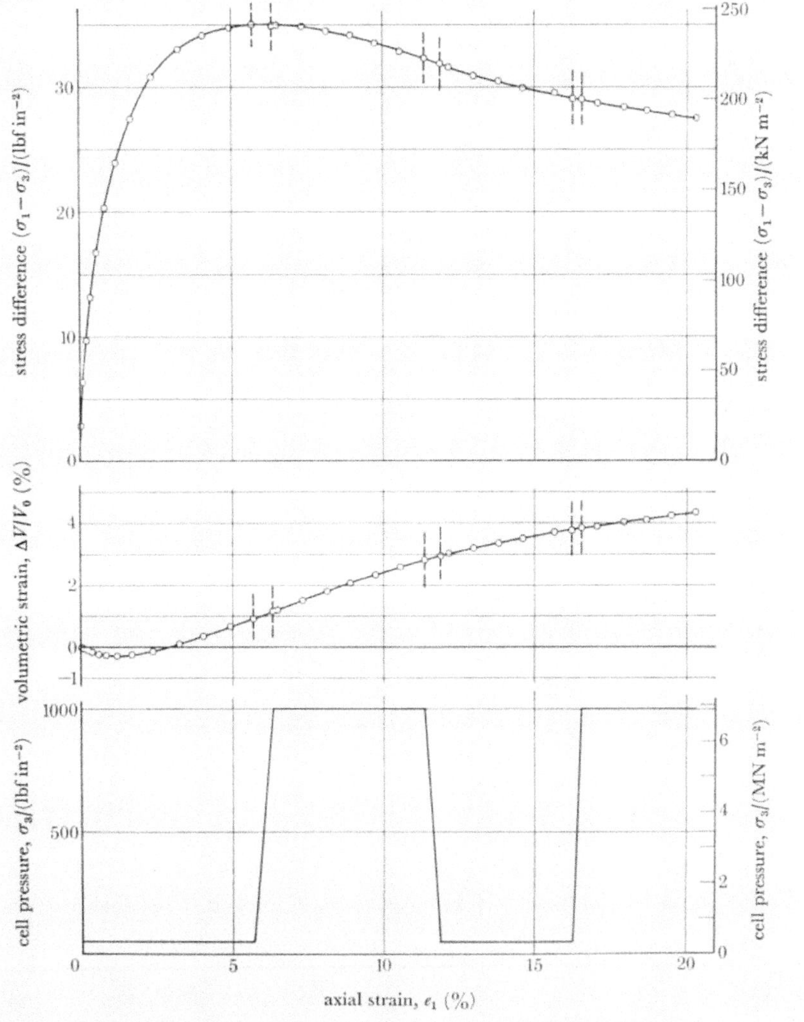

FIGURE 24. Crushed marble tested with $(\sigma_3 - u) = 69.0$ kN m⁻² (10 lbf in⁻²).

specified increase in pore pressure, of between 46.4 % and 88.3 %, whereas no detectable increase can be observed in figure 23. The validity of this equation is thus undisputably disproved. Skempton's effective stress equation predicts an increase of between 0.75 % and 1.42 %, and, though much less at variance with the observed stress–strain curve, must be considered to be significantly inconsistent with the experimental result.

Only the Terzaghi equation predicts to within the accuracy of the experimental observations.

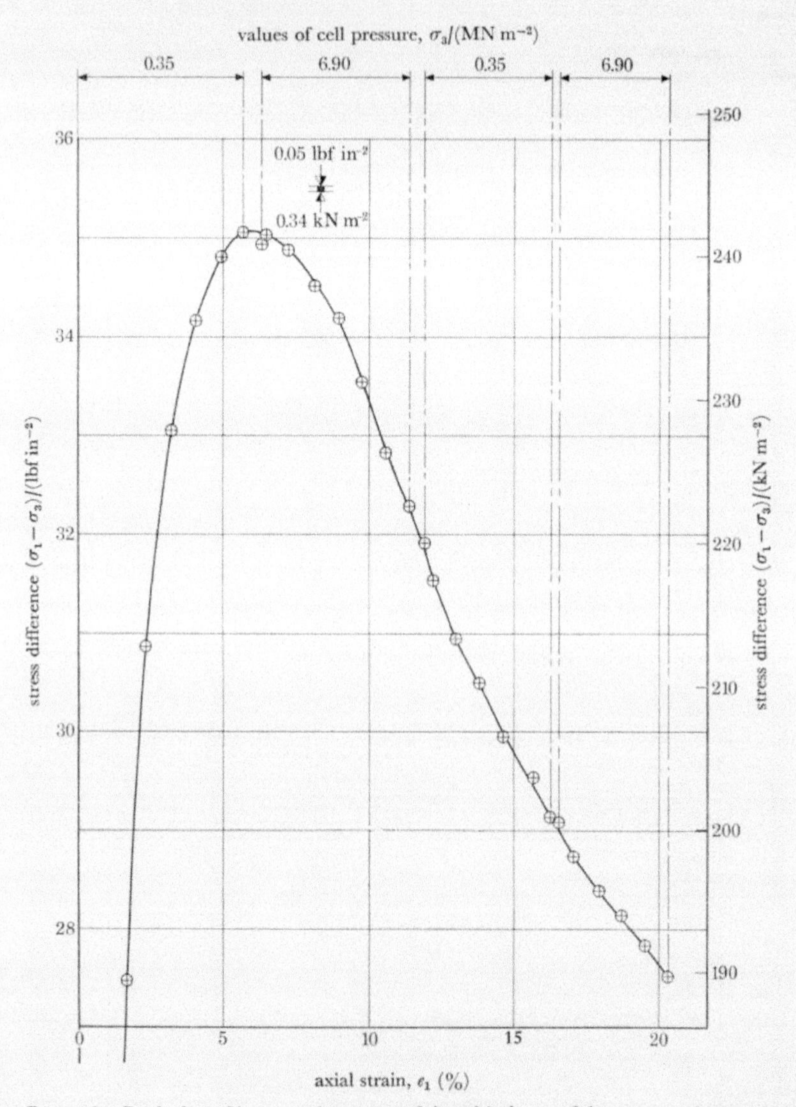

FIGURE 25. Crushed marble – an enlargement of the critical part of the stress–strain curve shown in figure 24.

(c) *Crushed marble*

Part of the testing facilities required for the high pressure cell was in use on another project at this stage of the programme, and the multistage test on crushed marble was therefore carried out in the 0–6.9 MN/m² capacity cell. Particular care was therefore taken to eliminate stray sources of scatter from the values defining the stress–strain curve.

The test was run with $(\sigma_3 - u)$ equal to 68.9 kN/m² (10.0 lbf/in²) and a variation of σ_3 between 349 kN/m² (50 lbf/in²) and 6.895 MN/m² (1000 lbf/in²). The test results are presented in figure 24 and the critical section of the stress–strain curve is given to an enlarged scale in figure 25. Inspection of the curve in figure 25 suggests that any consistent jump in the curve of more that 0.34 kN/m² (0.05 lbf/in²) would be readily discerned, i.e. a jump of 0.14 %.

Table 7. Predicted changes in strength for a change in pore pressure $\Delta u = 6.55$ MN/m², the value of $(\sigma_3 - u)$ being constant and equal to 68.9 kN/m²; crushed marble

		intergranular stress equation (2)		Skempton's effective stress equation (5)		Terzaghi effective stress equation (1)	
assumption about contact area a	value of a	$\frac{\Delta(\sigma_1-\sigma_3)_f}{\text{kN/m}^2}$	$\frac{\Delta(\sigma_1-\sigma_3)_f}{(\sigma_1-\sigma_3)_f}$ (%)	$\frac{\Delta(\sigma_1-\sigma_3)_f}{\text{kN/m}^2}$	$\frac{\Delta(\sigma_1-\sigma_3)_f}{(\sigma_1-\sigma_3)_f}$ (%)	$\frac{\Delta(\sigma_1-\sigma_3)_f}{\text{kN/m}^2}$	$\frac{\Delta(\sigma_1-\sigma_3)_f}{(\sigma_1-\sigma_3)_f}$ (%)
without junction growth, and with $b = 1$	2.63×10^{-4}	6.04	2.50	1.03	0.43	0	0
without junction growth, and with $b = 3$	8.78×10^{-5}	2.02	0.84	0.34	0.14	0	0
with junction growth, probable value	3.39×10^{-4}	7.78	3.22	1.32	0.55	0	0

The predicted values are based on two sets of data. The intrinsic parameters k and ψ may be assumed to approximate to those of calcite given by Skempton (1960) as 186.3 MN/m² (1900 kgf/cm²) and 8° respectively. Taking the value of $(\sigma_n - u)$ at an axial strain of 5.69 % as 112.85 kN/m² we obtain a value of the contact area a of 0.0263 %, using Skempton's expression (equation (13)). For $b = 1$ the same value is obtained from equation (12), but with $b = 3$ this value would fall to 0.00878 %. Micro-indentation hardness tests on the actual material gave a value of $H_v = 2.354$ MN|m². The value of the coefficient of friction μ may then be obtained from the relationship obtained by Skinner (1975):

$$\mu = \tan\psi + \frac{k}{H_v \sin\psi}. \tag{34}$$

This gives a value of 0.7, which is close to the range of values obtained for calcite under saturated conditions by Horn & Deere (1962), namely 0.6–0.68.

Assuming junction growth as in the case of quartz sand we obtain a value of σ_{nj} of 330.0 MN/m² and thus a value of $a = 112.85 \div 333.0 \times 10^3$ (= 0.0339 %). The results of micro-indentation hardness tests on calcite by Brace (1963) indicate that the effect of the magnitude of interparticle force is in this case relatively unimportant.

The magnitudes of the predicted changes in strength on the basis of these contact areas are given in table 7. The validity of intergranular stress equation is again disproved. Skempton's effective stress equation, assuming junction growth, predicts a jump of about four times the minimum magnitude which could be detected, and must therefore be considered to be significantly at variance with the experimental results. The Terzaghi equation again gives a correct prediction.

(*d*) *Braehead silt*

Two multistage tests were carried out on Braehead silt in the 0–6.9 MN/m² capacity cell with $(\sigma_3 - u)$ equal to 68.9 kN/m² (10.0 lbf/in²) and a variation of σ_3 between 349 kN/m² (50 lbf/in²) and 6.895 MN/m² (1000 lbf/in²). The results are presented in figures 26 and 27. Inspection of the curves does not show any significant change in shearing resistance following the changes in pore pressure.

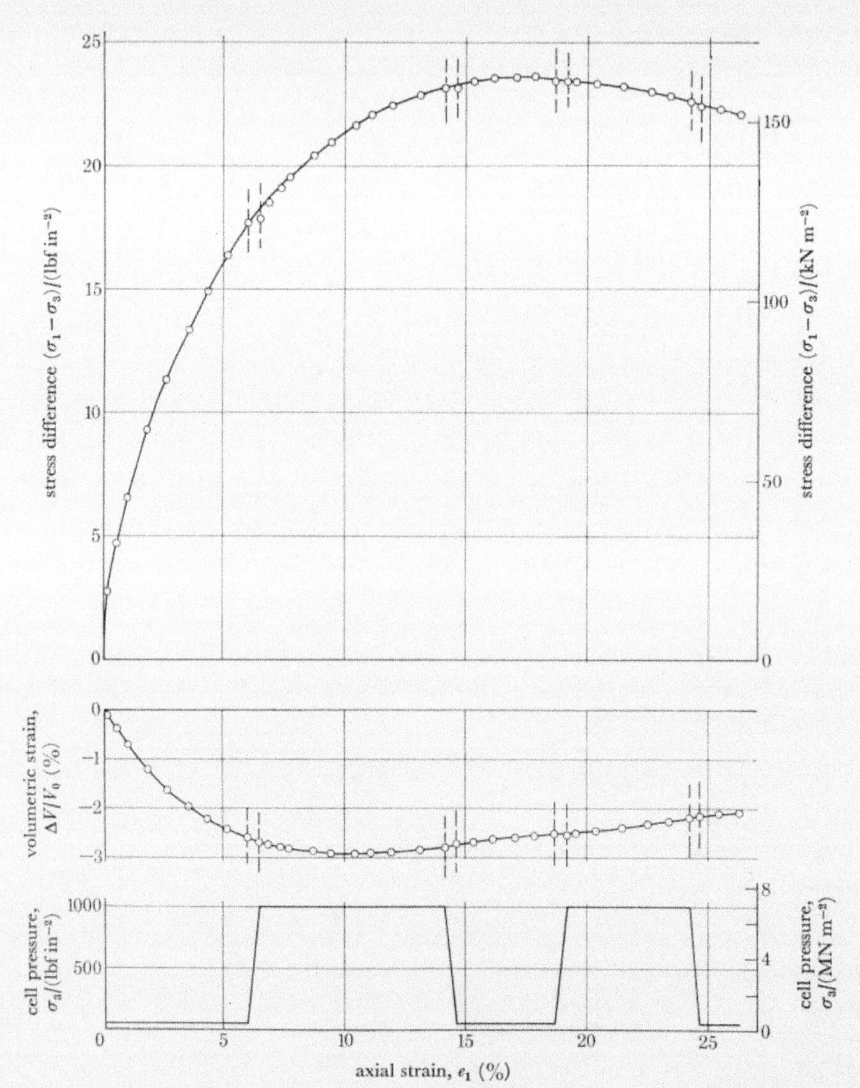

FIGURE 26. Braehead silt tested with $(\sigma_3 - u)$ = 69.0 kN m⁻² (10 lbf in⁻²).

The theoretical prediction of the magnitude of strength changes on the basis of the two equations involving the magnitude of the contact area a is subject to some uncertainty, as the predominant mineral is quartz and the average interparticle force is now at least two orders of magnitude smaller than the minimum load used in the micro-indentation tests by Brace (1963). In the absence of the data for evaluating σ_{nj}, table 8 (which refers to the test in figure 27 at 15 % strain) is restricted to estimates based on equations (12) and (13), which do not take account of junction growth or of the magnitude of the interparticle force. However, these values serve to show that a pore pressure change of this magnitude (6.55 MN/m²) is insufficient to provide

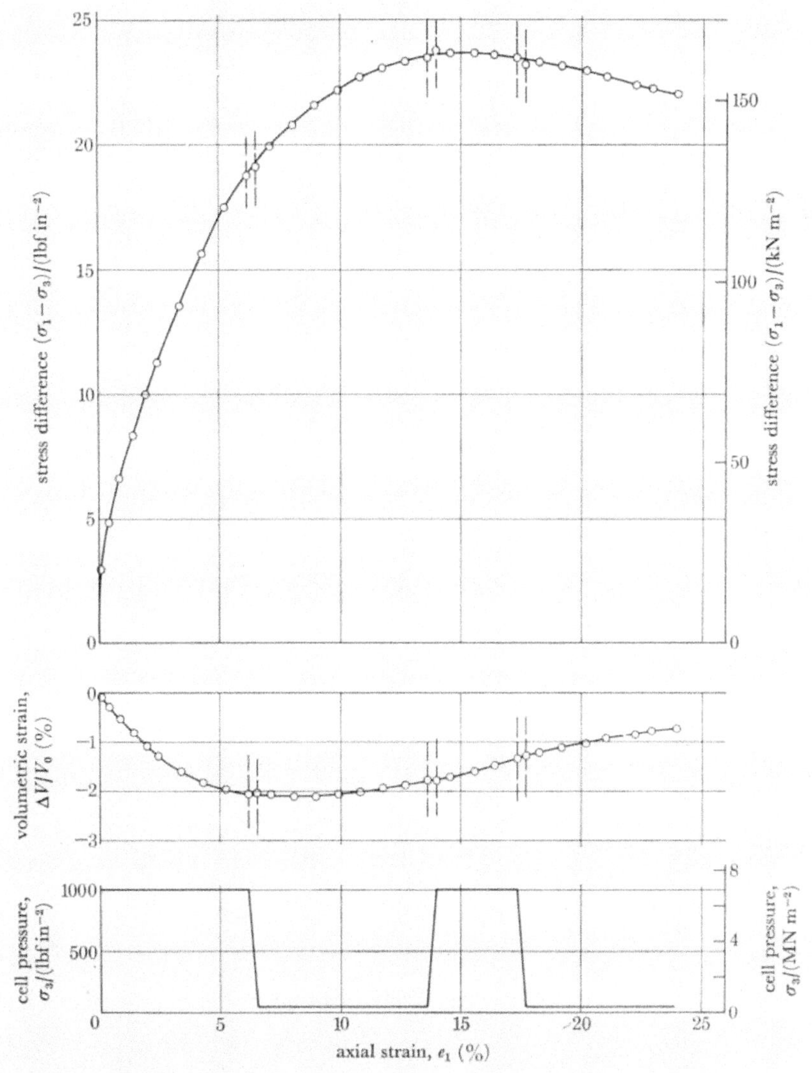

FIGURE 27. Braehead silt tested with $(\sigma_3 - u) = 69.0$ kN m⁻² (10 lbf in⁻²).

critical data for discriminating between the different theories in the case of materials composed predominantly of quartz particles.

The test results, however, do serve to indicate that no other physical phenomenon of greater significance than the interparticle contact area effect enters into the effective stress relationship as the particle size decreases and the material begins to show plasticity in the sense indicated by the Atterberg limits (the *liquid limit* being 29 % and the *plastic limit* 23 %).

TABLE 8. PREDICTED CHANGES IN STRENGTH FOR A CHANGE IN PORE PRESSURE $\Delta u = 6.55$ MN/m², THE VALUE OF $(\sigma_3 - u)$ BEING CONSTANT AND EQUAL TO 68.9 kN/m²; BRAEHEAD SILT

		intergranular stress equation (2)		Skempton's effective stress equation (5)		Terzaghi effective stress equation (1)	
assumption about contact area a	value of a	$\dfrac{\Delta(\sigma_1-\sigma_3)_f}{\text{kN/m}^2}$	$\dfrac{\Delta(\sigma_1-\sigma_3)_f}{(\sigma_1-\sigma_3)_f}$ (%)	$\dfrac{\Delta(\sigma_1-\sigma_3)_f}{\text{kN/m}^2}$	$\dfrac{\Delta(\sigma_1-\sigma_3)_f}{(\sigma_1-\sigma_3)_f}$ (%)	$\dfrac{\Delta(\sigma_1-\sigma_3)_f}{\text{kN/m}^2}$	$\dfrac{\Delta(\sigma_1-\sigma_3)_f}{(\sigma_1-\sigma_3)_f}$ (%)
without junction growth, and with $b = 1$	4.24×10^{-5}	0.66	0.40	0.30	0.19	0	0
without junction growth, and with $b = 3$	1.41×10^{-5}	0.22	0.13	0.10	0.06	0	0

6. CONCLUSIONS

The experimental results demonstrate, with a greater degree of accuracy than has been achieved in any previous investigation, the validity of the principle of effective stress as applied to the shear strength of cohesionless particulate materials. The results indicate that for the range of effective stresses and pore pressure changes investigated the principle of effective stress can be expressed to a high degree of accuracy by the simple Terzaghi equation $\sigma' = \sigma - u$. For the Ham River sand the maximum value of the minor principal effective stress was 6.91 MN/m² (1002.6 lbf/in²) and the maximum change in pore water pressure was 41.4 MN/m² (6000 lbf/in²) which is equivalent to a height of 4.2 km of water.

The investigation also demonstrates quite clearly that the often-made assumption that shear strength is controlled by the intergranular stress is invalid. In the case of lead shot this assumption would have led to gross errors of the order of 46–88 % and here there can be little doubt about the magnitude of the contact area a.

The expression derived by Skempton (1960) to take account of the intrinsic angle of friction of the solid material forming the particles gives a significantly less satisfactory agreement with the experimental data than the Terzaghi equation for lead shot and for calcite. For quartz sand the conclusion to be drawn is less clear, owing to the probable sensitivity of the value of the contact area a at a given value of $(\sigma_n - u)$ to the magnitude of the interparticle force. Changes in pore pressure an order of magnitude greater than were used in the present series of tests would be necessary to give a really convincing demonstration of the status of this expression for hard particles such as quartz. However, in as far as the expression predicts jumps in the stress–strain curve as indicated in tables 3 and 4, it is not supported by any positive evidence.

One reason for the apparent lack of validity of Skempton's expression for effective stress may be its dependence on the Caquot–Bishop–Horne type of relation between ϕ' and the coefficient of interparticle friction μ. More recent work (Skinner 1969 and 1975, and unpublished radiographic work) indicates that the mechanism of failure in random particulate materials is more complex than assumed in any current analysis and that no unique $\phi' - \mu$ relation exists. A rigorous analytical solution to this problem is not yet forthcoming.

The only data available in 1960 for studying the influence of contact area on the strength of porous materials subject to high pore-water pressures was from tests on strongly cohesive materials such as intact marble, limestone and concrete. The critical data (Skempton 1960) was provided by unjacketed samples (in which $\sigma_3 - u = 0$) tested under a range of cell pressures in rather cruder apparatus than used in the present series of tests. Unjacketed samples are subject to brittle failure at small strains, cohesion making the main contribution to compression strength. In the authors' view more relevant data would be obtained by running constant $(\sigma_3 - u)$ tests in the pressure range in which plastic failure occurs (for example $\sigma_3 - u = 50$–70 MN/m² on the basis of Karman's (1911) classic tests on marble). Here friction plays a more important part in the resistance to shear (whether considered in terms of the Mohr–Coulomb or the modified Griffith failure criterion (see, for example, McClintock & Walsh 1962; Hoek 1965)) and the shape of the stress–strain curve is sufficiently flat-topped to permit multistage tests, with due allowance in the rate of strain for the low permeability of the sample.

The early stages of the investigation were supported by the Department of Scientific and Industrial Research and by the Civil Engineering Research Association (now S.R.C. and C.I.R.I.A. respectively). Mr F. Winser contributed to the success of the experimental investigation by very high precision work on the high pressure cells and Mr David Evans assisted in the assembly of the various pressure systems.

References

Aldrich, M. J. Jr. 1969 Pore pressure effects on Berea Sandstone subjected to experimental deformation. *Bull. Am. Geol. Soc.* **80**, 1577–1586.

Bishop, A. W. 1953 (Private communication to Dr A. S. Laughton) (see Skempton 1960).

Bishop, A. W. 1954 Correspondence on shear characteristics of a saturated silt, measured in triaxial compression. Penman, A. D. M. (Geotechnique, 3, No. 8, 312–328). Letter to Geotechnique. *Geotechnique* **4**, 43–45.

Bishop, A. W. 1955 Lecture delivered in Oslo, entitled 'The principle of effective stress'. Printed in *Teknisk Ukeblad, No.* 39 (1959), 859–863.

Bishop, A. W. 1958 Test requirements for measuring the coefficient of earth pressure at rest. *Proc. Brussels Conf. Earth Pressure Problems* **1**, 2–14.

Bishop, A. W. 1965 Discussion. *Proc. 6th Int. Conf. Soil Mech.* **3**, 306–310.

Bishop, A. W. 1966 Strength of soils as engineering materials. 6th Rankine Lecture. *Geotechnique* **16**, 89–130.

Bishop, A. W. 1969 Discussion. *Proc. 7th Int. Conf. Soil Mech.* **3**, 182–186.

Bishop, A. W. 1973 The influence of an undrained change in stress on the pore pressure in porous media of low compressibility. (Technical note.) *Geotechnique* **23**, 435–442.

Bishop, A. W. & Blight, G. E. 1963 Some aspects of effective stress in saturated and partly saturated soils. *Geotechnique* **13**, 177–197.

Bishop, A. W. & Eldin, A. K. G. 1950 Undrained triaxial tests in saturated sands and their significance in the general theory of shear strength. *Geotechnique* **2**, 13–32.

Bishop, A. W. & Henkel, D. J. 1962 *The measurement of soil properties in the triaxial test*, 2nd ed. London: Edward Arnold.

Bishop, A. W., Kumapley, N. K. & El-Ruwayih, A. 1975 The influence of pore-water tension on the strength of clay. *Phil. Trans. R. Soc. Lond.* A **278**, 511–554.

Bishop, A. W., Webb, D. L. & Skinner, A. E. 1965 Triaxial tests on soil at elevated cell pressures. *Proc. 6th Conf. Soil Mech.* **1**, 170–174.

Bowden, F. P. & Tabor, D. 1942 Mechanism of metallic friction. *Nature, Lond.* **150**, 197–199.

Bowden, F. P. & Tabor, D. 1950 *The friction and lubrication of solids. Part I.* Oxford University Press.

Bowden, F. P. & Tabor, D. 1954 *The friction and lubrication of solids. Part I.* Corrected impression. Oxford University Press.

Bowden, F. P. & Tabor, D. 1964 *The friction and lubrication of solids. Part II.* Oxford University Press.

Brace, W. F. 1963 Behaviour of Quartz during indentation. *J. Geol.* **71**, 5, 581–595.

Bruggeman, J. R., Zangar, C. N. & Brahtz, J. H. A. 1939 Memorandum to Chief Designing Engineer: notes on analytic soil mechanics. *U.S. Dept. of the Interior, Tech. Mem.*, No. 592.

Caquot, A. 1934 *Equilibre des Messifs à Frottement Interne, Pulvérulents et Cohérentes.* Paris: Gauthier-Villars.
Hoek, E. 1965 Rock fracture under static stress conditions. Ph.D. thesis, University of Capetown. *CSIR Report MEG 383*, Pretoria, South Africa.
Horn, H. M. & Deere, D. U. 1962 Frictional characteristics of minerals. *Geotechnique* **12**, 319–335.
Horne, M. R. 1969 The behaviour of an assembly of rotund rigid, cohesionless particles. III. *Proc. R. Soc. Lond.* A **310**, 21–34.
Hubbert, M. K. & Rubey, W. W. 1959 Rôle of fluid pressure in mechanics of overthrust faulting. *Bull. Am. Geol. Soc.* **70**, 115–206.
Jaky, J. 1948 Pressure in silos. *Proc. 2nd Int. Conf. Soil Mech.* **1**, 103–107.
Karman, Th. von. 1911 Festigkeitsversuch unter allseitigem Druck. *Ziet. Vereines Deutsch, Ing.* **55**, 1749–1757.
Laughton, A. S. 1955 The compaction of ocean sediments. Ph.D. thesis. University of Cambridge.
McClintock, F. A. & Walsh, J. B. 1962 Friction on Griffith cracks under pressure. *4th U.S. Nat. Congress of Appl. Mech. Proc.* 1015–1021.
Procter, D. C. & Barton, R. R. 1974 Measurements of the angle of interparticle friction. *Geotechnique* **24**, 581–604.
Scott, R. F. 1963 *Principles of soil mechanics.* Addison Wesley.
Skempton, A. W. 1960 Effective stress in soils, concrete and rocks. *Conf. on Pore Pressure and Suction in Soils.* London: Butterworths.
Skinner, A. E. 1969 A note on the influence of interparticle friction on the shearing strength of a random assembly of spherical particles. *Geotechnique* **19**, 150–157.
Skinner, A. E. 1975 The effect of high pore water pressures on the mechanical behaviour of sediments. Ph.D. thesis, University of London.
Taylor, D. W. 1944 Cylindrical compression research program on stress-deformation and strength characteristics of soils. *M.I.T. 10th Progress Report to U.S. Engineers Department.*
Taylor, D. W. 1948 *Fundamentals of soil mechanics.* New York: John Wiley.
Terzaghi, K. 1923 Die Berechnung der Durchlässigkeits-Ziffer des Tones aus dem Verlauf der hydrodynamischen Spannungserscheinungen. *Sitz. Akad. Wissen. Wien Math-naturw. Kl. Abt. IIa* **132**, 125–138.
Terzaghi, K. 1925 *Erdbaumechanik auf bodenphysikalischer Grundlage.* Leipzig: Deuticke.
Terzaghi, K. 1936 The shearing resistance of saturated soils and the angle between the planes of shear. *Proc. 1st Int. Conf. Soil Mech.* **1**, 54–56.
Tombs, S. G. 1969 Strength and deformation characteristics of rockfill Ph.D. thesis, University of London.

THE BEHAVIOUR OF A SOFT ALLUVIAL CLAY REVEALED BY LABORATORY TESTS, FIELD TESTS AND FULL SCALE EMBANKMENT LOADING

LAURENCE D. WESLEY ME, MSC(ENG) PHD AND
RICHARD S. PUGH BSC, MSC, PHD, CENG, FICE
Retired Geotechnical Engineers

1 INTRODUCTION

This paper describes laboratory tests, field vane tests, and full scale trial embankments carried out by the authors between 1972 and 1975 under the supervision of Professor A.W. Bishop. The work was in connection with the raising of the earth embankment flood defences along the Essex shoreline of the Thames Estuary in response to the massive tidal surge that occurred in 1953. This tidal surge was the largest in recorded history and resulted in 839 breaches of the flood defences in Essex, resulting in 108km of failed embankments and flooding a total area of 21,900 hectares. One hundred and twelve people lost their lives in Essex, out of a national total in excess of 300, and 21,000 were made homeless in what, at the time, was probably the largest peace-time natural disaster known in Great Britain.

The laboratory testing was carried out by the first author and the field work by the second author with both being fully described in the respective theses of the two authors (Wesley, 1975, and Pugh, 1978). Separate papers for publication were planned by both authors to be co-authored with Professor Bishop, and some limited work was done on preparing these papers. Unfortunately, and sadly, the illness Professor Bishop suffered at the time and his early death some years later brought his involvement in these papers to an end. Intentions to complete the papers have taken second place to other commitments ever since, but retirement and the publication of the Bishop biography and this commemorative volume provided the opportunity and the motivation to write this paper. The question naturally arises as to whether tests done over forty years ago still have relevance today; in response to that question the following comments are made.

Firstly, with respect to the laboratory testing, the special triaxial apparatus designed and built for the tests has been described in detail by Bishop and Wesley (1975), and some material from the thesis has been included in papers published by other authors in the years following its completion. These include Jamiolkowski (1979) and Parry and Wroth (1981). The latter made the following observation when discussing specific aspects of the author's thesis:

> "*There is little published evidence of undrained triaxial compression and extension tests with pore pressure measurements on natural soft clay deposits. One reason for this may be the need to test at low stress levels consistent with field stresses. A notable exception is the work of Wesley (1975) on soft clay from Mucking Flats adjacent to the River Thames.*"

The situation has changed somewhat since that statement was made, the most notable example being the testing that followed the creation of the Bothkennar soft clay test site in Scotland. The testing of the particular clay at that site was very extensive and is described in detail in the Eighth Geotechnique Symposium in Print (1992). However, soft clays are not all identical and the Bothkennar clay has some characteristics not found in other clays, including those of southern England such as the Mucking clay described in this paper. Other significant features of the Mucking investigation were the following:

(a) the laboratory tests included some that were not carried out on the Bothkennar clay; these included plane strain tests and tests on samples at varying inclinations to the vertical.

(b) most of the triaxial tests were quadruple, that is, for any one particular type of measurement four identical samples were used. This was done primarily to maximise the reliability of the results, especially as there was an adequate supply of large diameter samples to make this possible.

Secondly, with respect to the field tests and trial embankments, these were unique for the following reasons;

(1) several full scale trial embankments were built and equipped with a very comprehensive array of monitoring equipment. This was in contrast to the Bothkennar investigations which did not include any trial embankments. Only with full scale trial embankments is it possible to relate the laboratory tests to field behaviour.

(2) in situ vane tests were carried out with both the Geonor vane and a vane four times the size of the Geonor vane. This was far larger than any previous vane tests, and tests were carried out at different depths and varying strain rates

(3) as with the laboratory tests, many of the vane tests were repeated to produce a high degree of reliability

PART I (Wesley) LABORATORY TESTS INVESTIGATING THE INFLUENCE OF ANISOTROPY AND STRESS PATH

1 SITE DESCRIPTION

The Mucking site lies on the north coast of the Thames estuary close to the town of Stanford le Hope in Essex. The coastline here is protected from sea flooding by a levee some 3 to 4 metres in height. The test pit from which samples were obtained was about 100m on the land side of this levee. The ground here is flat and the surface little more than a metre above mean sea level. The seaward side is inundated at high tide, but consists of mudflats several hundred metres wide at low tide.

The strata here consist of recent alluvial deposits nearest the surface, followed in succession by Thames gravel, London clay, and Woolwich and Reading beds. The upper part of the alluvial deposits, from which samples were obtained, consists predominantly of clay and silty clay having a thickness from 5 to 9 metres. Geological records show the relative sea level has risen about 30 metres over the past 10,000 years, and the upper clay layer formed over the past 5,000 years (Greensmith and Tucker, 1973 and Khan et al, 2014). The material being deposited to form the clay and silty clay layers was carried to the site by the River Thames, and it would be expected therefore that the fine fraction would consist of material originating mainly from London clay beds.

Evans (1965) suggests that deposition occurs mainly in the zone limited by high and low tide levels, and that streams may run across it during low tide which create quite deep channels. While this is somewhat hypothetical, it is a possible explanation for shear surfaces found in several of the shallow samples. Minor slips could have occurred along the banks of such drainage channels.

2 FIELD SAMPLING AND SAMPLE PREPARATION IN THE LABORATORY

2.1 *Test pit and cylinder samples*

The tests described here were nearly all carried out on large "block" samples taken from an investigation pit. They were obtained using large steel cylinders 25cm in diameter and 30cm high. The wall thickness was 6.2mm with a sharp cutting edge at one end. The wall thickness was turned down to 4.2mm for a distance of 7.5cm behind the cutting edge to help minimise sample disturbance. The cylinders were pushed into the soil using a dead weight pressing on a plate resting on the upper rim of the cylinder. When full the cylinders were dug out by hand. The top and bottom surfaces were carefully trimmed flush with the cylinder ends and the ends sealed with aluminium foil and rubber pads. These were held in place with metal plates and tie bolts. A total of 36 cylinders was obtained, details of which are given in Table 1. Generally two samples were obtained from each depth but from one level (3.33m) ten were obtained to enable an extensive testing programme to be carried out on soil expected to have the same properties.

The samples were extruded in a vertical position using a hydraulic jack and plate below the sample while the cylinder was held in place by a retaining collar at the top. Generally, the samples extruded cleanly, with a smooth outer surface. However, there were a few cylinders in which corrosion occurred between the soil and cylinder surface resulting in drag and slight disturbance of the samples.

Table 1 Details of test pit cylinder samples.

Depth to centre of sample (m)	**Number of samples**	**Identification numbers**
0.45	1	T8
0.75	1	T7
0.93	2	T9 and T16
1.23	2	T10 and T17
1.53	2	T18 and T19
1.83	2	T11 and T12
2.03	2	T13 and T14
2.43	2	T21 and T22
2.73	2	T22 and T23
3.03	2	T34 and T35
3.33	10	T24 to T33
3.53	1	T42
3.63	2	T36 and T37
3.93	2	T38 and T39
4.23	2	T40 and T41
4.43	1	T43

2.2 *Extruding the samples and preparing smaller samples for laboratory tests*

Cutting up the large blocks into samples suitable for triaxial or oedometer tests was done entirely with a wire saw. The soil was of an ideal consistency to be sliced easily with a wire saw in this way. Normal procedure was to extrude only 10cm of the cylinder at a time and complete tests on this layer before extruding the next 10cm. The principal difficulty that arose was in separating each 10cm layer after a horizontal cut had been made with a wire saw using the rim of the cylinder as a guiding surface. This was overcome to a large extent by placing spatula blades in the saw cut as it was formed; these kept the surfaces apart and prevented them sticking together again.

Two, and sometimes four, vertical cuts were then made in the 10cm thick layer to provide small blocks that could be trimmed to the specific shape and dimensions needed for triaxial or oedometer tests. The blocks not for immediate testing were sealed and stored until required. The soil remaining in the cylinder was sealed again with the rubber and metal plates at each end.

2.3 *Visual description of the soil*

The soil can generally be described as soft gray clay over the full depth investigated. It is somewhat darker in colour near the surface, presumably due to a higher organic content. The organic content is very small and becomes insignificant after a depth of about 2 metres. From this depth it appears to be confined to traces of dark material remaining in what appear to be fine root holes running vertically through the soil. The frequency of these decreases with depth. These root holes are the main structural feature apparent from visual examination. There are no traces of horizontal bedding or any other bedding. A further structural feature of some significance is the existence of distinct "fissures" or slip surfaces in the soil above a depth of about 2.2m. In the author's view these are clear slip surfaces on which definite shear movement has occurred. The most likely explanation is that given above, namely failures of the banks of minor water courses running through the soil during its deposition process.

2.4 *Preparation of samples for triaxial and plane strain tests.*

Most of the triaxial testing was on 3.8cm by 7.6cm cylinders, and the samples were prepared using a vertical lathe and a wire saw. This proved a very easy and quick method, although as discussed later it did not seem to involve a consistent degree of disturbance. The plain strain tests were carried out on samples having two dimensions identical to the triaxial samples and a length of 20.3cm in the plane strain direction.

Preparation of the plane strain samples was much more difficult and time consuming than the triaxial samples. The difficulties are firstly, trimming an oblong block of soft clay with precise dimensions, and secondly surrounding it with a rubber membrane. A special trimming mould with spacers and "templates" was used. The templates were thin rigid metal plates 7.6cm by 20.3cm and were used to support the long sides of the sample and act as guides during the trimming operation. The mould and spacers enabled the sample to be cut cleanly with the wire saw to the required dimensions. The most delicate part of the operation was enclosing the sample in the special rubber membrane. This was done by first folding the membrane down below the bottom platten and then placing the sample on the platten. Two thin plastic plates 3.8cm by 7.6cm were then placed at each end of the sample to prevent disturbance as the membrane was pulled up over the sample. Screwing the cover plate to clamp the membrane in place required considerable care but presented no real problems. A full description of the plane strain apparatus is given by Atkinson (1975).

3 INFLUENCE OF MEMBRANE AND STRAIN RATE IN TRIAXIAL TESTING.

3.1 *Tests to investigate the membrane influence*

Some limited testing was done in an attempt to determine the magnitude of correction, if any, that should be made to take account of the membrane used in the triaxial tests. Bishop and Henkel (1962) suggest that the correction needed with 7.6cm x 3.8cm samples was 4.1kPa at a strain of 15%. This was based on tests on remoulded samples. If the correction is assumed to be proportional to the strain then the correction would be about 0.7kPa.

An attempt to verify this figure was made by conducting a series of undrained triaxial tests on 7.6cm by 3.8cm samples. Samples with membranes were first tested in the usual manner with

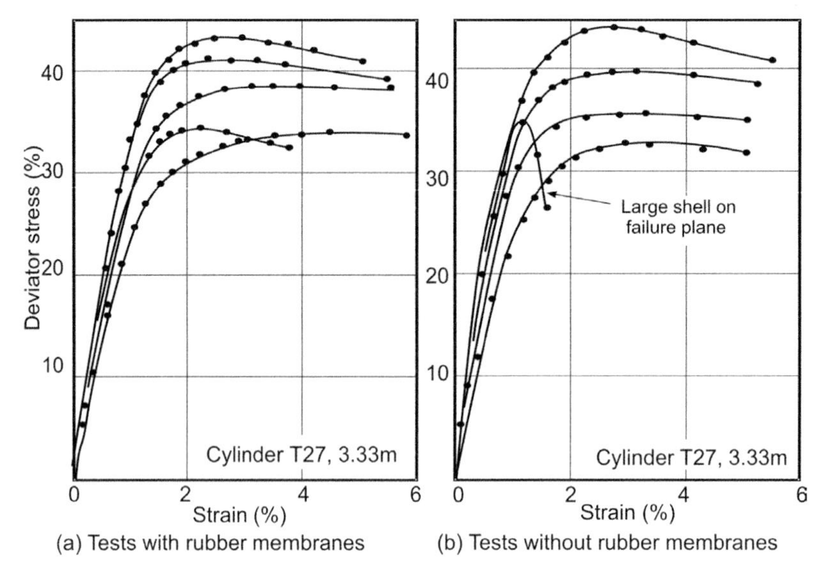

Figure 1. Unconsolidated undrained (UU) triaxial tests with and without rubber membranes.

a cell pressure of 210kPa. Samples without membranes were then tested using liquid paraffin as the cell fluid in place of water and a pressure of 210kPa. This technique appeared satisfactory although there was always the danger that paraffin would enter the sample through small root holes or other discontinuities in the soil. The results suggest that this did not occur. Five samples were tested with and without a rubber membrane. The results are shown in Figure 1. The two sets of results are seen to be almost identical except for one sample from the second set which appeared to fail prematurely. An examination of the sample showed that the failure had been influenced by the presence of a large shell on the sample. If the sample with the anomalous result is ignored then the average strength from the two sets of tests is almost identical. For this reason no membrane corrections were made to the results.

3.2 *Tests to investigate the strain rate influence.*

The strain rate used for most of the tests in this paper, especially those in which the pore pressures were measured, was normally about 0.33% per hour, and strain to failure averaged about 2%. This corresponds to a time to failure of 6 hours. Most tests were conducted over a 20 to 24-hour period, so the strain limit was usually between7% and 8%. The influence of strain rate was investigated by conducting UU tests at 10 and 100 times faster, and 10 times slower than this standard rate. Three or four samples were tested at each rate. The results are summarised in Table 2 and Figures 2 and 3.

Figure 2 illustrates the strong influence of strain rate on the undrained strength. The average strength increase per tenfold decrease in time to failure is of the order of 15%, which is much larger than the 5% increase reported by Bishop and Henkel (1962) from tests on remoulded clays. However, it is similar to results obtained by Berre and Bjerrum (1973) from tests on undisturbed Drammen clay. The question arises as to whether the strength increase with strain rate results from

Table 2 Strain rate effect in unconsolidated undrained (UU) triaxial tests

Series	Number of tests	Strain rate % / minute	Time to failure (min)	Average failure deviator stress kPa	Average strain to failure %
1	4	0.79	3.8	42.8	3.2
2	2	0.074	41	34.8	4.2
3	3	0.0059	513	32.4	2.6
4	4	0.00048	6160	26.5	3.0

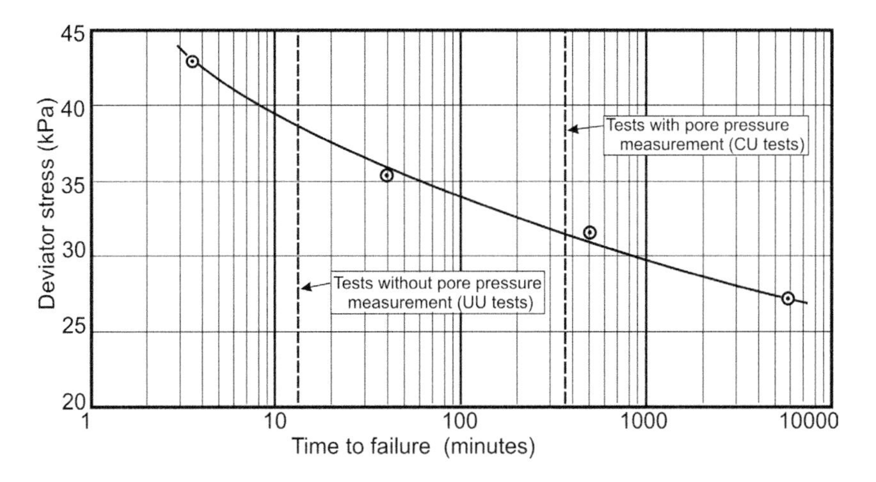

(Above) Figure 2. Influence of strain rate on undrained strength in unconsolidated undrained triaxial tests.

(Right) Figure 3. Effective stress paths for tests at varying strain rates following isotropic consolidation to the average in situ effective stress (CIU tests).

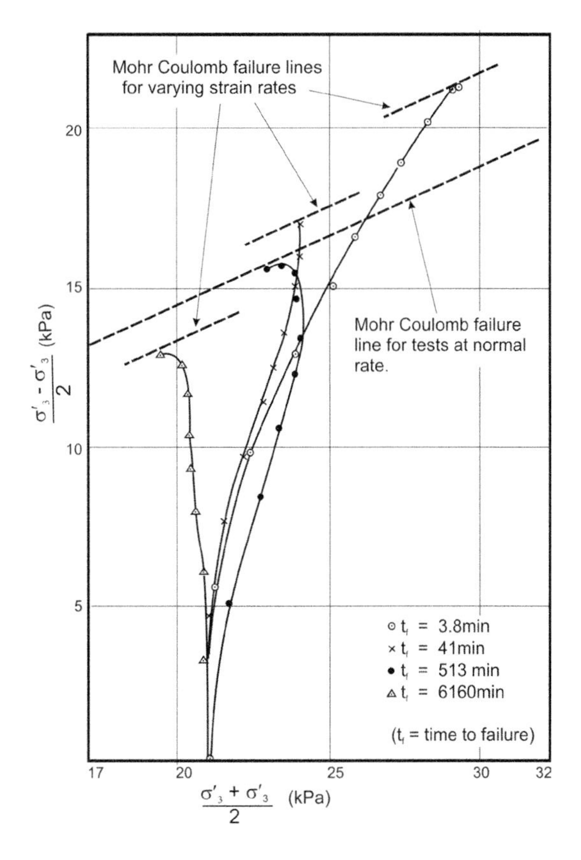

a change in the pore pressure response during tests or to an increase in the Mohr-Coulomb failure line. Figure 3 showing the results of undrained tests on samples consolidated to the average in situ effective stress (CIU tests) suggests that it is a combination of these two effects. Insufficient testing was done to determine the complete Mohr-Coulomb line for each strain rate; Figure 3 shows the Mohr-Coulomb failure line for tests done at the normal strain rate and the likely positions of the failure line for each different strain rate. The effective stress path from the test with the shortest time to failure (3.8min) should be treated with some caution as this may not have allowed sufficient time for the pore pressure to be uniform over the full length of the sample. This would mean the pore pressure measured at the base to be less than that at the centre of the sample;

4 BASIC LABORATORY TESTS

Basic laboratory tests were carried out on samples from each depth at which they were obtained. These included natural water content, Atterberg limits, and clay fraction. The results of these tests, along with visual descriptions of the samples, are shown in Table 3 and in the upper part of Figure 4. It is evident that there is considerable variation in the properties; the most significant being that the clay content and plasticity decreases with depth and there is a steady increase in liquidity index and sensitivity. Below 3m the natural water content is very close to the liquid limit and the soil consequently becomes highly sensitive.

Table 3. Results of basic soil tests at a range of depths.

Sample No	Depth (m)	Water Content (%)	Density (gm/cm^3)	Atterberg limits			Clay Fraction (%)	S_u (kPa)		S_r
				LL	PL	PI		Und.	Rem.	
T8	0.45	67.3	-	138	61	77	61	50.9	62	1
T9	0.93	76.3	1.50	111	40	71	61	15.1	6.3	2.5
T18	1.53	85.8	1.51	114	37	77	62	15.0	3.9	4
T13	2.13	73.8	1.57	113	39	74	66	12.4	7.4	2
T22	2.73	79.9	1.53	94	32	62	54	13.6	3.1	4.5
T25	3.33	51.1	1.73	54	25	29	30	18.3	3.5	5.5
T38	3.93	63.5	1.62	66	26	40	34	19.3	2.1	9
T43	4.43	752	1.57	73	29	44	37	20.8	2.4	9

The soil descriptions in Figure 4 are based on a visual examination of the soil and the use of manipulation tests ("quick" behaviour and dilatancy) to detect whether the soil showed any evidence of silt characteristics. No such behavioural characteristics were found in any of the samples, and it was not possible to detect changes in plasticity from manipulation tests, or from handling and trimming the samples. The Atterberg limits are shown in the plasticity chart in Figure 5. All of the soils have liquid limits above 50 and plot above the A-line, except for the sample with the highest liquid limit. This sample was from the shallowest depth and is believed to have a significant organic content which accounts for its position on the plasticity chart. Thus,

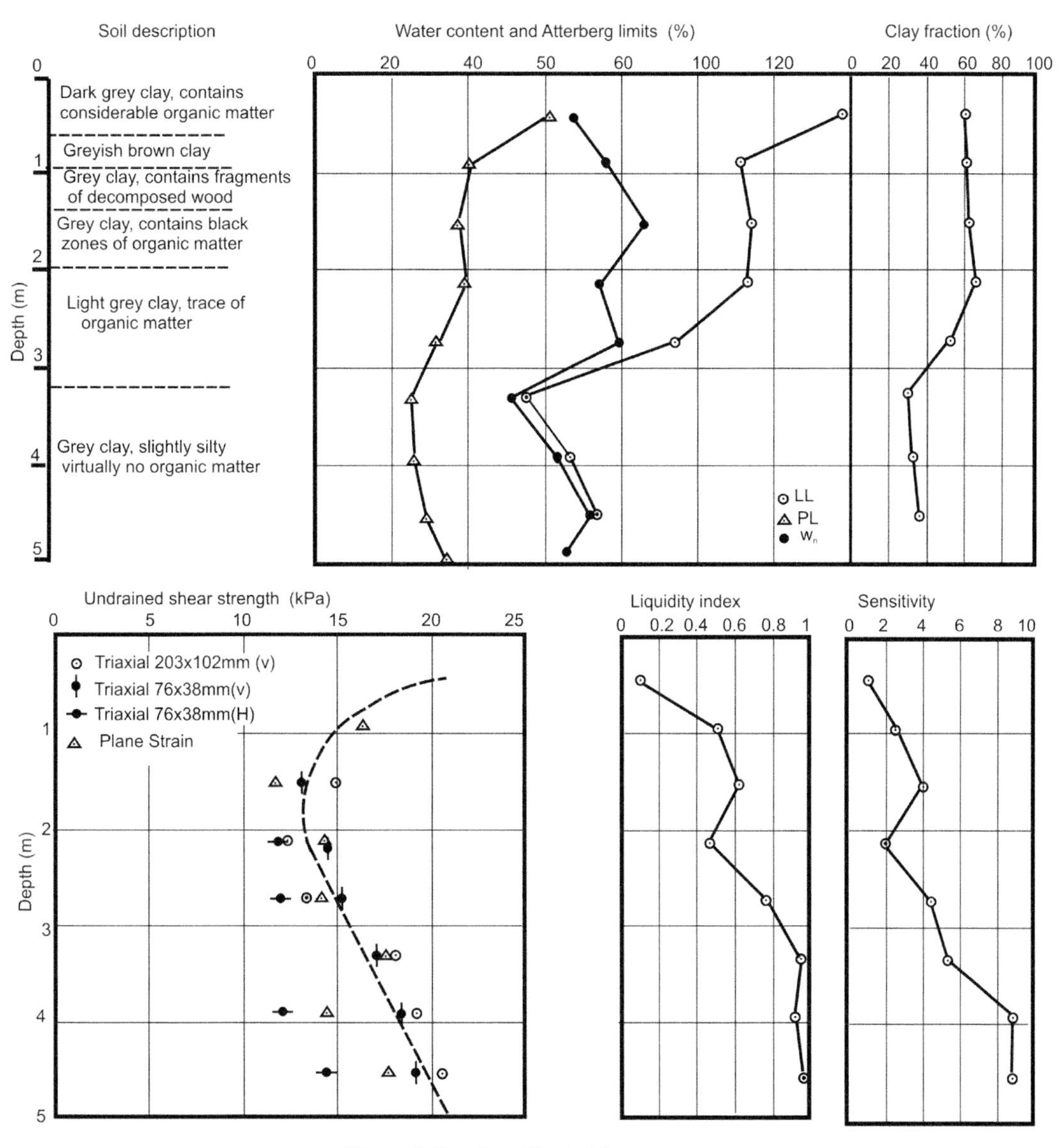

Figure 4. Results of basic laboratory tests.

according to both visual behavioural characteristics and position on the plasticity chart, the soil is classified as high plasticity clay. The lower part of Figure 4 also shows the values of undrained shear strength measured in a variety of ways that are fully described later in this paper. These tests included conventional triaxial tests on samples trimmed in both vertical and horizontal direction, and plane strain tests. At shallow depths the strength is strongly influenced by drying from evaporation at the ground surface

The clay fractions given in Table 3 were from normal hydrometer sedimentation tests. The activity obtained from the tests in Table 2 is about 1.1, which is comparable to the value of

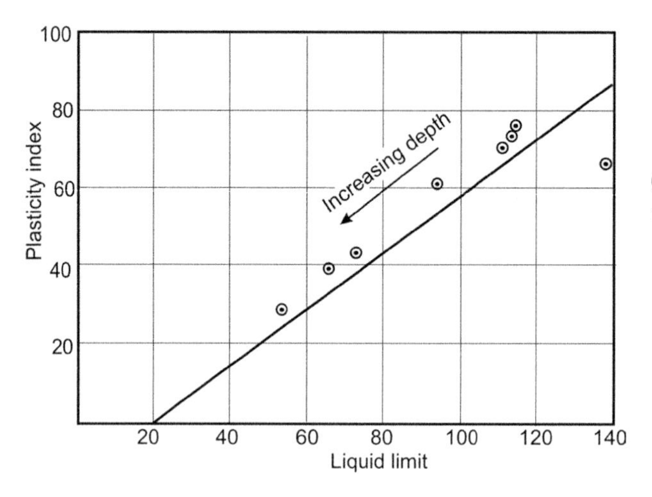

(Left) Figure 5. Mucking clay soils on the plasticity chart.

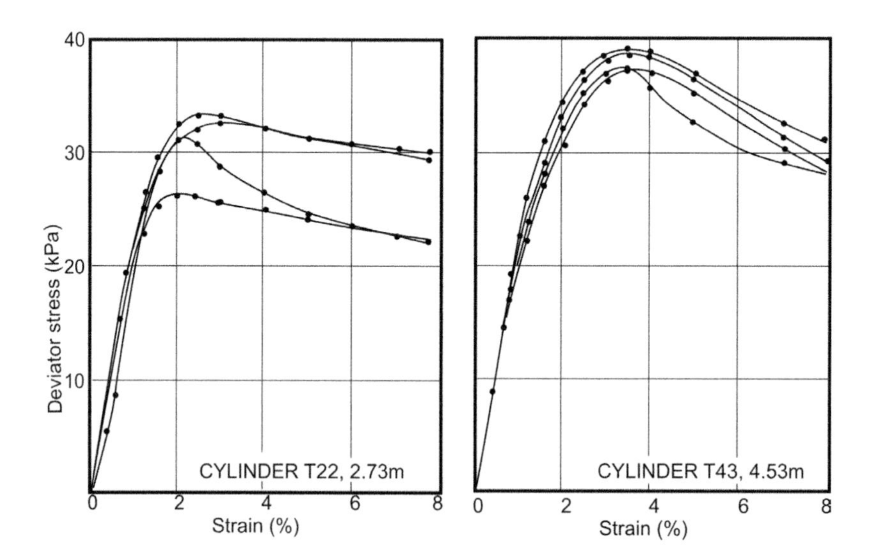

(Below) Figure 6. Variability in unconsolidated undrained (UU) triaxial tests on 7.6cm by 3.8cm samples.

0.95 given for London clay by Skempton (1953). This supports the observation made earlier that the most probable origin of the clay fraction of the Mucking clay was from erosion and re-deposition of London clay

5 UNCONSOLIDATED UNDRAINED (UU) TRIAXIAL AND PLANE STRAIN TESTS

The study of the stress deformation behaviour of the soil began with a series of essentially conventional triaxial tests. These were carried out on 7.6cm by 3.8cm samples at a strain rate of 0.2% per minute, this rate being chosen to be close to normal commercial testing rates. However, in addition to these conventional tests, a number of triaxial tests were carried out on samples at variable inclinations to the vertical. Also a limited number of plane strain tests were carried out. For the earlier part of the tests, strain readings were taken at intervals of 0.2%, this being increased in the latter part of the tests.

The results given earlier in Figure 1 show considerable scatter even when the samples were carefully chosen to be as close as possible to each other within the block from which they were taken. Figure 6 also illustrates the extent of scatter in results from two different cylinders. There is considerable scatter in results from Cylinder T22, but much less scatter in those from Cylinder T43. A decision was made therefore to carry out each experiment as four identical tests on samples taken as close as practical to each other. A set of four of the "Bishop-Wesley" triaxial devices was made, so that the four tests could be conducted simultaneously. With the plane strain tests it proved impractical to carry out four tests, partly because only one apparatus was available, and partly because trimming and setting up the samples was very time consuming. For these reasons the plane strain tests were usually, but nor always, restricted to two samples.

To investigate the anisotropy of undrained strength with respect to inclination of the sample, a series of UU triaxial tests was carried out on samples trimmed from Cylinder T39 (3.93m). Only two tests were carried out at each inclination, and the results, which were very similar, were averaged for each pair and are shown in Figure 7. These show a steady decrease in strength as the inclination to the vertical increases, in agreement with Lo (1965) and Berre and Bjerrum (1973). The results in Figure 7 are not as tidy as was hoped with these tests, and for this reason further tests to investigate anisotropy were carried out on several additional samples making use of four identical tests in each case. The results of these tests, after averaging the data are shown in Figure 8. These confirm the trend evident in Figure 7.

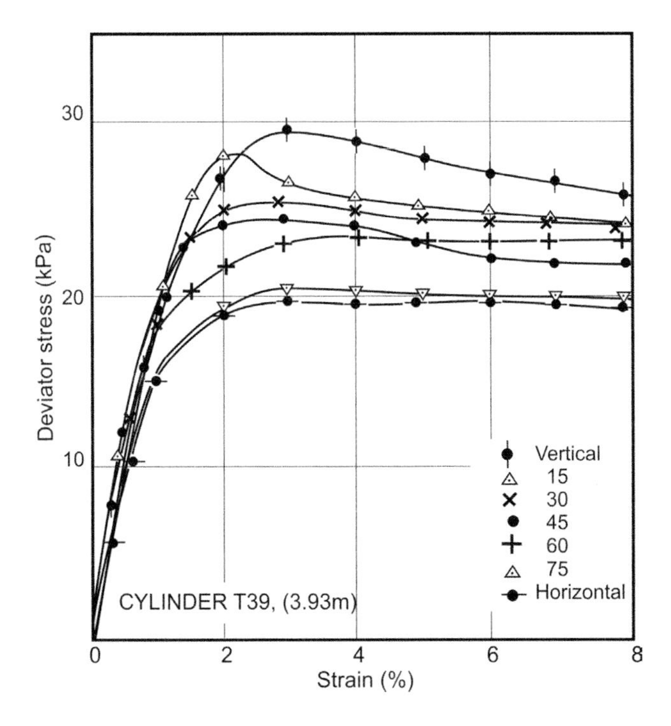

Figure 7. Undrained triaxial tests at varying inclinations

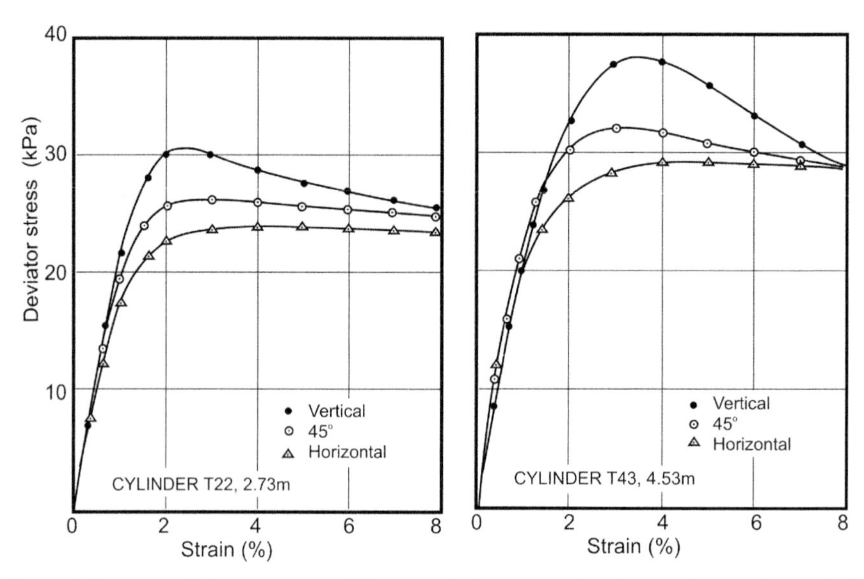

Figure 8. Undrained triaxial tests in different cylinders at varying inclinations to the vertical.

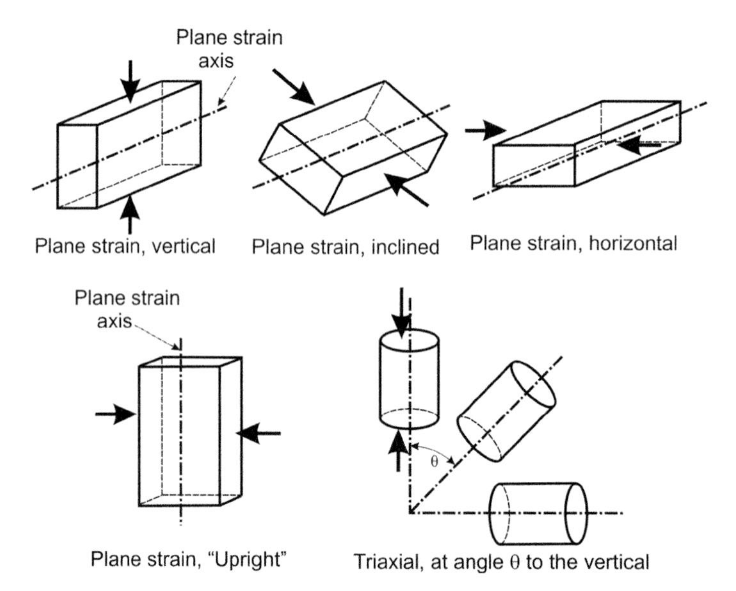

Figure 9. Shape and orientation of samples used in undrained compression tests.

With respect to Figures 6 to 8, it is important to recognise that these tests are all unconsolidated undrained (UU) tests in which pore pressure was not measured. In a later section of this paper (Section 9), undrained tests are carried out after consolidation of the samples to the in situ effective stress state. These are a special case of consolidated undrained (CU) tests. The CU term usually refers to a set of tests at different consolidation pressures in order to measure the effective stress parameters c' and ϕ', but in this case refers to a range of tests of various types after consolidation to the same effective stress, that is, the effective stress in the ground. Pore pressure was measured in all these tests.

The important point to recognise is that variations in the measured strength of soils arise from two causes, the first being intrinsic variability in its natural state in the ground and the second is variability in the degree of disturbance to which the samples are subject during sampling and preparation for testing. Hight (2018) refers to these respectively as natural and induced variability. This issue is discussed in detail in Section 7. A series of UU plane strain tests have also been carried out, with samples at vertical, horizontal and 45° orientation. In addition one "horizontal" test was carried out with the plane strain axis in the vertical direction. For want of a better term these are called upright samples (or tests). The shape and orientation of the samples are indicated in Figure 9.

The results of the plane strain tests are illustrated in Figure 10. The two tests carried out at each orientation showed very good agreement and the curves in Figure 10 are averages from the two. The most significant features of the results are the following:

1. The general pattern of behaviour is similar to that from the triaxial tests, with the vertical samples showing the highest strength, and a steady decline to the minimum strength with the horizontal samples.
2. The values of maximum and minimum strength are also very similar: the respective values being about 32kPa and 20kPa.
3. The only clear difference is that the deviator stress graph shows a much sharper peak than that from the triaxial tests.

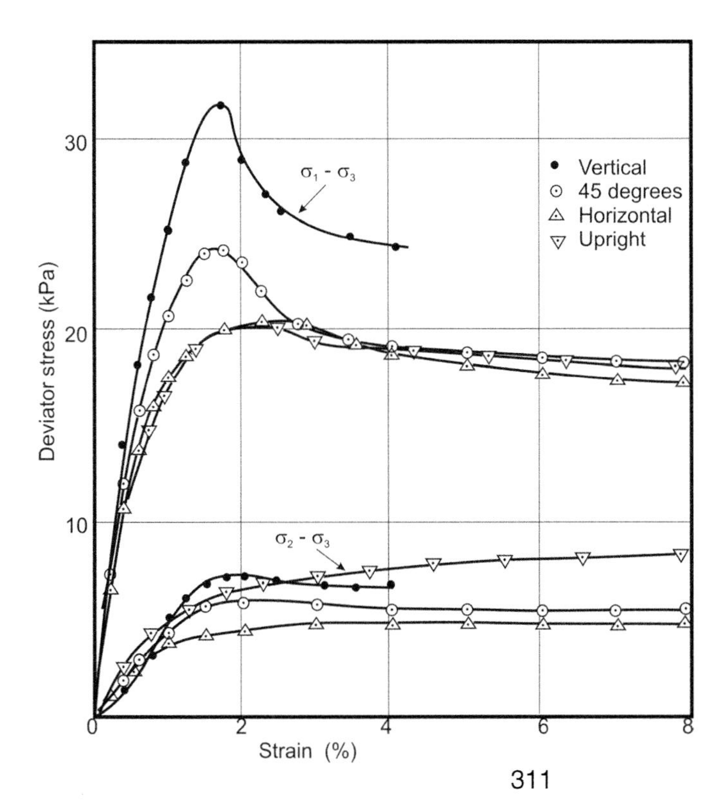

Figure 10. Undrained plane strain tests at varying orientations

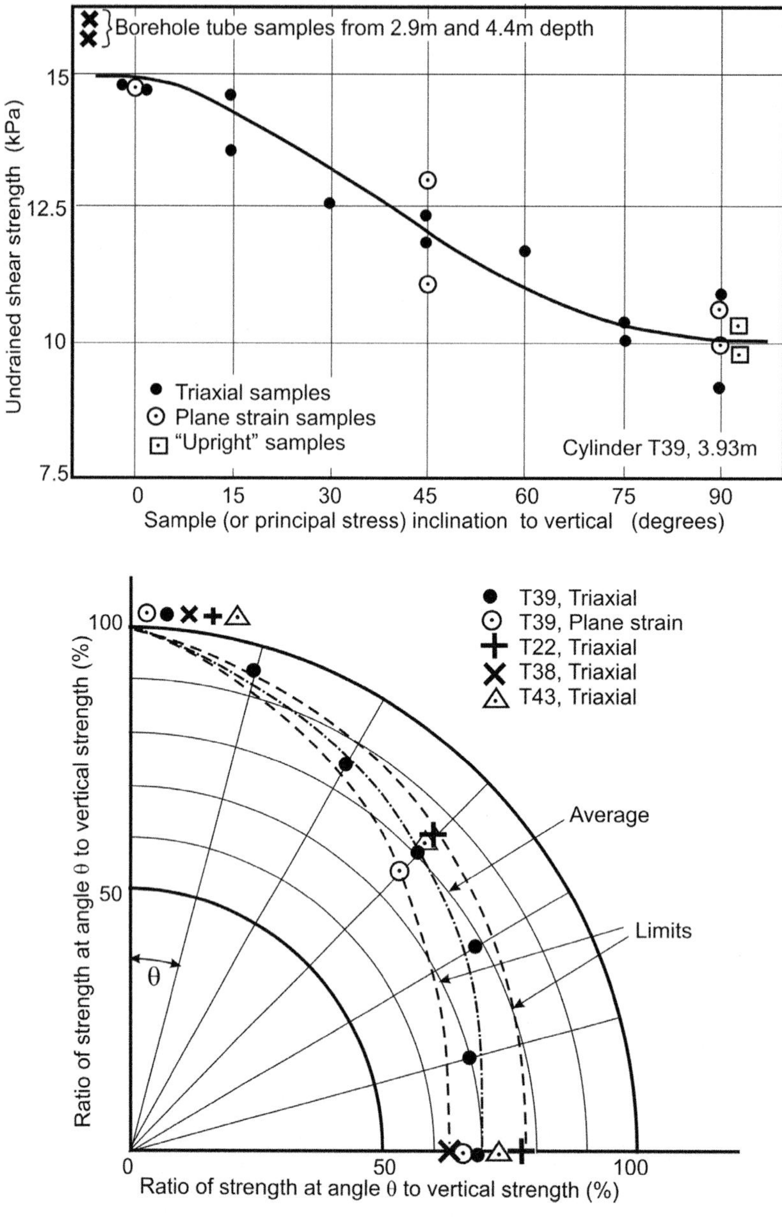

Figure 11. Shear strength ratio in triaxial and plane strain tests at varying inclinations.

This difference in behaviour is believed to be due to the difference in the failure mode in the two types of test. With the triaxial samples failure was a combination of "plastic" bulging and shear on specific slip planes, although with some samples these planes were scarcely visible. With the plane strain samples failure was invariably on a distinct plane having a "strike" in the direction of the plane strain axis, and a "dip" from the top corner of the sample to the bottom on the other side. The dip angle was generally slightly greater than 60°.

Figure 10 also shows the values of the intermediate principal stress (σ_2), plotted in terms of $\sigma_2 - \sigma_3$). The curves average about 5kPa, and do not show distinct peaks. The value from the "upright" test is somewhat different to the other three, including the horizontal sample in which the direction of the major principal stress is the same. The difference presumably arises because of a difference in stiffness in the vertical direction compared to that in the horizontal direction. This difference in stiffness is discussed in a later section

The results of this series of undrained triaxial and plane strain tests are summarised in Figure 11. They are presented in two forms. The upper one shows the undrained shear strength plotted against the angle of inclination to the vertical (θ). As indicated earlier the strength declines steadily form just above 30kPa for the vertical samples to about 20kPa for the horizontal samples. The actual ratio for the strength averaged from all tests was 64%. The data are also plotted in a polar diagram in the lower part of Figure 11. This is similar to that of Lo (1965).

It is not possible on the basis of undrained tests to determine the specific cause of the anisotropy associated with the direction of the principal stresses. It could be caused by a difference in the Mohr-Coulomb parameters c' and ϕ', or a different pore pressure response due to differences in stiffness in the vertical and horizontal directions. This is investigated closely later in this paper.

6 SAMPLE DISTURBANCE AND PORE PRESSURE CHANGES DURING SAMPLING AND PREPARATION FOR LABORATORY TESTS

In the undrained tests already described, the pore pressure was not measured and the effective stress was not known, either at the start or during the tests. In tests described in later sections (CU tests), the pore pressure was measured immediately after setting up the sample and throughout subsequent testing. This section examines only the pore pressure change that occured during sampling and setting up samples. If the field sampling operation is perfect, the pore pressure immediately after sampling and removal from the sample tube can be estimated theoretically making use of the pore pressure coefficients A and B, and the in situ earth pressure coefficient K_o. Because the soil is fully saturated the coefficient B = 1. This gives the value of the negative pore pressure, u_s (soil suction or pore water tension) in the soil sample as:

$$u_s = K_o \sigma'_v + A(1 - K_o) \sigma'_v$$

In the present case the values of σ'_v can be calculated from the following assumptions:

Depth = 3.33m (for Cylinders T24 to T33).
Water table depth = 1.0m
Soil density = 1.52gm/cm^3
Pore pressure coefficient A = 0.19 (from later tests)
At rest soil pressure coefficient K_o = 0.55.
This gives a value of u_s = 17.0kPa.

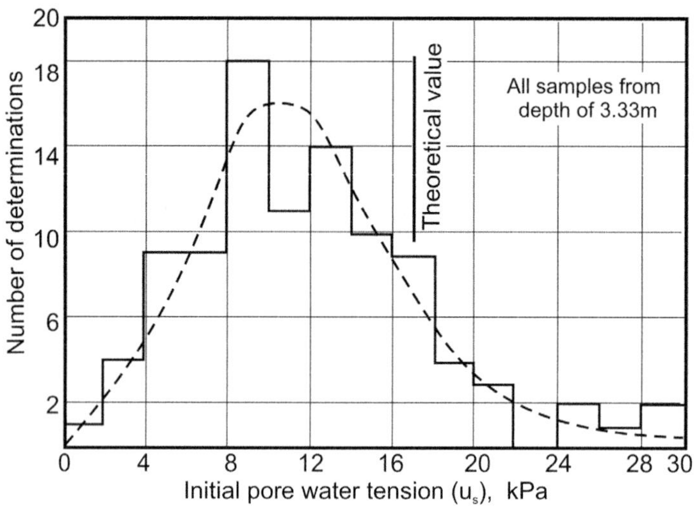

(a) Pore water tension in samples following application of cell pressure.

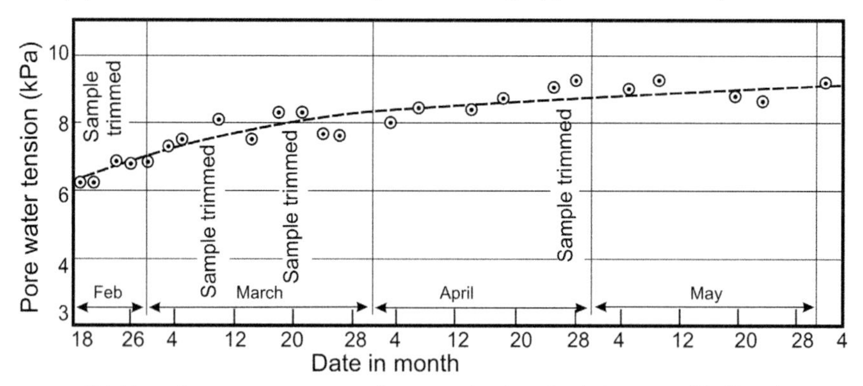

(b) Negative pore pressure (pore water tension)change with time in samples sealed with plastic sheeting and wax.

Figure 12. Initial pore water tension and changes when sealed in plastic and wax.

The pore water tension, u_s, was determined each time a soil sample was set up in a triaxial cell (or plane strain apparatus) ready for testing. The cell pressure was applied with all outlets from the sample closed, and the pore pressure (u_s) measured once equilibrium was achieved. The difference between these gives the value of the pore water tension. The measured values are illustrated in Figure 12(a) as a histogram plot. Also shown is the theoretical value of 17kPa. The difference between the actual value and the theoretical value can be regarded as a measure of the degree of disturbance, and the ratio u_s/u_{st} where u_{st} = the theoretical value, used to express this. There is wide scatter in u_s with an average value of 11.5kPa, so that the ratio u_s/u_{st} is close 65%. This is higher than most of the values from the Bothkennar tests (Hight, et al, 1992), where measurements were made on samples obtained by three different methods. Two methods involved special large diameter Canadian samplers developed by Laval and Sherbrooke universities, and the third was conventional thin walled sampling tubes. The results showed considerable scatter, with the ratio u_s/u_{st} ranging from 20% to 100%. Most values were below 50%.

The difference in values is not surprising. The Bothkennar samples were taken using special equipment from depths up to 17m while the Mucking samples were taken from a pit down to a depth of 5m. As explained above, the Mucking samples were taken in large diameter tubes and immediately sealed very firmly on site, so that disturbance opportunities were limited to the actual sampling operation and the later trimming and setting up of samples in the laboratory. In contrast the Laval samples were extruded from the large diameter tube on site and then sealed for further handling and transport, while the Sherbrooke samples were taken through drilling mud that had to be removed when the sample reached the surface, so water was available to the sample from the drilling mud.

The principal causes of sample disturbance are damage to the fabric or structure of the soil which occurs during sampling and handling and changes in its water content, either by drying or absorption. With some of the present samples, soil surfaces formed with the wire saw appeared shiny as though a thin film of water covered the surface. This could be due to the release of free water stored in the tiny root holes or cavities formed by the occasional shells that were present. In general, it is more likely that there will be loss of water due to evaporation during trimming or slow loss through sealing material.

With the Mucking samples, one such cause was the rubber membrane. These were normally dry when installed around the sample, but slightly damp when removed. This would account for a loss of about 0.5gm in the weight before and after testing. Other possible causes were loss of moisture through the plastic and wax sealing used on blocks extruded from the cylinders, and evaporation during handling and trimming the samples. Some simple experiments were conducted to examine these factors.

Blocks of soil extruded from the cylinders were used in stages between which they were wrapped in plastic sheeting and sealed with wax. To measure possible changes in pore water tension, transducers were installed through the sealed surface of one of these blocks. This was a block from Cylinder T30 having a diameter the same as the cylinder (25cm) and a thickness of 20cm. The result is shown in Figure 12(b). The initial suction was about 6kPa, and this rose to just over 9kPa over the three month period of the experiment.

To determine the influence of evaporation, two identical 7.6cm by 3.8cm cylindrical samples were set up side by side, one on weight scales and the other with a pore pressure transducer embedded in it. Readings were taken over a period of about 5 hours. The experiment was carried out on samples from Cylinders T30 and T42. The results are shown in Figure 13. With both samples the rate of change of pore water tension and evaporation is surprisingly large. The rate of evaporation and consequent weight loss was almost identical for the two samples but the tension changes were somewhat different. There was an increase in the pore water tension of about 40kPa in a time of 3.5 hours to 7 hours. The tests could not be continued beyond a tension of about 45kPa as the pore pressure transducers ceased to function beyond this value.

The time to trim triaxial samples and set them up in the apparatus was quite short due to the ease with which samples were trimmed with a wire saw. Unless there was an unforeseen

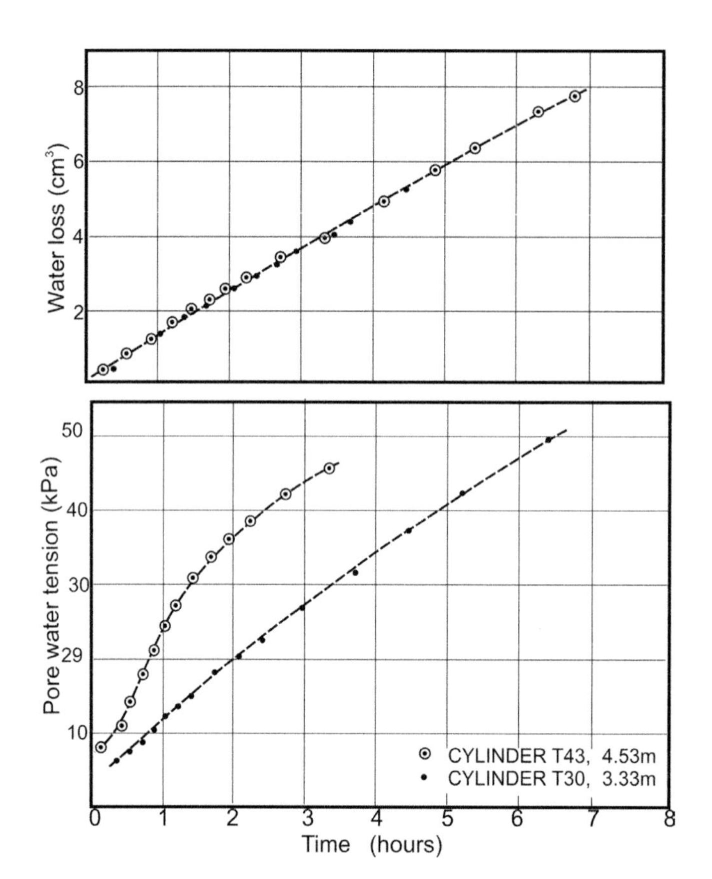

Figure 13. Weight loss and pore water tension changes in triaxial samples exposed to laboratory atmosphere.

interruption the time was less than 30 minutes. This meant that the increase in pore water tension during sample preparation was unlikely to be more than about 5 to 10kPa.

So far in this section only the sources of induced variability have been discussed, and it is not possible to know the extent to which these are responsible for the range of strengths obtained in the quadruple tests on samples from almost identical locations in the ground. Some light is thrown on this question by the consolidated undrained tests described in the next section. Figure 15(b) shows the effective stress paths from undrained tests on samples consolidated to effective stresses ranging from about 2kPa to 20kPa. From about 5kPa to 20kPa the stress paths tend to converge towards a similar failure stress, but there is still a small but significance difference in the peak value.

A simple investigation of the natural variability of the Mucking clay was made by measuring the water content of samples taken from a regular grid of points on both horizontal and vertical planes in one of the cylinder samples. The results are given in Figure 14, and show a surprisingly wide variation in water content. However, in accordance with expectations, the water content is reasonably constant on the horizontal cross section and highly variable on the vertical cross section. It appears therefore that variations in the undrained strength of the soil may be due as much to natural variability as to induced variability. It was not possible to determine the precise origin of this natural variability, but it is perhaps to be expected in an alluvial deposit where the material deposited by the river or stream is likely to vary with weather conditions. The nature of the material deposited and the rate of deposition can be expected to vary significantly between wet and dry seasons and during severe storm events. These two causes of variability are not entirely independent of each other. Coarser material and fine material are likely to have different undrained strengths and coarse material is more likely to lose moisture from evaporation during sample trimming in the laboratory.

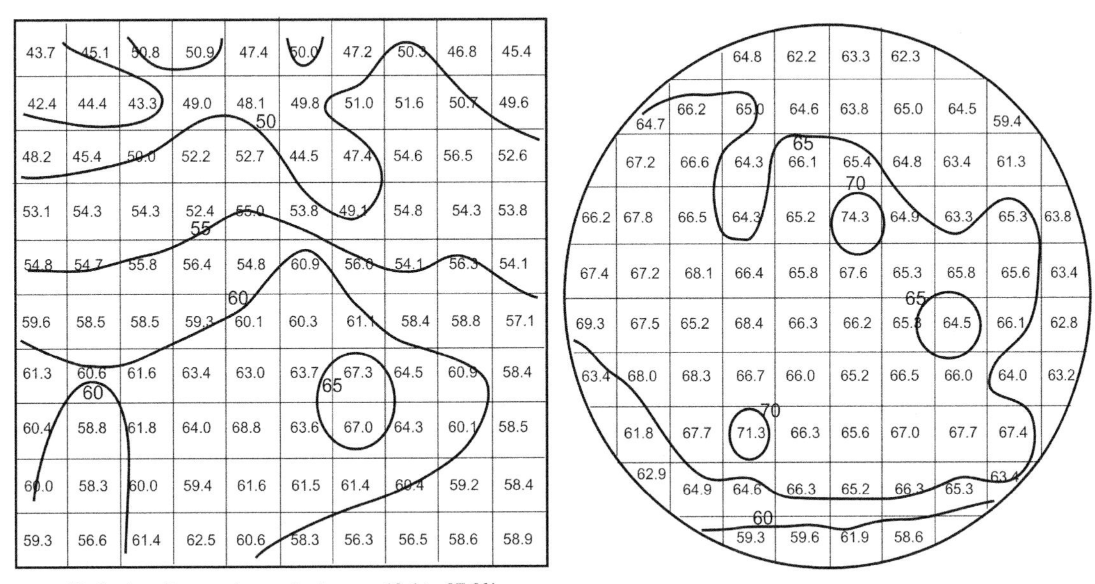

Vertical section - water content range 42.4 to 67.3%
Horizontal section, water content range: 59.3 to 74.3 %

Figure 14. Natural water content on a grid of points on horizontal and vertical cross sections.

7 A SERIES OF CONSOLIDATED UNDRAINED (CU) TRIAXIAL TESTS.

These tests are a series of conventional consolidated undrained triaxial tests extending over a range of consolidation pressures from just over 3kPa to just over 300kPa. A back pressure of 200kPa was used in all tests to ensure there was no air in the samples or measuring apparatus.

Measurements of the coefficient of consolidation were made by both oedometer and pore pressure dissipation tests prior to commencing these tests; based on these a strain rate of 0.33% per hour was used. As the tests were taken to a strain of about 8%, this meant the duration time for each test was about 24 hours. Allowing for consolidation time, this made possible a set of four tests every two days.

Figure 15 shows curves of deviator stress and pore pressure change against strain for consolidation pressures of 3.4kPa, 50.4kPa, and 152.1kPa. The results conform to normal behaviour for soft normally consolidated natural clays. At the lowest stress level, the pore pressure hardly rises above zero and then declines to a slightly negative value, while at the highest level the pore pressure rises at the same rate as the deviator stress and slightly exceeds

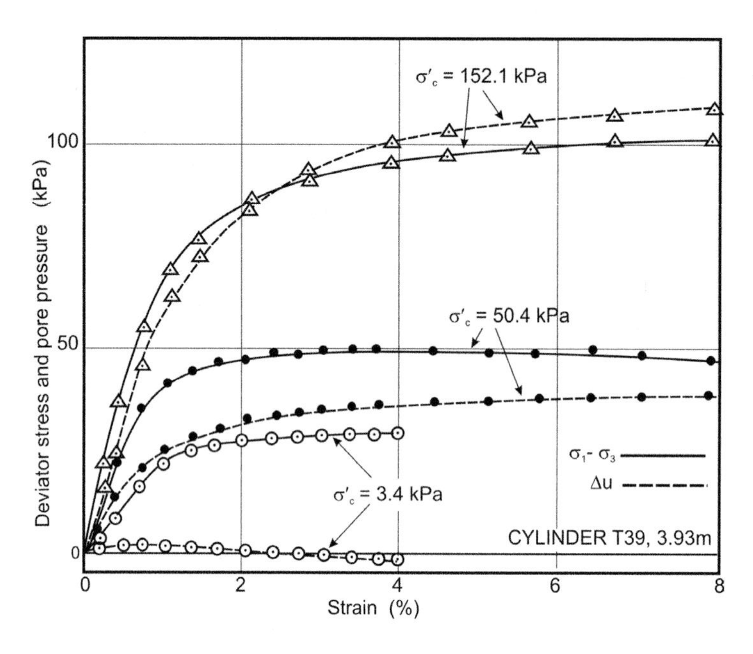

(Left) Figure 15. Deviator stress and pore pressure curves from consolidated undrained triaxial tests.

(Below) Figure 16. Stress paths in consolidated undrained triaxial tests

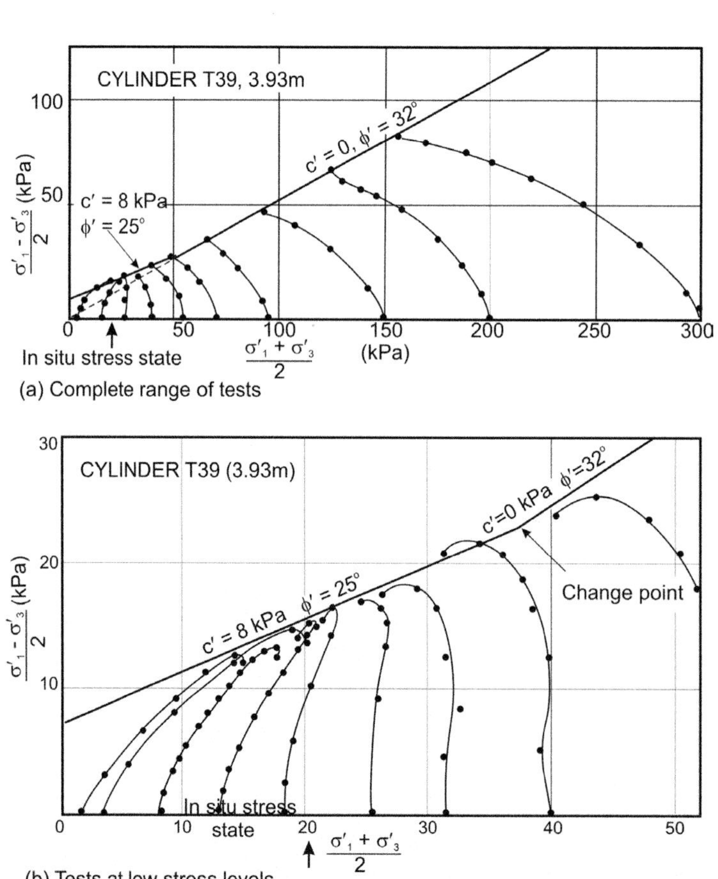

it near the end of the test. In other words the pore pressure parameter A is almost zero at the low consolidation pressure and close to unity at the high pressure.

Figure 16 shows the effective stress paths and the Mohr Coulomb failure lines. The upper figure shows stress paths from the full range of consolidation pressures, but does not include all the tests at low consolidation pressure, as including these would have made the drawing too cluttered. For this reason the results of all the tests at low pressures are shown in the lower figure. The Mohr Coulomb failure envelope has been drawn as two distinct straight lines. However, this is an idealisation as there is probably a gradual steepening of the envelope as the pressure increases until it finally reaches the "virgin" line passing through the origin. The additional strength in the lower stress range is considered to be due to some form of bonding that develops between the particles as the deposit ages. The question of the pre-consolidation (or yield) stress is addressed in a later section.

Figure 17 relates the undrained shear strength and initial stiffness modulus to the effective consolidation pressure (σ_c'). The stiffness modulus has been taken as the initial slope of the deviator stress versus strain curve. Both graphs show a relatively flat portion which begins to rise at σ_c' values somewhat in excess of the in situ vertical stress. At higher stress levels the points can be expected to lie on a straight line through the origin as the behaviour approaches that of a normally consolidated clay. That this is not the case is probably attributable to the

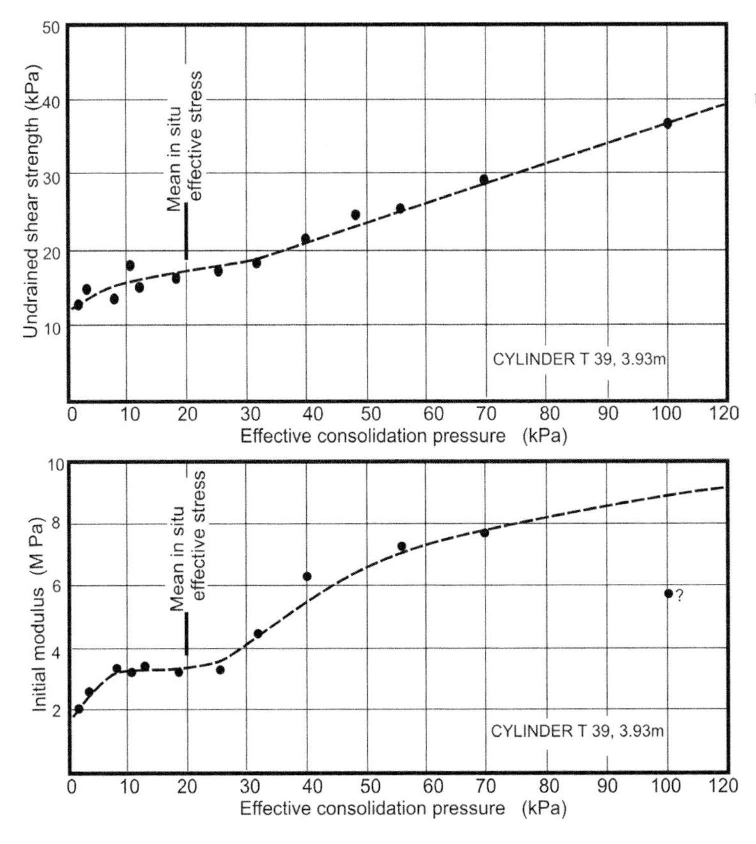

Figure 17. Undrained shear strength and undrained modulus versus consolidation pressure.

rather short consolidation time allowed in the tests. Had a much longer time been allowed both the undrained shear strength and initial modulus would probably have been significantly higher. The trend in the initial modulus is not well defined at higher stress levels due to some rather erratic scatter in the values.

The relationship between undrained shear strength and consolidation pressure in Figure 16 raises the question of how best to measure the undrained shear strength of undisturbed samples of clay, namely whether it is preferable to do a strictly undrained test or to consolidate the sample to its in situ stress and then do the undrained test. This issue is addressed in the following section

8 INFLUENCE OF STRESS PATH ON UNDRAINED BEHAVIOUR AFTER CONSOLIDATION TO IN SITU STRESS STATE

The tests described in this section had three main objectives:

(a) to compare unconsolidated undrained (UU) behaviour with that after consolidation to the in situ stress level.

(b) to compare behaviour after isotropic consolidation and after consolidation to the in situ K_0 stress state (anisotropic consolidation).

(c) to compare behaviour in extension loading to that in compression loading.

All of the tests were carried out on samples from Cylinder T29 at a depth 3.33m. Consolidation of samples to the average in situ stress level (isotropic consolidation) is straightforward as it simply involves raising the cell pressure and allowing drainage from the sample. Consolidation to the actual (Ko) stress state is more complex as it involves applying different vertical and horizontal effective stresses. This was achieved making use of the mercury pot pressure system and the motor driven gear boxes to raise and lower the pots. The actual sequence of applying the stresses could be done in variety of ways. The method chosen was that which would involve least volume change or distortion of the samples and consisted of following stages:

(1) The sample was set up, the cell pressure applied, and the pore pressure measured. The difference between the two gave the initial effective stress.

(2) The vertical stress was then increased until it reached the required K_0 value of 0.55.

(3) The vertical and horizontal stresses were then raised simultaneously in accordance with the adopted K_0 value until the in situ state was reached.

The first stage involved leaving the sample for 24 hours to ensure full pore pressure equalisation would occur. The stresses in the next two stages were applied very slowly to ensure drainage would occur as the stresses were applied. The time required was between 1 and 2 days, depending on the initial pore water tension. The actual shear tests were carried out at a controlled strain rate of 0.33% per hour and taken to a strain of 8%. The duration of each test was thus about 24 hours. The tests on four identical samples were carried out simultaneously.

To carry out the "K_0" tests it was first necessary to estimate the in situ stress state. This involved making an assumption regarding the K_0 value. Using the empirical relationship $K_0 = 1 - \sin \phi'$

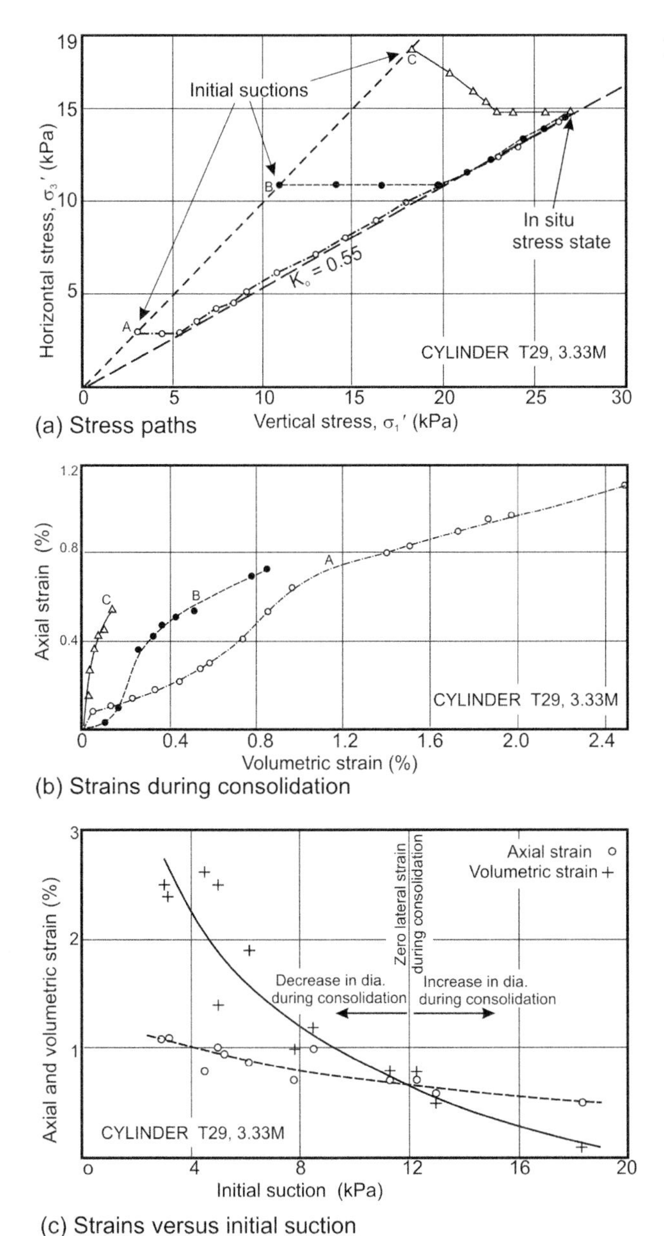

Figure 18. Anisotropic consolidation to the in situ stress state.

gives $K_o = 0.47$, which would be valid at the time of deposition of the soil, but soil creep with time is likely to cause an increase in the horizontal stress. The value of K_o can be related to the over-consolidation ratio (OCR) for soils that are genuinely over-consolidated. However the Mucking clay shows an OCR of between 1.5 and 2.0 which is not due to stress history, but rather to the growth with time of structural bonds referred to as "ageing" or hardening. Taking account of these factors a K_o value of 0.55 was adopted. The vertical and horizontal effective stresses are then 26.7kPa and 14.7kPa respectively and the average in situ effective stress is 20.7kPa.

Typical examples of the stress paths during anisotropic consolidation are shown in Figure 18, along with measurements of the axial and volumetric strains that occurred during this process. As shown earlier in Figure 12(a) the initial pore water tension ("suction") varies widely despite identical preparation and test procedure, and that was the case with the present tests. The pore water tension was generally below that expected theoretically. This meant that the initial effective stress was less than the in situ value and consolidation to the in situ stress state involved increasing both the vertical and horizontal effective stress. This was the case with Samples A and B in Figure 18. Sample C had an initial effective stress higher than the in situ horizontal pressure, and the consolidation stage involved lowering the horizontal stress and raising the vertical stress. The stress paths do not exactly follow the intended path; this is due to small errors in the gear box settings controlling the mercury pots.

The graphs in Figure 18 (b) show that Sample A suffered volumetric strain about double the axial strain, which means the diameter of the sample was decreasing, while Sample C shows an opposite trend: the sample diameter is increasing. Figure 18 (c) shows the total volumetric and axial strains from all the samples subject to K_o consolidation plotted against the initial pore water tension (suction). Those samples with low initial suction undergo large volumetric strains, over double the axial strains, while those samples with initial suctions of about 12kPa undergo equal volumetric and axial strains. This behaviour indicates that the volumetric strain in these samples is very sensitive to stress changes in the horizontal direction, suggesting the soil skeleton may be much less stiff in the horizontal than the vertical direction. This issue is discussed later in this paper.

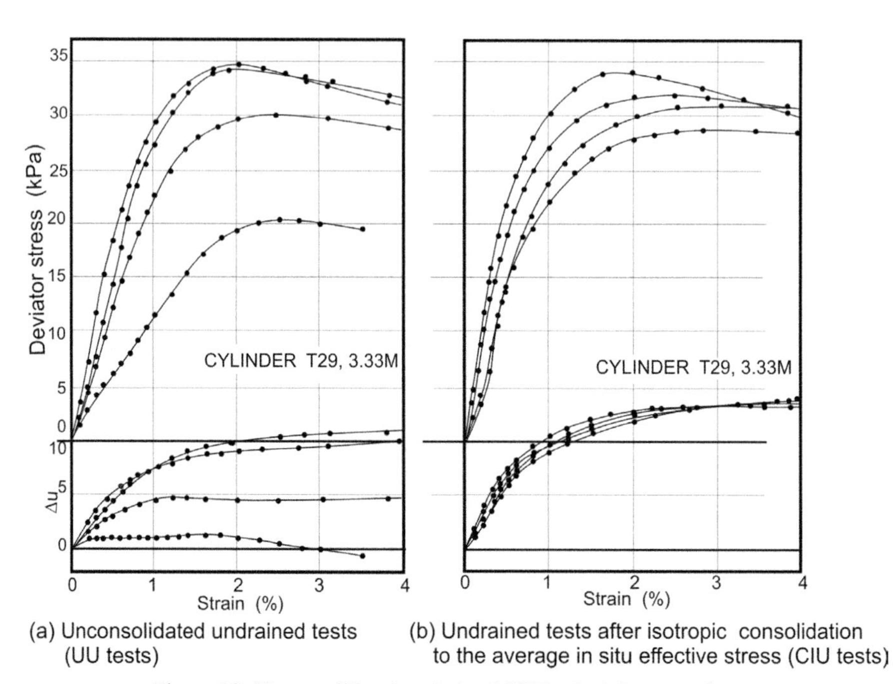

Figure 19. Unconsolidated undrained (UU) triaxial tests and tests after consolidation to the in situ mean stress state (CIU tests).

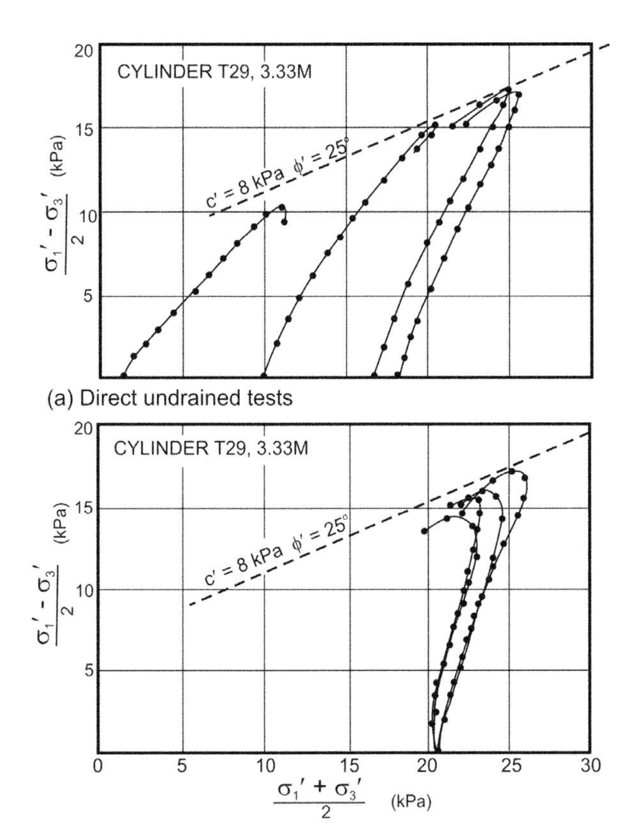

Figure 20. Stress paths in the tests shown in Figure 18.

Figures 19 and 20 show two sets of identical undrained triaxial tests carried out on samples from the same cylinder of clay. The first set is unconsolidaed undrained (UU) tests, and the second set is after isotropic consolidation to the average in situ stress state, namely 20.7kPa. These will be designated CIU tests The purpose of the tests was to investigate the merits of first consolidating samples to their in situ effective stress state prior to undrained testing, compared to normal practice of directly testing the samples. Figure 19 shows graphs of deviator stress and pore pressure versus strain. The samples tested directly show wide scatter while those tested after consolidation to the in situ stress level show significantly more consistent behaviour.

Figure 20 shows the stress paths in all the tests, indicating that the scatter in the unconsolidated undrained tests is directly related to the scatter in initial effective stress caused by the widely varying values of initial pore water tension. As mentioned earlier, the initial pore water tension is believed to indicate the degree of disturbance that occurs during sampling and preparation of the test samples. The conclusion to be drawn from this is very clear – the most consistent and reliable estimates of undrained strength are to be expected from samples consolidated to the in situ stress level. It is important to note that the method of consolidation referred to here as K_o consolidation is not true K_o consolidation. True K_o consolidation implies zero strain in the horizontal direction, which would mean continual adjustment of the horizontal stress (the cell pressure) to maintain this condition. This would be neither very practical nor desirable, since it

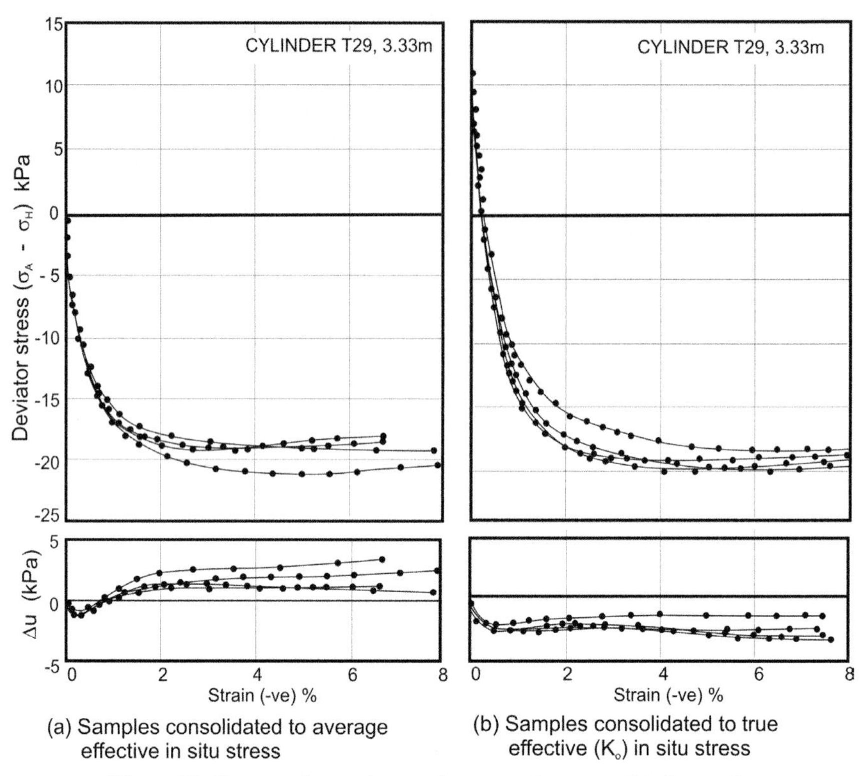

Figure 21. Compression and extension tests after consolidation to the average confining stress and the estimated K_o stress state.

would result in widely different values of horizontal stress depending on the initial pore water tension.

We will now examine the different behaviour between compression and extension tests. Figure 21 shows triaxial extension tests on samples consolidated to the average in situ stress state and to the K_o state. Four tests of each type were carried out, as the graphs show. The deviator stress is plotted as $\sigma_A' - \sigma_H'$ in order to avoid any ambiguity. In extension tests the vertical stress is less than the horizontal stress so the deviator stress defined in this way becomes negative. It should be understood that extension tests are not tensile tests, as both vertical and horizontal stresses remain positive.

These tests show very consistent behaviour between the four samples within each set of four, and that behaviour in the two types of test is very similar. Important points from the tests are.

(1) The peak value of the deviator stress is very similar, and is reached at a strain of about 4%.

(2) The deviator stress does not decline once this peak value is reached, and remains steady until a strain of about 8%, at which value the tests were stopped. This is in contrast to the behaviour in compression tests which generally show a peak deviator stress followed by a slight decline (see Figure 18(b) for example).

(3) The strength measured in extension tests is well below the strength measured in coventional compression tests. The peak deviator stress in extension is close to 20kPa while that in compression is close to 31kPa.

The graphs in Figure 21 have been averaged and the results from both test types shown in Figure 22. The results of all these tests are summarised in Figure 23 in the form of stress strain graphs and stress paths in terms of effective stress.. Because each set of four tests all showed consistent behaviour with very small scatter, only the averaged curves are shown. Also shown in Figure 23(b) are the theoretical stress paths according to elastic theory. These are shown for both isotropic and anisotropic cases. It is shown later in this paper that vertical stiffness of the soil skeleton is double the horizontal stiffness, as indicated in the figure. These theoretical paths are not greatly different and are a reasonable indication of behaviour, at least at low stress levels. However, soils are not elastic materials and close agreement is not to be expected.

The important points to note from these tests and Figure 23 are the following:

(1) To measure undrained shear strength in triaxial tests it is preferable to first consolidate

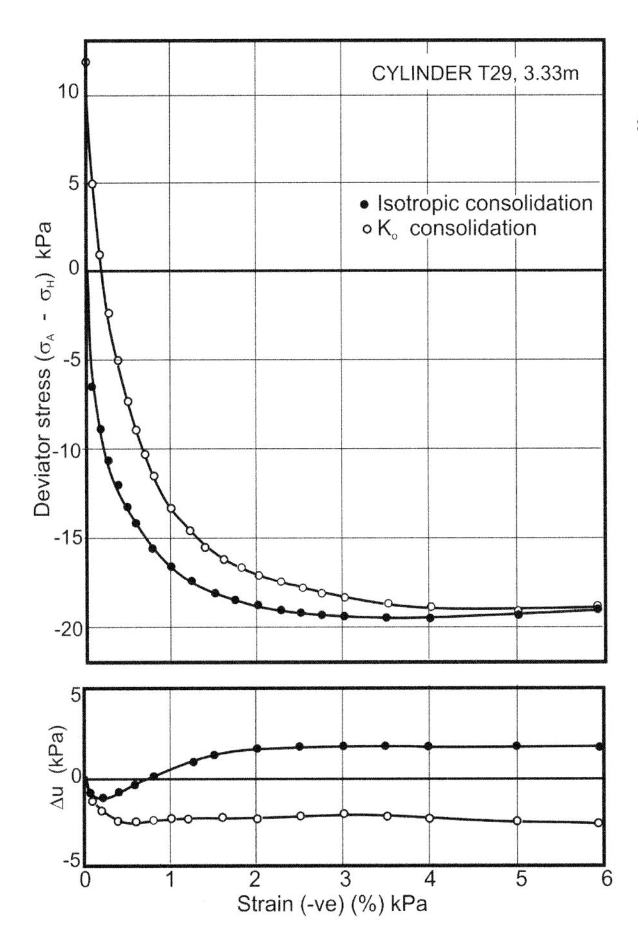

Figure 22. Extension tests after consolidation to the average in situ stress state and the K_0 state (averaged from four tests of each type).

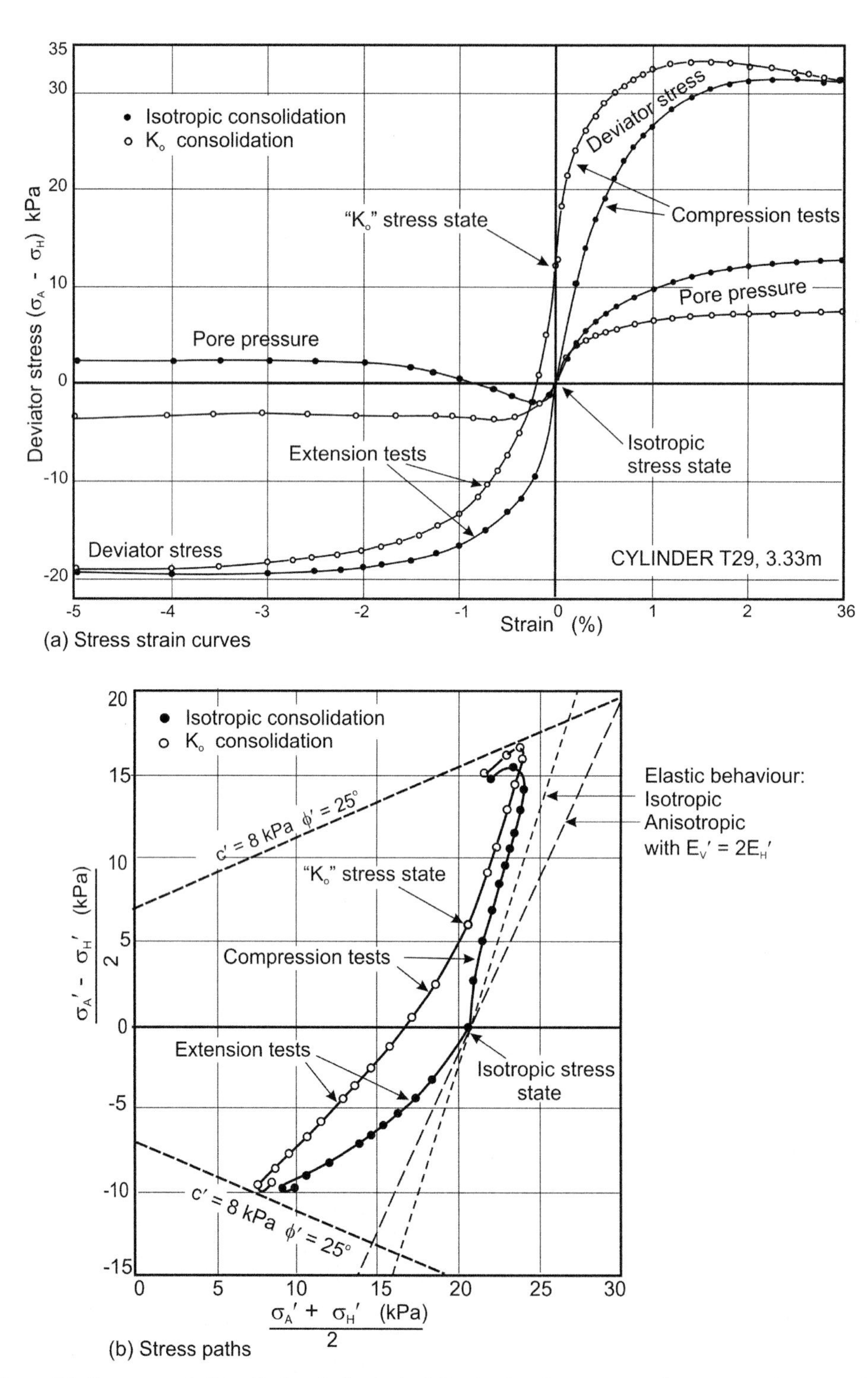

Figure 23. Summary of all vertical undrained triaxial tests showing stress-strain curves and stress paths

the sample to the in situ stress state, rather than the conventional practice of unconsolidated undrained measurement.

(2) Initial consolidation to the K_o state produces somewhat different behaviour from isotropic consolidation to the average in situ stress. The difference is not dramatic. The peak deviator stress is slightly different but becomes almost identical at a strain of 6%.

(3) The extension tests show a peak strength that is only about 60% of the peak strength in compression tests. This is due to the different stress paths resulting from different skeleton stiffness in the horizontal and vertical direction referred to above. This is discussed in more detail later in this paper

9 THE EFFECTIVE STRENGTH PARAMETERS (c′ AND Φ′)

The consolidated undrained tests illustrated in Figures 15 and 16 showed typical behaviour of soft clay in such tests and indicated the values of the effective strength parameters c′ and ϕ′. Earlier tests (Figures 6, 8, and 10) indicated that the soil is anisotropic when subject to consolidated undrained triaxial tests at varying inclinations to the vertical. Vertical samples showed the highest undrained strength and horizontal the lowest. The objective of this section was to investigate further the issue of anisotropy. Various types of test were carried out after consolidating the soil to its average in situ effective stress. The first set was plane strain tests on vertical and horizontal samples, similar to earlier tests (Figure 10), but with pore pressure measurement. Four identical samples were tested in each case. The results were averaged and are presented in graphical form in Figure 24.

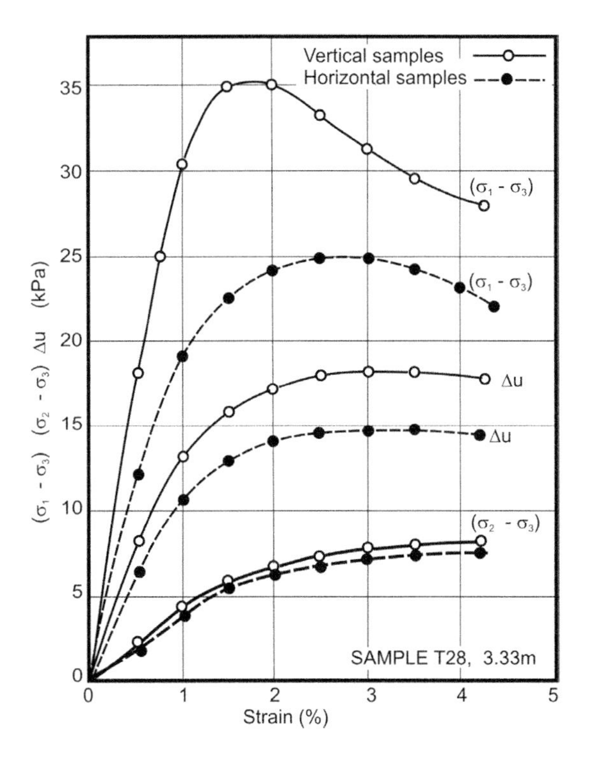

Figure 24. Vertical and horizontal plane strain tests after consolidation to the in situ stress (20.7kPa).

The behaviour is similar to that in Figure 10 although the samples were from different cylinders taken at different depths. The vertical samples show a sharp peak deviator stress at a strain close to 2%, while the horizontal samples show flat deviator stress-strain graphs without a clearly defined peak. The intermediate principal stress (σ_2) plotted as (σ_2-σ_3) is relatively small and does not vary much between vertical and horizontal samples. It was not possible to carry out extension tests with the plane strain apparatus as these would involve intermediate stresses lower than the cell pressure and each end of the sample would move away from the end platens. In other words there would be deformation along the plane strain direction and the essential condition of plane strain behaviour would be violated.

The second set of tests was both compression and extension triaxial tests on samples prepared with inclinations varying at 15° intervals from vertical to horizontal. The aim of these tests was to investigate whether there was any evidence of anisotropy with respect to the effective stress parameters c′ and ϕ'. If such anisotropy existed its most likely origin would be a difference in strength between horizontal and vertical planes arising from either the deposition process that formed the soil or the difference in the natural stress state on the two planes. Tests on horizontal and vertical samples would be unlikely to indicate such anisotropy as the form of failure would involve failure surfaces that cross the horizontal and vertical planes rather than lie along them. Tests on samples cut with inclinations of 30° or 60° would be the most likely to induce failure on these planes

Several samples were tested at each inclination and the results averaged to give the graphs shown in the following two figures. Figure 25 shows typical stress paths for the tests at 30° intervals. In order to avoid confusion only the results at 30° intervals are included. The Mohr-Coulomb failure lines shown in this figure are obtained from the whole range of test types carried out and presented in later figures. The stress paths indicate that in compression tests the undrained shear strength varies from a maximum in vertical samples to minimum in horizontal samples, while in extension tests the reverse is true.

Figure 26 shows the results plotted in a similar manner to the earlier tests in Figure 11. Included in Figure 26(a) is the average graph from the earlier tests, which were all UU triaxial compression tests, while the current tests are CIU tests. The present tests show a similar strength variation for compressive tests as the earlier tests, namely the horizontal strength is about 70% of the vertical strength.

There is thus no substantial evidence from these tests that the effective stress strength on horizontal or vertical planes is any different from other planes and thus no indication of anisotropy with respect to the parameters c′ and ϕ'. There is, however, an indication in the different modes of failure in the extension tests that the strength on vertical planes is less than that on horizontal planes. The failure modes are shown in Figure 26(b). With vertical samples the mode of behaviour was uniform "necking" of the sample as indicated in the figure. With horizontal samples, however, the behaviour was quite different and involved failure on distinct planes which in the ground would have been vertical planes. The inclination of the failure planes was between 60° and 70° to the vertical plane, which in extension tests was the plane of the major principal stress. This is a little higher than the theoretical value according to the relationship:

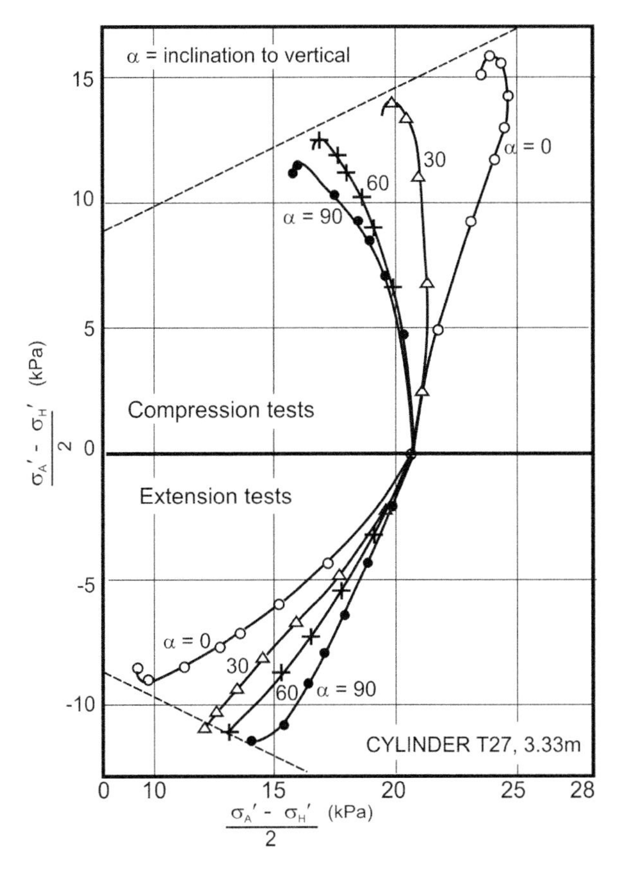

(Left) Figure 25. Stress paths in compression and extension CIU triaxial tests at varying inclinations.

(Below) Figure 26. Triaxial compression and extension tests at angles 0, 30, 60, 90 degrees to the vertical

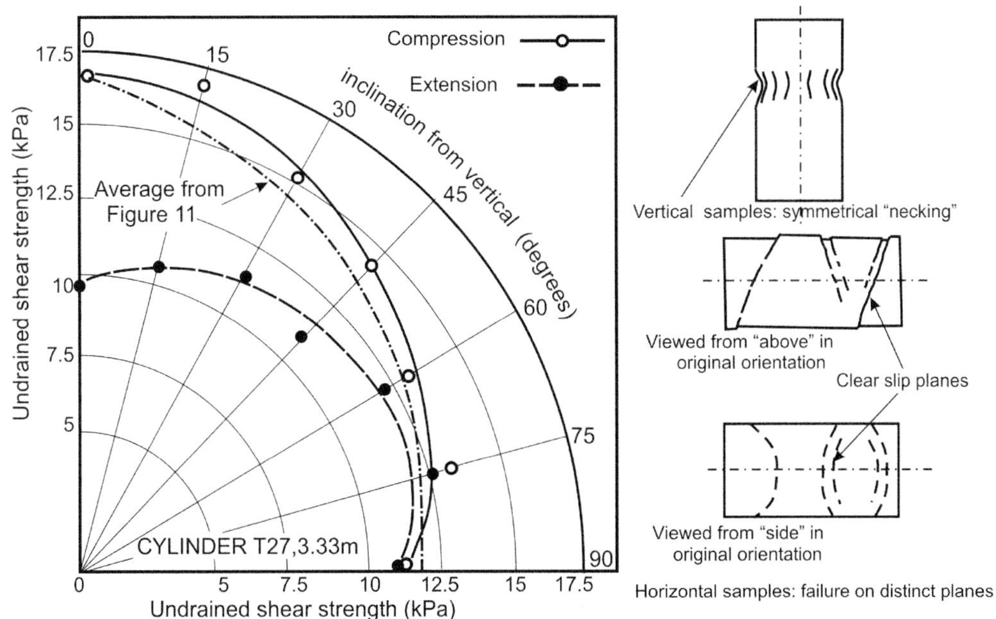

(a) Undrained shear strength versus sample inclination

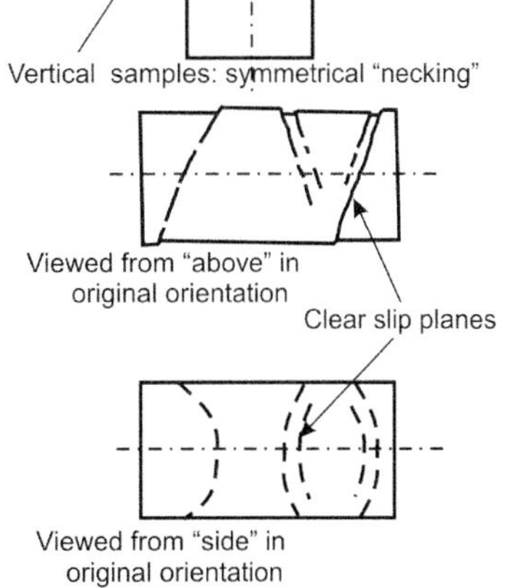

(b) Form of failure

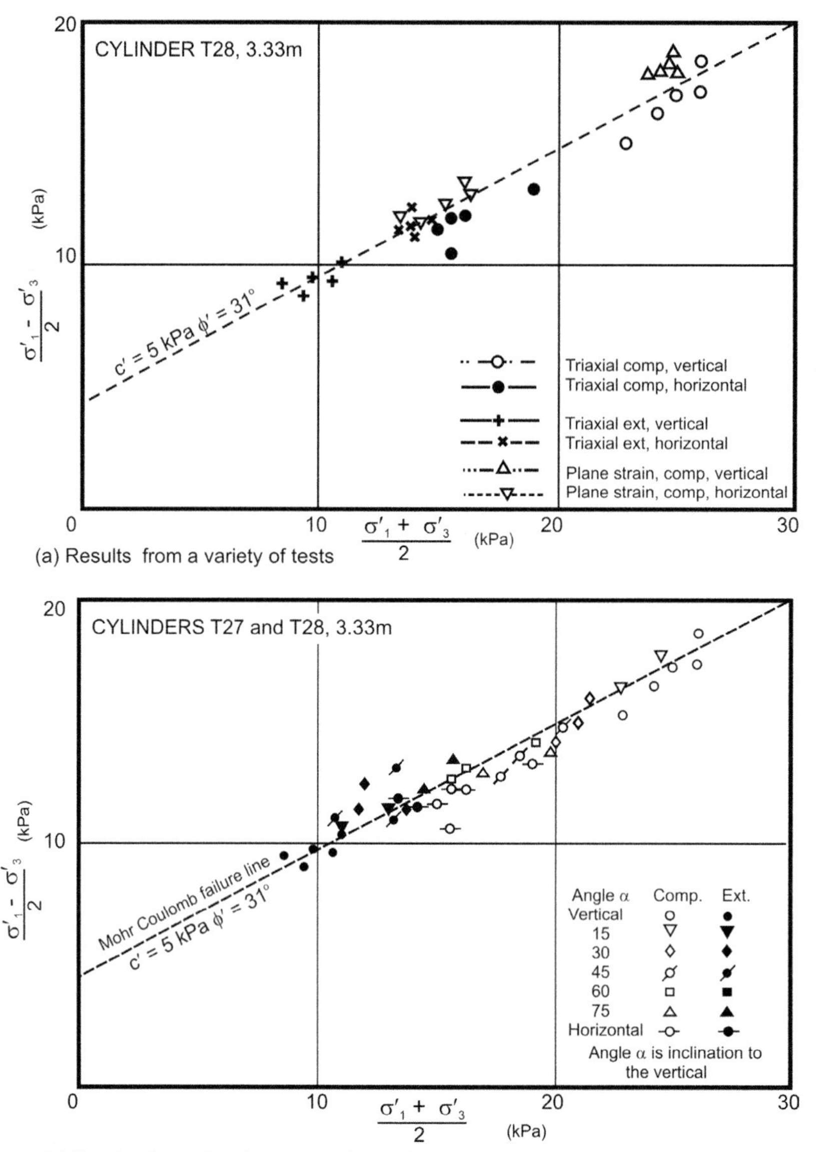

Figure 27. The Mohr Coulomb failure line and the effective stress parameters from all tests.

$$\alpha = 45 + \phi'/2$$

where α is the angle to the major principal stress plane. Thus, although this behaviour during extension tests suggests a lower strength on vertical planes than horizontal planes, this difference must be very small as it is not reflected in the triaxial tests.

The effective stress parameters are examined in Figure 27 by plotting the failure values from all tests in which the samples were initially consolidated to the in situ stress state. The upper figure shows values from triaxial compression and extension tests and plane strain compression tests. The fitted line is slightly different from the line shown in Figure 15 from earlier tests. The points all lie close to this line except for the plane strain tests, which lie a little above the line. The lower figure shows the values from triaxial compression and extension tests at varying inclinations. These have been shown separately from the other data to avoid the confusion caused by a mass or data points. Some of the small differences may be due to the fact that the tests in Figure 27(a) are all from Cylinder T28, while those in Figure 27(b) include the tests on inclined samples which came from Cylinder T27. Both cylinders are from the same depth, namely 3.3m.

The failure line in Figure 27(a) confirms the general validity of the Mohr Coulomb failure criterion, especially with respect to the influence or otherwise of the intermediate principal stress. In a compression triaxial test the intermediate principal stress is the same as the minor principal stress (the cell pressure) while in an extension triaxial test the intermediate principal stress is the same as the major principal stress, which is now the cell pressure. The vertical stress thus becomes the minor principal stress. The results in Figure 26 do not provide an explanation for the different failure modes between extension and compression triaxial tests. Thus if there is a difference in effective stress strength between vertical planes and horizontal planes it is very small.

10 SOIL SKELETON STIFFNESS IN THE HORIZONTAL AND VERTICAL DIRECTIONS

A series of tests was carried out to investigate the effective stiffness of the soil structure in the horizontal and vertical direction. The first tests were conventional oedometer tests and a triaxial test on a vertical sample in which no horizontal deformation was permitted. This involved adjusting the cell pressure as the test proceeded in order to maintain the zero lateral strain condition.

In addition to these tests further oedometer tests were carried out on horizontal and vertical samples. The results are shown in Figure 28. The upper graph shows similar, but not identical behaviour in the two types of test. Both indicate a yield pressure of about 40kPa but the oedometer test shows higher compressibility than the triaxial test, the most likely explanation for which is the greater disturbance in the oedometer sample. The lower graph shows results of two oedometer tests on both horizontal and vertical samples. This indicates that at low stresses the compressibility in the horizontal direction is about double that in the vertical direction, and the yield stress is both lower and less well defined in the horizontal than the vertical direction. At higher stresses the compressibility tends to become uniform in both directions.

Further measurements of soil skeleton stiffness are shown in Figure 29. The upper graph shows the results of drained triaxial tests on vertical and horizontal samples, and the lower graph shows the strains occurring during isotropic consolidation in a triaxial cell. Both confirm the indication in Figure 28 that at low stresses the compressibility in the horizontal direction is approximately double that in the vertical direction, while at higher stress levels it becomes very similar. .

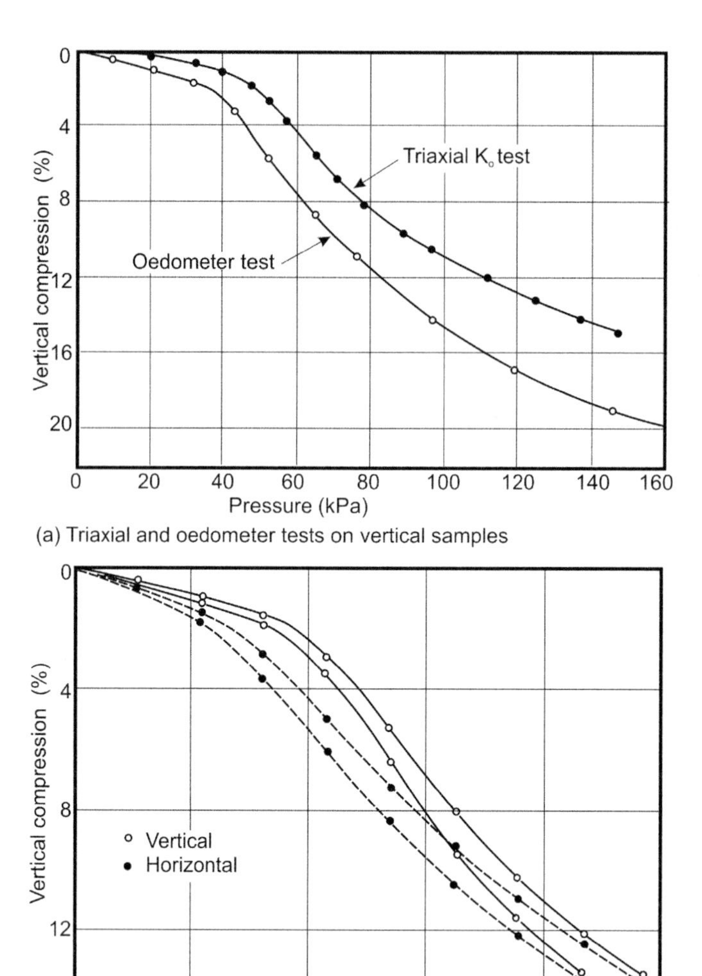

Figure 28. Soil skeleton stiffness in the vertical and horizontal directions.

11 BEHAVIOUR IN DRAINED TRIAXIAL TESTS

To ensure adequate pore pressure dissipation these tests had to be carried out at a slow strain rate, and the time needed for the tests varied between a few days and just over three weeks. Because of this long test duration it was only possible to carry out a limited number of tests. A primary aim of this study was to investigate the behaviour of the soil at the stress level under which it existed in its natural state in the ground. For this reason triaxial tests were carried out after consolidating the soil to its K_0 stress state in the ground. The soil was then subject to drained loading following four different stress paths as follows:

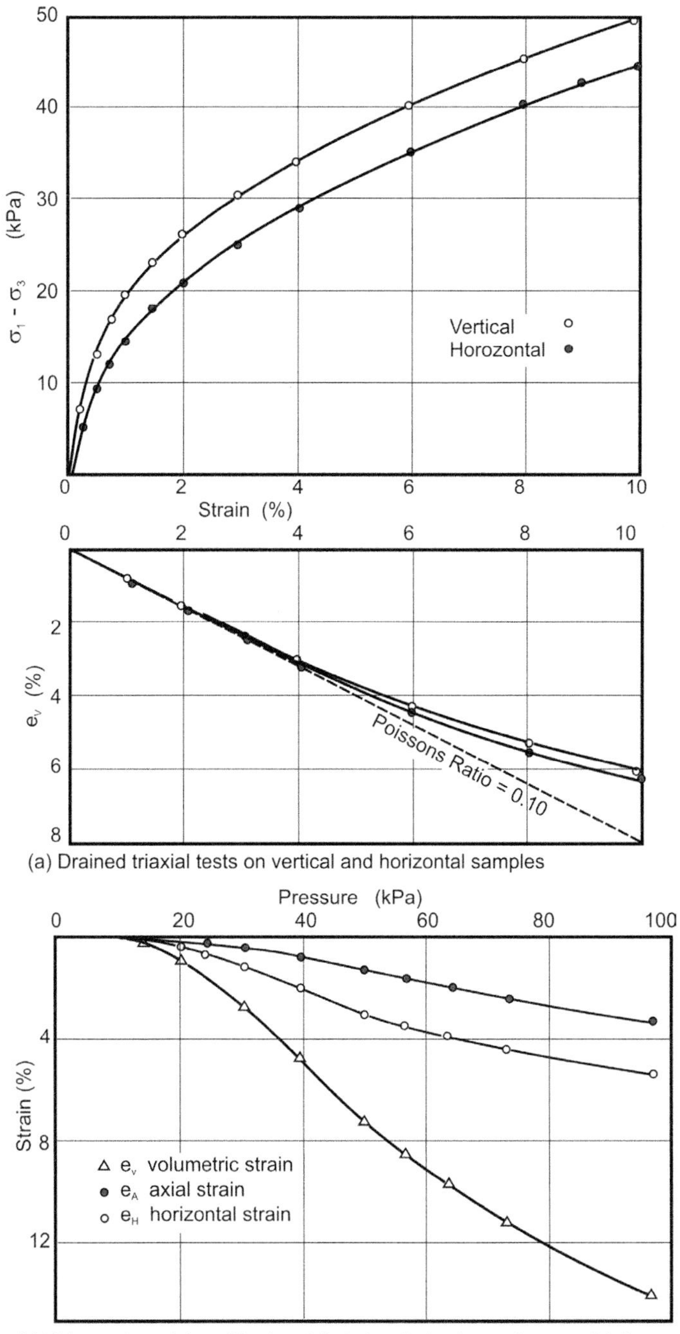

Figure 29. Further measurements of soil skeleton stiffness.

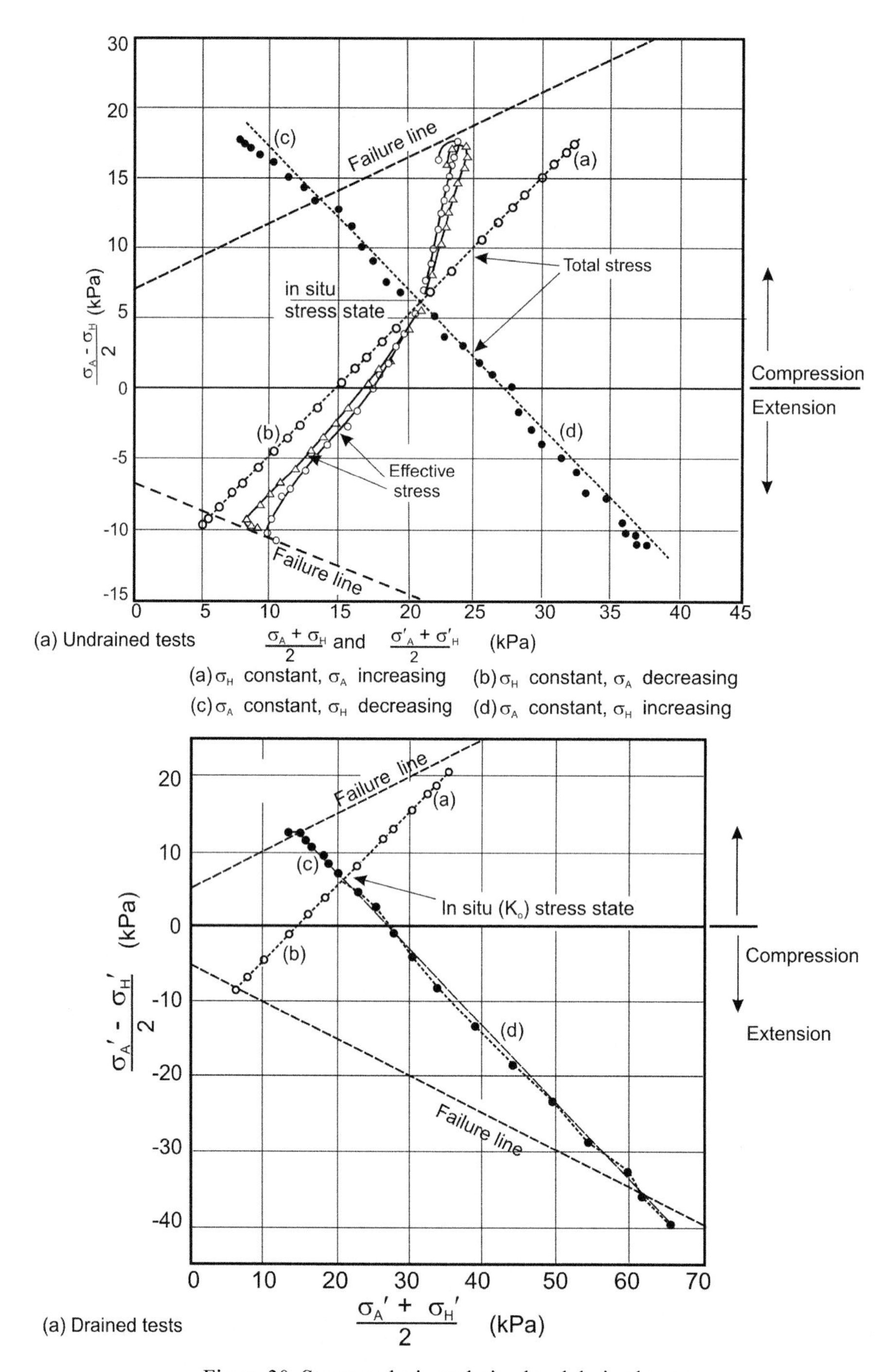

Figure 30. Stress paths in undrained and drained tests

(a) Constant horizontal stress, σ_H, and increasing axial (vertical) stress, σ_A, (compression).
(b) Constant horizontal stress, σ_H, and decreasing axial stress, $\sigma_{A,}$ (extension)
(c) Constant axial stress σ_A, and decreasing horizontal stress σ_H (compression
(d) Constant axial (vertical) stress $\sigma_{A,}$ and increasing horizontal stress σ_H (extension)

In the following discussion tests will be identified using these letters

Figure 30 shows these stress paths and the results of the triaxial tests for both undrained and drained loading. The stress path in Cases (a) and (b) is straightforward as it involves a constant cell pressure, and an increase or decrease in the vertical stress. Thus the intended stress path is exact, as Figure 30 indicates. The stress path in Cases (c) and (d) is more complicated as it involves holding the vertical stress constant while decreasing or increasing the cell pressure. Thus the actual stress path "wanders" a little from the intended path and corrections are made to correct this as the test proceeds.

For undrained loading, Figure 30(a) shows that provided the applied deviator stress is in the same direction the behaviour of the soil will be identical. In other words the effective stress will

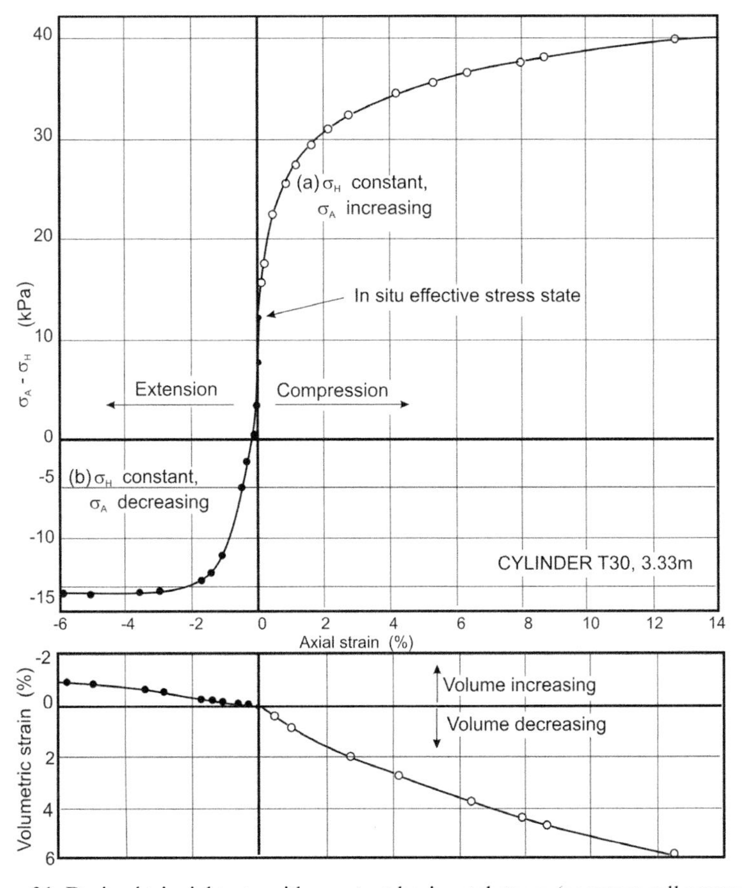

Figure 31. Drained triaxial tests with constant horizontal stress (constant cell pressure)

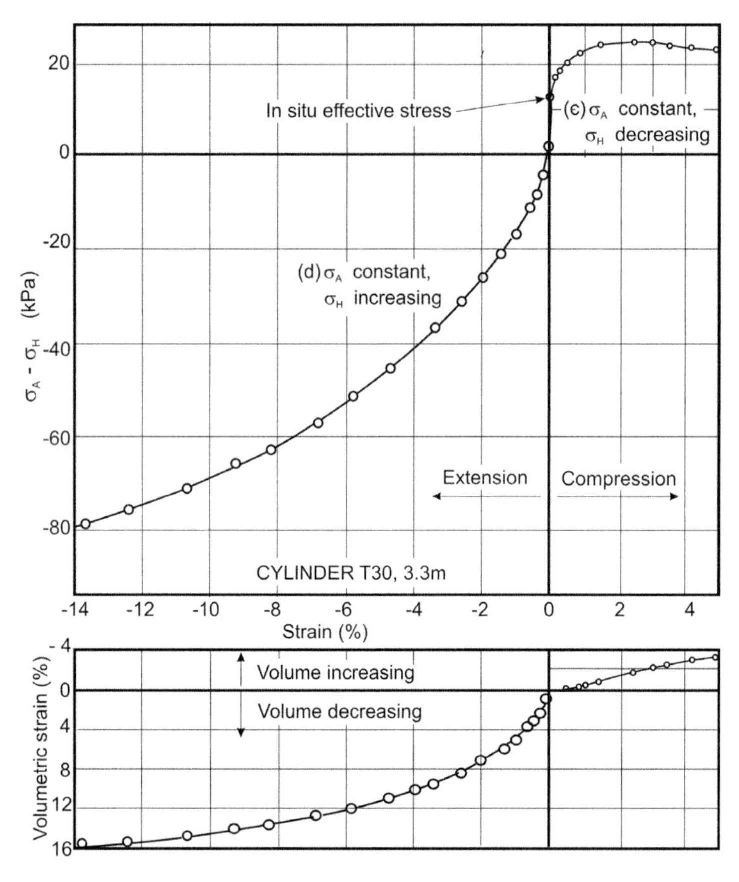

Figure 32. Drained triaxial tests with constant vertical stress.

be the same and lead to the same failure stress. The slight difference in the actual stress paths is to be expected from tests on natural soils.

For drained loading, Figure 30 shows very different behaviour, which is not at all surprising as the effective stresses to which the samples are subjected are very different. This behaviour is illustrated more fully in the following figures. The significant point to note from Figure 30 is that the failure stress state is very close to that expected from the Mohr-Coulomb failure line determined from undrained tests. However, it should be noted that for Cases (a) and (d), the deviator stress had not reached a peak value at a strain of 14% at which the tests were terminated. At this strain the samples were very distorted and the calculated failure stress can only be regarded as rather approximate.

Figures 31 and 32 show the complete deviator stress and volume change curves for the tests. Figure 31 gives the curves for Cases (a) and (b), that is for a constant horizontal stress and increasing and decreasing vertical (axial) stress, while Figure 32 shows the curves for Cases (c) and (d), that is for constant vertical stress and decreasing or increasing horizontal stress. When the mean stress level is decreasing, as in Cases (c) and (b), the soil quickly reaches its failure state, at a strain of approximately 3%. This is not very different from the failure strain in undrained triaxial tests, as shown in Figures 6 to 8. When the mean stress level is increasing, as in Cases (a) and (d) the increase in deviator stress is more gradual, and fails to reach a clear peak at the strain of 14% at which the tests were terminated.

The behaviour in Figure 32 is to be expected as it closely parallels that in active and passive pressure on retaining walls. Case (c) is, in effect, the active case as the vertical stress is constant and the horizontal stress is reduced to induce failure, while Case (d) is the passive case as the cell pressure pushes against the soil to move it towards the passive failure state. The volume changes that occur during shearing are also to be expected. Where the average effective stress is increasing as in Cases (a) and (d) the volume decreases significantly, while in the tests with the average stress decreasing there is a small increase in volume.

An attempt was made, using elastic theory, to relate behaviour in undrained tests to that in drained tests but without convincing results. The only situation where elastic theory was a useful indictor of soil behaviour was in indicating effective stress paths in undrained tests as described earlier and illustrated in Figure 22.

12 THE UNDRAINED RESIDUAL STRENGTH

A number of tests have been carried out to investigate the rate of decline of undrained strength with displacement and to measure the undrained residual strength. The tests were also a "challenge" to find out whether such tests were actually feasible using the Imperial College-Norwegian Geotechnical Institute ring shear apparatus (Bishop et al, 1971). Tests with this apparatus, and the simplified Bromhead version, are almost always drained tests. The author is unaware of any undrained tests on soft clays. The challenges involve the following:

(a) Trimming the annular sample of soft clay and installing it in the apparatus without unacceptable disturbance
(b) Maintaining the undrained condition during the test.
(c) Ensuring that shear takes place at the centre of the sample and not between the soil and the upper or lower porous confining stones.

Trimming and setting up the sample proved not to be problematical. The same procedure as for much harder clay was used. This involved sharp circular cutting rings for the inside and outside of the sample after which the sample was easily installed in the apparatus. Challenges (b) and (c) proved more difficult. To obtain sensible results, and especially to prevent shear between soil and confining stones, it was clearly necessary to apply a small vertical stress, and also to carry out the test very quickly. The first test was carried out with a normal stress of 50kPa, the total in situ stress. This did not result in a sensible result as the strength decline after passing the peak was small and steadily increased with displacement. It was clear that consolidation of the sample was occurring rapidly.

Two more tests were done with variations in the vertical stress and rate of rotation. In the second test shearing occurred between the sample and stones. The third test (which was to have been the last), however, produced a satisfactory result. The displacement curve printed out by the chart recorder as the test proceeded looked convincing and examination of the sample after the test showed clearly that shearing had occurred within the soil mid-way between the porous stones. Details of the test are as follows.

(1) The initial vertical stress was 18kPa
(2) The test was immediately started with a rotation speed of 5°/minute (0.53cm/min).
(3) After 3 minutes the vertical stress was reduced to 9kPa.
(4) At a displacement of 2cm the speed was increased by a factor of 5 to 2.58cm/min.
(5) At a displacement of 10cm the speed was again increased by a factor of 5 to 13.3cm/min.
(6) The test was stopped at a displacement of 120cm (16 minutes).

The printout curve is shown in Figure 33, together with the essential results of associated laboratory tests. These tests were:

- a ring shear test on an undisturbed sample with a pre-cut plane at its centre.
- ring shear tests and vane tests on remoulded samples
- an undrained triaxial test on a remoulded sample

The stress displacement curve in Figure 33 for the undisturbed soil shows that the shear resistance rises rapidly to a peak at a displacement of 1.5cm, and then falls fairly rapidly at first and then at a steadily declining rate to become constant at a displacement of about 20cm. The graph then remains approximately constant with a little "up and down wander" until the test was stopped at a displacement of 120cm, (1.2m). The peak strength recorded in the test is only a little below that obtained in unconsolidated undrained triaxial tests, as shown in Figures 18 and 23. The shear stress in those figures is half the deviator stress. Examination of the surface on which shear occurred showed it to be smooth and flat and covered with a thin "paste" of fully remoulded soil. The surface was very flat and not "slickensided" as is normally the case with stiff over-consolidated clays. Samples of the intact soil before and after the test and the past on the shear plane were taken and measurements of water content made. The results were:

Intact soil before test. w = 53.8%
Intact soil after test: w = 53.4%
Paste from shear surface w = 48.8%

The almost identical values of water content on the intact soil shows that no significant consolidation was taking place during the test. This showed that the applied vertical stress had been judged correctly; large enough to ensure adequate contact between the soil and the porous stones and confining rings, but not so high as to cause consolidation.

The test on the sample with the pre-cut plane showed a very slight, gentle peak followed by a slight decline to a value not much different to that from the undisturbed sample. The tests on remoulded samples showed a residual strength a little lower than that from the undisturbed sample, as also shown in Figure 35.

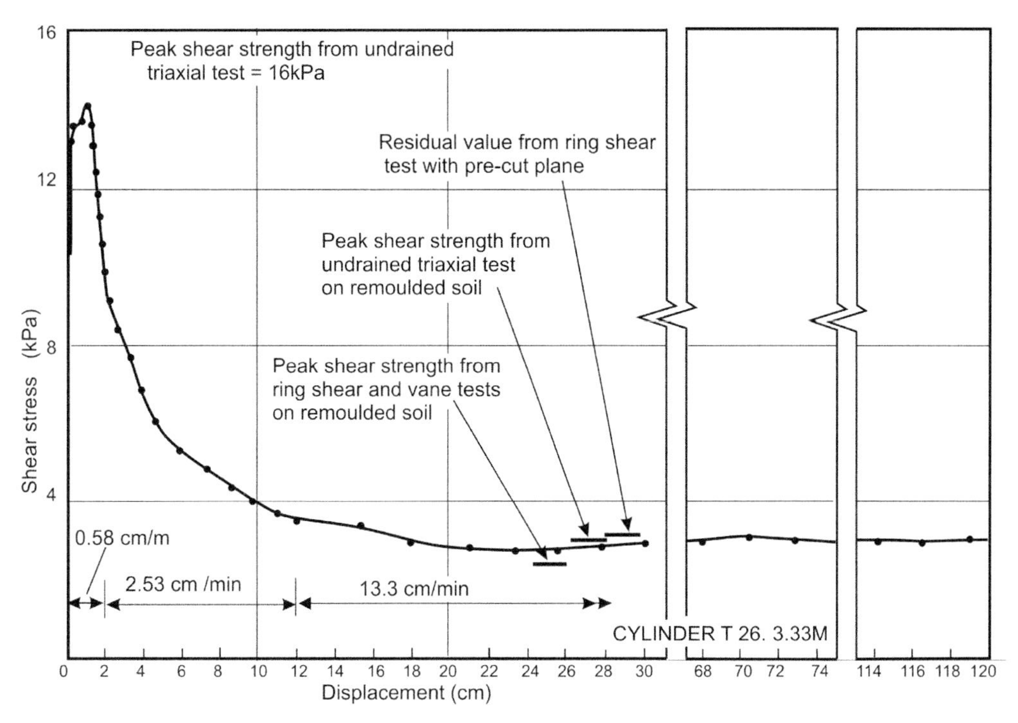

Figure 33. Undrained residual strength from ring shear tests.

The peak strength in Figure 33 was just under 16kPa, and thus close to the peak value from undrained triaxial tests, as shown for example in Figures 19 and 23(a). Also, the value of the residual strength is reasonably close to that expected from the sensitivity of the soil, which was just under 6. The undrained residual strength is just over 3kPa, corresponding to a sensitivity of 4.6. However, there is little doubt that while the ring shear test gives a reliable value of the residual strength, its value of peak strength is unlikely to be as reliable as that from a triaxial test. If the value from the triaxial test is used the sensitivity becomes 5.3. Perhaps the most useful information to come from these tests is that it is possible to determine the stress displacement curve of soft undisturbed clays for the undrained condition provided the apparatus is available as well as large samples and an experimentalist with the necessary skill and patience.

13 CONCLUDING OBSERVATIONS

13.1 *The two options for measurement of the undrained strength*

Various observations have been made over the years as to whether measurement of the undrained shear strength of clay by laboratory test is best by maintaining the water content at its in situ value (in as much is this is possible), or by first reconsolidating the sample the in situ effective stress state prior to carrying out the undrained test. The author's view, backed up by the test results in Figures 19 and 20, is that consolidation to the in situ effective stress will produce the most consistent and reliable results, although the difference with tests not consolidated is quite

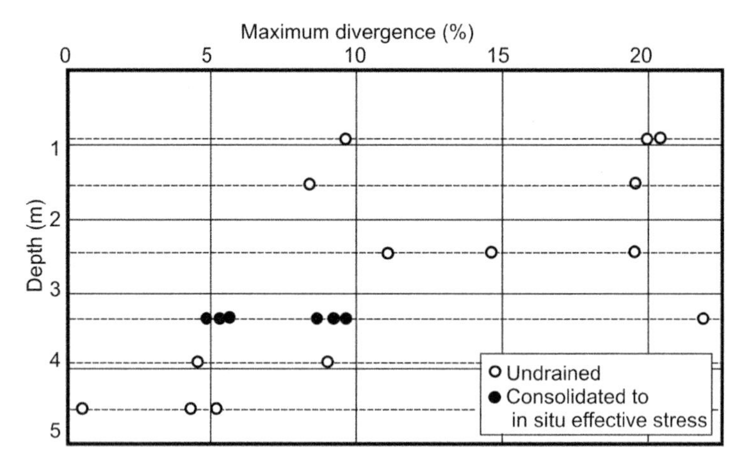

Figure 34. Scatter in triaxial tests.

small. It is also clear from Figures 21 to 23 that there is a small but significant difference in behaviour when consolidation is to the average in situ stress and the K_o state.

13.2 *The benefits of quadruple tests*

It is already apparent from the test results presented above that quadruple tests provide a greater degree of reliability than that from single tests. A further attempt has been made to assess the advantages of quadruple tests. For all the quadruple tests, the average value was calculated and the maximum value above or below this determined. The difference has been designated the "Maximum divergence" and is shown in Figure 35, expressed as a percentage. The scatter of values is surprisingly wide, but some significant trends emerge. The variability above a depth of about 3m can be partly attributed to the presence of fissures in some samples that were not found deeper down. The tests on samples consolidated to the in situ effective stress are more consistent than the direct strictly undrained tests

13.3 *Degree of disturbance in the Mucking samples and those from the Bothkennar site.*

If loss of pore water tension is used as a guide to degree of disturbance, then the samples from Mucking were less disturbed than those from Bothkennar, except for the Laval samples. This is not surprising in view of the much greater depth of the Bothkennar samples, and the smaller opportunity for disturbance with the Mucking samples. The latter remained tightly sealed in the cylinders in which they were taken during handling and transport from the site to the laboratory. However, it should be recognised that the degree of disturbance that occurs in taking samples and preparing them for tests is likely to be as much dependent on the skill of the experimentalist as on the technique itself. Bishop, in one of his many letters to Bjerrum, expresses his envy of the situation at the NGI with its team of permanent skilled experimentalists, while he relies on post graduate students who (he states) frequently lack such skills.

13.4 *The Mohr-Coulomb failure line.*

The triaxial tests do not show any clear evidence of dependence of the parameters c′ and ϕ′ on the type of test used to measure them. There is a slight indication in Figure 26(a) that plane strain tests result in a marginally higher Mohr-Coulomb failure line. However, this could be the result of a small but significant membrane influence. The type of membrane used in these tests was somewhat different to that in the triaxial tests, and could have influenced the strength though it seem unlikely. The only significant difference in behaviour between plane strain and triaxial compression tests is in small difference in the shape of the stress strain curves as seen by comparing Figures 10 and 8. The plane strain pre-failure curve is slightly steeper and there is a shaper peak at failure than in the triaxial tests.

13.5 *The nature of anisotropy*

While there is no evidence of anisotropy with respect to the effective stress parameters, c′ and ϕ′, there is clear anisotropy with respect to the stiffness of the soil skeleton, at least at the low stress level relevant to practical situations. The horizontal stiffness is approximately half that of the vertical stiffness resulting in higher pore pressures when loaded horizontally than vertically and thus lower undrained shear strength.

Part 2 (Pugh) In Situ Vane Tests And Full Scale Embankment Loading

14 FIELD SHEAR STRENGTH TESTING

14.1 *Introduction.*

Field measurements of the undrained shear strength of the Mucking clay were carried out using both a conventional Geonor vane (height 130mm x 65mm diameter) and a large vane apparatus (height 508mm x 254mm diameter) that was purpose-built to investigate the effect of size and strain rate on the field undrained shear strength. The extent of the field vane testing was as follows, while the test locations are shown on the site plan of Figure 35:

21 Geonor vane tests located on the instrumented section lines of the trial embankments

8 vertical and 9 horizontal Geonor vane tests adjacent to, and from within respectively, the test pit from which the large cylinder samples for the laboratory testing program were obtained

36 large vane tests, of which 35 were successfully completed, performed in nine boreholes

4 Geonor vane tests performed adjacent to the large vane test boreholes

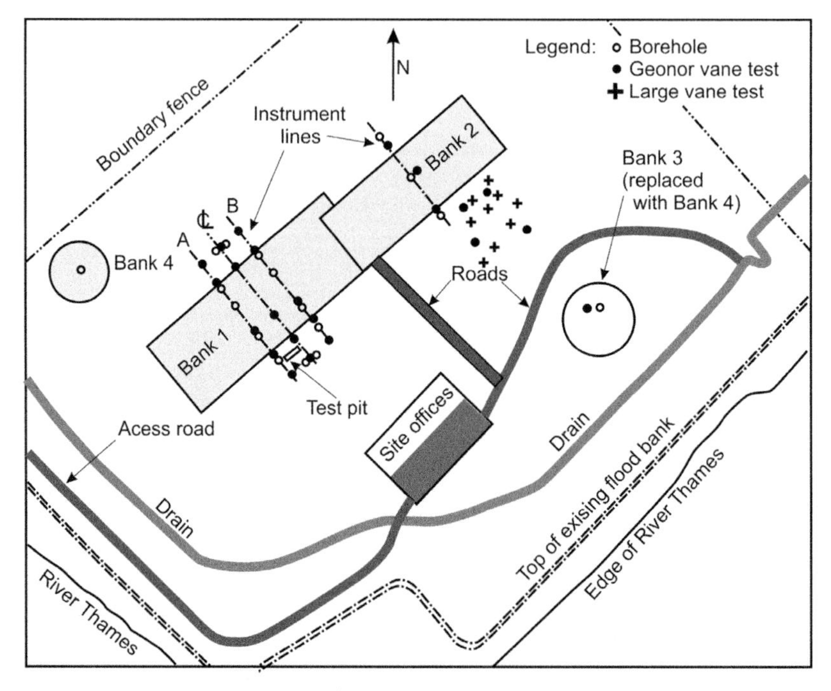

Figure 35. Site plan

14.2 *Geonor vane tests.*

In theory, the shear surface generated by a horizontal vane test should more closely represent the direction (horizontal/circular) and form (active/passive) of shearing associated with the failure of an embankment on very soft clay than those (vertical/circular-active/passive) associated with a conventional vertical test. Comparison of the results of the vertical and horizontal tests (Figure 36(a)) shows, however, that the combined results produce a remarkably consistent shear strength/depth profile. Also shown on Figure 36(a) are the undrained shear strength/depth profiles for the laboratory triaxial tests performed on vertical and horizontal samples (Figure 11). This shows that from 1.5m to 3.25m depth the vane shear strength/depth profile is parallel to and above that for the vertical triaxial tests, while below 3.5m the vane shear strengths are very similar to the vertical triaxial strengths, albeit slightly lower.

At the time the vane tests were carried out Bjerrum (1972, 1973) had recently proposed a correction factor to be applied to the results of Geonor vane tests for use in stability analyses. The correction factor (Figure 37) is a function of the plasticity index and combines a strain-rate reduction in shear strength with an increase in respect of anisotropy. For the Mucking site (Figure 37) the overall correction factor varies from about 0.7 at 1m depth to 0.85 at 3.5m and 0.7 at 6m depth. Within this range the anisotropy correction (increase) is only 0.05 to 0.1, or about 1 to 2kPa in undrained shear strength, which would not be evident within the typical scatter of the field data, as is evident from Figure 36(a). It is thus unsurprising that there is no clear distinction between shear strengths measured in the vertical and horizontal vane tests. With respect to the estimated strain rate reduction, Figure 37 indicates the correction factor to be between about 0.65 and 0.75, with a mean of 0.7.

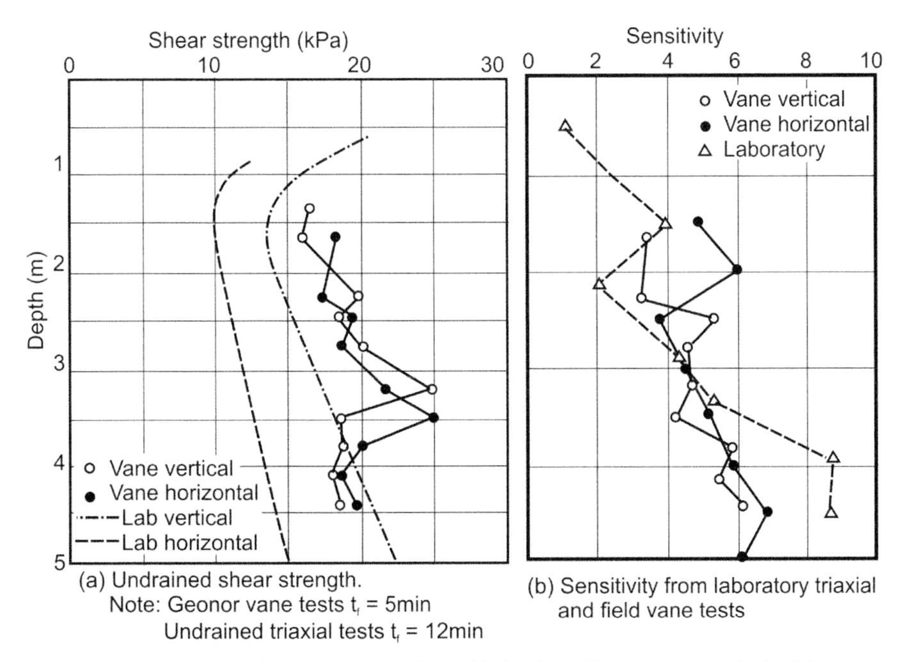

Figure 36. Undrained shear strength and sensitivity from Geonor vane and triaxial tests.

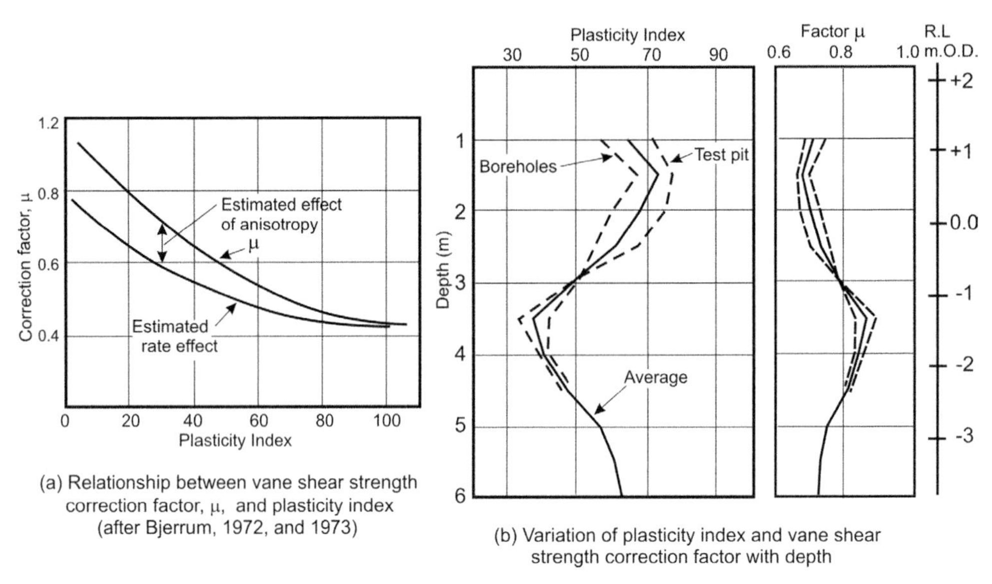

Figure 37. Bjerrum's (1972, 1973) vane test correction factors

Figure 36(b) shows the sensitivity of the Mucking clay as measured in the laboratory and in the vertical and horizontal vane tests. The three sets of data are consistent in showing the sensitivity to increase from about 4 at 1.5m depth to 5 at 3.25m depth. Below 3.25m depth the laboratory sensitivity increases sharply to 9 at 4m depth and below. This increase in sensitivity at 4m depth was not recorded by the Geonor vane tests, almost certainly because of the inherent internal friction in the torque measurement system preventing the Geonor vane from accurately measuring very low remoulded shear strengths. The increase in sensitivity below 3.25m depth is the result of an increase in the liquidity index to close to unity which, in turn, is the result of decreasing plasticity and clay content (Figure 4).

Despite the variations in plasticity, liquidity index and sensitivity, the laboratory measured undrained shear strengths of the Mucking clay below the desiccated crust increase linearly with depth, being a function of the in-situ vertical effective stress. Both the vertical and horizontal vane test shear strengths below 3.25m depth correspond closely to the laboratory results, indicating that the vane tests at depth were effectively undrained. The vane test shear strengths above 3.25m depth also increase linearly with depth, indicating they are also a function of the in-situ effective stress. The higher measured shear strengths (+4kPa on average) above 3.25m depth were attributed to the tests being partially/fully drained. This conclusion is supported by the soil descriptions (Figure 4) which showed there to be traces of organic matter above 3.25m but none below, with the traces of organic matter being decayed remnants of roots, with a system of vertical root holes being the dominant fabric of the upper clay.

14.3 *Large vane tests*

The identical depth sequence of the four large vane tests carried out in each of nine boreholes is shown in Figure 38, which also shows the depths of the four test levels (designated Levels 1 to 4).

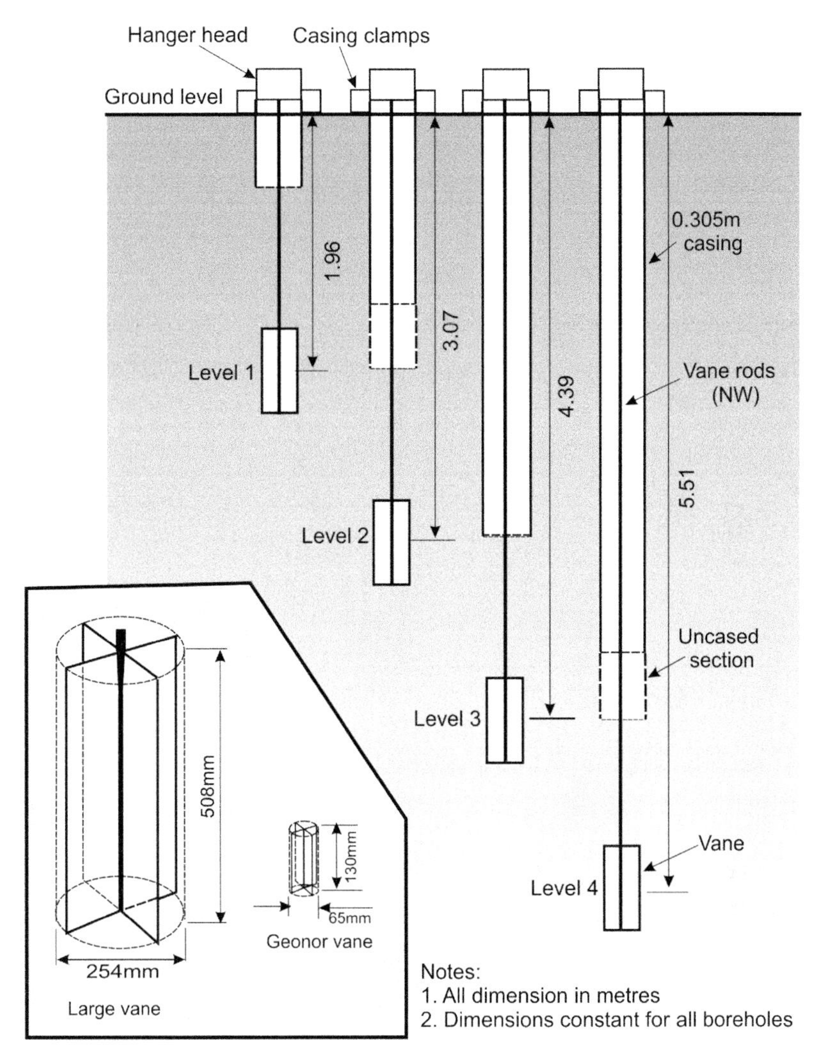

Figure 38. Sequence of large diameter vane tests

At each test level five different rates of vane rotation were used, resulting in times to failure of between 1.25 and 2.75 minutes for the fastest tests and between 750 and 1260 minutes for the slowest. A single test was carried out at each level for the highest and lowest rotation rates, while two tests per level were carried out at the second and third slowest rates. For the second fastest rate, corresponding to average times to failure of between 6.42 minutes and 11 minutes, considered to be the "standard" rate of rotation, three tests were carried out at each location. Despite being a prototype, the large vane apparatus performed exceptionally well, with 35 out of a planned 36 tests successfully completed (only two of the three planned standard tests were successfully completed at Level 4). The multiple tests carried out at the same strain rate in different boreholes produced highly repeatable results at each test level, giving a high degree of confidence in the veracity of the results.

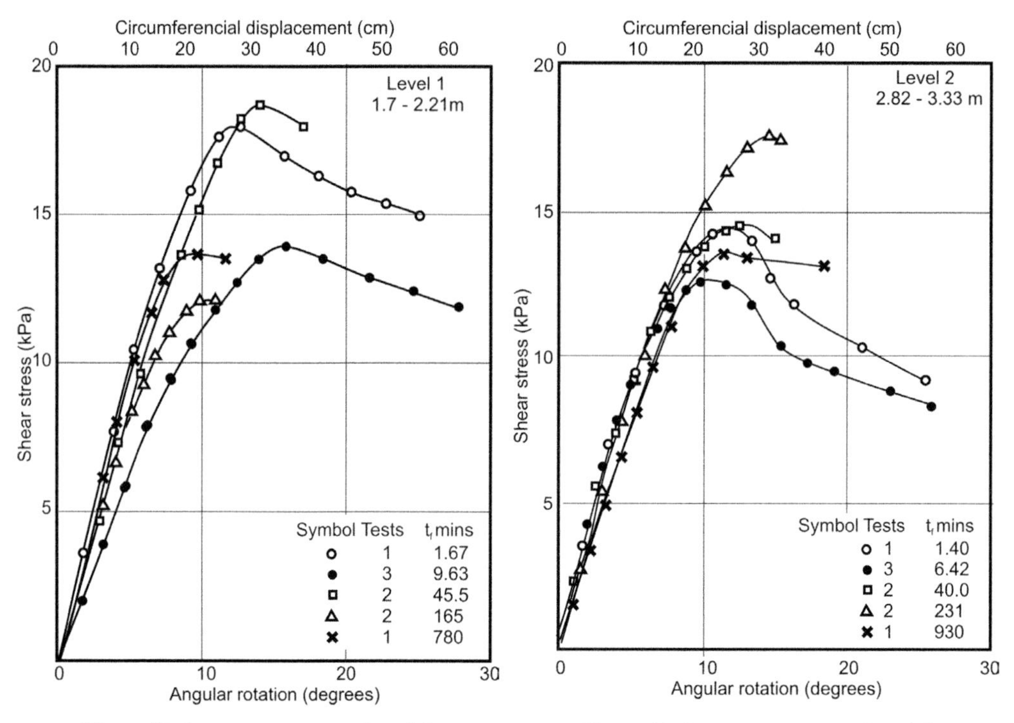

Figure 39. Large vane tests at Level 1

Figure 40. Large vane tests at Level 2

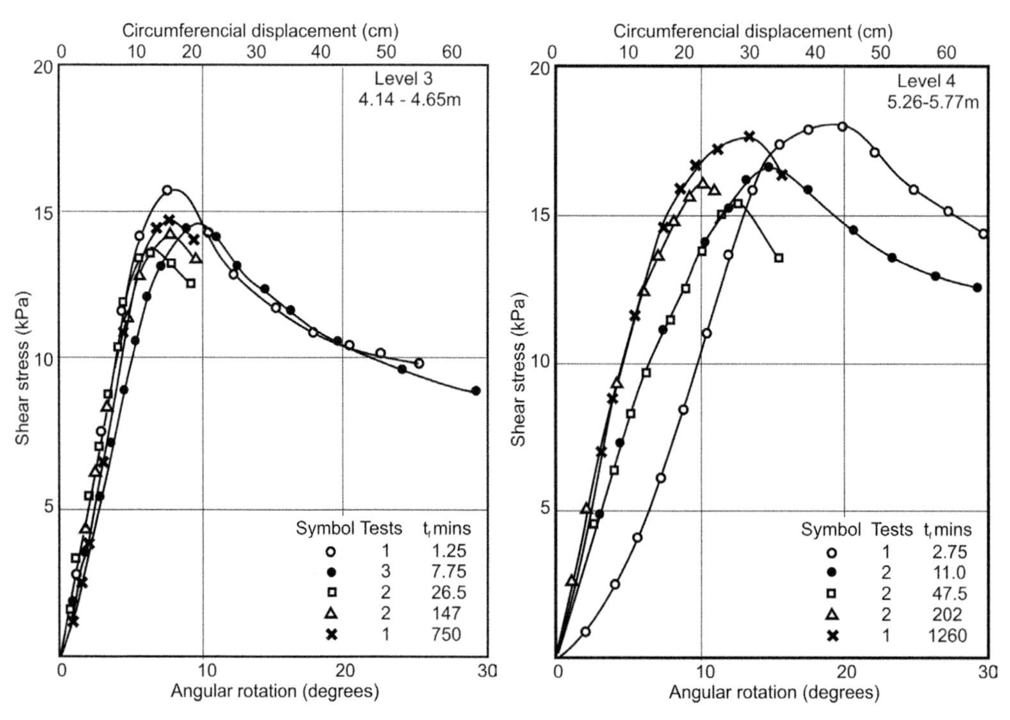

Figure 41. Large vane tests at Level 3

Figure 42. Large vane tests at Level 4

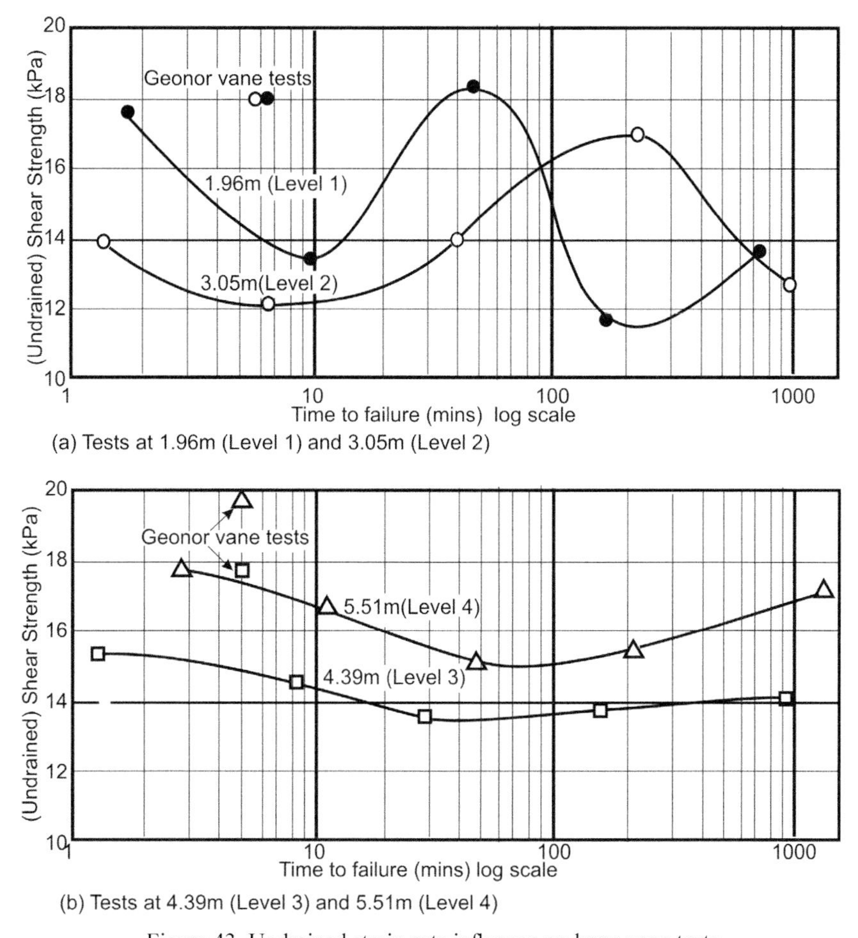

Figure 43. Undrained strain-rate influence on large vane tests

The shear stress/angular rotation relationships for the large vane tests, averaged for each strain rate, are presented in Figures 39 to 42. This data is re-plotted as shear strength versus log time to failure in Figures 43(a) (Levels 1 and 2) and 43(b) (Levels 3 and 4). Figure 43 shows that, as for the Geonor vane shear strength/depth plot of Figure 36(a), the large vane shear strength/log time plots for Levels 1 and 2 are completely different in form to those for Levels 3 and 4. Figure 43(b) shows that at Levels 3 and 4, up to a failure time of about 40 minutes, the results exhibit decreasing shear strength with time, indicating a field undrained strain rate effect. With increasing time to failure the shear strengths gradually increase, suggesting that the effect of drainage increasingly offsets the effect of the undrained strain-rate reduction. Also shown on Figure 43(b) are the Geonor vane shear strengths (t = 5 minutes) for Levels 3 and 4, being about 2.5 to 3kPa higher than the equivalent large vane values. As the Geonor vane tests at Levels 3 and 4 were effectively undrained, as were the large vane tests at t = 5 minutes, it is possible that the lower large-vane shear strengths were the result of a degree of installation disturbance in the more sensitive clay below 3.25m depth.

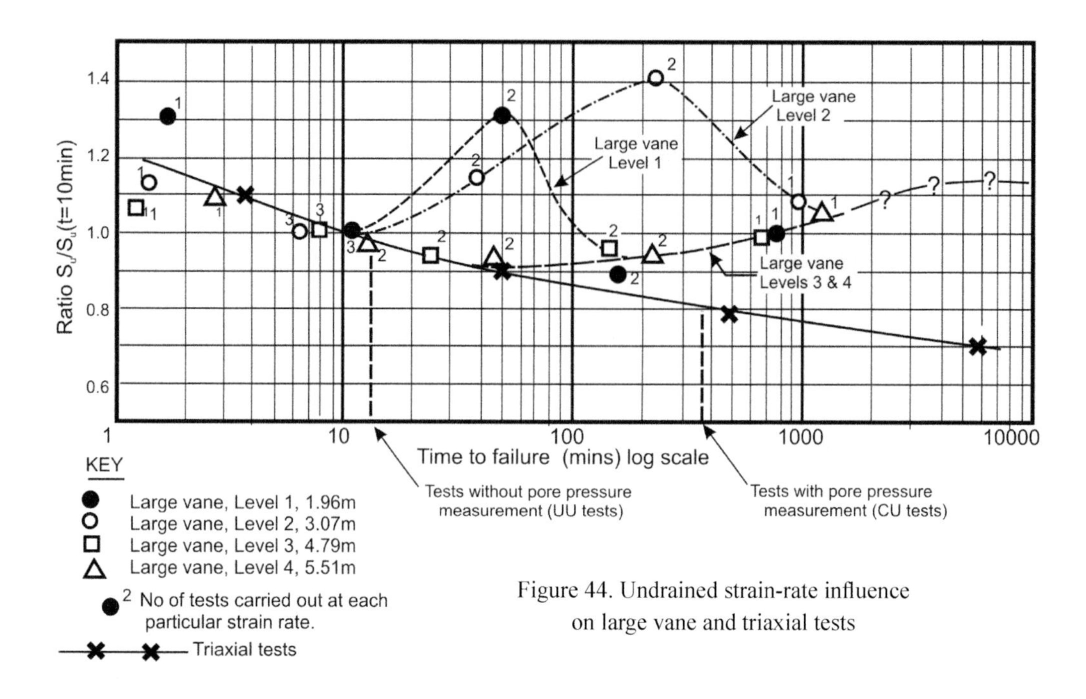

Figure 44. Undrained strain-rate influence on large vane and triaxial tests

At Levels 1 and 2 (Figure 43(a) the shear strength/log time relationships are again similar, but in this case there is an initial drop in strength with time (undrained strain-rate effect) followed by a rapid increase (partially to fully drained) and ultimately a further decrease (drained strain-rate effect) suggesting that at these levels none of the tests beyond t = 10 minutes were undrained. In the case of Levels 1 and 2 the Geonor vane shear strengths are 4kPa to 6kPa higher than the equivalent (t = 5 minutes) large vane tests, also suggesting that the Geonor vane tests were drained. The very rapid drainage apparent at large vane-test Levels 1 and 2 is hardly surprising given the very short (c.a. 0.9m-see Figure 38) vertical drainage path lengths occasioned by the vertical root holes.

In Figure 44 the large vane-test data of Figure 43 is presented as shear strength, normalised with respect to the shear strength for a failure time of 10 minutes, versus log time to failure. The shear strength/log time data for the laboratory triaxial tests is also shown for comparison purposes. For Levels 3 and 4 there is a single normalised shear strength/log time relationship, exhibiting an initial strain-rate drop in normalised strength with time, identical to that from the triaxial tests. Beyond about t = 40 minutes the normalised shear strengths exhibit a gradual increase with time, reflecting partial to full drainage. At Level 1 the normalised shear strengths (undrained) initially reduce, but beyond t = 10 minutes increase (fully drained) before falling back (drained strain-rate effect) to join the relationship for Levels 3 and 4 at about t = 150 minutes. The relationship for Level 2 is similar to Level 1, but does not rejoin that for Levels 3 and 4 line until t = 1200 minutes.

Figure 44 clearly shows that the large vane tests at the standard rate of rotation (average time to failure 9 minutes) were undrained. With increasing time to failure the large vane tests went from being undrained to fully drained, at which point the difference between the drained shear strength and the undrained shear strength at the same time to failure was, on average, 4kPa.The Geonor vane tests were effectively undrained below 3.25m depth, but effectively drained above. The average difference between the Geonor vane drained shear strengths above 3.25m depth and the corresponding undrained shear strengths from the vertical triaxial tests was also 4kPa. This suggests that, the reduction in the drained shear strength with log time is similar to that for the undrained shear strength.

15 THE TRIAL EMBANKMENTS

A total of four trial embankments were constructed at Mucking as shown in Figure 35. Bank 1 was constructed to failure in two stages, Bank 1/1 and Bank 1/2. Bank 1/1 was rectangular in plan (90m x 34m) and designed to induce a plane strain failure confined to the "normally consolidated" clay beneath the upper desiccated zone and to the south side of the embankment. This was achieved by the excavation of two 1m deep trenches through the desiccated zone, together with the sectional geometry of the embankment (Figure 45). The central 10m wide trench was excavated prior to construction while the toe trench (2m wide at the base, 4m wide at the surface) was excavated when the embankment reached a height of 2.75m.

Bank 1/1 had reached 4m high by 1730 hours on 9 October 1973 (51 days after start of construction) when work finished for the day. The only visible cracks were in the base of the toe trench. At 1930 hours, just after an inspection had revealed nothing untoward, the movement alarm bell was triggered and automatic and manual photographic recording commenced, being some 15 minutes before visible vertical displacement of the crest was observed.

Bank 1/2 was also rectangular in plan (90m x 47.5m) and was designed to induce a plane strain failure passing through the desiccated zone into the underlying clay to the northern side of the embankment. This was achieved solely by the sectional geometry (Figure 45). Fill height had reached 4.5m over the central area of the crest, grading down to 3.75m at the east and west ends, by 1700 hours on 16 November 1973 (89 days after start of construction) when work stopped for the day. At 1705 hours the alarm sounded and automatic photographic recording commenced. An inspection of the embankment revealed no visible signs of cracking of the fill or heave of the foundation. Inspection of the crest was hampered by the recently placed fill not having been raked flat and the crest was raked and smoothed and the alarm re-set. Observation of the mercury pots in the triggering device and subsequent photographic and photogrammetric data indicated that significant movement of the embankment occurred during this period. At 2240 hours the alarm was again triggered, accompanied from the outset by visible downward displacement of the crest.

Bank 2 was a rectangular in plan (75m x 25m) embankment designed to load the clay, including the desiccated zone, in plane strain conditions to the stress levels corresponding to a working factor of safety. The embankment was 2.8m high with a computed factor of safety of 1.5 (in terms of effective stress) at the end of construction. Following completion of the embankment on 5 February 1974 monitoring of the instrumentation (Figure 46) continued until September 1975.

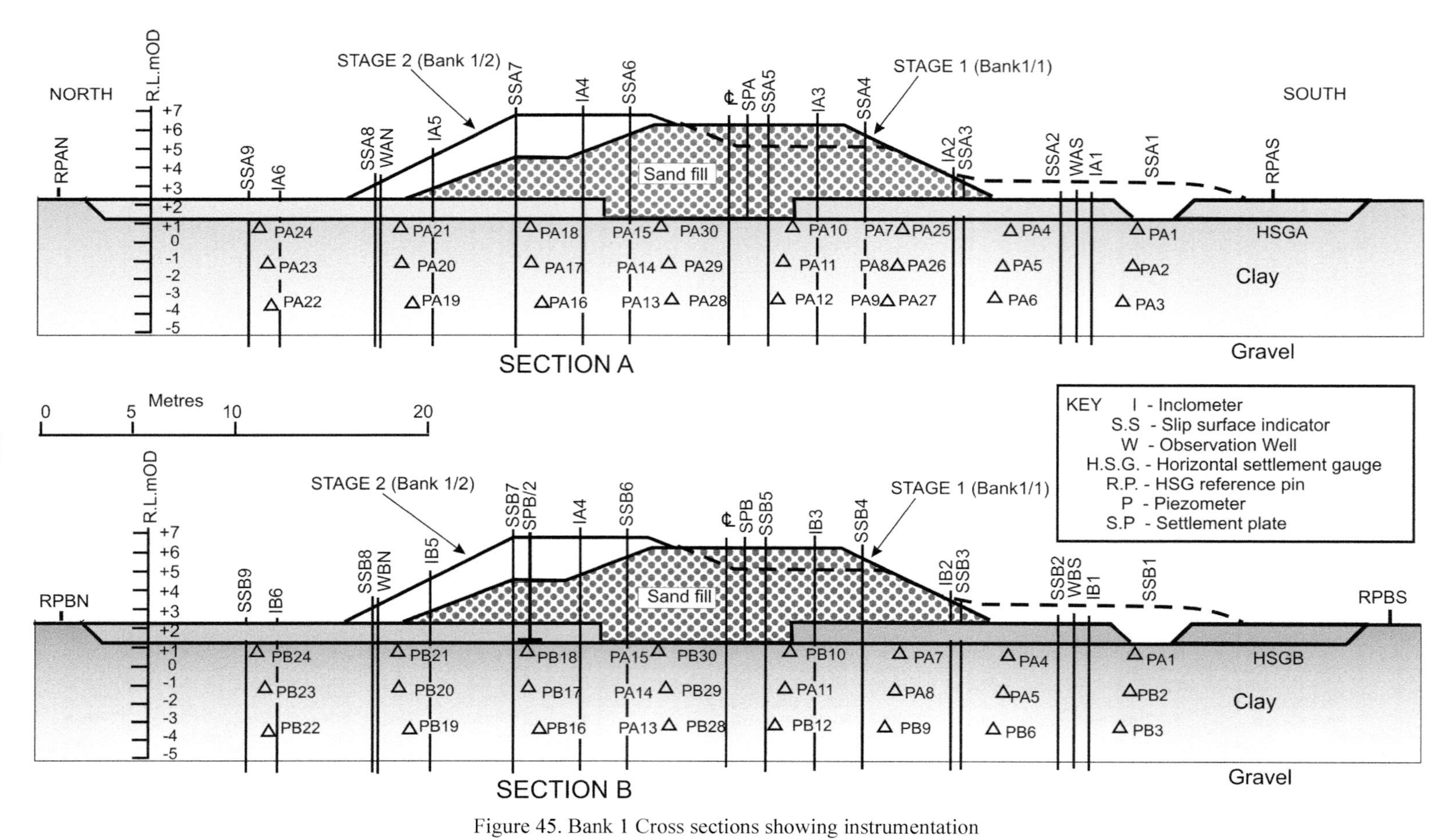

Figure 45. Bank 1 Cross sections showing instrumentation

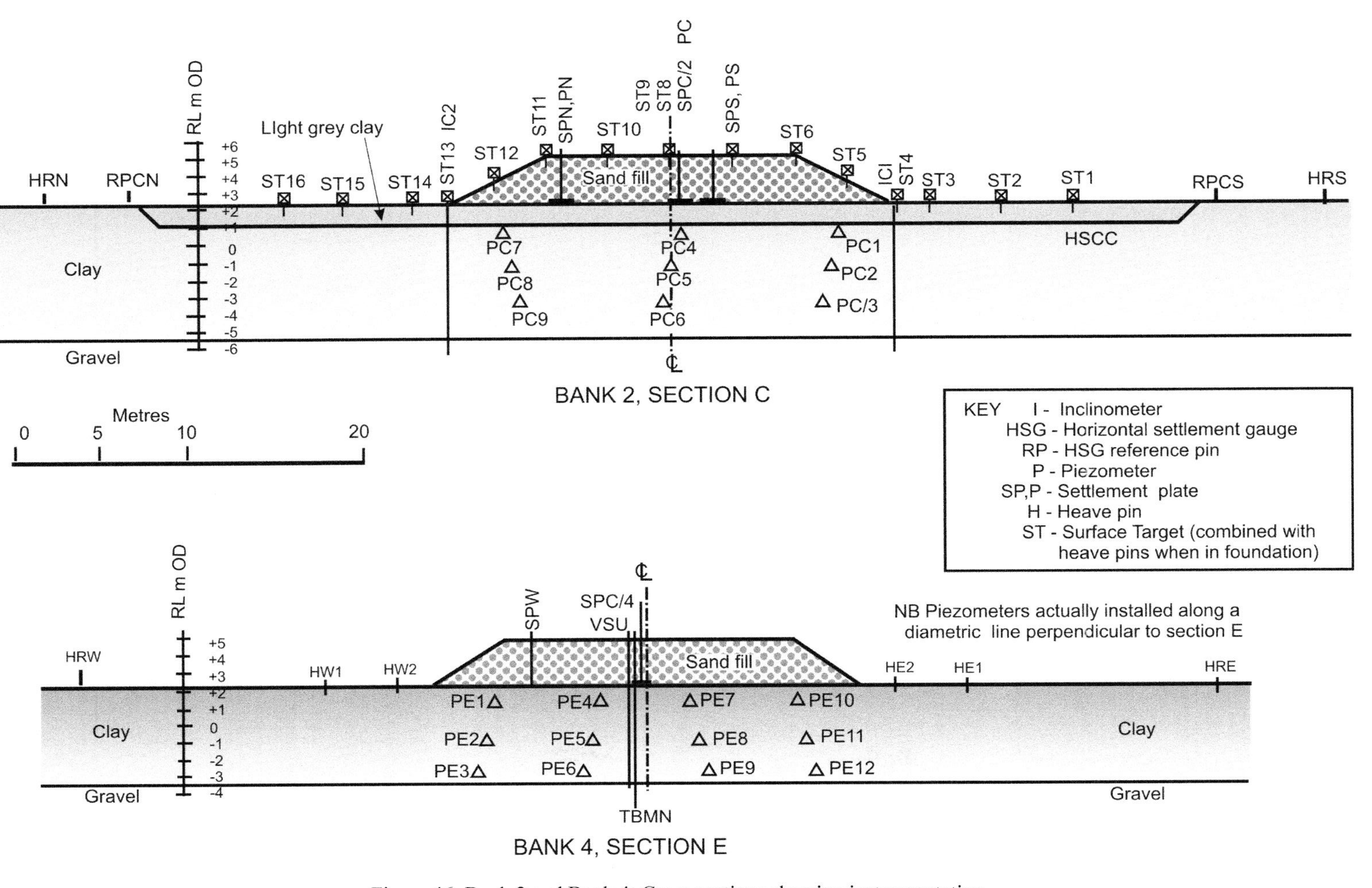

Figure 46. Bank 2 and Bank 4: Cross sections showing instrumentation

Bank 3 was circular in plan and designed to enable the study of the field consolidation characteristics of the Mucking clay. The intention was to construct it to a safe height of 4m (assumed factor of safety 1.28) based on stability analyses using laboratory undrained shear strengths from quick undrained tests on conventional piston samples. It failed at the design height of 4m because of both an overestimation of the operational undrained shear strength and the presence of a nearby drainage ditch.

Bank 4, which replaced Bank 3, was 24m in diameter and safely constructed to 2.5m in height, with the adjacent section of drainage ditch replaced by a culvert. It was completed on 13 December 1973 and the instrumentation (Figure 46) monitored until August 1975.

16 STABILITY ANALYSES

16.1 *Methods and extent of analyses*

Two methods of limit equilibrium analysis were used, both based on the method of slices. Bishop's Routine Method (1955) was used for circular arc failure surfaces and Sarma's Method (1973) for general (non-circular) surfaces. The circular arc surfaces were analysed using a computer program written by Professor E.N. Bromhead of Kingston University while the general surface analyses used a program written by Professor S.K. Sarma of Imperial College.

Back-analyses of the Bank 1 failures, using circular arc surfaces, were carried out employing both total and effective stress parameters; slip surfaces restricted to fit, as closely as possible, the field observations, and unrestricted (within the limitation of emerging within the top part of the embankment) surfaces. End of construction stability of Bank 2 was analysed in terms of both total and effective stresses using unrestricted slip surfaces while post-construction stability was only assessed in terms of effective stress.

Back-analyses of the Bank 1 failures, using general slip surfaces, were also carried out in terms of total and effective stress, but in this case only surfaces approximating to the field observations were considered for Stage 1. For Stage 2, however, unrestricted surfaces were considered. Stability of the critical surfaces for Bank 1, Stages 1/2 (Figure 45) was also investigated in conjunction with the post-failure profiles, assuming a state of limiting equilibrium to have been re-established, in order to assess the field residual strength parameters and hence the brittleness index (Bishop,1967 and 1971) in terms of both total and effective stresses.

16.2 *Shear strength parameters for stability analyses*

The large vane tests clearly demonstrate that the reduction in undrained shear strength with time to failure was the same for both laboratory and field conditions. The Geonor vane tests also indicate that, for undrained conditions, the measured vane shear strengths were equivalent to vertical triaxial test values. However, the combined effects of drainage during vane testing and possible installation disturbance mean that the best estimate of the short-term undrained strength is given by the triaxial tests on specimens from the undisturbed block samples. The strength/depth profiles for the vertical triaxial tests, the Geonor vane tests and the large vane tests are shown in Figure 47.

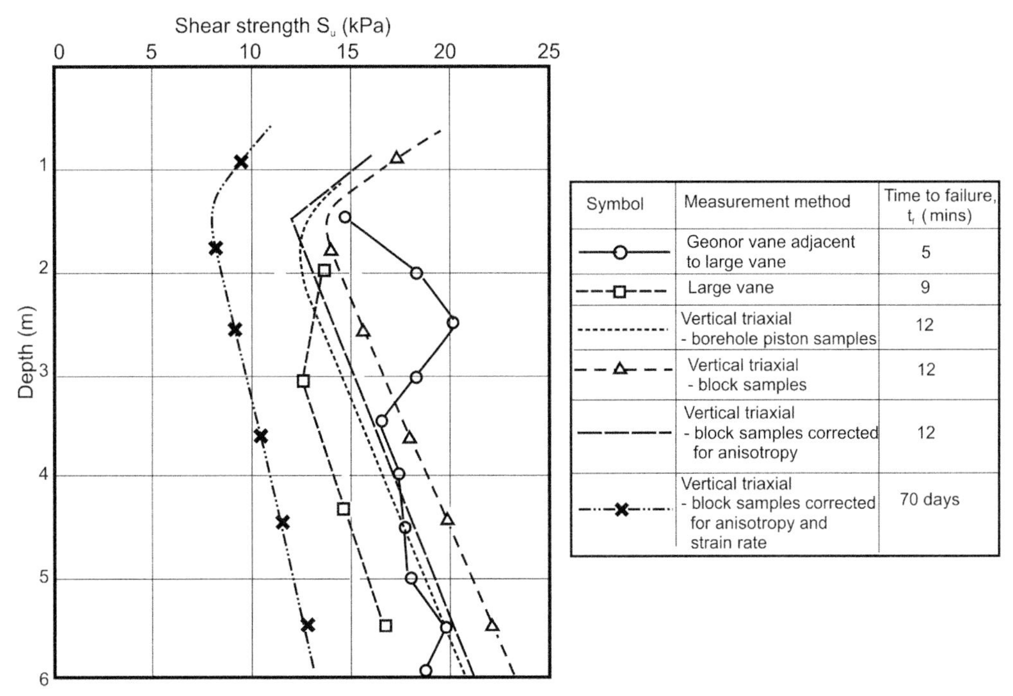

Figure 47. Undrained shear strength versus depth profiles for stability analysis

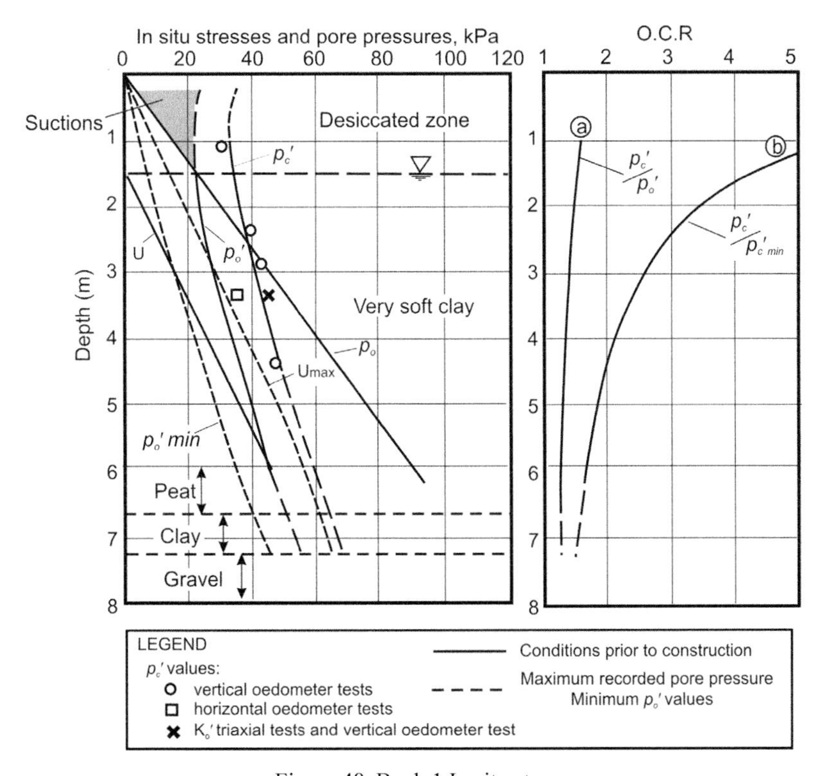

Figure 48. Bank 1 In situ stresses

In order to convert the laboratory vertical shear strength/depth profile of Figure 47 into a design profile for total stress stability analyses it was firstly necessary to consider the effect of undrained strength anisotropy on the shear strength mobilised on a total stress "failure" surface. Figure 47 shows the vertical triaxial strength/depth profile corrected for anisotropy (reduction of 0.87) based upon a simple average of the triaxial tests on vertical, 45^0 and horizontal samples (see Figure 11). This corrected profile is, however, for a time to failure of 12 minutes, compared to 70 days in the field, being the mean time to failure for Bank 1/1 and Bank 1/2. The design profile was thus based on a further strain-rate reduction of 0.65 (see Figure 2) giving an overall reduction to the laboratory vertical undrained strength profile of 0.57, as shown on Figure 47. Also shown on Figure 47 is the shear strength profile for commercial piston samples. Comparison of this profile with that for the vertical triaxial tests on the block samples indicates that, on average, the strength of the piston samples was about 2kPa lower than that for the block samples. This strength reduction was considered to be attributable to greater sample disturbance.

The laboratory consolidated undrained triaxial tests on specimens from the block samples indicated that the effective stress shear strength parameters of the Mucking clay were $c' = 8$kPa, $\phi' = 25^0$ at low consolidation pressures, changing to $c' = 0$, $\phi' = 32^0$ as the consolidation pressure exceeded the apparent pre-consolidation pressure (see Figure 16 and Figure 48). The laboratory testing program also demonstrated that the undrained strength anisotropy was excess pore-pressure related, there being no anisotropy in terms of the effective stress-strength parameters. There was, however, a significant effective stress strain-rate effect (as also suggested by the large vane tests) which manifested itself as decreasing c' with increasing time to failure, with φ' constant (Figure 3 and Figure 49). The laboratory strain rate-data (Figure 49) indicates that the design effective stress parameters (average time to failure 70 days) should be $c' = 5.5$kPa, $\phi' = 25^0$.

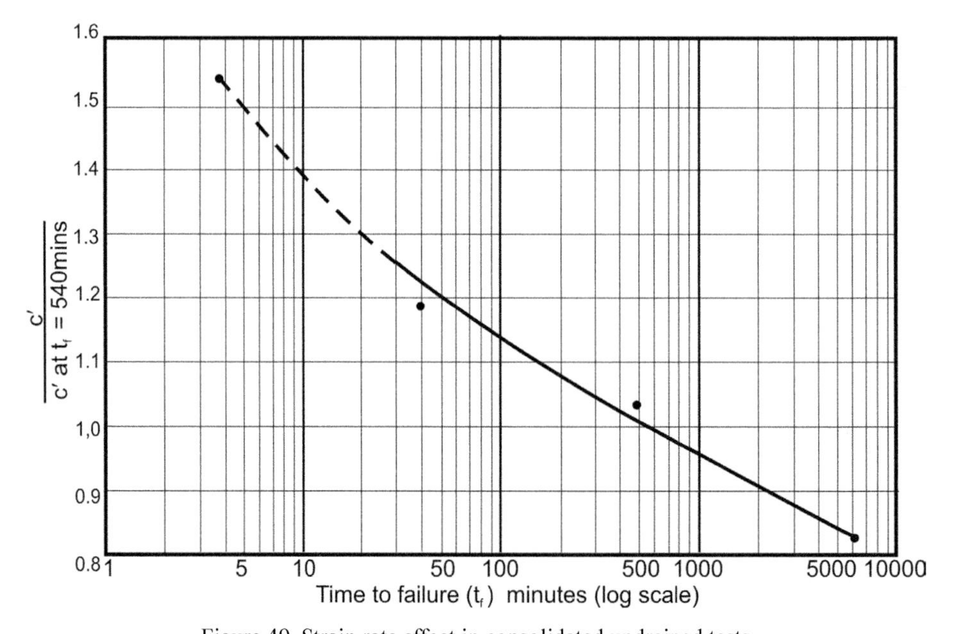

Figure 49. Strain rate effect in consolidated undrained tests

16.3 *Bank 1, failure: total stress analyses*

16.3.1 *Bank 1/1*

Figure 50 shows the limits of the zone of extreme deformation for Bank1/1 defined by the foundation instrument readings. These limits, taken together with the locations of the rear scarp of the slip surface and its emergence within the toe trench, indicated that the design of Bank 1/1 had achieved its aim of confining the slip surface to the "normally consolidated" clay beneath the desiccated crust.

A particular problem in the back analysis of the embankment failures at Mucking, and indeed for inland sites generally, is the appropriate value of the undrained strength of the upper levels of the clay which are subject to seasonal variations in moisture content and hence undrained shear strength. The Mucking site is in south-east England where pore pressures in this zone typically vary between large suctions (as a result of evapotranspiration during summer) and small positive values (as a result of moisture recovery during winter), accompanied by, respectively maximum and minimum values of the undrained strength. Construction of Bank 1 commenced in early October 1973 and, as such, the measured undrained strength of 50kPa was likely to have approximated to the maximum value.

Initial stability analyses were carried out using non-circular slip surfaces restricted to the deformed zone and the end points of the observed surfaces (Figure 50) in combination with the vertical triaxial undrained shear strength profile corrected for anisotropy (Figure 47). For these initial analyses the assumption was made of a "weak crust" (Zone 3, Figure 50) with an undrained strength of 25kPa based on the laboratory triaxial test profiles of Figure 47. The weak

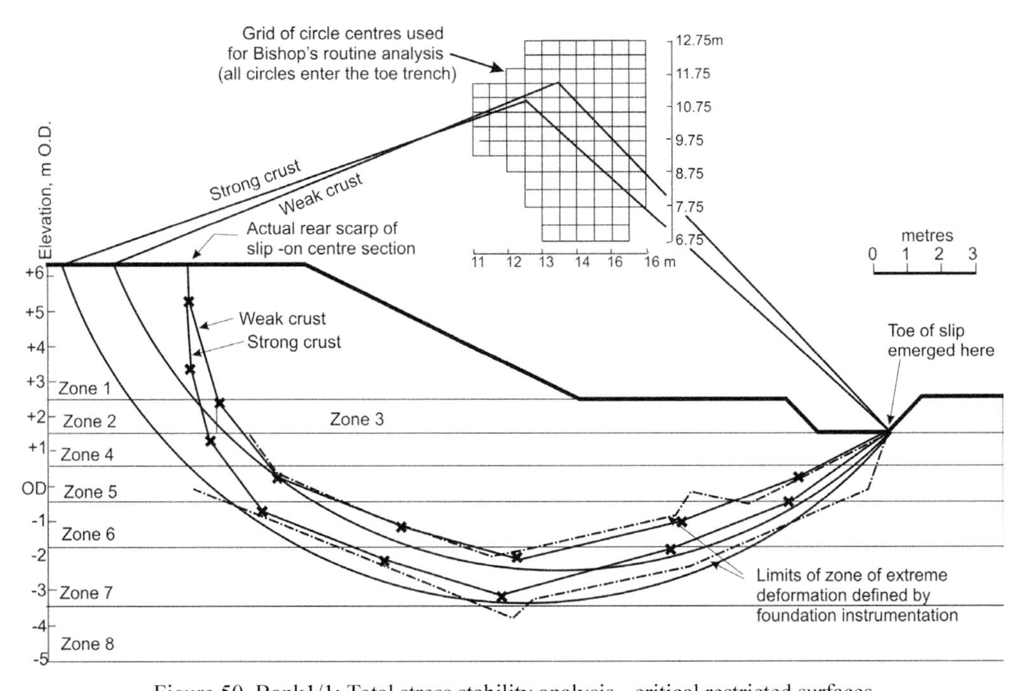

Figure 50. Bank1/1: Total stress stability analysis - critical restricted surfaces

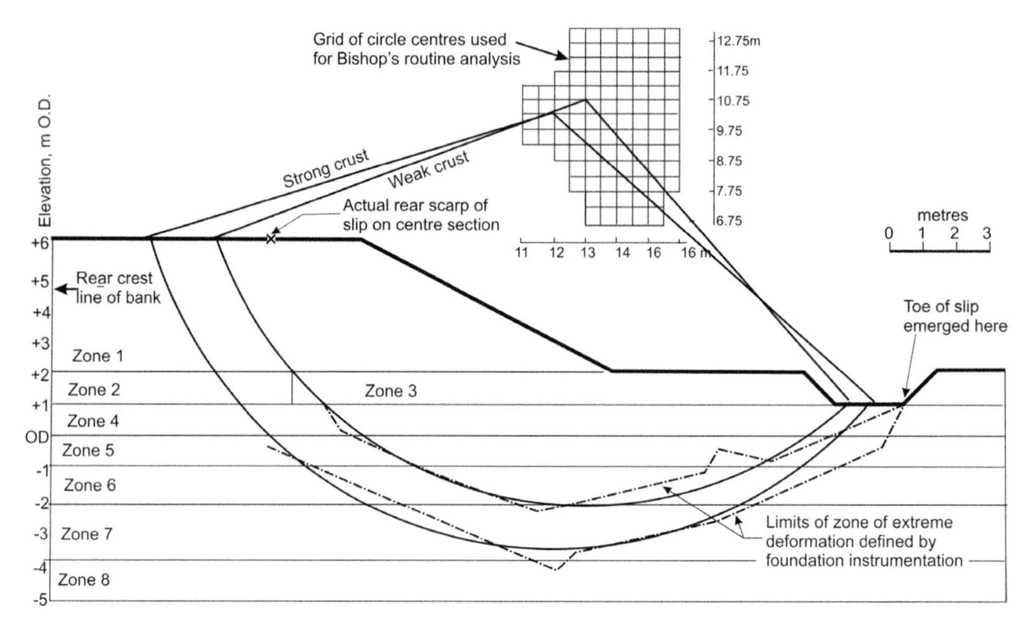

Figure 51. Bank 1/1: Total stress stability analysis - critical unrestricted surfaces

crust analyses produced a critical slip surface which passed through the crust, contrary to the field observations. Further analyses were then carried out with increasing crust strength until the critical surface was forced into the central trench. The required undrained strength to force the slip surface into the trench (noting that this was not a true $\phi_u = 0$ analysis) was 45kPa and therefore in excellent agreement with the field observations. The critical surface was associated with a 3m deep tension crack, supporting the field observation that formation of a tension crack preceded limiting equilibrium conditions.

Stability analyses using circular arc failure surfaces were then carried out, initially with restricted failure surfaces and then followed by unrestricted surfaces (Figure 51- genuine $\phi_u = 0$ analyses). The critical surfaces were also in good agreement with the observed zone of extreme deformation for both the restricted and, surprisingly, the unrestricted ($\phi_u = 0$) surfaces. Finally, the undrained shear strength of the clay was reduced (except for the derived shear strength (45kPa) for the strong crust analyses) until a factor of safety of unity was achieved on both unrestricted ($\phi_u = 0$) critical surfaces.

16.3.2 *Bank 1/2*

Figure 52 shows the limits of the zone of extreme deformation for Bank 1/2 defined by the foundation instrument readings. Using the shear strength profile back analysed from Bank 1/1 (with both weak and strong crust assumptions) a single critical non-circular surface was located. Contrary to Bank 1/1, Bank 1/2 was not considered to have cracked completely prior to reaching a state of limiting equilibrium and thus only a small tension crack was introduced in the back analyses. The critical surface (Figure 52) passed from the observed rear scarp into the uppermost part of the deformed zone (avoiding the central trench) indicating that the design of Bank 1/2 had also achieved its purpose. Unrestricted circular and non-circular slip surfaces were then considered, leading to the critical surfaces (both weak and strong crust assumptions) shown in Figure 53. The shear strength/depth profiles (weak and strong crust) derived from the back analyses of Bank1/1 both led to factors of safety close to unity for the critical unrestricted surfaces.

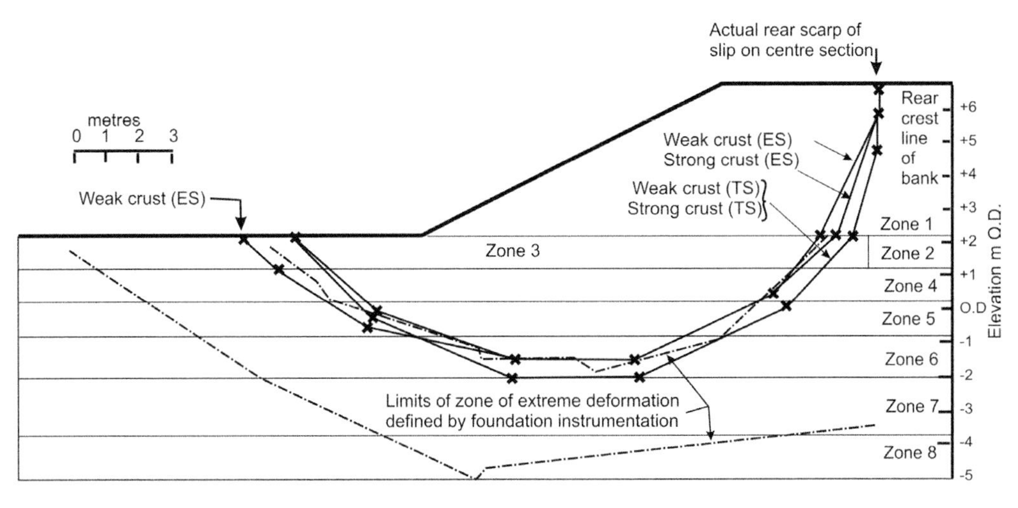

Figure 52. Bank 1/2: Total and effective stress stability analysis - critical restricted surfaces

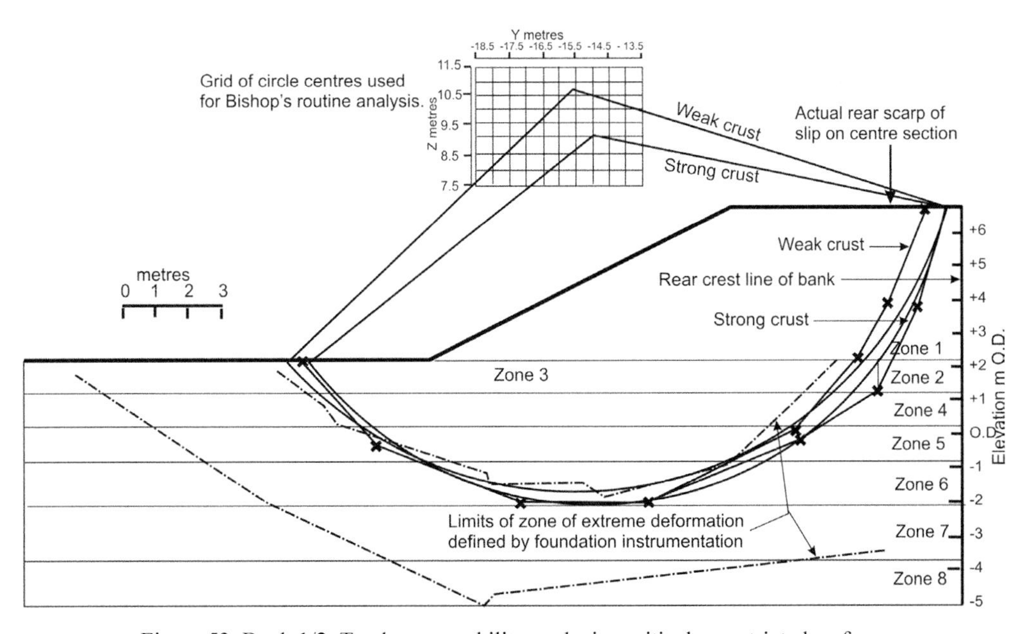

Figure 53. Bank 1/2: Total stress stability analysis - critical unrestricted surfaces

16.3.3 *Summary of results*

Stability analyses were carried out for all the critical surfaces using the shear strength profiles of Figure 47, together with the corrected Geonor and large vane strength and derived profiles. The shear strength profiles are summarised in Figure 54 and the results presented in Table 4.

The relevant data in Table 4 for the back analyses of the field undrained shear strength is provided by the unrestricted circular arc ($\phi_u = 0$) stability analyses. For the analyses using the various laboratory shear strength profiles the factors of safety were near identical for Banks 1/1 and 1/2. Restricted circular arc failure surfaces resulted in safety factors that were only slightly higher (+0.01 to +0.04) than the unrestricted values whereas those for the restricted non-circular surfaces were significantly higher (+0.08 to +0.14). The highest safety factor was given by the vertical laboratory-test data (1.37), reducing to 1.30 for the piston samples, 1.25 for the vertical triaxials corrected for anisotropy and 0.81 for the design profile (i.e. corrected for both anisotropy and strain rate). The vertical triaxial shear strength was considered to represent, as far as it is possible, the true undisturbed laboratory quick undrained shear strength of the Mucking clay. The slight drop (0.07) in safety factor using the piston sample data was considered to be the result of slight sample disturbance.

Table 4: Results of total stress stability analyses for critical surfaces

Soil Properties	**Factor of Safety**					
	Circular arc surfaces				**Non-circular surfaces**	
	Unrestricted		**Restricted**		**Restricted**	
	B1/1	**B1/2**	**B1/1**	**B1/2**	**B1/1**	**B1/2**
Laboratory Data						
Block samples, vertical specimens	1.37	1.37	1.38	1.41	1.50	1.51
Corrected for anisotropy	1.25	1.28	1.27	1.32	1.36	1.40
Design profile	0.81	0.83	0.83	0.86	0.89	0.91
Piston samples	1.30	1.30	1.31	1.33	1.42	1.44
Field Vane Data						
Geonor Vane	1.39	1.44	1.41	1.53	1.55	1.58
Geonor vane corrected (x μ)	1.10	1.21	1.12	1.30	1.22	1.36
Large Vane	1.12	1.20	1.15	1.27	1.23	1.38
Large vane corrected (x μ)	0.91	1.04	0.93	1.14	0.99	1.20
Derived Data						
Based on corrected vertical profile	0.99	1.01	1.01	1.08	1.06	1.15

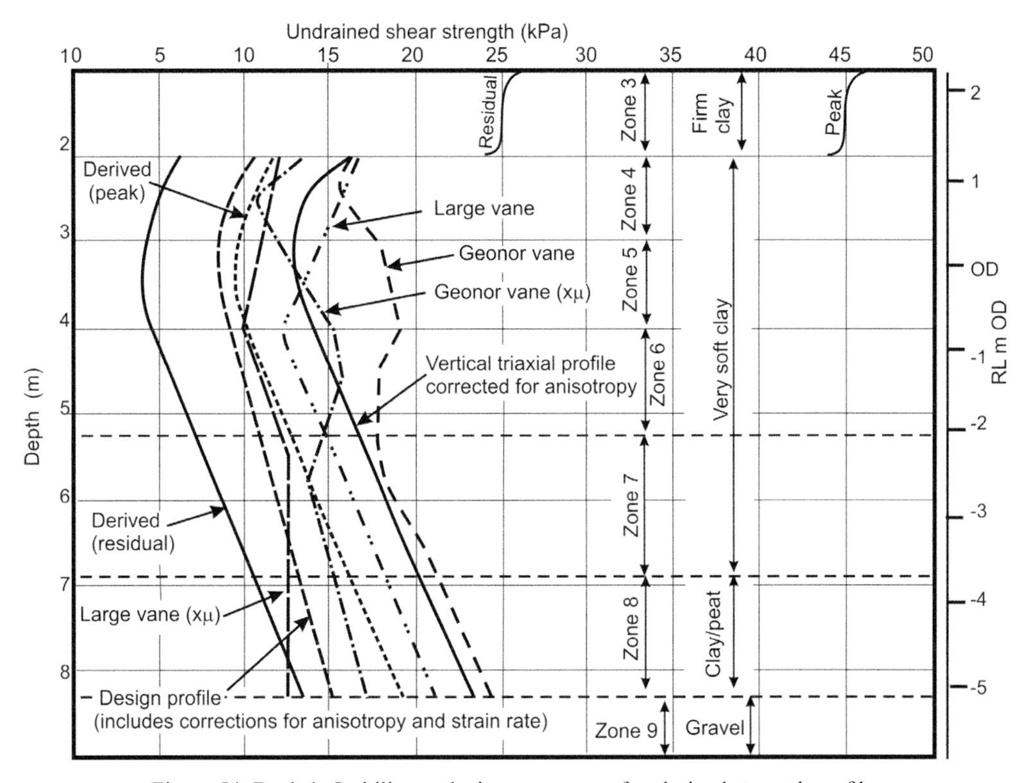

Figure 54. Bank 1: Stability analysis – summary of undrained strength profiles

For the field vane-test data the factors of safety for Bank 1/2 were between 0.05 and 0.13 higher than for Bank 1/1. This variation was considered to be largely the result of the irregular shear strength profiles for the vane tests. Again, the restricted circular arc surfaces resulted in slightly higher safety factors (+0.02 to +0.10) and the restricted non-circular surfaces in significantly higher values (+0.08 to +0.18). The highest factors of safety were for the Geonor vane test data (1.39 to 1.44) and were comparable to those for the vertical triaxial data. However, the Geonor vane test data corrected for plasticity produced safety factors (1.10 to 1.21) which were significantly lower than those for the laboratory triaxials corrected for anisotropy. The large vane test data produced even lower safety factors (1.12 to 1.20), reducing to between 0.91 and 1.04 when corrected for plasticity. Thus, of all the data sets, the large vane tests at the standard rate of rotation, corrected for anisotropy and strain rate, came closest to a correct prediction of the failure heights of Banks 1/1 and 1/2.

16.3.4 *Peak undrained shear strength under embankment loading*

The derived undrained strength profile is compared to the laboratory and field test data in Figure 54. The average undrained strength around the critical unrestricted circular arc failure surface for Bank 1/1 (Figure 51) for each of the various strength profiles is shown on Table 5. This failure surface did not pass through the desiccated crust. Table 5 also gives the correction factors

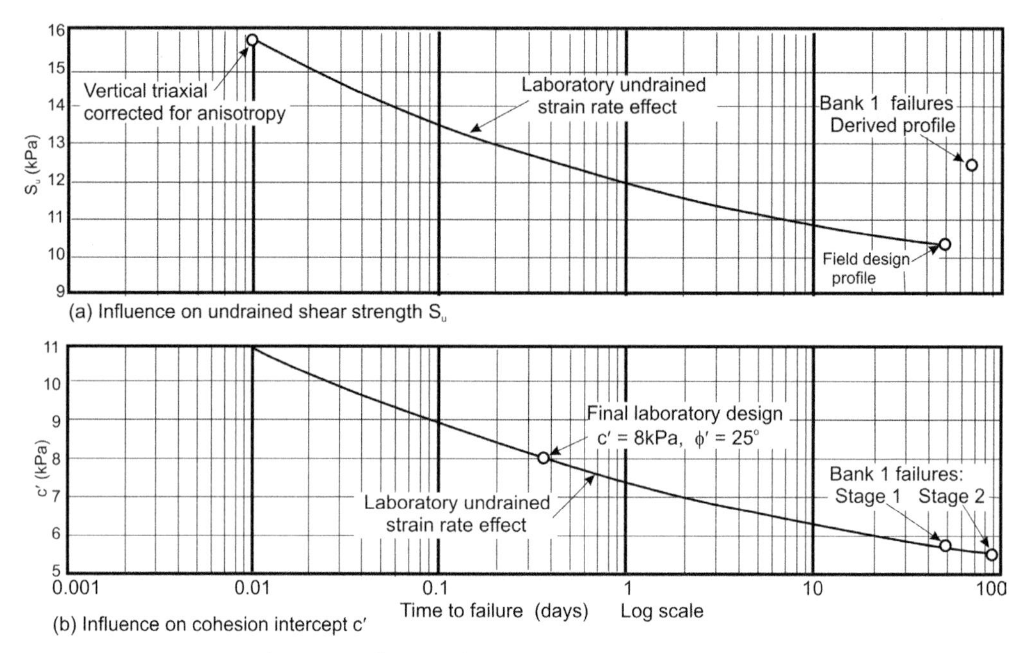

Figure 55. Influence of time to failure on laboratory and
full scale field measurements of undrained shear strength

that would have to be applied to the shear strength profiles in order to produce a safety factor of unity together with the factors applied to the laboratory and vane test profiles.

Figure 55(a) depicts the laboratory undrained strain-rate curve, on which the back-calculated undrained strength for Bank 1/1 has been plotted, at a mean time to failure for Banks 1/1 and 1/2 of 70 days. The laboratory shear strength is the average for the analysis using the laboratory vertical triaxial undrained strength profile corrected for anisotropy (× 0.87). The required strain-rate correction from the 12 minute laboratory tests to the 70 day field failure (× 0.79) is significantly less than predicted (× 0.65) by the design profile. While this discrepancy was initially considered to possibly be the result of the differing stress-strain-time paths followed in the laboratory and the field, subsequent effective stress analyses confirmed the field observation that significant pore-pressure dissipation took place during the very early stages of construction. Beneath the crest and shoulder of Bank 1/1 the average vertical effective stresses increased by a maximum of about 15kPa, from about 25kPa to 40kPa, the latter closely corresponding to the pre-consolidation pressure (Figure 48). Reference to Figure 16 indicates that such an increase in effective stress, followed by undrained shear, would increase the undrained strength by 4kPa. The back-calculated increase in undrained strength of 2.2kPa (see Figure 55(a)) suggests, however, that the long-term undrained strain-rate effect significantly reduced the strength gain in the early stages of construction.

The average undrained strength on the critical slip surface, using the vertical triaxial data, was 18.2kPa (Table 5). If 4kPa is added to this figure, to allow for early stage consolidation, the vertical triaxial strength becomes 22.2kPa. Correcting this figure for anisotropy (× 0.87)

Table 5: Average undrained shear strengths on critical slip surfaces for Bank 1/1

Data Source	Factor of Safety	S_u kPa	Correction Factors	
			Required	Applied
Laboratory				
Vertical specimens	1.37	18.20	0.69	-
Corrected for anisotropy	1.25	15.85	0.79	0.87
Design profile	0.81	10.20	1.22	0.57
Field Vane				
Geonor vane	1.39	18.30	0.68	-
Geonor vane corrected (x μ)	1.10	13.95	0.90	0.76
Large vane	1.12	14.85	0.84	-
Large vane corrected (x μ)	0.91	11.65	1.07	0.76
Derived Data				
Based on corrected vertical profile	0.99	12.5	-	0.79

and undrained strain rate (× 0.65) produces a predicted field design value of 12.7kPa, compared to 12.2kPa from back-analyses of the Bank 1/1 failures. This suggests that the peak undrained shear strength of the Mucking clay could have been predicted by measuring the undrained shear strength in undrained triaxial tests on specimens from high quality undisturbed block samples consolidated to the pre-consolidation pressure. The resulting strength profile would then require correction for anisotropy and strength loss appropriate to the predicted time of construction. The back analyses of the stability of Bank 1 also confirmed that the average undrained shear strength mobilised on the critical slip surfaces corresponded to the laboratory peak values and, therefore, that progressive failure was not an issue for the Mucking clay.

Turning to the Geonor vane tests, the critical slip surface produced a near identical safety factor and average shear strength to the vertical triaxial tests, despite the Geonor vane profile predicting higher strengths above 3.5m depth (+ 4 kPa, result of drainage) and lower strengths below 3.5m depth (–4 kPa, result of disturbance). Applying Bjerrum's (1972, 1973) correction factor (× 0.76) resulted in a safety factor of 1.10 and an average shear strength of 13.95kPa. This implies that the case histories considered by Bjerrum (loc cit.) were clays which, like Mucking, underwent rapid consolidation and strength gain at effective stresses less than the pre-consolidation pressure.

The large vane tests at the standard rate of rotation were undrained and the critical slip surface produced a safety factor of 1.12 and an average shear strength on the failure surface of 14.85kPa. Applying Bjerrum's (1972) correction factor reduced these values to, respectively 0.91 and 11.65kPa. It is considered that this slight under-prediction of the shear strength may have been the result of a degree of installation disturbance, particularly in the lower, more sensitive, clay. The large vane tests demonstrated that the laboratory-measured undrained strain rate could be reproduced in field tests. They also predicted the rapid pore-pressure dissipation that would occur, particularly in the upper 3.5m, at in-situ stress levels. In the context of the

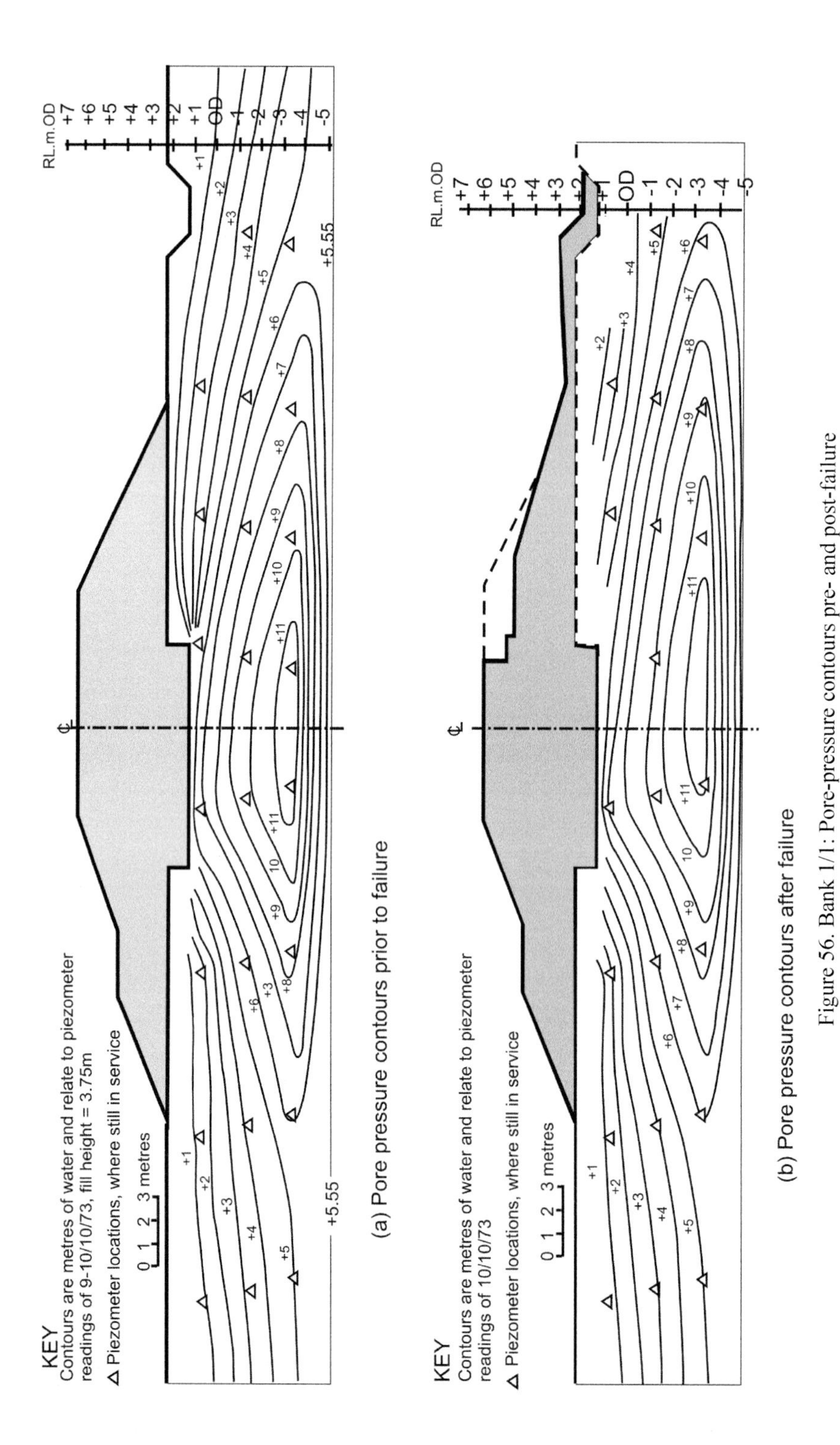

Figure 56. Bank 1/1: Pore-pressure contours pre- and post-failure

full-scale field behaviour it is interesting to note (Figure 43) that the difference between the minimum (undrained) and the maximum (partially/drained) shear strengths measured in the upper levels of the clay was 4.5 to 5kPa.

16.4 *Bank 1, failure: effective stress analyses*

16.4.5 *Bank 1/1*

As a result of the laboratory testing program being carried out over a three year period in parallel with the field trials, the stability analyses in terms of effective stress were performed using preliminary design parameters of $c' = 8$kPa and $\phi' = 26.5^0$. Figure 56(a) shows the pore-pressure contours used for the initial stability analyses, being based on the last set of observations prior to failure. These analyses were restricted to the observed end points of the actual slip surface and the observed zone of extreme deformation, as shown in Figure 57, and resulted in the critical slip surface passing through the desiccated crust (Zone 3). In order to force the critical restricted surface into the central trench it was necessary to increase c' for Zone 3 to 45kPa, being the back-calculated value of the undrained shear strength (S_u). It was thus decided to treat the desiccated crust in terms of total stress with $S_u = 45$kPa and $\phi_u = 0$, representing a strong crust analysis. It should be noted that because the strong crust analyses produced a critical surface passing through the central trench, the analysis for the critical surface was, *de facto*, an effective stress analysis. As a result the analyses of the critical "weak" and "strong" crust critical surfaces yielded similar safety factors.

Circular arc surfaces, restricted to the observed toe of the slip and the zone of deformation, were then considered for both the weak and strong crust assumptions, yielding the critical

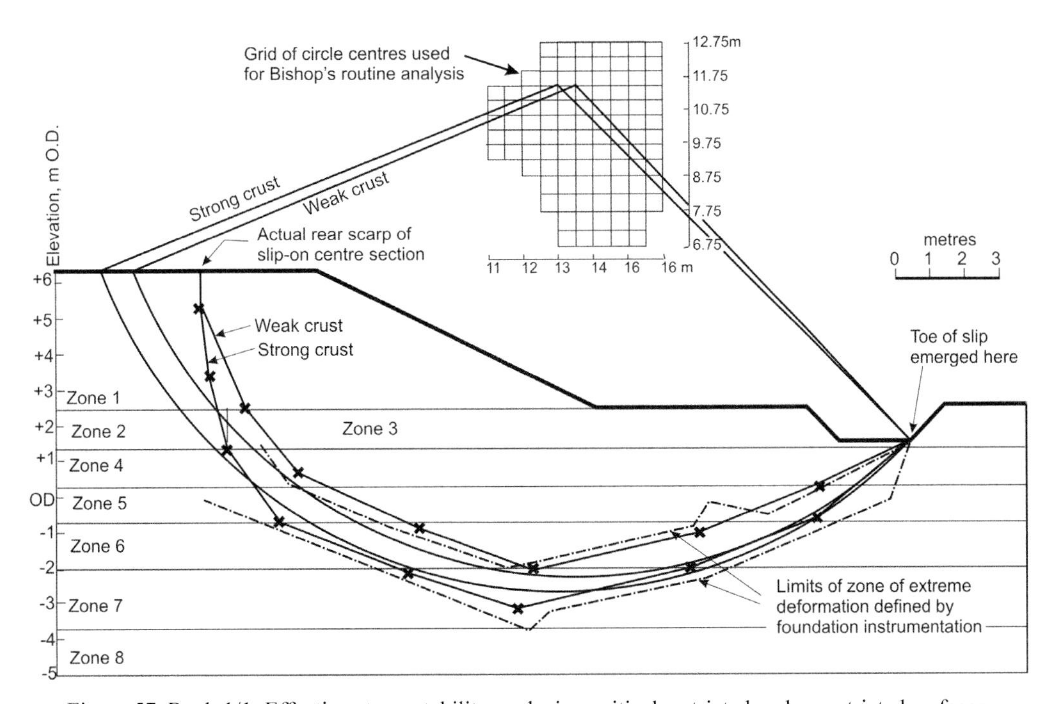

Figure 57. Bank 1/1: Effective stress stability analysis – critical restricted and unrestricted surfaces.

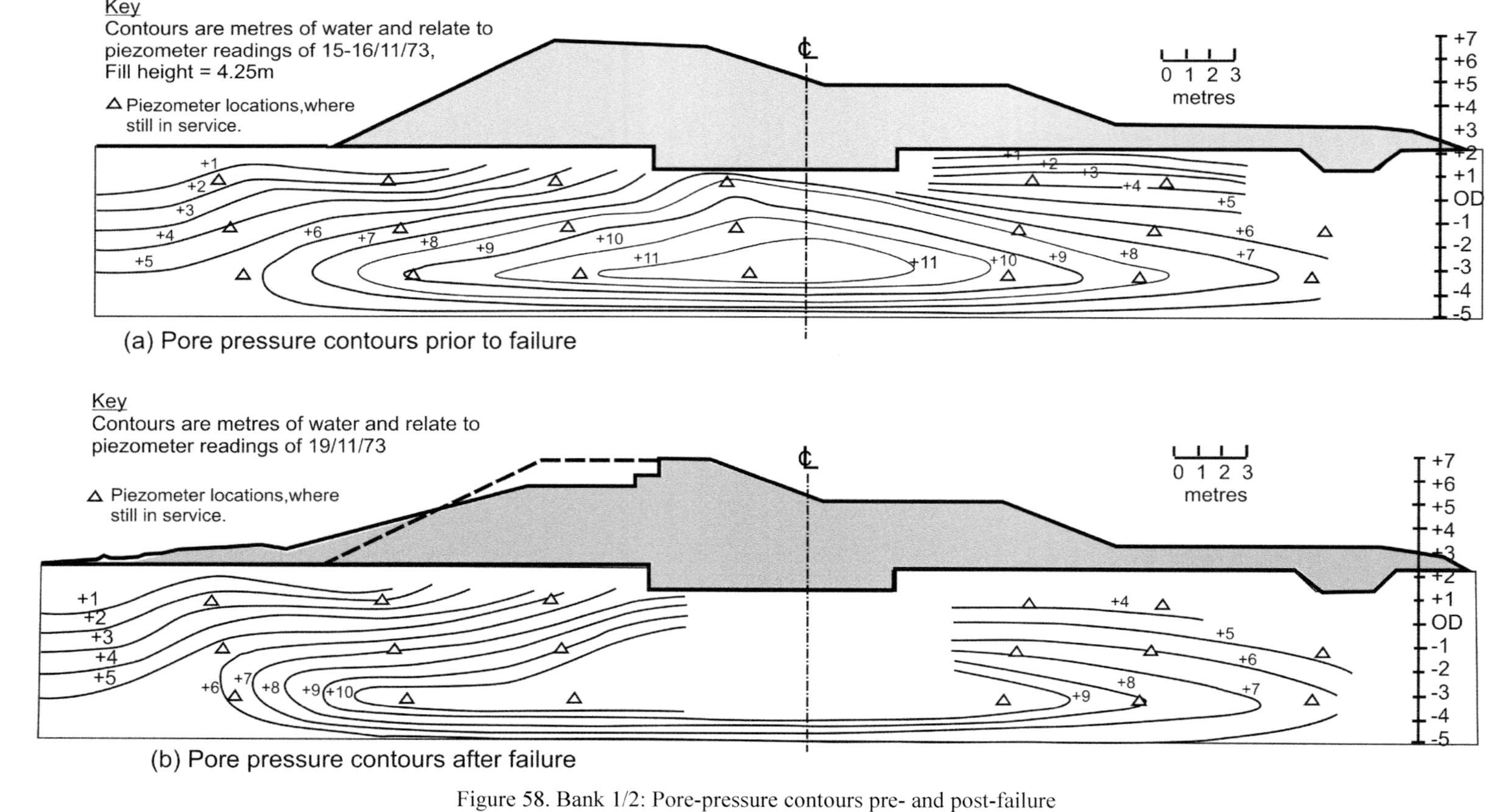

Figure 58. Bank 1/2: Pore-pressure contours pre- and post-failure

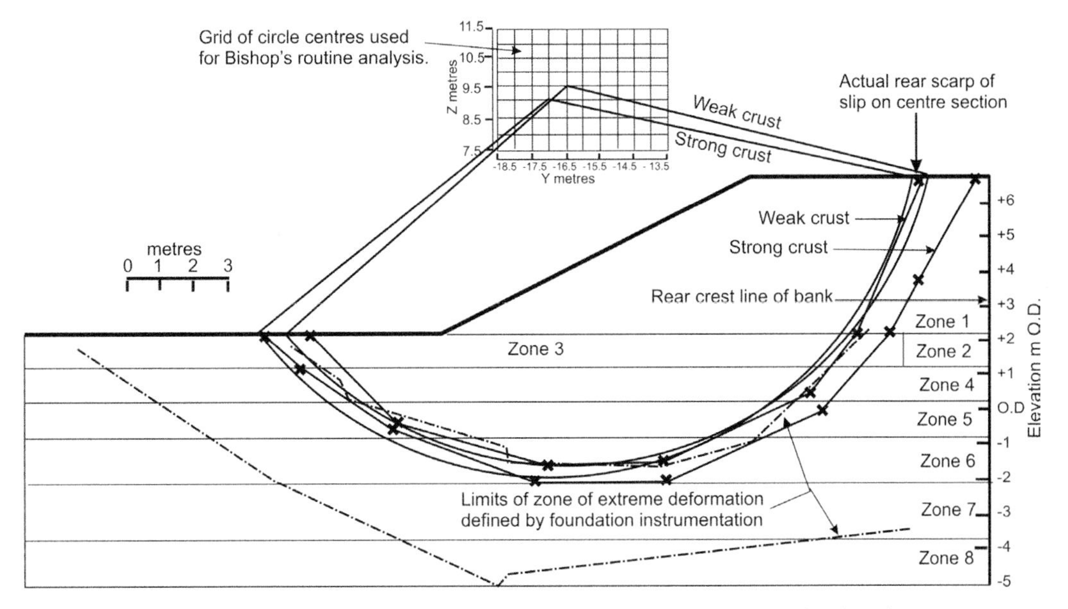

Figure 59. Bank 1/2: Effective stress stability analysis – critical unrestricted surfaces

surfaces shown in Figure 57. These same surfaces were also found to be critical when the stability analyses were extended to unrestricted circular arc surfaces. Analyses of the critical non-circular surfaces were then repeated using pore pressures extrapolated to a fill height of 4m, being the height at failure.

16.4.6 *Bank 1/2*

Figure 58(a) shows the pore-pressure contours used for the initial stability analyses, being based on the last set of observations prior to failure. As for Bank 1/1 the initial analyses were for restricted non-circular surfaces, with both weak and strong crust assumptions. Figure 52 shows that, as for Bank 1/1, the critical surfaces were close to the upper limit of deformation defined by the foundation instrument readings. Analyses were then extended to unrestricted circular arc and non-circular surfaces, with both weak and strong crust assumptions, for which the critical surfaces (Figure 59) were essentially the same as those for the restricted surfaces. The major difference between the effective stress analyses for Bank 1/1 and 1/2 was that, unlike Bank 1/1, the strong-crust analyses for the critical surfaces were not true effective stress analyses. Finally, the non-circular surfaces were analysed using pore pressures extrapolated to a fill height of 4.5m, being the height at failure.

16.4.7 *Summary of results*

Very consistent results were obtained from the effective stress analyses for Bank 1/1, for which none of the critical surfaces passed through the crust, and from the effective stress analyses for Bank 1/2. Results from the Bank 1/2 analyses incorporating a strong total stress crust were

Table 6. Bank 1: Effective stress stability analyses – summary of results.

Soil Properties	**Factor of Safety**			
	Circular arc surfaces		**Non-circular surfaces**	
	Unrestricted		**Restricted**	
	B1/1	**B1/2**	**B1/1**	**B1/2**
Initial Laboratory Design $c' = 6.5$kPa, $\phi' = 26.5^0$	1.25	1.39	1.23	1.31
Derived for Bank 1/1 $c' = 3.75$kPa, $\phi' = 26.5^0$	1.04	1.19	1.01	1.09
Derived for Bank 1/2 $c' = 2.75$kPa, $\phi' = 26.5^0$	0.96	1.12	0.93	1.00
a) Measured pore-pressures at 3.75m height (Bank 1/1) and 4.25m height (Bank 1/2)				
Initial Laboratory Design $c' = 6.5$kPa, $\phi' = 26.5^0$			1.09	1.11
Derived for Bank 1/1 $c' = 5.5$kPa, $\phi' = 26.5^0$			1.01	1.03
Derived for Bank 1/2 $c' = 5.25$kPa, $\phi' = 26.5^0$			0.98	1.01
b) Pore pressures extrapolated to 4.0m height (Bank 1/1) and 4.5m height (Bank 1/2)				

inconsistent with the effective stress analyses. As such, the results summary of Table 6 is restricted to the true effective stress analyses.

For Bank 1/1 the initial laboratory design parameters produced a safety factor of 1.23 using the measured pore pressures, reducing to 1.09 when pore pressures extrapolated to maximum fill height (4.5m) were used. A reduction of only 1kPa was required to c' to achieve a factor of safety of unity using the extrapolated pore pressures. There was also excellent agreement between the safety factors on critical restricted non-circular surfaces and critical unrestricted circular arc surfaces as well as between the critical failure surfaces and the field observations.

For Bank 1/2 the initial laboratory design data, in conjunction with the measured pore pressures, yielded a safety factor of 1.31, being somewhat higher than the 1.25 recorded for Bank 1/1. However, when the extrapolated pore pressures were used the safety factor reduced to 1.11, being in excellent agreement with the equivalent 1.09 for Bank 1/1. As for Bank 1/1, there was excellent agreement between the locations of the critical restricted and unrestricted non-circular surfaces and the unrestricted circular arcs, all corresponding to the upper limit of extreme deformation defined by the foundation instrumentation.

16.4.8 *Peak effective stress shear strength parameters under embankment loading*

Using the critical non-circular slip surfaces for Bank 1/1 and Bank 1/2, parametric studies were carried out to determine the influence of c', ϕ' and pore pressure on the factors of safety. The results of these parametric studies are presented in Figure 60, which enables the final laboratory

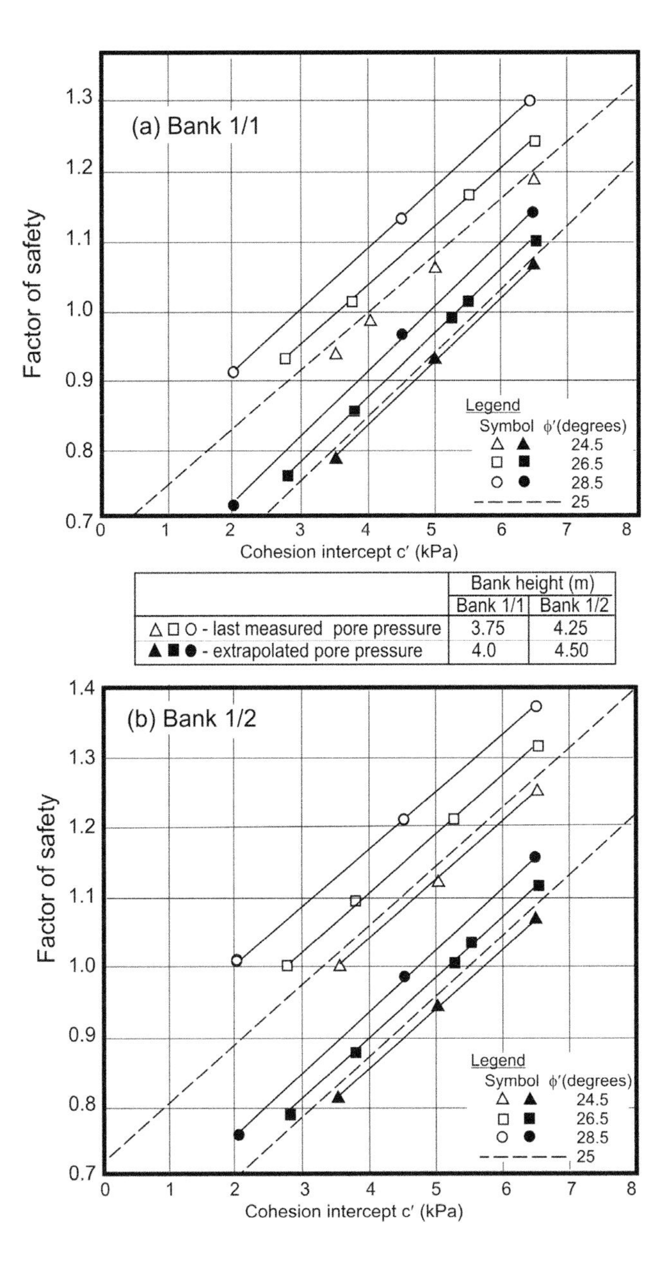

Figure 60. Influence of c′ ϕ′ and pore pressure on the factors of safety for Bank 1/1 and Bank 1/2

design parameters of c′ = 8kPa, ϕ′ = 25^0 to be related to the back-calculated field effective stress parameters. Using the extrapolated pore pressures at failure the factor of safety of Bank 1/1 with c′ = 8kPa, ϕ′ = 25^0 is 1.21 and for Bank 1/2 is 1.22. These results are not only in excellent agreement with each other but also in good agreement with the respective values based on the laboratory total stress design profile (Table 4), being 1.25 and 1.28.

The back-calculated values of c′ for a factor of safety of unity with ϕ′=25^0 and extrapolated pore-pressures are 5.7kPa for Bank 1/1 and 5.4 kPa for Bank 1/2. These data are compared to the laboratory values extrapolated on a strain-rate basis (Figure 49) in Figure 55(b), wherein it can be seen that the back-calculated values of c′ are entirely consistent with the values predicted by the laboratory tests. Reference to Figure 55 shows that the drop in S_u from a time to failure of 12 minutes to a time of 70 days is 5.65kPa, compared to 5.5kPa for c′ over the same period. This suggests that the undrained strain-rate reduction in the field shear strength is primarily the result of a partial breakdown in the lightly over-consolidated clay mineral structure of the Mucking clay, coupled with associated secondary pore pressure rises. .

16.5 *Bank 1: Post-failure total and effective stress analyses*

Figure 61 shows the post-failure profiles for Bank 1/1 and Bank 1/2 together with the critical restricted failure surfaces, which all conform to the observed end points of the slip surfaces and the zone of extreme foundation deformation. Both total and effective stress analyses were carried out, initially using, respectively, the derived peak undrained shear strength profile (Figure 54) and peak effective stress parameters c′ = 5.5kPa, $\phi' = 25^0$.

The effective stress analyses were carried out using the pore pressures measured post-failure, noting that for Bank 1/1 (Figure 56(a)) the pore pressures were measured immediately post-failure while for Bank 1/2 they were not measured until three days post-failure. Both total and effective stress analyses were carried out with weak and strong crust assumptions and, as can be seen from Figure 61, the critical failure surfaces for each of the desiccated crust assumptions were very similar for both the total and effective stress analyses. In this respect it should be noted that the total stress analyses of the actual failure surfaces were not true $\phi_u = 0$ analyses.

The initial analyses were highly problematic in terms of the necessary gross simplification of the zoning of the clay layers with respect to the total stress analyses, particularly in the case of the desiccated crust (Zone 3) for Bank 1/2. In addition, the effective stress analyses were subject to uncertainties regarding the locations of the remaining piezometers and the delay in reading the Bank 1/2 piezometers post-failure. Referring to Figure 61 it can be seen that the only analyses that are not influenced by the crust are the effective stress analyses for Bank 1/1, for which the immediately post-failure pore pressures were available. These analyses, as shown in Table 7, were very consistent with safety factors of 1.32 and 1.36. These are considered to be the definitive post-failure safety factors based on peak strength parameters. The equivalent effective stress analyses for Bank 1/2 yielded a higher safety factor of 1.65 for the weak crust (i.e. true effective stress) analysis, being considered to reflect the delay in the post- failure pore-pressure measurements. It thus seems likely that, in effective stress terms the post-failure Bank 1/1 and Bank 1/2 profiles had safety factors of between about 1.3 and 1.4. As can be seen from Table 7, the total stress analyses were very consistent for Banks 1/1 and 1/2 irrespective of the shear strength assumptions for the crust, but yielded safety factors considerably higher than the effective stress analyses. This was considered to be the result of the analyses not being for the critical $\phi_u = 0$ surface.

The shear strength parameters in terms of both total and effective stress were then reduced to achieve a safety factor of unity for each of the critical surfaces in order to obtain the average shear and normal effective stresses for each post-failure slip surface. The same eight critical surfaces were then analysed in combination with pre-failure geometries and peak shear strength parameters

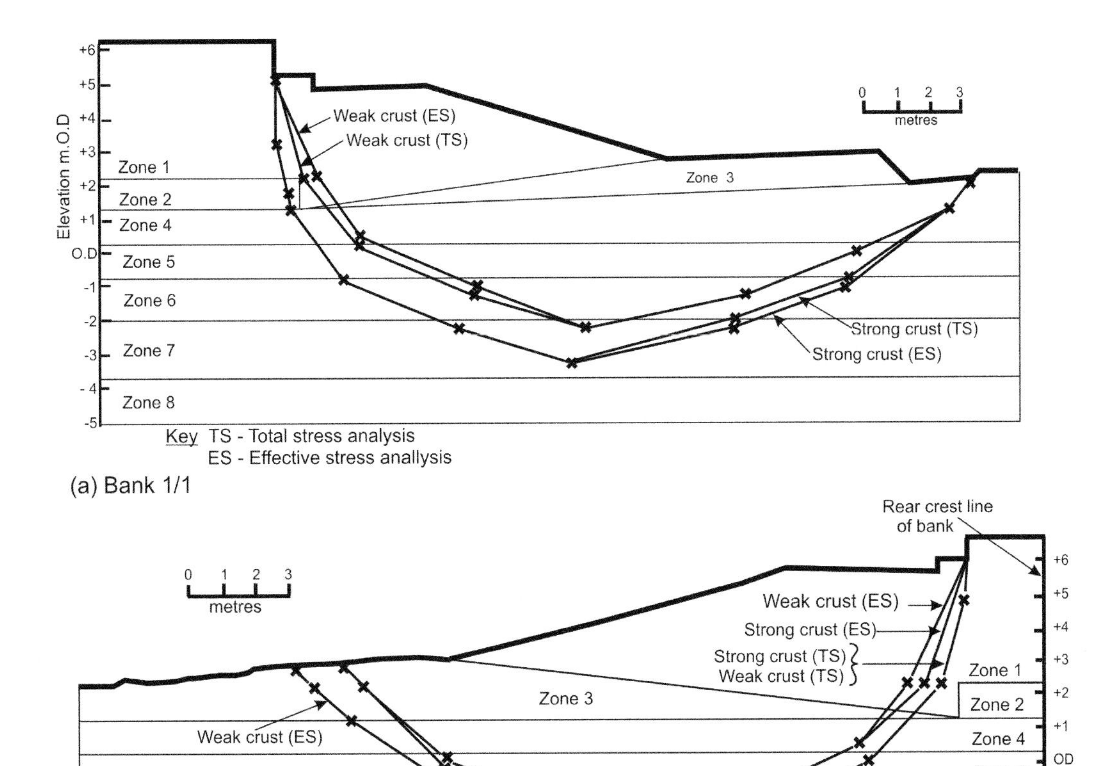

Figure 61. Bank 1: Post-failure total and effective stress stability analysis – critical restricted surface

Table 7: Results of post-failure stability analyses using peak shear strength parameters

Analysis	Factor of Safety			
	Bank 1/1		Bank 1/2	
	Weak crust	Strong crust	Weak crust	Strong crust
Effective stress	1.32	1.36	1.65	1.85
Total stress	1.63	1.85	1.65	1.83

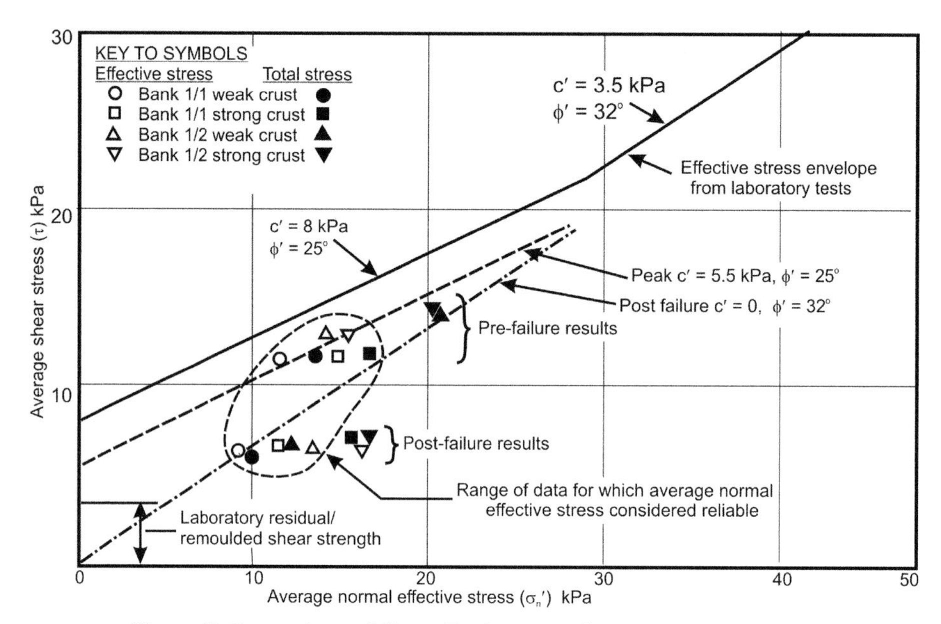

Figure 62. Pre- and post-failure effective stress shear strength parameters

and, for the effective stress analyses, with pore pressures extrapolated to the fill heights at failure. The average shear and normal effective stresses on the critical total and effective failure surfaces for both the pre- and post-failure geometries are presented in Figure 62.

As expected, the effective stress data for the pre-failure conditions conform to the peak field effective stress parameters of c′ = 5.5kPa, ϕ′ = 25^0. Perhaps surprisingly, the Bank 1/1 total stress data is also in good agreement. The only data that doesn't conform is that for the total stress analyses of Bank 1/2, because of the higher than average normal effective stresses, almost certainly the result of underestimated post-failure pore pressures. The peak shear stresses, however, are remarkably consistent, varying between 11.3 and 13.3kPa.

The post-failure effective shear and normal stress data for Bank 1/1 conform to c′ = 0, ϕ′ = 32^0 suggesting that, post-failure, the lightly over-consolidated Mucking clay becomes effectively "normally consolidated", as indicated by comparison with the effective stress envelope derived from the laboratory tests (Figure 62). As was the case for the peak shear stresses, the post-failure shear stresses were very consistent, being within the narrow range 6.0 to 6.9kPa. This consistency extended to the brittleness index (Bishop, 1967, 1971) which was an average of 46% in terms of both total and effective stress.

The continued shearing of the clay as Banks 1/1 and 1/2 moved from pre- to post-failure profiles resulted in both the undrained shear strength and the apparent cohesion reducing by 5.5kPa. This suggests that brittleness, in terms of both total and effective stress, is largely the result of a breakdown of the lightly over-consolidated clay mineral structure coupled with associated secondary pore pressure rises.

The results of laboratory ring-shear tests show (Figure 32) that the undrained residual strength of the Mucking clay was about 3kPa. A similar value for the undrained residual strength was obtained from ring-shear tests on specimens with a pre-cut plane. Similar values were, however, also recorded for the peak undrained shear strengths of remoulded specimens in laboratory triaxial, ring-shear and vane tests. The laboratory tests thus confirmed that the there was no difference between the ultimate residual and remoulded undrained shear strengths of the Mucking clay. The average peak undrained strengths from the laboratory triaxial and ring-shear tests was 15kPa, resulting in a laboratory brittleness index of 20% compared to the field value of 46%. The field brittleness, in terms of both total and effective stress, was largely the result of a complete loss of apparent cohesion caused by the breakdown of the lightly over-consolidated structure of the clay. It is thus clear that, for the Mucking clay, increasing shear strains beyond the scale of the field failures results in increasing excess pore pressures until ultimately the undrained shear strength drops to the residual/remoulded value (Figure 62). During this post-failure shearing the clay is effectively normally consolidated and the increased undrained brittleness is a function of the excess pore pressures.

16.6 *Bank 2, Post-construction stability, pore pressures and displacements*

It will be recalled from section 14 that Bank 2 (Figure 35) was designed to load the clay, including the upper desiccated zone, in plane strain conditions with a working factor of safety of 1.5. The embankment was constructed to a height of 2.8m (see Figure 46) with both horizontal and vertical displacements and pore pressures monitored during construction and in the long term. Stability analyses were carried out using Bishop's routine method for the end-of-construction situation and 5, 50, 200 and 600 days thereafter.

The pore pressures relevant to each of these analyses are presented as contours of the pore-pressure ratio (r_u) in Figure 63, which shows that the maximum pore pressures occurred 5 days after the end-of-construction and did not dissipate to the end-of-construction values until 50 days later. The back-analyses of the Bank 1 failures in terms of effective stress enabled the analyses of Bank 2 to incorporate the field-derived time-dependent values of apparent cohesion, as shown in Figure 55(b). For comparison purposes a second set of analyses were carried out using the Bank1/1 field-derived effective stress parameters $c' = 5.5$kPa, $\phi' = 25^0$.

The results of the analyses are shown as factor of safety versus time after the start of construction (log scale) in Figure 64.The factor of safety for the actual time-dependent c' can be seen to drop from about 1.6 at the end-of-construction to a minimum of 1.5 some 35 days later and thereafter rise to nearly 1.9 some 600 days after the end of construction. Also evident from Figure 64 is that the adoption of a constant (70 day) value of $c' = 5.5$kPa results in an initial under-estimate of the safety factor, followed by an over-estimate beyond about 90 days from the start of construction.

Finally, Figure 65 shows the horizontal displacements recorded by the inclinometers at the opposite toes of the embankment. The maximum horizontal displacements rise from between about 30mm at the end-of-construction to about 100mm some 570 days later. With respect to the previous discussion on the strain-rate effect together with field residual and remoulded shear strengths, it is of interest that these post-construction horizontal displacements and strains (creep) were initially accompanied by increasing excess pore pressures and that it was not until some 50 days after the end of construction that the pore-pressures returned to the end of construction values

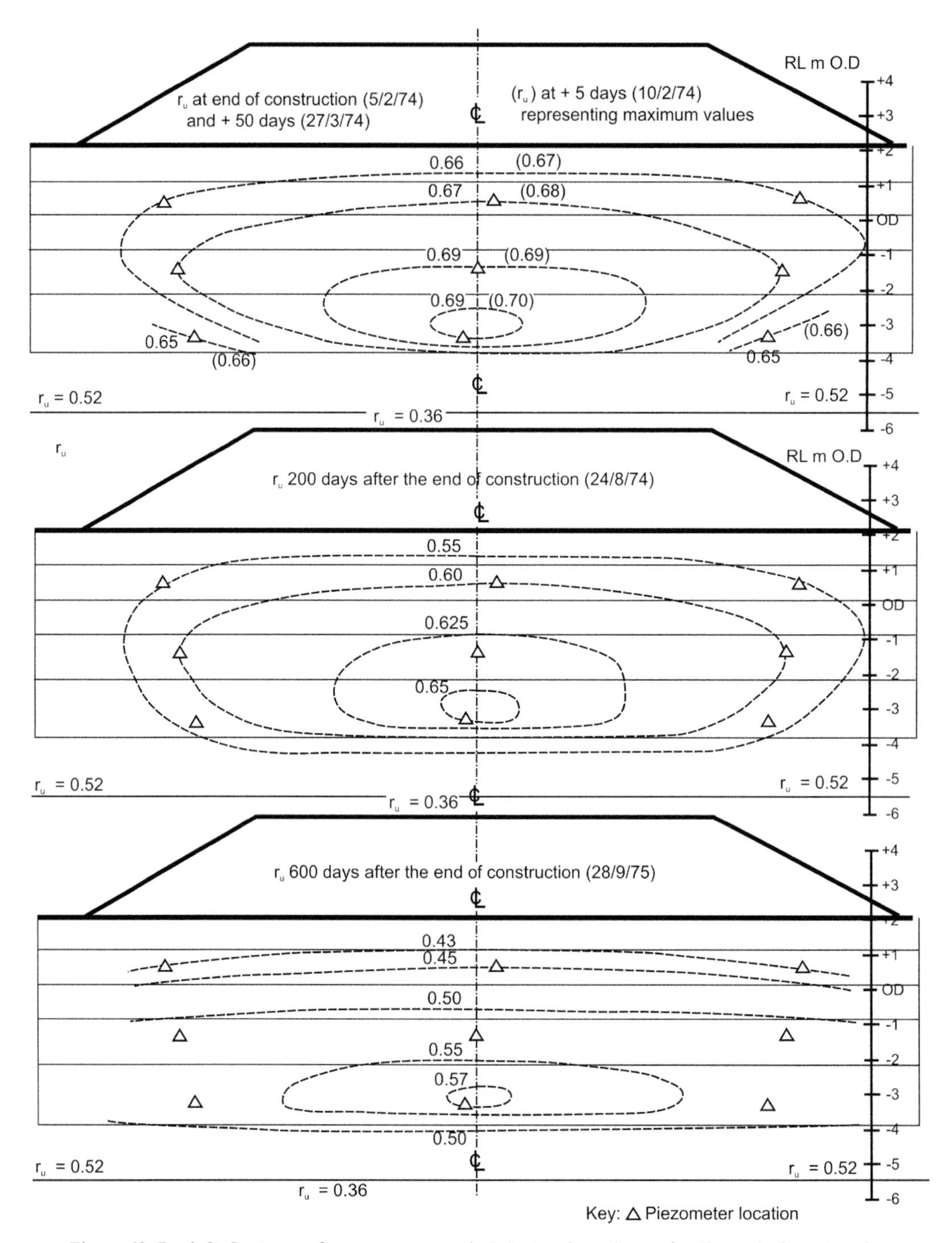

Figure 63. Bank 2: Contours of pore-pressure ratio (r_u) at various times after the end of construction

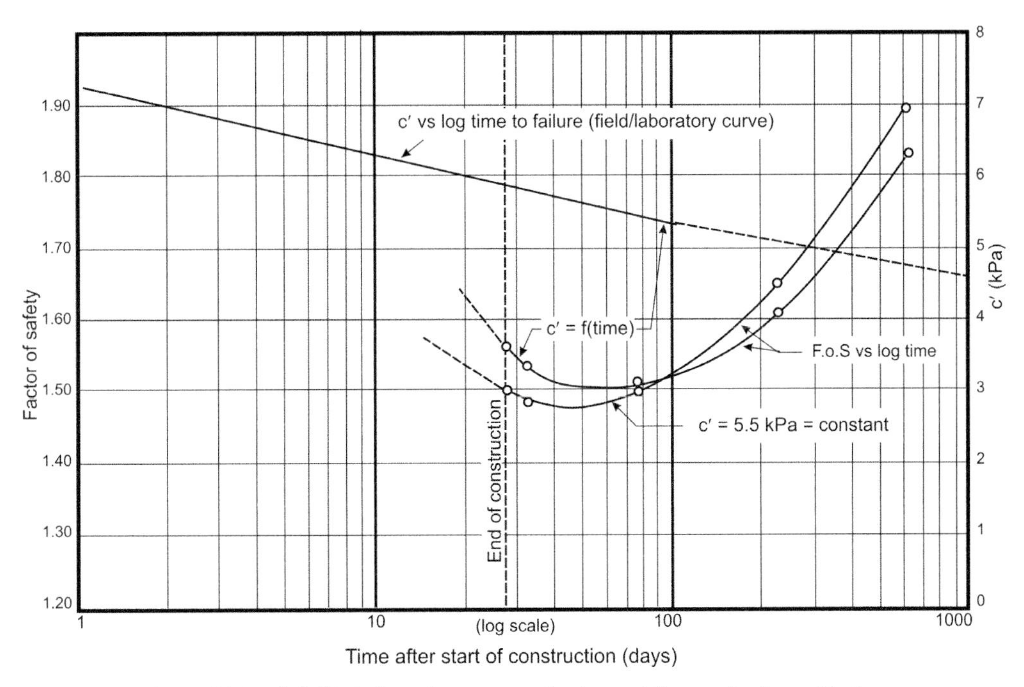

Figure 64. Bank 2: Variation of apparent cohesion and factor of safety with time

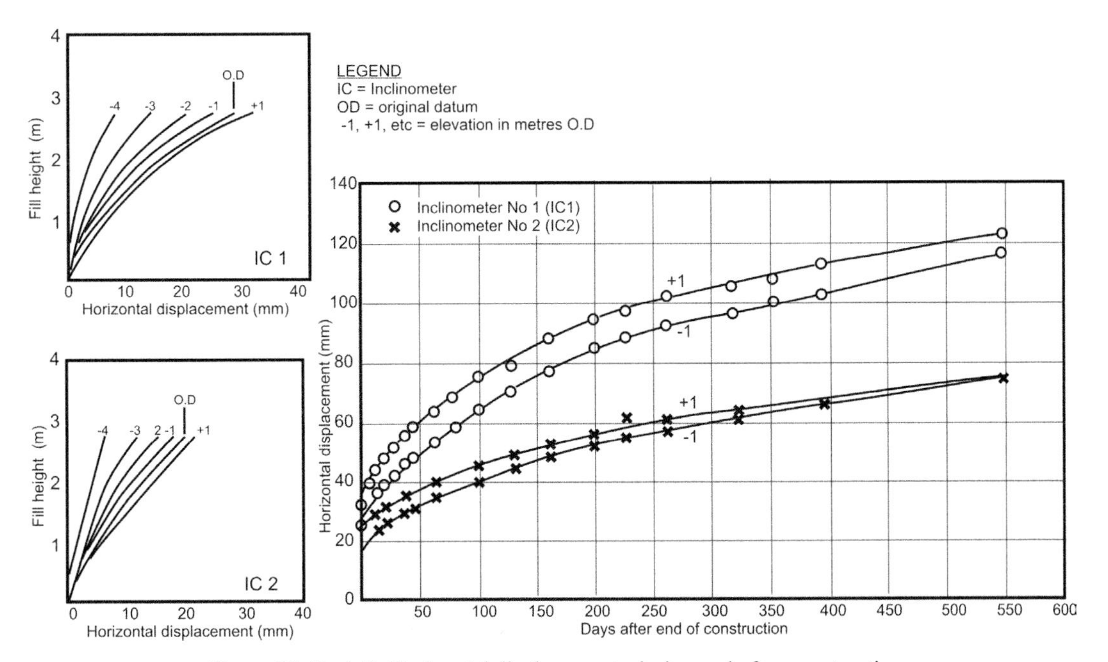

Figure 65. Bank 2: Horizontal displacements during and after construction

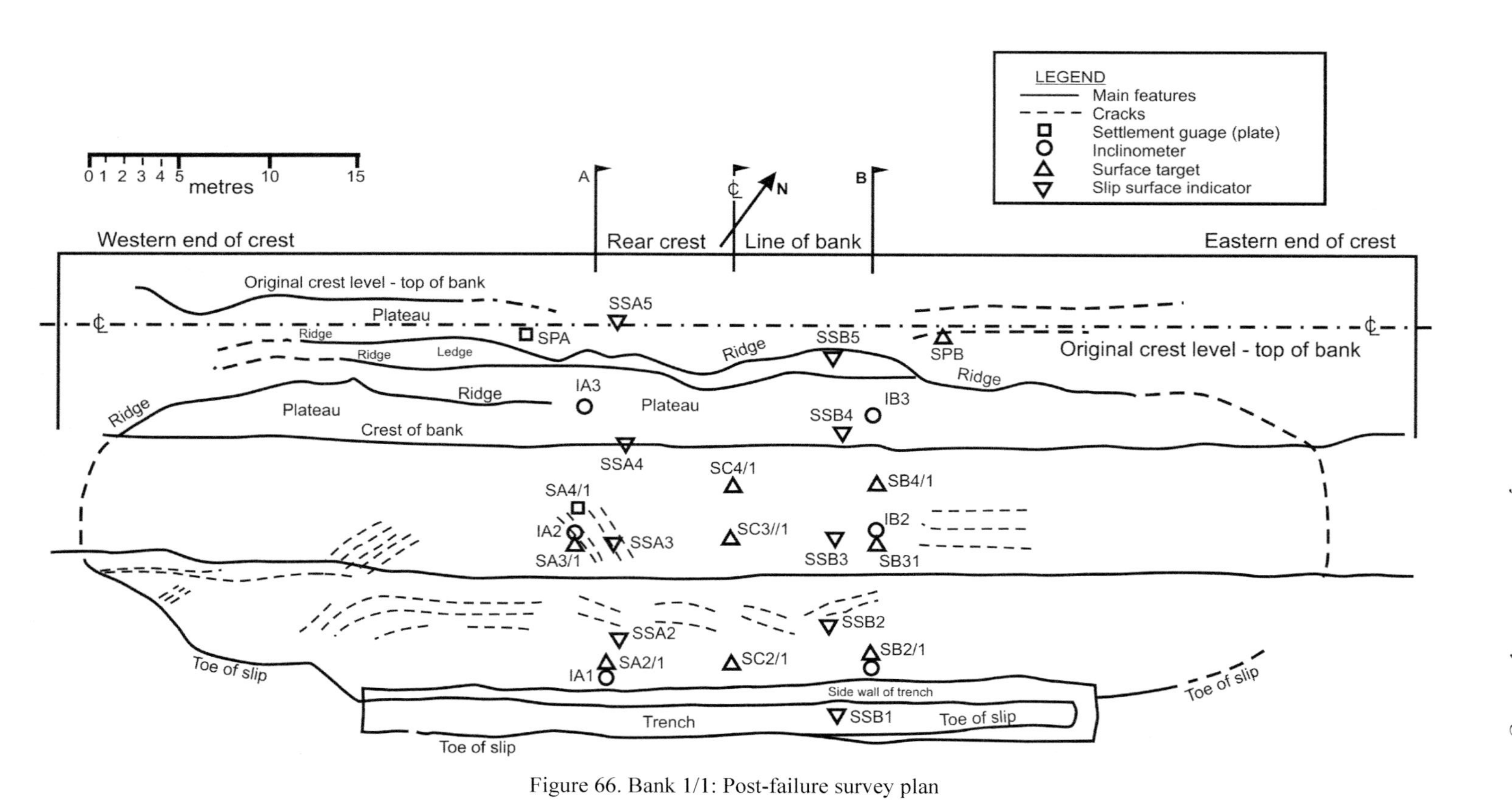

Figure 66. Bank 1/1: Post-failure survey plan

17 PRE- AND POST-FAILURE FOUNDATION MOVEMENTS

The post-failure survey plan for Bank 1/1 is presented in Figure 66. A crack in the fill, up to 5mm wide, opened up some 15 minutes prior to any discernible relative vertical displacements occurring across it. This crack (Figure 67) had widened to a maximum of 100mm before vertical movements were observed, this being similar to the observed behaviour for other embankments on soft clay in the absence of significant desiccation in the near surface clay (Bjerrum, 1972). Some 2.5 hours prior to failure, tensile horizontal strains of about 4% had developed within the near surface foundation clay adjacent to the central trench; the indications were, therefore, that this figure was nearer 10% when relative vertical displacements commenced. Monitoring of the ground surface beyond the embankment toe suggested that, during this period of horizontal movement of the embankment, heave movements of similar magnitude were being experienced beyond the toe.

The initial failure was followed by the development of secondary failure surfaces on the top of the embankment, to the rear of the main surface. These secondary surfaces developed subsequent to the period of maximum velocity and when maximum vertical displacement was about 750mm. Although the failure of the embankment was more pronounced towards the west end (Figure 66) it extended along virtually the entire 74m length. The visual evidence suggested that the failure of the embankment, similarly to the construction, was under essentially plane strain conditions. Maximum displacements along the length of the embankment were up to 15% of those along the section lines, although generally much less.

The geometry of the slip, particularly near the embankment toe where the slipped mass was disrupted by tension cracks – and so indicating internal distortion, taken together with the rear scarp of the slip suggested a non-circular failure surface. The tension cracks in the fill and the foundation were restricted to a plan area extending about 4m from the toe (see Figure 66) where they were well developed and clearly visible. The main scarp of the slip was a vertical face about 1m high, located above the central trench. The shear surface within the foundation was clearly defined and planar where it entered the toe trench. The foundation instrumentation (see Figure 45) indicated the rupture zone to be relatively narrow, being about 500mm wide, and composed of two inclined zones meeting beneath the embankment toe and with their extremities within the toe and central trenches respectively (see Figure 67, which also shows the coordinates used in the non-circular stability analyses described in section 3). Maximum horizontal shear strains recorded by the inclinometers also coincided with this zone, while marked increases in pore pressure post-failure were also observed within or close to the rupture zone.

In summary, the evidence suggests that failure was preceded by the formation of a vertical tension crack through the fill, the tensile horizontal strains being about 4% at this stage. A large part of the foundation beneath the embankment is anticipated to have been in a state of local active failure prior to formation of the crack, which possibly precipitated local passive failure beyond the toe. It would appear probable that during the period of horizontal spreading of the embankment, coupled with heave of the foundation beyond the toe, the areas of local failure were coalescing into a continuous rupture zone, the bounds of which are delineated in Figure 67.

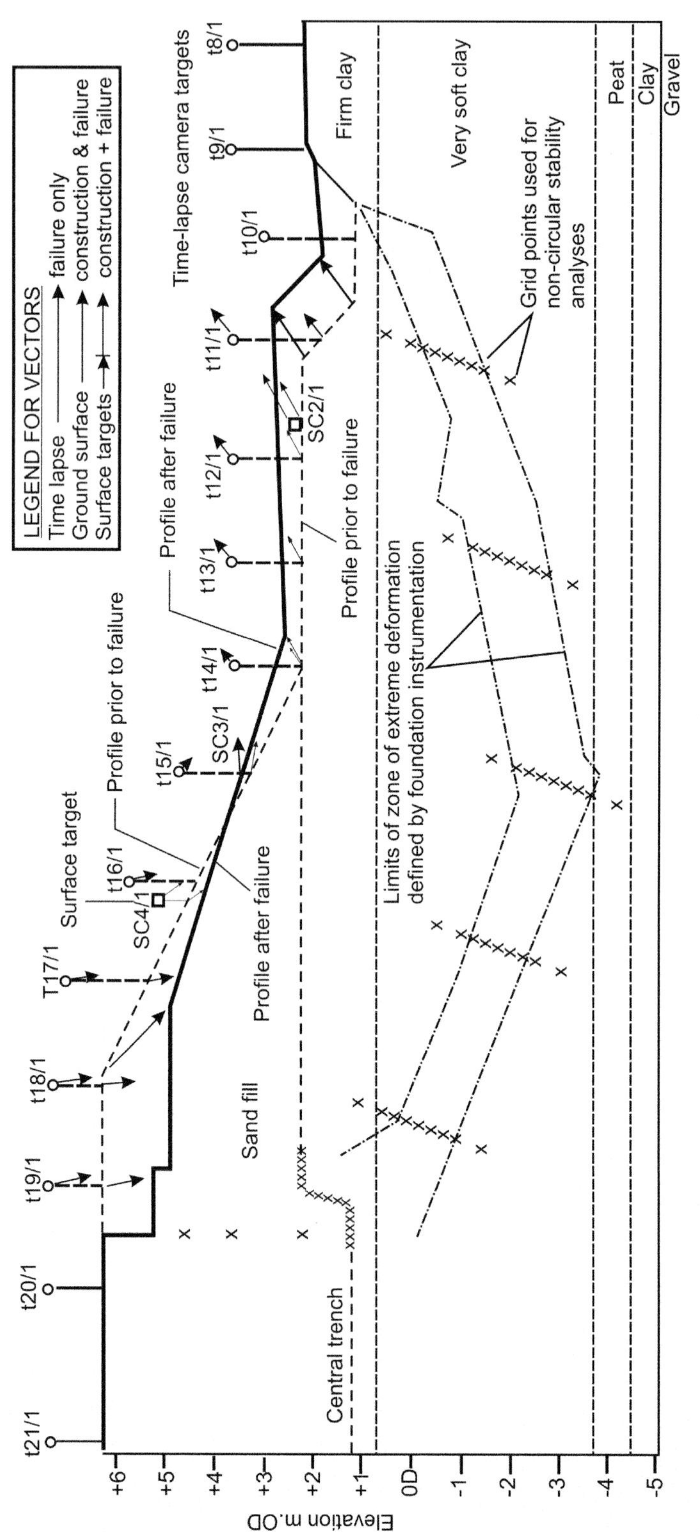

Figure 67. Bank 1/1: Pre- and post-failure profiles and surface movement vectors

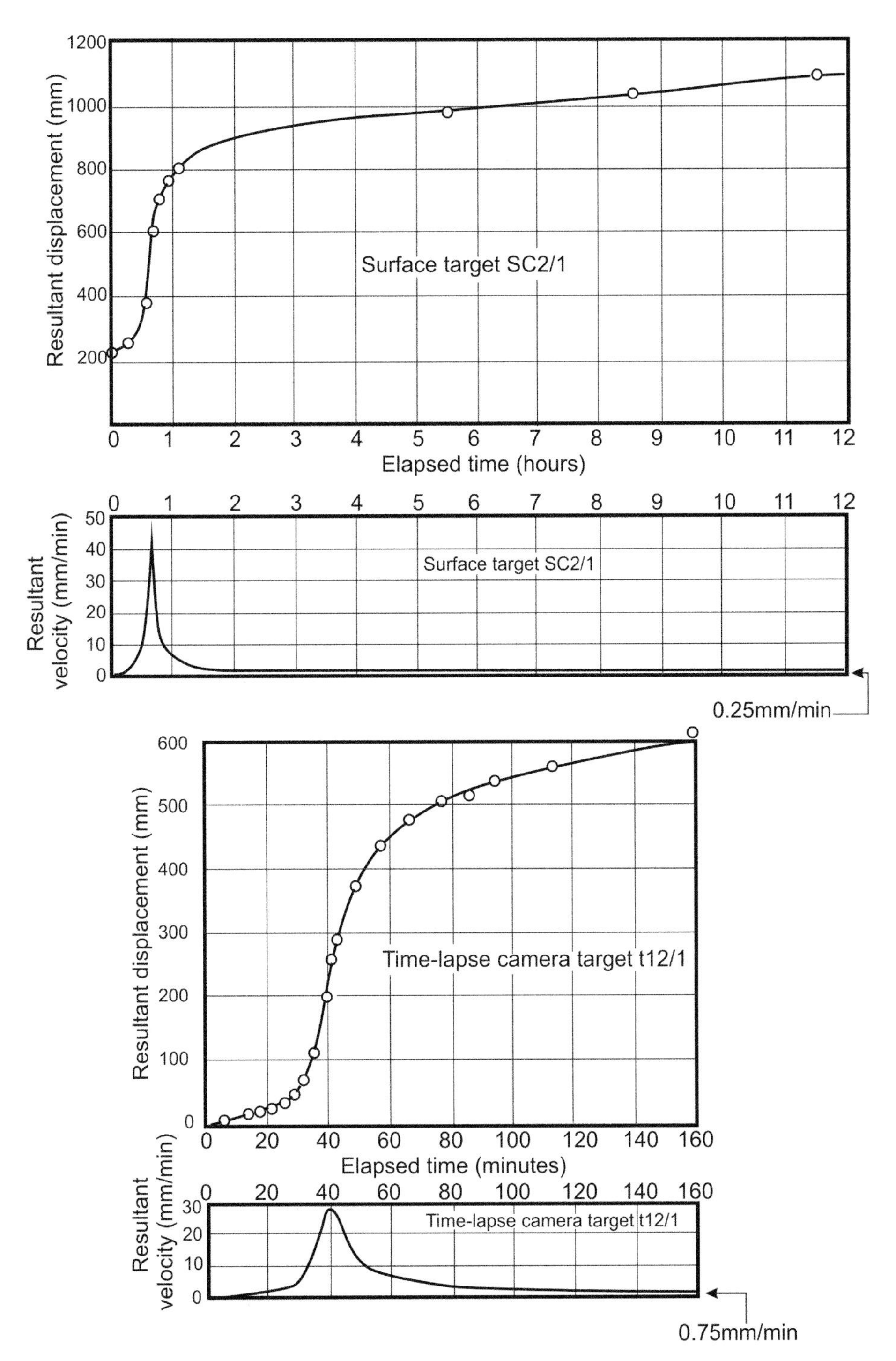

Figure 68. Bank 1/1: Resultant displacements and velocities of surface and camera targets versus time from start of failure

Recording of surface movements during failure was carried out using surface targets, time-lapse photography targets and the ground survey as shown in Figure 67. The resulting displacement vectors were fully supportive of the orientation and location of the postulated rupture zone. It is of interest to note that the time-lapse photography targets beyond the embankment toe all moved, initially, vertically upwards by some 30 to 40mm.

Figure 68 shows plots of displacement and velocity versus time from the formation of the tension crack for a surface target and a time-lapse photography target located adjacent to each other between the embankment toe and the toe trench. As can be seen the resultant displacements were between 600 and 900mm. Peak velocities of 30 to 45mm/minute occurred about 40 minutes after formation of the tension crack and some 25 minutes after commencement of major movement. Movements slowed to 0.75mm/minute after 2.5 hours but were still 0.25mm/hour after 12 hours.

18 THE DIRECT USE OF FIELD MEASUREMENTS TO CONTROL CONSTRUCTION

The "undrained" behaviour of the Mucking clay under initial construction loading was characterised by small displacements and strains, coupled with elastic pore-pressure responses. Initial conditions beneath the centre of Bank 1/1 were indicative of zero lateral yield i.e. K_0 conditions. Pore-pressure dissipation, leading to increases in the average effective stresses, resulted in plastic yielding associated with increased displacements, strains and pore pressures per unit height of fill. Local failure was indicated by further increases in the rates of displacement, strain and pore-pressure generation and, as described in section 16 above, appeared to progress outwards from the crest of Bank 1/1 towards the toe trench. The unloading following the Bank 1/1 failure produced elastic pore-pressure responses and virtually no recoverable displacements. Bank 1/2 produced similar behaviour to Bank 1/1 except that the tensile strains developed beneath the embankment prior to failure were much larger.

It was hoped that the full-scale embankment trials at Mucking would enable the direct use of field instrumentation to provide advance warning of potential instability during construction of the main works. Plots of excess pore pressure versus fill height for Bank 1 indicated early yielding of the foundation with signs of an increase in pore-pressure response as local failure commenced. However, the interpretation was not easy, the only definitive indication of local failure being provided by those piezometers beyond the embankment toes; even in this case the distinction between yield and failure was problematic. Piezometers 1–3 (Figure 45) indicated local failure at the toe of Bank 1/1 at 3.75m height while piezometers 22–24 (Figure 45) likewise indicated local failure of Bank 1/2 at 4.25m height. However, measurements of excess pore pressure at embankment toe locations tend to be less accurate because of the low absolute values and, as a result, pore-pressure monitoring was not considered a viable method of construction control.

Only horizontal displacements were considered for construction-control purposes, in order to avoid the problem of consolidation associated with the interpretation of vertical displacements. Figure 69 shows horizontal displacements divided by fill height plotted versus time from the start of construction for Banks 1/1, 1/2 and 2. The data in each case is from inclinometers at the embankment toes and at a depth of about 1m below original ground level. All three plots exhibit a distinct change in behaviour at 3.25m height for Bank 1/1, 3.8m for Bank 1/2 and 2.8m for Bank 2.

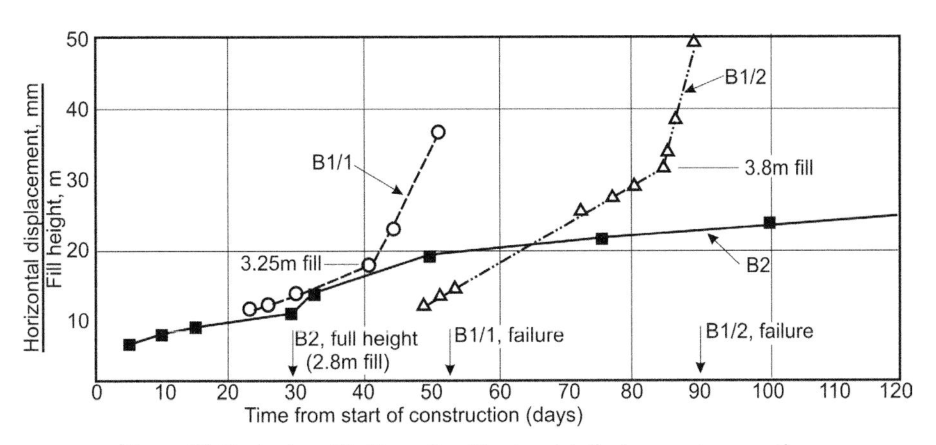

Figure 69. Banks 1 and 2: Normalised horizontal displacement versus time

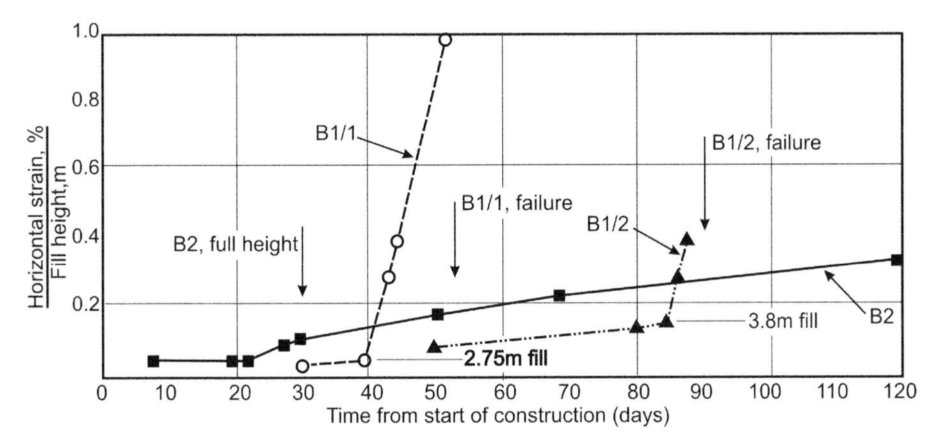

Figure 70. Banks 1 and 2: Horizontal strain/fill height versus time

Horizontal displacements from extensometers beneath the embankments (Figure 45) enabled horizontal strains to be monitored across the entire embankment widths. Figure 70 shows horizontal strains divided by fill height plotted versus time from the start of construction for extensometer locations beneath the embankment crests at about 1m below original ground level. The strains are tensile and suggest local failure developed beneath the crest of Bank 1/1 at 2.75m height, Bank 1/2 at 3.8m and Bank 2 at 2.7m. Similar plots beneath and beyond the embankment toes showed a change from tension beneath the embankment crest to compression beyond the toe. Local failure at the embankment toe corresponded to a fill height of 3.5m for Bank 1/1 and 4.25m for Bank 1/2. At the toe of the ensuing failure surfaces, local failure coincided with overall embankment failure.

It is clear from Table 8 that, for construction control purposes, local failure must be detected by monitoring horizontal displacements beneath the embankment crest in order to permit continuing observations and, if necessary, a halt in construction. The observations for Bank 2

Table 8. Embankment heights for local failure

Location	Embankment height, m-Local failure		
	Bank 1/1	**Bank 1/2**	**Bank 2**
Embankment crest	2.75	3.8	2.7
Embankment toe	3.5	4.25	
Toe of slip	4.0	4.5	

suggest that commencement of local failure beneath the crest coincided with a factor of safety of about 1.5

19 CONCLUDING OBSERVATIONS

19.1 *The effect of strain-rate and strain on the peak, field-residual and laboratory-remoulded shear strengths.*

The results of the field trials showed that the field undrained shear strength under full-scale loading conditions could be accurately determined from the results of laboratory vertical triaxial tests on specimens cut from undisturbed block samples, corrected for the effects of anisotropy and strain rate determined from laboratory testing. With respect to anisotropy, a simple average of the vertical, horizontal and 45° inclined tests at each test level was found to provide a realistic correction, while for strain rate it was necessary to reduce the undrained strength from the value corresponding to failure time in the laboratory to that corresponding to failure time in the field.

In terms of effective stress, the laboratory tests showed that the effective stress parameters were isotropic and therefore that no correction for anisotropy was necessary for the field situation. As a result the field effective stress parameters under full-scale embankment loading could be obtained by simply reducing the apparent cohesion from the value corresponding to failure time in the laboratory to that corresponding to failure time in the field, again as determined from laboratory tests.

The required strain-rate reduction in shear strength from the laboratory to the full-scale field situation was the same in terms of both total stress (reduction in undrained shear strength) and effective stress (reduction in apparent cohesion). It was thus considered that the observed field strain-rate reduction in shear strength was primarily the result of a partial breakdown in the clay-mineral structure of the lightly over-consolidated Mucking clay, coupled with associated secondary pore-pressure rises.

Field observations showed that significant pore-pressure dissipation occurred during the early (pre-yield) stages of embankment construction, such that vertical effective stresses rapidly approached the apparent pre-consolidation pressure. The back-analyses of embankment stability in terms of total stress confirmed that the correct laboratory undrained shear strength for stability analyses would be given by consolidated undrained tests on specimens consolidated to the apparent pre-consolidation pressure. While this rapid pore-pressure dissipation at effective stresses less than the apparent pre-consolidation pressure was greatly facilitated by the fabric of the Mucking clay, in the author's experience it is a common feature of the behaviour of soft alluvial clays.

Post-failure effective shear and normal stress data from the back-analyses of the post-failure embankment profiles conformed to the laboratory "normally consolidated" effective stress failure envelope. The analyses showed that continued shearing as the embankments moved from the pre- to post-failure profiles resulted in both the undrained shear strength and the apparent cohesion reducing by the same amount. In the case of the effective stress analyses the apparent cohesion was reduced to zero. This suggests that the field brittleness of the Mucking clay in both total and effective stress terms is largely the result a complete breakdown of the lightly over-consolidated clay-mineral structure coupled with associated pore-pressure rises.

Laboratory ring shear tests on Mucking clay showed that the ultimate residual and remoulded shear strengths were the same and only about 50% of the field residual strength. It thus seems clear that increasing strains beyond the field residual state results in increasing pore pressures until, ultimately, the undrained shear strength drops to the remoulded value. During this post-failure shearing the clay is effectively normally consolidated and the increased undrained brittleness is a direct function of increased pore pressures.

Construction of an embankment to a height commensurate with a safety factor of 1.5 and the development of local failure beneath the crest enabled long-term monitoring of pore-pressures and horizontal strains beneath the embankment toe. The results showed that the post-construction horizontal displacements and strains (creep) were initially accompanied by increasing excess pore pressures and that it was not until some 50 days after the end of construction that pore pressures returned to end of construction values. It was the author's view at the time of the trials, and remains so at the time of writing, that the onset of yielding marked the onset of significant creep strains and that creep is associated with pore-pressure generation. As such it is a significant factor in the strain rate effect.

19.2 *Practical application*

The extensive laboratory testing and field trials conducted on the Mucking clay provide a detailed review of the behaviour of the Mucking clay with respect to shear strength. The field trials also provided a detailed record of the deformations and pore pressures for four trial embankments (Pugh, 1978). It was beyond the scope of this paper to present the results of the field instrumentation but, taken together with the results of the laboratory testing presented here, they represent a comprehensive and high quality resource against which to test the accuracy of current advanced numerical modelling techniques.

With respect to shear strength, the data presented enables a complete understanding of the full-scale behaviour of the Mucking clay relevant to both failure and post-failure conditions. However, its practical application would require a quantity and quality of testing rarely available in the commercial world. Over the forty plus years that have passed since the Mucking trials the author has been involved in the successful design of embankments and reclamations on similar soft clays both in the UK and overseas, without coming close to the quality of the Mucking data. In practice, laboratory undrained shear strengths for total stress stability analyses might be from quick undrained tests on, at best, piston samples or, at worst, thick walled, driven tube samples.

Field measurements might be indirect via cone penetration tests or field vane tests. The evidence is that, in the hands of experienced geotechnical engineers, successful outcomes can be achieved using all of these data sources, the secret being to know the applicability of the particular data source to the stability analysis of embankments on soft clay.

A perfect example of the above is the use of undrained shear strengths from quick undrained triaxial tests on thick walled, driven tube samples. Such tests were carried out on the Mucking clay. The resulting strength/depth profile had an average undrained strength of 12.5kPa, being exactly the same as the average shear strength on the critical slip surfaces, but only 0.69 of the average shear strength for the vertical triaxial tests. The reason for this reduction is sampling disturbance: the greater the disturbance the greater the strength reduction; and the reason for the successful prediction of a safety factor of unity is that the disturbance reduction coincidentally equals the consolidation strength increase less the combined anisotropy and strain rate reductions.

The average undrained strengths from the piston samples were about 2kPa lower than those for the vertical triaxial tests on block samples, this being a disturbance reduction, and as a result the strength/depth profile closely resembles that for the vertical triaxial tests corrected for anisotropy. As such, undrained strengths obtained from piston samples would require a reduction of 0.79 to account for consolidation strength gain and strain-rate reduction.

The Geonor vane test results approximated to those for the vertical triaxial tests, although they were higher in the upper half of the soil profile and lower in the lower half. When corrected for plasticity they overestimated the field shear strength by 10%. The large vane tests which, unlike the Geonor vane tests, were truly undrained tests, when corrected for plasticity underestimated the field shear strength by 10%.

In conclusion, the author's view is that, for the routine design of embankments on lightly over-consolidated soft clays, laboratory undrained strength testing should be on piston samples, consolidated to the apparent pre-consolidation pressure and with a strain-rate reduction applied for design purposes. The piston sampling should be supplemented by in-situ testing using either a shear vane or cone penetrometer for calibration purposes. It is appreciated that the recommendation of consolidation to the pre-consolidation pressure appears to be at variance with the conclusion in Part 1 that consolidation to the situ effective stress is preferable. However, the latter is based entirely on laboratory testing on its own, while the former takes into account the field factors described earlier in Section 15.3.4.

20 FINAL REFLECTIONS

This paper began with mention of Professor Bishop and it is appropriate to bring him back in making these final reflections. The above accounts of investigations carried out under his supervision give an impression of the sort of man he was. Among the many outstanding features of this investigation for which he was directly responsible are the following:

1. The size and number of block samples that were taken for the triaxial and oedometer tests.
2. The design of the triaxial cell for stress controlled triaxial tests. Although the first author came up with the basic design he was not without concerns as to whether it would really

work as he hoped. Bishop, on the other hand, took a quick look at the design concept and was satisfied immediately that it would function as expected.

3. The design of the large vane apparatus. A vane of this size had never been used before, yet his design worked faultlessly from its first trial.
4 The number, layout and instrumentation of the full scale trial embankments.

As described in the book "The Bishop Method" it was always Bishop's aim to relate the findings of experimental investigations to field behaviour. More specifically, he sought information, whether it be theoretical or experimental, that would be useful to designers. This has largely been done in the conclusions to Part 2; some additional comments are the following:

(a) Choice of the method of stability analysis – total or effective stress?

It has been shown here that both total and effective stress methods can produce reliable results (ie factors of safety) provided the necessary data are available. This data includes the undrained shear strength (S_u) and the effective stress strength parameters (c' and ϕ'), and the pore pressure. The strength parameters can be determined from appropriate tests, but the pore pressure can only be predicted by theoretical methods, none of which are very reliable. The measurement of the undrained strength is much simpler (though not without uncertainties) than measurement of the parameters c' and ϕ', especially at the low stress levels involved in stability analysis of soft clays. This means that a total stress analysis is generally much simpler and more reliable than an effective stress analysis.

(b) Determination of the strength parameters

This issue has been well covered in the conclusion to Part 2. However, with respect to the influence of rate of loading (or time to failure) the following point needs to be emphasised. The strain rate correction is not an argument for building an embankment as rapidly as possible. This is because, as shown in Figures 63 to 65, after completion of construction there are still large ongoing horizontal creep strains and associated pore pressure rises. Thus, the critical point for bank stability on soft clays is not necessarily at the end of construction: it may well be some time later, depending on whether the pore pressures generated by creep exceed the rate at which pore pressures can dissipate by the normal consolidation process.

(c) Consolidation and its associated strength gain – staged construction.

The measurements of pore pressures during, and on completion of construction, show clearly that for the Mucking clay considerable consolidation occurred during construction of the embankment, at least in the top several metres. This was a significant finding, although it is probably the case with many embankments built on soft clays. Even if little pore pressure dissipation occurs during construction, it will certainly occur with time and the associated increase in strength is often used by designers as a means of building higher embankments than would otherwise be possible. Prediction of both the rate of pore pressure dissipation and the associated increase in soil strength is subject to considerable uncertainty. The rate of consolidation is often under-estimated which is naturally of benefit to the designer. Estimating the increase in strength is not straightforward because, as shown in Figure 17(a), there is not a fixed relationship between undrained strength (S_u) and consolidation pressure (p_c'). For design purposes the S_u/p_c' relationship needs to be established both below and above the pre-consolidation pressure and the relationship between the pre-consolidation pressure and depth also requires definition. Neither of these is easy.

Monitoring during the construction of embankments on soft clays is largely outside the scope of this paper; however, based on their own experience the authors would make the following recommendation:

If the successful completion of the embankment relies on increased strength of the soft clay due to consolidation then this increase should be measured directly, rather than by relying on either settlement or pore pressure records. The original strength of the clay can be measured with in situ vane or cone penetrometer tests (CPTs) and these tests repeated as the embankment is built. This may mean having to pre-bore the fill, especially if it consists of well compacted granular material. Only in this way can the increase in strength be known with certainty.

Another "reflection" of interest is that in the introduction to this paper mention was made of papers to be co-authored with Bishop. Part of Bishop's introduction to the joint paper with Wesley is still in the latter's hands and contains the following interesting observation:

Skempton and Bishop (1950) had pointed out the important effect of rate of loading, which in most construction operations is some 10,000 times smaller than that in routine soil tests. On the basis of tests by Taylor (1943) and Casagrande and Wilson (1949) it appeared to Skempton and Bishop that the undrained strength relevant to full scale analysis would fall to a value 15 to 30% lower than normally measured in the laboratory.

Thus, three decades before the time of the tests described in this paper there was already literature on this issue. It was clearly of concern to Bishop and to further investigate the issue he instigated both the laboratory and field tests described here. Hopefully, the data presented in this paper would meet with his approval.

The authors recognise that a literature review is not included in their paper, and they may have overlooked important papers published since completing their PhDs. However, their principal aim has been to put their work on paper so that it is available to those with an interest in this subject, and apologise if they have overlooked the important relevant work of others. .

Finally, the authors have enjoyed putting this paper together, and found revisiting their PhD's some four decades after writing them an interesting and demanding experience. They were impressed with both the scope and meticulousness of their research, which was due as much to their supervisor as to them. Perhaps they feel some regret that they did not get their work published in the years immediately following its completion. On the other hand, perhaps their years of practical experience have enabled them to better appreciate the relevance or otherwise of their research.

21 REFERENCES

Atkinson, J.H.(1975). Anisotropic elastic deformations in laboratory tests on London clay. *Geotechnique* **33**, No 2, 357–374.

Berre, T. and Bjerrum, L. (1973). Shear strength of normally consolidated clays. *Proc. 8th Int. Conf. Soil Mech.* Moscow 1.1, 39–49.

Bishop, A W (1955) The use of the slip surface in the stability analysis of slopes. *Geotechnique* 5, No 1, 7–17

Bishop, A.W. (1967) Progressive failure-with special reference to the mechanism causing it. *Proc. Geotechnical Conf. on the shear strength properties of natural soils and rocks.* Oslo.

Bishop, A.W. (1971) The influence of progressive failure on the choice of stability method. *Geotechnique* 21, No 2, 168–172

Bishop, A.W., and Henkel, D. J. (1962). The measurement of soil properties in the triaxial test. *Edward Arnold, London* 1957.

Bishop, A.W., Green, G.E., Garga, V.K., Andresen, A., and Brown, J.D. (1971) A new ring shear apparatus and its application to the measurement of residual strength. *Geotechnique* 21, 273–328.

Bishop, A.W. and Wesley, L.D. (1975) A New Hydraulic Triaxial Apparatus. *Geotechnique* 25. No 4, 657–670.

Bjerrum, L. (1972) Embankments on soft ground. *Proc. ASCE Speciality Conf. on performance of earth and earth supported structures.* Purdue University, Lafayette, Indiana.

Bjerrum, L. (1972) Problems of soil mechanics and construction on soft clays and structurally unstable soils (collapsible, expansive and others). *8th ICSMFE,* 3, 111–159.

Boehler, J.P. and Giroud, J. P. (1971). Measurements of soil anisotropy. *Proc. 9th Annual Symp. on Eng. Geology and Soil Engineering.* Idaho .175–187.

Evans, G. (1965) Intertidal flat sediments and their environments of deposition in the Wash. Quart. Journal of the Geological Soc. London, Vol. 121, 209–245.

Hight D.W. (2018) Personal communication.

Greensmith, J. T., and Tucker, E. V. (1973).Holocene transgression and regressions on the Essex Coast, outer Thames Estuary. *Geologie en Mijnbouw.* **52**, No 4, 193–202.

Jamiolkowski, M. Lancellotta, R. Pasqualini, and Marchetti, S. (1979). Design parameters for soft clays. *Proc. 7th Euro. Conf. on Soil Mechanics and Foundation Engineering,* Brighton **5**, 27–57.

Kane, N.S., Vane.C.H., Horton, B.P., and Fackler, S. (2011). A new record of Holocene sea-level change in the Thames Estuary and its implications for geophysical modelling. British Geotechnical Society publication.

Khan, S.N., Vane, C.H., Horton, B.P., Hillier, C., Riding, J.B., Kendrick, C. (2015). The application of δ^{13}C, TOC, C/N geochemistry to reconstruct Holocene relative sea levels and paleo-environments in the Thames Estuary, UK. *Journal of Quaternary Science* 30, 5, 417–433.

Lo, K.Y. (1965). Stability of slopes in anisotropic soils. *ASCE Journal Soil Mech. and Found. Eng.* Vol. 91, SM4, 85-106.

Parry, R.H.G., and Wroth, C.P. (1981) Shear stress-strain properties of soft clay. *Soft Clay Engineering Ed E.W.Brand and R.P Bremmer.* Elsevier, Amsterdam, 1981

Pugh, R.S. (1978) The strength and deformation characteristics of a soft alluvial clay under full scale loading conditions. *PhD thesis,* Imperial College University of London

Sarma S.K. (1973) Stability of embankments and slopes. *Geotechnique*, 23, 3, 423–433.

Wesley, L.D. (1975). Influence of anisotropy and stress path on the behaviour of a soft alluvial clay. PhD Thesis, *Imperial College*, University of London.